D0620885

INTRODUCTION TO ENVIRONMENTAL FORENSICS

INTRODUCTION TO ENVIRONMENTAL FORENSICS

edited by

Brian L. Murphy and Robert D. Morrison

San Diego San Francisco New York Boston
London Sydney Tokyo

This book is printed on acid-free paper.

ACADEMIC PRESS
A Harcourt Science and Technology Company
Harcourt Place, 32 Jamestown Road, London NW1 7BY, UK
http://www.academicpress.com

ACADEMIC PRESS
A Harcourt Science and Technology Company
525 B Street, Suite 1900, San Diego, California 92101-4495, USA
http://www.academicpress.com

ISBN 0-12-511355-2

Library of Congress Catalog Number: 2001090246

A catalogue record for this book is available from the British Library

Typeset by Charon Tec Pvt. Ltd, Chennai, India
Printed and bound in China by RDC Group Limited
02 03 04 05 06 07 RD 9 8 7 6 5 4 3 2 1

CONTENTS

CONTRIBUTORS

Dr. Shelley Bookspan is currently a consultant in Santa Barbara, California. In 1982, she was among the founders of PHR Environmental Consultants, Inc., and she is the past president of this history-based consulting company which specializes in conducting research and investigations to ascertain and clarify the factual circumstances underlying pollution liability disputes. Dr. Bookspan's doctorate, from the University of California, Santa Barbara, is in history, with emphasis on American history, the history of technology, and urban history. She also holds a master's degree in city planning from the University of Pennsylvania and a master's degree in history from the University of Arizona. For the past four years, Dr. Bookspan has also served as the Editor in chief of an internationally circulated journal published by the University of California Press in Berkeley, California, called *The Public Historian*. Prior to that time, she was Contributing Editor for the Environment of *The Real Estate Law Journal*. She has written multiple essays elucidating the application of history to environmental cases. shelleybookspan@juno.com

Dr. Judith C. Chow is a Research Professor in Desert Research Institute's Division of Atmospheric Sciences specializing in aerosol measurement method development and applications. At DRI, Dr. Chow established an environmental analysis facility that quantifies mass, elemental, ionic, and carbon concentrations on filter samples. She has prepared guidance documents on aerosol measurement for the USEPA, authored the 1995 Air and Waste Management Association's Critical Review of $PM_{2.5}$ measurement methods, and has completed more than 100 research publications on sampling and analysis methods and interpretation of results. She is a member of the National Research Council's Committee on Research Priorities for Airborne Particulate Matter and was the 1999 recipient of DRI's Alessandro Dandini Medal of Science and the 2001 recipient of the University of Nevada Board of Regent's Researcher of the Year Award. Dr. Chow received her ScD degree in Environmental Science from the Harvard School of Public Health in 1985. judyc@dri.edu

Julie Corley received an MA from Arizona State University in 1992. Her work in the field of history has included extensive research in issues surrounding land-use history. She has been employed by PHR Environmental Consultants since 1992, and is currently the Director of Research and West Coast Operations A25/PHR. jcorley@theitgroup.com

Dr. James I. Ebert is an archeologist, an anthropologist, and an environmental and forensic scientist, and is Chief Scientist at Ebert & Associates, Inc., an Albuquerque, New Mexico firm specializing in forensic, environmental and archeological applications of photogrammetry, photointerpretation, remote sensing, and digital mapping technologies. He studied archaeology at Michigan State University where he earned his BA degree, and then at the University of New Mexico, where he was awarded his MA and PhD, and founded Ebert & Associates in 1983. He is a Certified Photogrammetrist (American Society of Photogrammetry and Remote Sensing), a fellow of the American Academy of Forensic Sciences, and a member of the New York State Police Medicolegal Investigations Unit. Dr. Ebert has conducted an ongoing program of archaeological and environmental research at Olduvai Gorge in Tanzania since 1989 with Rutgers University's Olduvai Landscape Palaeoanthropology Program. jebert@ebert.com

For the past 30 years, **Dr. Robert Ehrlich** has been a leader in the application of image analysis and pattern recognition procedures in the earth and environmental sciences. His MS and PhD degrees are from Louisiana State University. Dr. Ehrlich has been editor of the *Journal of Mathematical Geology* and has published more than 100 papers on the application and development of numerical techniques. He was an Associate Professor of Geology at Michigan State University from 1965 to 1974, and Professor of Geology at the University of South Carolina from 1974 to 1997. Dr. Ehrlich was also the Director of the Center for Industrial Imaging at the University of Utah from 1995 to 1999, and is currently Vice-President of Residuum Energy Incorporated and heads Residuum's Salt Lake City office. bobehrlich@home.com

Stephen Emsbo-Mattingly holds an MS in Environmental Science from the University of Massachusetts (1994). He has 12 years of environmental analytical chemistry experience, including as a former Quality Assurance Director and Laboratory Director for environmental chemistry contract laboratories. As a consultant, he has managed environmental forensic projects throughout the US. His specialty is the characterization of coal- and petroleum-derived tars, tar by-products and the environmental behavior and analytical chemistry of PCBs. He has developed and applied numerous hydrocarbon methods for volatile and semi-volatile petroleum constituents. Mr. Emsbo-Mattingly is the current President and former Chairman of the Technical Committee of the Independent Testing Laboratory Association (Boston). emsbo-mattinglys@battelle.org

Dr. William Full has spent his career in academia and industry focusing on computer applications and algorithm development in the geosciences. He has a BS in mathematics from the University of Notre Dame, and MS and PhD degrees in Geology from University of Chicago and the University of South Carolina, respectively. In 1982 he received the Andre Borisovich Vistelius Research Award for Young Geomathematicians from the International Association of Mathematical Geology. He has published many papers on information entropy and multivariate analysis, and he was the principal developer of the polytopic vector analysis procedure. Dr. Full was on the faculty of the Department of Geology at Wichita State University for 15 years, and is currently a Distinguished Professor at the University of Rome, Italy, and President of WEF consulting in Florence, SC. nw.full@worldnet.att.net

Dr. Thomas D. Gauthier is a senior environmental chemist at Sciences International, Inc. in Bradenton, Florida. He has over 12 years of consulting experience and works on projects involving the transport and fate of organic chemicals in the environment, historical dose reconstruction, source identification, and the statistical analysis of environmental chemistry data. Dr. Gauthier received his BS in chemistry from Merrimack College and PhD in analytical chemistry from the University of New Hampshire. He is a member of the American Chemical Society and is a frequent reviewer for *Environmental Science and Technology*. tgauthier@sciences.com

A.J. Gravel is the Managing Director of A2S/PHR a wholly owned subsidiary of The IT Group. A2S/PHR, specializes in providing attorneys and businesses with historical research, investigations, and management consulting services in support of environmental and product liability litigation. Mr. Gravel has extensive experience managing environmental investigations that have dealt with numerous CERCLA related issues such as PRP identification; corporate succession and asset searches; land use and business operation histories; past industrial chemical generation, usage, and disposal analysis; and government involvement and war claims. Additionally, his product liability experience includes conducting historical research and investigations relating to alternate causation, state-of-knowledge, state-of-practice, level-of-awareness, and regulatory development and regulatory oversight issues. Mr. Gravel earned a Bachelor of Science degree from Springfield College, Springfield, Massachusetts and a Masters of International Management from the University of Maryland, College Park, MD. agravel@theitgroup.com

Dr. Glenn W. Johnson received his MS from the University of Delaware and his PhD from the University of South Carolina. Both graduate research programs focused on the application of multivariate techniques to geological and environmental chemical data. He spent seven years between degrees as an environmental consultant with Roux Associates, Inc. and McLaren/Hart Environmental Engineering Corp., where his work focused on investigation of contaminated sites, environmental forensics and associated litigation support. Dr. Johnson is currently a Research Assistant Professor at the Energy and Geoscience Institute at the University of Utah, where he continues research in environmental chemometrics, and teaches in the Department of Civil and Environmental Engineering. He is also President of GeoChem Metrix, a consulting firm in Sandy, Utah. gjohnson@egi.utah.edu

Kevin J. McCarthy received a bachelor of sciences degree in 1986 from Westfield State College in Massachusetts. He has over 14 years experience in the field of petroleum environmental chemistry. His experience is in the detailed chemical analysis and chromatographic interpretation of petroleum products and wastes, and their characterization using advanced instrumental methods and chemometric data analysis techniques (forensic chemical fingerprinting). His expertise includes the molecular-level characterization of petroleum and petroleum products and investigations of the chemical alteration of petroleum due to physical and biological weathering. He has participated in developing specialized methodologies for the analysis and characterization of oil and petroleum products. mccarthy@battelle.org

Dr. Robert D. Morrison has a BS in Geology, an MS in Environmental Studies, an MS in Environmental Engineering and a PhD in Soil Physics from the University of Wisconsin at Madison. Dr. Morrison has worked as an environmental consultant for over 30 years on issues related to soil and groundwater contamination. Dr. Morrison specializes in the forensic review and interpretation of scientific data in support of litigation involving soil and groundwater contamination. Dr. Morrison is Editor in Chief of the *Journal of Environmental Forensics* and has served on the Editorial Boards of *Ground Water* and *Groundwater Monitoring and Remediation.* Dr. Morrison has provided testimony in over 40 cases throughout the world with some claims in the billions of dollars. He has also served as a confidential consultant in a number of other cases. bob@rmorrison.com

Dr. Brian L. Murphy has an ScB from Brown University and MS and PhD degrees from Yale University. All his degrees are in physics. He has spent most of his career as an environmental consultant, including as General Manager for Physical Sciences at what is now ENSR Corporation, Founding President of Gradient Corporation, and most recently as Vice President of Sciences International. He also has had Visiting Instructor positions at Harvard School of Public Health and the University of South Florida. He is an Associate Editor of the *Journal of Environmental Forensics.* Dr. Murphy has testified as an expert on air exposure and dose reconstruction in toxic torts; on determining site histories using soil and groundwater data in insurance litigation; and on various risk assessment topics. He also has served as a consulting expert in these areas. blmurphy@home.com

Dr. R. Paul Philp is Professor of Petroleum and Environmental Geochemistry at the University of Oklahoma. His major areas of research interest in environmental geochemistry are centered on the use of carbon and hydrogen isotopes for the correlation of chemicals spilled in the environment with their suspected sources. Dr. Philp obtained his PhD from the University of Sydney, Australia and subsequently held research positions at the University of Bristol, England, University of California, Berkeley, and with CSIRO, Sydney before joining the faculty at the University of Oklahoma in 1984. He has authored or coauthored over 240 articles and books on various aspects of geochemistry and has lectured extensively on petroleum and environmental geochemistry in Southeast Asia, South America, Europe and Africa. In May 1998 he was awarded the DSc degree from the University of Sydney on the basis of his research in geochemistry over the past 20 years. pphilp@ou.edu

Dr. Scott A. Stout received MS and PhD degrees in Geology from Penn State University (1985 and 1988). He is an organic geochemist with more than 15 years of coal and petroleum industry experience. His particular expertise surrounds the compositions of coal-, petroleum-, gasoline-, and other fuel-derived sources of contamination and the synthesis of geochemical data for property management decisions and environmental torts. Dr. Stout is a registered geologist and member of the American Chemical Society and the American Academy of Forensic Sciences and an Associate Editor for the *International Journal of Environmental Forensics.* stouts@battelle.org

Dr. Allen D. Uhler holds MS and PhD degrees in Chemistry from the University of Maryland at College Park (1981 and 1983). He has 18 years experience in the field of environmental chemistry, with an emphasis on environmental forensics. His particular expertise is the analysis of petroleum hydrocarbons and other trace organic compounds in waters, soils, and sediments, the use of data analysis techniques to elucidate relationships among samples and suspected sources. Dr. Uhler is a member of the Society of Petroleum Engineers, serves on the editorial board for the *International Journal of Environmental Forensics* and is the Environmental Forensics Feature Editor for *Contaminated Soil, Sediment, and Water.* uhler@battelle.org

Dr. John G. Watson is a Research Professor in Desert Research Institute's Division of Atmospheric Sciences specializing in the characterization, source apportionment, and control of suspended particles that cause adverse health effects and haze. Dr. Watson has developed theoretical and empirical models that, when coupled with appropriate measurements, quantify contributions for pollution from different sources. With his colleagues, he has applied these methods in urban and regional particle studies to solve problems of excessive concentrations, visibility impairment, and deposition. Dr. Watson obtained his PhD degree in Environmental Science from the Oregon Graduate Institute in 1979 and was awarded the Howard Vollum Prize for Distinguished Achievement in Science and Technology in 1989, DRI's Alessandro Dandini Medal of Science in 1992, and the Air and Waste Management Association's 2000 Frank A. Chambers Award for major contributions to the science and art of air pollution control. Dr. Watson has served as chair of the Publications Committee for the Journal of the Air and Waste Management Association. johnw@dri.edu

INTRODUCTION

'Forensic' is related to 'forum' and refers to any public discussion or debate. In the United States 'forensic' most often refers to courtroom or litigation proceedings. However, environmental forensics may also provide the fact basis for mediated or negotiated transactions or for any public inquiry related to environmental matters, common situations in many countries.

Questions that environmental forensics seeks to answer are:

- Who caused the contamination?
- When did the contamination occur?
- How did the contamination occur? (For example, was it an accidental spill or a series of routine operating releases?)
- How extensive is the contamination?
- Are the test results valid? Is there evidence of fraud?
- What levels of contamination have people been exposed to?

Chapter 1 discusses the common contexts in the United States in which environmental forensic investigations occur. These contexts include liability allocation at hazardous waste sites where multiple parties are involved, site assessments for property transfers, insurance litigation, and toxic torts.

Environmental forensic investigations frequently deal with the historical release of contaminants. Generally there are two sources of information in conducting an investigation, namely:

- The documentary record, including statements by witnesses or other knowledgeable individuals.
- Measurement or sampling data.

This book addresses witness statements only as one form of the documentary record. Chapters 2 and 3 address the first component of a forensic investigation. The phrase 'documentary record' is a broad description encompassing many types of information such as aerial photography, fire insurance maps, and electronic information copied from computer hard drives. Chapter 2 describes techniques used to acquire the historical documents necessary to reconstruct a narrative site history. Chapter 3 discusses the acquisition and interpretation of the photographic images that are frequently used to identify potential contaminant sources and to chronicle waste handling practices.

Chapters 4 through 12 deal with using measurements for forensic purposes. In a forensic investigation any datum or limited subset of data should be viewed with suspicion until its veracity is confirmed. Chapter 4 contains a discussion of potential biases that can be introduced into soil gas, soil and groundwater data, both in field sampling and in the laboratory.

The most common environmental forensic measurements are of chemical concentrations. However, the use of isotope measurements, as described in Chapter 5, is becoming an integral component of many environmental forensic investigations.

Forensic techniques are categorized in different ways, with the most common being by contaminant type or medium. The two most common contaminant types are petroleum hydrocarbons and chlorinated solvents. Forensic techniques for these classes of contamination are described in Chapters 6 and 7. Properties of automotive fuels and chlorinated solvents that are useful in forensic investigations are detailed in several appendices.

Taking the alternative point of view, by medium, Chapter 8 describes soil and groundwater models used for environmental forensic investigations. Chapter 9 discusses air modeling for forensic purposes.

Chapters 10 through 12 introduce statistical aspects associated with an environmental forensic investigation. Chapter 10 summarizes statistical tests for comparing data sets and evaluating temporal or spatial relationships. Chapters 11 and 12 present advanced pattern recognition techniques, of increasing utility with today's greater computing power. Chapter 11 discusses particulate pattern recognition techniques used for source identification. Because many of the techniques discussed in Chapter 11 are based on air data, they apply to currently operating, as opposed to historical sources. Chapter 12 focuses on several techniques, including principal components analysis and polytopic vector analysis. These techniques have been successfully used in a variety of media, including identifying sources of sediment contamination.

There are a limited number of ways that chemical or isotope concentration data can be manipulated; essentially there are only the following four families of techniques:

- Tracer techniques based on the presence or absence of a particular chemical.
- Ratio techniques where the relative amounts of two or more chemicals are compared. (The pattern recognition techniques described in Chapters 11 and 12 are primarily examples of ratio techniques.)

- Trend techniques where the spatial or temporal variation of a concentration, or a ratio, is of interest.
- Quantity techniques that depend on the integrated concentration over space or time, i.e., the mass of a chemical, to provide forensic information.

A forensic investigation may involve any or all of these families. For example, identifying the source of and age dating a hydrocarbon spill may be of interest. The presence or absence of lead, methyl tertiary butyl ether (MTBE) or other additives, for example, may provide crucial information. The ratio of different hydrocarbons may distinguish different fuels or brands. The spatial variation of a plume or its growth over time may assist in both source identification and age dating. Finally, the total mass or volume of petroleum hydrocarbons in the environment may be compared to inventory or leak detection records for source identification.

The most successful forensic investigations rely on a toolbox of techniques. A case relying on the results of a single forensic technique exclusive of other available tools often topples when contrary evidence is introduced. When forensic evidence is arrayed as multiple, but independent lines of evidence, a stronger scientific case, less susceptible to scientific challenge, emerges. This book is intended to provide you with your own toolbox of forensic techniques.

BRIAN L. MURPHY

ROBERT D. MORRISON

CHAPTER 1

APPLICATIONS OF ENVIRONMENTAL FORENSICS

Brian L. Murphy

1.1 INTRODUCTION

There are two separate motivations for environmental forensic studies. First, such studies may be performed for the sake of obtaining knowledge of historical emissions to the environment or historical environmental processes and for no other reason, what might be termed purely 'research' or 'academic' studies. Second, such studies are carried out to determine liability in a variety of contexts. This latter purpose is the focus of this chapter.

Our discussion of liability-driven forensic studies is based on liability under United States laws. However, we do not focus on the law itself, although some of the references given discuss various legal issues. Rather we focus on how the

Table 1.1
Liability context and related forensic issues.

Context	Forensic Issues
Cost allocation at Superfund sites	Responsibility for waste streams or areas of contamination Contribution of waste streams to remedial cost or to the need for a remedy
Site investigation for property transfer	Existence and extent of contamination Cost of remediation Existence of other responsible parties
Insurance litigation	Policies 'triggered' by a release Characteristics of a release or of contamination relative to contract language Equity of liability allocation
Toxic tort	Probability that chemical exposure caused manifest or latent injury

legal requirements concerning liability translate into technical issues and questions, which can be answered using forensic methods.

We discuss liability in four different contexts as shown in Table 1.1. This table also lists key forensic issues for each context. In the remainder of this chapter we describe how measurements of chemical concentrations or other properties combined with the forensic techniques described later in this book can be used to illuminate these issues.

These four contexts probably represent a good proportion of the situations in which the tools of environmental forensics are employed to allocate liability. However, they do not represent all such situations. We have selected these four contexts because they are the most structured and universally applicable in the United States.

Other situations where the tools of environmental forensics have been employed, but which are not discussed below, include identification of the source of maritime oil spills and identification of air pollution sources, including in international or transboundary air pollution situations. Hydrocarbon pattern recognition, discussed in Chapter 6, and dispersion or receptor modeling, discussed in Chapter 11, are the most relevant tools for these situations.

1.2 LIABILITY ALLOCATION AT SUPERFUND SITES

The Comprehensive Environmental Response, Compensation and Liability Act or CERCLA, commonly referred to as Superfund, prescribes specific procedures for dealing with chemical release or disposal sites that are considered to be the most hazardous in the United States. States also have hazardous waste site remediation programs patterned to varying degrees after the federal program. Thus state lead occurs at many sites deemed less hazardous than those in the federal Superfund program, for example at dry cleaning facilities across the country. Also, because the federal Superfund legislation excludes petroleum release sites, state lead occurs at facilities such as gasoline stations and former manufactured gas plant sites.

State laws are often modeled after CERCLA, although as discussed below CERCLA does not provide much guidance on liability allocation. In any case, the discussion in this chapter of methods of liability allocation at federal Superfund sites is still relevant but the details may vary from state to state.

Potentially responsible parties (PRPs) at Superfund sites include present owners and operators, past owners and operators, waste generators, and transporters or arrangers for transport of waste. Costs borne by PRPs at CERCLA sites may be for site remediation or in payment for past, present, and future damage to natural resources. Payment for future damages arises when the site cannot be totally remediated so that the habitat is restored.

Superfund liability includes for actions that predate the CERCLA legislation. Furthermore, liability is perpetual, it cannot be circumvented by being assigned to someone else. Liability does not depend on fault but simply on being a member of one of the classes of PRP described above. Finally, Superfund liability can be 'joint and several,' that is in principle all liability may be borne by a single PRP irrespective of the relative degree of fault.

Two sections of CERCLA touch on allocation of liability among the PRPs. Section 107 provides for recovery of remediation costs. Plaintiffs in a recovery action may be the US Environmental Protection Agency (EPA) or states. Courts are divided on whether a PRP may be a plaintiff but the recent trend has been to deny Section 107 to PRP plaintiffs (Aronovsky, 2000). Although Section 107 specifies joint and several liability, this is discretionary with the court. In particular, where a PRP can demonstrate distinct harm or divisibility of harm, that party may be responsible just for their contribution to harm. A distinct harm arises, for example, when there are separate groundwater plumes or areas of surface soil contamination. A divisible harm might be where there are successive site owners conducting the same operation. The basis for divisibility in that case might be the relative number of years of operation.

Section 113 of CERCLA allows a party who has incurred response costs to seek contribution from other PRPs. This section also provides contribution protection for parties that have settled with the United States. Under Section 113 the liability of nonsettling PRPs is limited to their proportionate share. The nonsettling PRP's liability may be determined in either of two ways. It may be determined by subtracting out the amount of prior settlements or by subtracting out the proportionate share of the harm for the settling PRPs. As Ferrey (1994) points out, the results of these two approaches may be quite different.

No guidance in determining proportionate shares, beyond citing 'equitable factors' is found in CERCLA.

The equitable factors most often cited are the Gore factors proposed by then-Representative Albert Gore in 1980 but not enacted. They are: (1) the ability to distinguish the party's contribution to the nature and extent of the problem, (2) the degree of the party's involvement in the activities that caused the problem, (3) the degree of care exercised by the party, (4) the degree of cooperation of the party with governmental agencies, (5) the quantity of the hazardous waste involved, and (6) the toxicity of the waste. These factors have been found to be far from sufficient and in some cases may not be applicable at all. Furthermore, they are simply a list; they provide no conceptual framework for allocation.

Other factors that have been suggested include: (7) existing contracts between the parties, (8) the owner's acquiescence in the operator's activities, and (9) the benefit to the owner from the increase in land value due to remediation.

Thus CERCLA and its legislative history are not particularly helpful in specifying how liability is to be allocated. Courts have generally determined that there is a presumption of 'joint and several liability' unless harm is distinct or there is a reasonable basis for its division.

1.2.1 EQUIVALENCE OF 'HARM' AND RISK

What is 'harm' at a Superfund site? It seems logical that it be closely identified with the concept of 'risk'. A baseline risk assessment is conducted at all Superfund sites. Removal actions may precede completion of the risk assessment for urgent matters but the continued remediation of a site is based on a finding that the computed baseline risks, either human or ecological, are unacceptable. Therefore, it is logical to identify the risks at a site, including those requiring removal action with 'harm'. As discussed by Rockwood and Harrison (1993) several federal circuit court rulings also support this notion.

Thus, an argument can be made that risk assessment is the appropriate tool to use in apportioning liability. However, in fact risk is often not a consideration in apportioning liability. If the PRPs at a Superfund site agree on an allocation scheme, then that scheme is by definition satisfactory, assuming that there is no 'second guessing' by other parties such as insurers. PRPs often decide to allocate liability based on the contribution of each to the cost of the remedy. Of course, surrogate measures for estimating contribution to costs may be used, such as counting barrels or estimating plume areas.

There are several common situations where the harm due to multiple PRPs is not distinct but requires division. For example: (1) commingled groundwater plumes, (2) hazardous waste disposal sites with multiple users, and (3) successive site ownership. If these situations result in contamination by similar chemicals, a straightforward allocation based on contribution to the cost of a remedy may make sense. However, when one or more PRPs' wastes differ significantly from the others in the risk they pose, those PRPs may wish to consider a risk-based approach.

1.2.2 ALLOCATION PRINCIPLES

Economic principles of cost allocation, including the stand-alone cost method, have been discussed by Butler *et al.* (1993) and by Wise *et al.* (1997). The cost allocation matrix approach of Hall *et al.* (1994) is also based on determining contribution to the cost of a remedy. Marryott *et al.* (2000) present a stand-alone cost type model in which a weighted sum of contaminant mass in the plume and plume volume serves a surrogate for remediation costs. The basic equation for calculating stand-alone costs is:

$$f_i = \frac{SAC_i}{\sum SAC_i} \tag{1.1}$$

where SAC_i is the stand-alone cost for the waste stream due to the *i*th PRP generator/transporter.

This equation does not address how liability is to be allocated between the generator, transporter and site owner nor does it address orphan shares, such as from unidentified or defunct parties. It is solely an allocation by waste stream. Equation 1.1 states that each PRP pays in proportion to the cost that would have been incurred if there were no other PRPs at the site. Because of redundancy of cost items and economies of scale the total cost of a remedy will generally be less than the denominator of Equation 1.1 and hence each PRP will actually pay less than their computed stand-alone cost.

Risk-based allocation methods have been discussed by Murphy (1996, 2000) and by Mink *et al.* (1997). The risk contribution analogue to Equation 1.1 is:

$$g_i = \frac{SAR_i}{\sum SAR_i} \tag{1.2}$$

where g_i is the cost fraction for the *i*th generator/transporter based on stand-alone contribution to risk SAR_i. The analogy with stand-alone costs is incomplete; however, the total risk is equal to the sum of the individual PRP-caused risks rather than generally being less.

Of course cost allocation may be a mixture of cost-based and risk-based methods:

$$h_i = \alpha f_i + (1-\alpha)\, g_i \tag{1.3}$$

where α is a constant. As α decreases from 1, a 'contribution to the need for a remedy' component is mixed in with the 'contribution to the cost of a remedy.'

The kind of information needed to calculate f_i or g_i differs. For example, in computing stand-alone costs, well installation costs may vary as plume area and groundwater treatment costs may vary as contaminant mass in the plume. How long a pump and treat remedy needs to be maintained will depend on the ratio of individual chemical concentrations to acceptable levels in groundwater and on chemical properties which determine partitioning to soil. In computing stand-alone risks, concentrations and toxicities of specific chemicals will be required. Of course, as indicated above, it may be to the advantage of all PRPs to lower transaction costs by using surrogate quantities rather than attempting to collect the additional information necessary for refined or precise calculations.

For a cost-based allocation, typical forensic issues are:

- Attributing different groundwater plumes to individual parties or where plumes are inextricably commingled to two or more parties.
- For successive site owners determining when major releases occurred, or for contamination by chronic operating discharges determining relative production amounts or years of operation.
- At hazardous waste sites accepting waste from multiple parties determining waste stream volumes attributable to individual generators or transporters.

The additional information needed for a risk-based allocation is concentrations of specific chemicals in groundwater plumes, waste streams or historical releases.

Time is a missing factor in many allocations whether by risk or by cost. For example, a PRP's wastes in groundwater might not arrive at an extraction point for many years because of a slow groundwater velocity or retardation effects. If the remedy will not be relevant to that PRP's wastes until some possibly distant future time it can be argued that that PRP's contribution should be discounted to a smaller present value.

1.3 ENVIRONMENTAL SITE ASSESSMENT

As the term is used in this chapter, an environmental site assessment is conducted as a preliminary to a real estate transfer. (Similar tasks may be conducted as part of an internal management assessment, a process generally known as an environmental 'audit.' An audit may be concerned solely with compliance with applicable laws and regulations or it may include a more management-oriented review of responsibilities, organization, communications and measurement of progress.) The main purposes of an environmental assessment are to determine:

- whether contaminants are present on site,
- if present to determine the extent of contamination so that likely remediation requirements and costs can be estimated.

The American Society for Testing and Materials (ASTM) has published two 'Standard Practices' for conducting Phase I Assessments. Phase I is intended to assess the likelihood of site contamination. As the term is used in these standard practices a Phase I Assessment does not include any environmental sampling. These Standard Practices were originally developed to satisfy one of the requirements for the 'innocent landowner defense' under CERCLA.

ASTM Standard Practice E 1527 describes the four components of a Phase I site assessment as on-site reconnaissance, interview of site owners and occupants as well as local government officials, records review, and report preparation. This Standard Practice is intended to be conducted by an 'environmental professional.' ASTM Standard Practice E 1528 on the other hand may be conducted by any of the parties to a real estate transaction as well as an environmental professional. This Standard Practice is based on a 'transaction screen process.' The transaction screen process consists of the same three components prior to report preparation: a site visit, questions for the owner/occupants, and a records review. The difference is that the questions or issues to be addressed during the conduct of these components are all prescripted.

In the ASTM description, sampling of soils, groundwater, or other media would be a Phase II Assessment. ASTM has published a framework for the Phase II Assessment as Standard Guide 1903–97 and has published a number of standards dealing with sampling methods. These have been collected in a document: 'ASTM Standards Related to the Phase II Environmental Site Assessment Process.' A Phase II Assessment is generally necessary in order to determine the extent of contamination and hence the likely remedial requirements and associated costs. The Phase II Assessment would be guided by the results of Phase I.

The ASTM descriptions provide a framework but not one that should be followed slavishly. For example, at some sites the necessity of sampling certain locations and media may be evident and it may make the most sense to conduct sampling simultaneously with the components of a Phase I Assessment. Similarly, if one or more potential 'fatal flaws' are obvious, the Phase I or Phase II Assessments may focus solely on those areas.

As noted above, if contamination is found, an understanding of the extent and options for remediation to regulatory acceptable limits becomes important. If the cost of remediation and the uncertainties are determined this may become the basis for structuring a deal by allocating risks between the parties. Ideally, one would like a description of the complete spectrum of cost possibilities and their associated probabilities. An estimate of the expected time to regulatory closure and the associated uncertainties may also be factored in.

Environmental forensics enters into the site assessment process in several ways. First, in Phase I the site use history, as revealed by interviews and records, and visual clues during reconnaissance are combined with the analyst's knowledge of specific industrial operations to develop expectations of the presence and type of contamination. Secondly, in Phase II this information is augmented by sampling data to determine the extent of contamination. Finally, determining who is responsible for the contamination may involve other parties and hence introduce other remediation cost-sharing options. For example, groundwater contamination under a site may in fact originate from off-site sources.

1.4 INSURANCE LITIGATION

Insurance claims are based on the contract language between insurer and insured. Contracts until the mid-1980s were based on comprehensive general liability (CGL) policies. Subsequently, environmental impairment liability (EIL) policies were introduced to deal specifically with contamination and other environmental issues. Interpretation of the language is governed by state law and can vary greatly. However, the same phrases in the contract and the same issues produce the need for forensic information in any state in order to determine matters of fact.

Insurance coverage for damages associated with chemical contamination in the environment may depend, among other things, on the 'imminence' of off-site migration, whether coverage was 'triggered' during a policy period, and whether the release was 'expected and intended,' or 'sudden and accidental.' When multiple parties have contaminated a site, equitable cost sharing may also be a coverage issue (Murphy, 1993).

Parties may agree on the facts but still produce different descriptions for the same facts in order to construe the policy language most effectively. Several examples of this are noted below under the 'trigger of coverage' and 'sudden and accidental' sections. While there may be no 'correct' point of view, it may be useful to consider whether a particular point of view only arises in a litigation context and hence is not a customary point of view.

1.4.1 IMMINENCE OF OFF-SITE MIGRATION

Policies often apply only to third party property. However, if there is an imminent threat to off-site locations, coverage may exist for on-site cleanup. In some states, groundwater under a site is 'off-site.' To predict whether significant migration off-site is likely, soil leaching and soil erosion runoff models may be used. If the chemicals of concern are only slightly soluble in water and sorb appreciably to soils, then chemical transport through the vadose zone will be slow and concentrations reaching the water table may be below regulatory limits.

Groundwater transport models may be used to determine if a threat is imminent when groundwater contamination has not yet reached the property line and groundwater is considered off-site. Because of biodegradation in the plume as well as weathering and sequestration of mobile waste constituents in the source region, some plumes may reach a steady state before going off-site, or even recede over time. This is a common observation for BTEX (benzene, toluene, ethylbenzene, and xylene) plumes from gasoline spills (National Research Council, 2000).

1.4.2 TRIGGER OF COVERAGE

Some policies provide coverage only if in force when a claim is made. Coverage in other policies is triggered in environmental remediation cases by property damage. However, states differ in their determination of when damage actually occurs and if it can occur only once or can occur in a continuing fashion.

The possibility of triggering multiple policies in different time periods with multiple triggering events can lead to different interpretations of the same events. For example, a groundwater plume from a spill on one occasion may be stabilized and even shrinking, but since new water molecules are always entering the plume, some might argue that new damage is continually being done. Others would, of course, argue that the plume itself demarcates the extent of the damage.

Determining when policies are triggered often involves back-calculating a time of release or time to reach the water table as described in Chapter 8. In some cases, structures such as cesspools, french drains, and leaching pits, which were specifically designed and installed to facilitate disposal of wastes to groundwater, negate the need for model calculations. A one-time liquid release, which is large enough to penetrate to groundwater, will generally do so over a period of hours or days. A cumulative or 'drip' release will reach groundwater over a period determined by the drip rate. If the total quantity released is insufficient to reach the water table, the rate of contaminant travel will be controlled by the rate at which precipitation infiltrates the soil column and carries soluble waste components downward.

Reverse groundwater modeling, discussed in Chapter 8, can be used to determine the time when a property line was crossed or groundwater was first contaminated. However, there are always substantial uncertainties introduced by limited measurements in the subterranean environment. In addition, care must be used in defining the plume front; while the peak plume concentration may move with the retarded velocity, contamination in front of the peak moves more rapidly, up to and in theory even exceeding the groundwater velocity.

It may be possible to establish the time of release by linking the observed contamination to known process changes, such as a change in degreasing fluids from trichloroethylene to 1,1,1-trichloroethane (TCA). TCA releases can also be dated by the amount of the hydrolysis product, 1,1-dichloroethylene, present. This is described in Chapter 7.

1.4.3 EXPECTED AND INTENDED

It will generally be important to determine if the damage was 'expected and intended.' There can be issues which vary from state to state as to precisely what was 'expected and intended,' i.e., the release to the environment or the damage. Depending on the state, a 'reasonable man' standard may apply or it may be necessary to produce evidence of actual knowledge by specific individuals. Of course what is reasonable for an individual to know depends on their background and role in an organization. Expectations are different for an accountant and an engineer, whose job might require her to read professional literature in her field.

Thus in some cases it will be useful to compare facility practices with historical waste disposal practices as evidenced by the engineering literature for the appropriate time period. The following illustrate the type of information that can be found.

- 'The old fallacy of the speedy self-purification of streams was once pretty firmly fastened upon the engineering profession itself, and it is only in relatively recent times that it has been wholly abandoned.' – Editor of Engineering News, Stream Pollution Fallacies, *Engineering News*, Vol. 42, No. 9, 1899.
- 'The discharge of manufactural waste into streams without purification and treatment has frequently resulted in serious pollution. Manufacturers are coming to realize the seriousness of the conditions and consequently much study is being devoted to methods of rendering the wastes innocuous before their discharge into bodies of water.' – Disposal Methods for Manufactural Wastes, *Engineering Record*, August 27, 1910.
- 'In the arid and semi-arid regions of the West, many large communities are virtually dependent upon groundwater supplies... Surveys show that refinery wastes in particular penetrate to considerable distances from sumps and stream beds.' – Burt Harmon, Contamination of Ground-Water Resources, *Civil Engineering*, June, 1941.

As evidenced by these examples, the early pollution incident literature tends not to be chemical specific. It also is concerned with levels of contamination much higher than the levels that can be recorded with present measurement technology.

Pollution control legislation, practices at other companies in the same field, or trade association publications may also be introduced to illustrate the state of knowledge or practice at a given time. Generally, the engineering literature will provide a picture of a more advanced state of knowledge at an earlier time than these other references. Chapter 2 describes some sources of historical information.

If documentary information as to practices at a particular facility is lacking, it may still be possible to discern historical waste disposal practices from the spatial location or 'footprint' of contamination at the site. For example, a groundwater plume emanating from a dry well could be linked to disposal of chemicals down a laboratory sink drain.

It may be important to distinguish contamination that arose from routine operational spills, which could be argued to be expected and intended, from such things as tank failures. Estimating the mass of contamination in soils and groundwater and characterizing the location relative to process areas can help in making such a distinction by determining the origin of contamination.

1.4.4 SUDDEN AND ACCIDENTAL

In the 1970s a clause was introduced to CGL policies that stated coverage for various kinds of releases would only apply if these were 'sudden and accidental.' Some states consider 'sudden' to have a temporal meaning while others consider it to be more akin to 'unexpected.' In the former case, it may be important to determine if a release was gradual or sudden in a temporal sense. However, even if the parties agree on the facts different interpretations can arise. For example, a leaking underground storage tank might be viewed as the result of years of electrochemical corrosion or it might be viewed in terms of a single instant when the tank is finally breached. Similarly, routine periodic degreaser cleaning and discharge to the environment might be characterized as a series of 'sudden' releases or as a chronic operating condition.

1.4.5 EQUITABLE COST SHARING

Equitable cost sharing becomes an issue if there are multiple PRPs at a site. An unfavorable cost allocation scheme may be a basis to dispute full policy coverage. For example, as discussed in Section 1.2, when there are wastes that differ greatly in toxicity or mobility, a scheme based solely on the quantity of waste will be unfair to the disposer of large volumes of innocuous waste and its insurers.

Equitable cost sharing requires that waste streams be identified with specific PRPs. Methods are available for unmixing commingled waste streams. These include isotope techniques, discussed in Chapter 5, as well as principal components analysis (PCA) and polytopic vector analysis (PVA), discussed in Chapter 12. When indemnification costs are presented or settlements proposed, the question may arise: 'Should these techniques have been used?'

1.5 TOXIC TORTS

In a toxic tort the issue is most often whether an injury was 'more likely than not' caused by exposure to chemicals or other substances (e.g., radiological or biological). The causation requirement may also be phrased as 'but for' the exposure the injury would not have occurred or that the exposure was a 'substantial contributing cause.' Environmental forensics enters because historical chemical concentrations in air, water, soil, or foodstuffs are needed to estimate exposure and dose.

Dose is exposure times some uptake rate, e.g., cubic meters of air inhaled or an average number of grams of fish eaten per day. Exposure is determined by the concentration in environmental media, air or fish in the above examples, and by the time period over which uptake occurs. Since chemical concentrations may vary with time, exposure may be characterized over various time periods, acute or peak exposure, subchronic, or chronic (long-term) exposure. The averaging time of interest depends on the specific health effect being investigated.

Proof of causation for the injury may proceed in at least three ways. (1) If a sufficient number of people have been exposed epidemiological evidence may be offered. (2) A differential diagnosis may be performed for specific individuals. (3) Although this is less frequent than the other two procedures, a risk assessment may be performed.

1.5.1 EPIDEMIOLOGY

Epidemiological information is often presented as an odds ratio for a specific type of injury. The odds ratio is the ratio of the number of observed cases in the putatively exposed community to the expected number of cases for a community of that size and demographic composition. Usually adjustments are made for age, smoking, ethnicity, etc. in determining the expected number of cases. It is often claimed that a probability of causation can be calculated directly from the odds ratio. If the odds ratio is OR, the probability of causation, P_c, is said to be:

$$P_c = \frac{\text{OR} - 1}{\text{OR}} \tag{1.4}$$

P_c is thus just the fraction of total number of cases represented by the 'excess' cases above background.

There are several things wrong with this argument and they show the role that historical exposure information, and hence forensic analysis, can play in assessing epidemiological evidence of causation: (1) association is being confused with causation; (2) no accounting is given of the number of different disease endpoints that were examined in order to find an OR > 2; and (3) in the above discussion the uncertainty associated with OR itself does not enter into determining P_c.

1.5.1.1 Association and Causation

Epidemiological evidence by itself describes 'association' not 'causation.' To move from association to causation the Hill criteria formulated by Sir Austin Bradford Hill are often invoked (Hill, 1965). These criteria are:

1 *Strength of the statistical association.* As noted above, this is often measured by an 'odds ratio' comparing exposed and unexposed populations.
2 *Consistency of the association.* Is the disease observed with similar exposures in other places and times? Do studies using a variety of techniques arrive at similar conclusions?
3 *Specificity of the association.* Does the disease have many causes? Can the chemical in question cause many diseases?
4 *Temporality.* Does exposure precede disease? Is the disease onset consistent with what is known concerning latency?
5 *Biological gradient of the disease with exposure.* Are data for the population under study consistent with a dose–response relationship? Do the data show increasing rates of disease with increasing dose?
6 *Plausibility.* Is a causal relationship between disease and exposure biologically plausible?
7 *Coherence.* Is a causal interpretation consistent with other scientific understanding?
8 *Experiment.* For example, if the suspected cause is removed, does the disease rate change?
9 *Analogy.* Does experience with similar situations give any guidance?

The Hill criteria are intended as different viewpoints for examining causality rather than constituting a pass/fail exam. An additional criterion, which is sometimes added, is:

10 *Elimination of confounders.*

Several of the Hill criteria are exposure related. The biological gradient criterion asks whether the number of cases increases with increasing exposure. The temporality criterion asks whether exposure preceded effect and if so whether it was by enough time to be consistent with what is known about disease latency. Consistency of the association is also exposure related. The injury may have been observed to occur elsewhere only when exposure was above some level. Both concentration in the exposure medium and averaging time enter into 'level of exposure.'

1.5.1.2 Texas Sharpshooter Effect

The second thing that is wrong with the simple odds ratio/probability of causation argument is that how many endpoints were looked at is an important consideration in interpreting the results. If enough disease endpoints are examined, an odds ratio greater than 2 may be found for some condition on a purely statistical basis. Restricting results to a 95% confidence level will not prevent this. At a 95% confidence level one test out of twenty will appear to be statistically significant, even if exposure conditions are actually identical in the two communities being compared. If a large number of disease endpoints are examined and only the high odds ratio and high confidence level cases are then presented this constitutes what is sometimes called the 'Texas sharpshooter' effect – where the bull's eye is drawn after the gun is fired!

1.5.1.3 Statistical Significance

The third thing that is wrong with an odds ratio greater than 2 simply equating to a causation probability greater than 50% has to do with the statistical uncertainty inherent in the odds ratio determination.

The basic concept is that even if one could do a series of identical studies with identical test populations and exposures there would be a distribution of odds ratios because of statistical fluctuations. Let x be the odds ratio and $f(x)$ the distribution of odds ratios. Then the probability of causation is:

$$P_c = 1 - \int_0^\infty \frac{f(x)}{x}\,dx \tag{1.5}$$

Charrow and Bernstein (1994) show that Equation 1.5 implies:

$$P_c < 1 - \frac{1}{\int_0^\infty x f(x)\,dx} \tag{1.6}$$

for any distribution $f(x)$ subject to the normalization condition:

$$\int_0^{\infty} f(x)\,\mathrm{d}x = 1 \tag{1.7}$$

If the odds ratio is identified with the expectation value of x:

$$\mathrm{OR} = \int_0^{\infty} x f(x)\,\mathrm{d}x \tag{1.8}$$

then it follows that:

$$P_\mathrm{c} < 1 - \frac{1}{\mathrm{OR}} \tag{1.9}$$

Thus just the fact that a distribution of odds ratios would occur if the epidemiological study could be repeated with different, but equivalent, populations causes the probability of causation to be overestimated by the simple expression $P_\mathrm{c} = 1 - 1/\mathrm{OR}$.

1.5.2 DIFFERENTIAL DIAGNOSIS

The second way of determining causation is through a 'differential diagnosis.' In a clinical setting this term means determining the underlying disease from among various possibilities through an analysis of symptoms. In a toxic tort context the term has taken on the meaning of determining the cause of a disease from among various possibilities. For example, if the claim is that a heart attack was chemically induced, among the factors that should be looked at as part of a differential diagnosis are the individual's weight, smoking habits, blood pressure, and age. This is in addition to whether the specific chemical is associated with heart disease and what the level of exposure was.

Causation is often considered in two parts. General causation addresses the question of whether the chemical in question is believed capable of causing the injury in question at any level of exposure. Specific causation addresses the question of whether the chemical caused the injury in the specific individual. The Hill criteria are used to support the general causation argument. Of course, for substances where the effect is well known, appeal to medical textbooks or government documents may be sufficient. Specific causation relies on the subset of the Hill criteria that may be applied to an individual rather than a population. These are the consistency of the association for the level of exposure, specificity of the individual's injury for the chemical, temporality (e.g., exposure preceding disease), experiment or whether symptoms are alleviated when the supposed chemical cause is removed. In addition, as indicated above, confounding factors or other potential causes are eliminated as part of a differential diagnosis.

Historical exposure is thus an essential part of a correct differential diagnosis and can enter into a causation analysis in a quantitative way. Most obviously, if we know the human exposure level at which disease is likely to occur. However, this condition is a rarity. More often we know the exposure level at which disease is not likely to occur. Thus one may compare the estimated historical exposure for an individual with chemical specific standards and criteria, both public health and occupational. Chemical exposure is unlikely to be a significant cause of disease if this exposure is less than these criteria. Of course, some caveats are necessary. Occupational criteria are generally less stringent than public health criteria and there may be issues in comparing a 'healthy worker' to the general population. Also criteria may not be based on carcinogenic effects, particularly for chemicals with limited evidence of carcinogenicity.

Similarly, one may investigate the significance of specific exposure levels by comparing the exposures or the concentrations involved with the exposures and concentrations of those same chemicals that people are normally exposed to through the natural or anthropogenic background, such as consumer products or urban air.

1.5.3 RISK ASSESSMENT

The basic risk assessment algorithm, risk = toxicity × dose, demonstrates the key role that dose and hence exposure plays. A risk assessment may be conducted at a hazardous waste site using EPA methods to calculate the lifetime risk of excess cancer, that is of a cancer which would not otherwise have occurred. EPA's criterion for cleanup at a Superfund site is a computed risk, ΔR, larger than 10^{-4} to 10^{-6}. Computed risks rarely approach the 'more likely than not' criterion of 0.50. This might be taken to imply that risk assessment is not a useful tool for demonstrating a causal toxic exposure. However, when there is a background rate for cancer of a certain type of R_0 and a computed chemical risk ΔR, the probability of causation for an individual who has cancer of that type can be written as:

$$P_c = \Delta R / (\Delta R + R_0) \quad (1.10)$$

Thus, even computed risks in the 10^{-4} to 10^{-6} range can be used to support a causation argument for an individual who already has cancer.

EPA risk assessment methods also can be used to calculate what is known as a 'hazard index' for noncarcinogens. The hazard index is simply the computed dose over some averaging time, usually 24 hours, divided by a reference dose. The reference dose is a dose at which no adverse effects are believed to occur. Thus computing a hazard index <1 means that no adverse effect would have been expected from the exposure. Computing a hazard index >1 leaves the question of a chemically caused adverse effect open. In that case an informed judgment requires reviewing the primary medical literature including the occupational or animal studies upon which the reference dose is based.

ACKNOWLEDGMENTS

Certain of the above material appeared in different form in *Environmental Claims Journal*, Vol. 5, No. 4, Summer 1993 and Vol. 8, No. 3, Spring 1996 © Aspen Law & Business, Inc. as well as in *Environmental Forensics*, Vol. 1, No. 3, September 2000 © Academic Press.

REFERENCES

Aronovsky, R.G. (2000) Liability theories in contaminated ground water litigation. University of Wisconsin, Environmental Litigation: Advanced Forensics and Legal Strategies, April 13–14.

Butler, J.C. III, Schneider, M.W., Hall, G.R., and Burton, M.E. (1993) Allocating superfund costs: cleaning up the controversy. *Environmental Law Reporter* 23, 10 133–10 144.

Charrow, R. and Bernstein, D. (1994) *Scientific Evidence in the Courtroom: Admissibility and Statistical Significance after Daubert*. Washington Legal Foundation, Washington, DC.

Ferrey, S. (1994) The new wave: Superfund allocation strategies and outcomes. *BNA Environmental Reporter* 25, 790–803.

Hall, R.M., Harris, R.H., and Reinsdorf, J.A. (1994) Superfund response cost allocations: The law, the science and the practice. *The Business Lawyer* 49, 1489–1540.

Hill, A.B. (1965) The environment and disease: association or causation. *Proceedings of the Royal Society of Medicine* 58, 295–300.

Marryott, R.A., Sabadell, G.P., Ahlfeld, D.P., Harris, R.H., and Pinder, G.F. (2000) Allocating remedial costs at Superfund sites with commingled groundwater contaminant plumes. *Environmental Forensics* 1(1), 47–54.

Mink, F.L., Nash, D.E., and Coleman, J.C. II (1997) Superfund site contamination: Apportionment of liability. *Natural Resources and Environment* 12, 68–79.

Murphy, B. (1993) Technical issues in Superfund insurance litigation. *Environmental Claims Journal* 5, 573–592.

Murphy, B. (1996) Risk assessment as a liability allocation tool. *Environmental Claims Journal* 8, 129–144.

Murphy, B. (2000) Allocation by contribution to cost and risk at Superfund sites. *Journal of Environmental Forensics* 1(3), 117–120.

National Research Council (2000) *Natural Attenuation Groundwater Remediation*, National Academy Press, Washington, DC.

Rockwood, L.L. and Harrison, J.L. (1993) The *Alcan* decisions: Causation through the back door. *Environmental Law Reporter* 25 ELR 10542.

Wise, K.T., Maniatis, M.A., and Koch, G.S. (1997) Allocating CERCLA liabilities: The applications and limitations of economics. *BNA Toxics Law Reporter* 11, 830–833.

CHAPTER 2

SITE HISTORY: THE FIRST TOOL OF THE ENVIRONMENTAL FORENSICS TEAM

Shelley Bookspan, A.J. Gravel, and Julie Corley

2.1 INTRODUCTION

There is a polluted site. Cleanup costs are in the millions of dollars. There have been many industrial occupants over a period of decades, but there are no obvious culpable ones. Under provisions of the federal Comprehensive Environmental Response, Compensation, and Liability Act of 1980 (CERCLA), and particularly as it was revised and reauthorized in the Superfund Amendments and Reauthorization Act of 1986 (SARA; 42 *US Code* Sections 9601, et seq.), current landowners who are not themselves accountable for all or any of the contamination found on their property can sue certain responsible parties for recovery of their cleanup costs. Such potentially responsible parties include those whose activities actually caused some or all of the contamination, no matter how long ago they occurred, or those parties' successors in interest (i.e., generators). Other potentially responsible parties are those who owned or operated the site when the contamination, or part of it, occurred (i.e., past owners or operators at the time of disposal). Others still are those who may have taken contaminating materials to the site for disposal, or who arranged for them to have been taken there (i.e., transporters or arrangers for disposal, respectively) (42 *US Code* Section 107).

CERCLA and analogous state level legislation, then, provide an array of prospective sources for financial assistance. Practically, however, how is it possible for the current landowners, faced with a cleanup bill, and their legal counsel to begin to unravel the necessary information about the site to ascertain whether there are other extant responsible parties? How is the context derived within which science-based environmental consultants can interpret the findings of field and laboratory studies to argue that a particular party's practices contributed to the environmental problem? A narration of the site's history may be what provides disputing parties, scientists, and adjudicators alike with clear and compelling information about the origin of the environmental problem. Certainly, it is not enough to find the pathways that chemicals have taken through the environmental media, nor enough to characterize the offending chemicals. In order to connect those findings to an existing party and to argue effectively that the party owes cost contribution, the scientists and legal specialists also need to define the universe of past owners, occupants, generators, transporters, and/or arrangers. They need to be able to link the contaminants and the site conditions today somehow with the activities of the past. They need a narrative of the site's history to make sense of their findings.

Compared with the seemingly arcane formulae and models of science experts, the 'stuff' of narrative history is ordinary documentation that most lay people can understand. The narrative history itself will become a source of information for the entire environmental liability investigation. Because it is developed after the fact by a nonparticipant, this narrative is called a secondary source of historical information. A secondary source is a verbal reconstruction of events over a period of time that reflects the historian's best interpretation of the facts as discerned from the primary sources of information. The primary sources in the field of history are original documents, or documentary materials that were produced at the time of an occurrence. They may reside anywhere, in public agency files or in public or private collections. An archive is specifically a place housing original documentation, and there are archives around the country which maintain distinct types of collections. (For guidance to the country's archives see, for example, the National Historical Publications and Records Commission (1988) or try logging onto the Library of Congress's National Union Catalog of Manuscript Collections: lcweb.loc.gov/coll/nucmc.html. Or, for another site identifying and describing some 4000 plus repositories of primary documentation, see: uidaho.edu/special-collections/Other.Repositories.html.)

Primary history sources can be verbal or graphic, and the ability to discern their meaning usually does not depend on a grasp of chemistry, engineering, hydrological, geological, or mathematical or other scientific jargon. This does not mean, however, that the work of history reconstruction is uncomplicated or lacking rigor. Ordinary documents – letters, testimony, photographs, maps, field reports – are the raw materials of a site history, indeed. Nonetheless, identifying them, locating them, and piecing them together in a way that coheres, assesses their relative merit, is true to their meaning, and suspends value judgment requires adherence to historical methodology. Like studies conducted using the scientific method, history conducted using the historical method must withstand tests that include peer scrutiny, reproducibility, and reliability (Bookspan and Corley, 2001).

Developing a work plan for a site history. How might a historian proceed to find and analyze documents relative to the history of a contaminated site for which responsible parties are being sought? Among historians who specialize in research and analysis for the purpose of clarifying the origin of specific circumstances of pollution, it is axiomatic that there is no one-size-fits-all formula for designing a site history work plan. A timeline or chronology serves as the backbone of a narrative, and there are a handful of standard types of sources for which anyone compiling a site timeline must look. Topographic maps and aerial photographs, for example, help define the period of development and thereby help define the research periods. Sanborn fire insurance maps (block-by-block insurance maps of American cities published by the Sanborn Map Company), if they exist for the subject property, can supply a snapshot of the on-site land uses at a given point or points in time. Building permits and city directory listings, if either of those exist for the subject site, may define some more of the bones of the land-use skeleton.

These are places to start, necessary but insufficient. A historian working a pollution liability claim or dispute needs to know much more about a site than simply what was on it, who was associated with it, and when, although these are vital data indeed. So vital, in fact, that such data can lead to the formation of a tailored work plan designed to unearth the idiosyncratic data on which a case may turn. For what gains the attention of the mediator, judge, or jury is not the listing in the city directory of a salvage yard. It is the fifty-year-old transcribed testimony of a neighbor who told the Planning Department that he saw 'dozens' of leaking drums buried on the site. It is not the aerial photograph that shows a sump, it is the letter of complaint about the 'rotten egg' smell from that sump, found within the microfilmed files of the County Board of Supervisors. More than likely, historians will use disparate sources, some common, some obscure, to develop a sequential picture of what happened on the site. It is rare that a single source, or even single type of source, of historical information is complete, unambiguous, or unequivocal.

The methodology guiding historians through their planning for and conducting of site research, then, starts with the basic sources and proceeds through a warren of avenues. The work plan the historian develops at the outset of the research is effectively a road map identifying the main highways through a selected set of prospective sources, including the type of document sought, the information it is expected to provide, and the place or places where the document might reside and be found. The use of key words to navigate through published indexes to libraries and to archival collections will aid in the construction of this initial work plan.

Internet searches designed to identify repositories of potentially relevant primary sources can also be of use, especially in this important first phase of the historical work. The scope and types of information and resources available on the Internet vary greatly. Some of the information is accurate, and a lot of it is not. The sheer amount of information has made it more difficult to determine the veracity of the data available. Nonetheless, certain searches can be particularly helpful when starting out on a project, for example:

- Manuscript collections pertaining to the companies and/or agencies of interest.
- Environmental Protection Agency (EPA) databases for information regarding sites under agency scrutiny that are located in the vicinity of the target site.
- Library and Special Collection Databases.
- Databases containing corporate information.

Very little of the information available online, however, is of the documentary type. For historical research the Internet best serves as a means to find places, outside of the office and inside files, shelves, or archival boxes, to look.

The initial work plan, or research map, starts the historians on the treasure hunt, but is essentially an uncharted guide that requires revision as they actually explore those routes. They may find dead ends, or new or forgotten trails, back roads, or shortcuts. The value of the work plan at the outset will be commensurate with the historians' ability, through an understanding of change over time, to step away from current infrastructure, systems, and values of the prevailing society in which environmental discharges now occur. Files from the 1920s, for example, will simply not contain the words 'environment' or 'ecology.' That does not mean that there are no relevant files, however. They may be found perhaps by use of engineering nomenclature rather than 'green' words, reflecting what were then the central values in that heyday of the civil and sanitary engineer.

Context, then, is key to a successful site history work plan. A work plan tailored to the site will reflect:

- the time periods of potential interest;
- the jurisdictions in which it falls and into which it has fallen over time;
- the types of chemicals currently found on or under the site;
- the types of land uses or businesses occupying the site and when;
- the remedial and legal issues at hand.

Although there may be some similar elements, a work plan designed for each of, say, the following situations, will vary greatly from each other. Each of the situations contains clues for crafting the plan. Table 2.1 describes three distinct site history problems for pollution liability matters and extracts the clues to the explanatory context and consequently leads to useful primary source documents therefrom.

So it is that specific historical questions as well as the locations of the sites and the types of chemicals (and their containers, if any) causing the problems will suggest to the historian an era of origin and a range of sources. Review of the standard kinds of historical sources, such as the topographic or Sanborn fire insurance maps, aerials, city directories, and building permits will help confirm the initial impression and, perhaps, provide some of the other clues, such as the name of a former site occupant, on which to build that in-depth work plan. For more detailed listings of sources of site history information and repositories in which to find them, see Bookspan (1991).

Hypothetical cases and workplans. To provide further illustration of the interrelationship between research goals, historical context, and historical research

Table 2.1

Three examples of pollution liability site-history problems.

Research Objectives	Contextual Elements Suggested	Some of the Possible Primary Sources Suggested
To name contributors to an acid sludge pit in Louisiana	Refinery wastes reflective of Louisiana industry? Off-site generation? Multiple contributors? Possible association with aviation gasoline production and wartime era land disposal? Pre-dating of specific state or federal environmental regulations	Contemporaneous petroleum industry trade literature, such as *Oil and Gas Journal*. Parish courts for early pollution or nuisance complaints. Local history ephemera, such as booster/Chamber of Commerce literature. World War II aviation gasoline program records in the National Archives
To identify owners or the successors thereto of a defunct manufactured gas plant in Pennsylvania	Era of operation the gaslight era, between 1880 and 1940? Utility selling gas service to the public in an era pre-dating specific federal environmental regulations? Owners possibly benefiting from extensive acquisitions and mergers in the utility industry in early twentieth century? Possibly coming under state regulation for rates because of early Pennsylvania state involvement? Possibly making enough change through its operation in town to receive local attention?	County recorded grant deeds. State Public Utilities Commission ownership files and records. *Brown's* directories, for utilities. *Moody's Industrial Securities* for acquisition and merger information. Securities and Exchange Commission filings for post-1930s corporate tracings
To identify parties contributing to an extensive trichloroethylene (TCE) plume in a California aquifer	Post-1930s and pre-1980s, reflecting the market for TCE versus carbon tetrachloride or trichloroethane? Multiple contributors? Metal working industries? Industries using heavy mechanical equipment? Cold war-related industries, as flourished in California? Local permitting for sewer use and special land uses beginning?	Various federal record groups within the National Archives for wartime, military-related industry records. Industrial waste disposal permits, conditional use permits, discharge regulations, at the state and county level for 1950s onward

methodology, provided below are three additional hypothetical situations in which the key question is who is responsible for a specific pollution problem. These hypotheticals represent each of the three basic situations in which potentially responsible parties (PRPs) may find themselves:

1 They may be involved in a single-party site with a history of preceding uses and users that may have contributed to the current problem.

2 They may be involved in a multi-party site, such as a regionally spread groundwater plume, where operations that resulted in chemical releases to the environment occurred *in situ.*

3 They may be involved in a multi-party site, such as a landfill, to which wastes generated off-site were transported and commingled over a period of time.

A detailed work plan follows each of these hypotheticals. While neither the situations nor the work plans account for all of the possible configurations of environmental events or available documentation, they are intended to show how the case historian interweaves goals, context, and sources to reconstruct history.

2.2 HYPOTHETICAL SITUATION 1: Single-party sites with a history of preceding uses and users that may have contributed to the current problem

Problem. The client is a law firm representing a multinational chemical company (Chemicals, Inc.) that, among other things, owns several paint and dye manufacturing plants in the northeast of the United States. Ownership derived from the acquisition of two paint companies, one of which dates to the 1890s and the other to 1910. Some of the facilities remain in use, although, of course, they have been rebuilt and expanded many times over the years. Others of the facilities, however, are not functional, and others still have been sold to new operators.

As part of its program for minimizing its environmental exposure and for obtaining, where possible, third-party contributions for the costs of environmental remediation, the company has hired the client law firm to undertake a cost recovery program on its behalf. That law firm, in turn, has asked the historical consultant to assist in developing the requisite information by which to evaluate the cost recovery potential for the various sites.

The goals of the research are as follows.

1 Insuring that all of the facilities that have been acquired over time are identified.

2 Identifying and analyzing those facilities that have the potential for federal government cost recovery through wartime associated ownership and/or operatorship.

3 Identifying and analyzing those facilities that have the potential for private third-party cost recovery through other theories, including prior or subsequent ownership and/or use; toll contracting; and off-site contributions.

Workplan. In the Chemicals, Inc. example it would be important to use a tasked approach that would allow the researchers to build on the research done in each task, ultimately to answer the questions posed in the research goals.

2.2.1 TASK 1: CORPORATE SUCCESSION RESEARCH

In this particular situation, prior to conducting what will undoubtedly be extensive research on ownership and operations issues, it is necessary to determine which company or companies to examine. To do this, it will be critical to gain as much insight as possible from the attorney(s) representing Chemicals, Inc. and company personnel regarding existing common knowledge about the company's corporate lineage. In addition, the research will focus first on the main company and, subsequently, on the companies it acquired over time. A study of Chemicals, Inc. may well lead to research of Paint Company 1, Inc. (Paint 1), Paint Company 2, Inc. (Paint 2), etc.

Primary sources. Because of its evidentiary value, the most desirable and useful information for the client will be obtained from primary, or documentary, sources. The three primary sources of corporate succession data for this case will include:

1 Company-specific collections: Researchers should always begin by conducting research and speaking with company personnel to determine if company-specific collections exist either internally or in the public domain. Often, company archives can be a rich source of relevant information. In addition, many company-specific collections exist in the public domain. Reference material to such collections should always be consulted to identify whether such collections of relevant materials exist.
2 Secretary of State filings: The next step in the corporate succession research is to conduct research in the appropriate Secretary of State office(s). Relevant information may be found in more than one state as companies sometimes reincorporate elsewhere over time for tax or other purposes. The types of information obtained should include:
   - articles of incorporation;
   - amendments;
   - name changes;
   - mergers;
   - acquisitions;
   - dissolutions;
   - foreign corporation registrations; and
   - certificates of good standing.

3 Security and Exchange Commission (SEC) filings: In addition to the Secretary of State, SEC filings can also be a rich source of corporate filings, if the company under scrutiny is or has been traded publicly. The purpose of federal securities laws, namely, The Securities Act of 1933, The Securities Exchange Act of 1934, and The Investment Act of 1940, was to require companies seeking to raise capital through the public offering of their securities, as well as companies whose securities are already publicly held, to disclose material financial and other information about their companies. Annual reports, 10K filings (forms filed annually by publicly traded companies with the SEC) and investment prospects are the main vehicles for this communication and often contain very useful information for this type of research. This information can be obtained through a variety of sources including several service bureaus (e.g., Disclosure, Inc.) that make a business of providing this information.

Published sources. In addition to primary sources, there are a variety of published contemporaneous sources to be researched for the corporate succession phase of the project. Some of the most useful information can be found in business- and industry-specific directories, and these can often be found in libraries, particularly business libraries at universities or community research libraries. These sources tend to be published by research services, business services, or educational institutions and are compiled by researchers or through voluntary submission by the companies listed in the publications. In contrast, other secondary sources such as trade literature, business magazines, newspapers, and the like often contain articles or profiles of companies that are authored by a variety of individuals or groups. Published sources for the Chemicals, Inc., corporate succession research may include but not necessarily be limited to:

- *Moody's* Manual of Corporations;
- *Standard & Poor's* Corporation Records;
- *Ward's* Business Directory;
- *Directory of Corporate Affiliations*;
- *Million Dollar Directory*;
- *Wall Street Journal* Index;
- *New York Times* Index;
- Chemical Manufacturers Association *Newsletter*;
- *Chemical Week* Magazine.

2.2.2 TASK 2: FACILITIES RESEARCH

Once the corporate succession work is done, the historians will compile two important pieces of information into one client work product: a detailed corporate history of Chemicals, Inc. and a list of facilities linked to Chemicals, Inc. through its predecessor companies. To compile the latter, many of the same sources, particularly company-specific records and directory information, will be used. This information will then be used to inform Task 3.

2.2.3 TASK 3: OWNERSHIP AND OPERATIONS RESEARCH

Taking the list of facilities compiled in Task 2, the historians can now focus on conducting ownership and operations history research.

2.2.3.1 Ownership History through Chain-of-Title and Tax Research

Chain-of-title research. In order to answer the question of ownership over time at each of the Chemicals, Inc. facilities, the researcher must determine the parcels of land related to each facility and conduct detailed chain-of-title research at the appropriate county, parish, city clerk's, or other office. In addition to researching the subject site, chain-of-title research to identify the owners of adjacent property may be conducted using the same time period criteria used in the Chemicals, Inc. research. This information may prove very useful in identifying off-site contributors to contamination of the subject property over time and in identifying parties from which Chemicals, Inc. can recover costs. In this phase of the work, it is critical to detail all of the transactions related to the subject properties, to analyze any subdivisions of property, and to note any lease agreements and/or easements recorded in the deeds.

Tax research. In addition to deed research, it is often helpful to conduct research in the local tax collector's office. This office generally holds all information related to taxes paid, taxpayer identification, parcel information, and property liens and foreclosures. These data can be very useful in documenting the history of a property and confirming the identity of PRPs for cost recovery.

Once the ownership research is complete, the data should be compiled, analyzed and recorded so they can be used to inform the operations history research.

2.2.3.2 Operations Research

Next, the researcher can focus on the operations portion of the project. To complete this phase of the work, the researcher or research team must now examine federal (if appropriate), state and local and industry information sources to piece together a history of operations at the Chemicals, Inc. sites. Research sources may vary depending on the facility's location and other factors such as types of operations and period of operation. However, based on the Chemicals, Inc. example, the following types of sources should be helpful:

Federal data sources. Since there is a question about the federal government's involvement with the Chemicals, Inc. facilities, the research team will analyze the chain-of-title and other relevant documents to determine any government ownership or other types of control, particularly during times of war. If evidence of ownership or government involvement at an operational level were found or suggested, a more detailed review of federal records would be initiated. Depending on the agency or branch of the military service involved, the researcher might examine records contained in:

1 The National Archives and Records Administration (NARA) archival research facilities, which house records accessioned by the Archives, deemed to have potential historical value, and maintained within Record Groups signified by their federal agency or department of origin. (Online indices to the collections held within the regional branches of the National Archives and within the main research facility in College Park, Maryland, may be found, respectively, at the following sites: nara.gov/regional/findaids/findaids.html and nara.gov/nara/nail.html.)
2 The Presidential Libraries.
3 The Federal Records Centers, also operated by NARA, which house old files of federal agencies not accessioned to the Archives and still under the originating agency or department's control.
4 Specific military collections such as the Air Force History Collection at Maxwell Air Force Base.

(See Hypothetical Situation 2, below, for further discussion of federal record sources.)

Site circumstances might suggest the utility of federal records other than ones relating to ownership or operations. For example, if issues of eminent domain because of the construction of an interstate were present, a review of Federal Highway Administration or US Department of Transportation records could be appropriate. Likewise, if a navigable waterway or dredging were at issue, a review of the records of the US Army Corps or Engineers or the US Coast Guard could be of use. In the Chemicals, Inc. example it is likely that World War II records reviewed would include but not necessarily be limited to:

- Chemical Warfare Service;
- War Production Board;
- Defense Plant Corporation;

- Office of Price Administration; and
- Ordnance Department.

2.2.3.3 State Level Research

In the Chemicals, Inc. example, the state agencies targeted for research would be determined based on the specifics of the facility. However, generally the state level research would include but not necessarily be limited to agencies, whose exact names will vary based on the state, such as the:

- State Archives;
- Attorney General's Office;
- State Department of Health;
- State Water Resources Board;
- State Library;
- State Department of Transportation;
- State Lands Bureau;
- State Historical Society;
- State Department of Environmental Conservation; and
- State Department of Public Works.

2.2.3.4 Local Level Research

In this area the same criteria for selection would apply, but it is likely that local level research would include but not necessarily be limited to agencies such as the:

- Department of Public Works;
- Clerks Office;
- Fire Department;
- Health Department;
- Library System;
- County Offices;
- County Court Records;
- Historical Society; and
- Local specialty collections.

2.2.4 TASK 4: WORK PRODUCT

Once all of the information is collected, it is analyzed and compiled in a 'user-friendly' format. At this point, the researcher, who has at least intellectually been sorting information along the way, must take the disparate pieces and weave together a coherent story about the company. In the Chemicals, Inc.

example, the work product will have several distinct components including:

1 Detailed ownership history of the real properties/facilities to be acquired: In the Chemicals, Inc. example it is likely that this portion of the work product would include a detailed ownership history for all of the properties associated with the acquisition. The ownership history would include a description of the lands over time and a detailed chronology of ownership past and present.
2 Detailed information on operations at the various sites: Based on the information gathered at the federal, state, and local levels along with industry and company-specific operations information, an operations history would be developed weaving all of the disparate pieces of information to tell a coherent story about the site. Some of the components of the site history might include manufacturing, building improvement, infrastructure improvement and waste generation, handling and disposal activities that have taken place at the site over time.
3 An evaluation of the federal government's involvement at Chemicals, Inc. facilities: In the Chemicals, Inc. example it would be critical to assess the federal government's involvement in site ownership and operations as would be done for any other PRP. For instance, if records showed that the government owned the property and controlled operations during World War II, by virtue of its exclusion from sovereign immunity under the Superfund law, the federal government would be liable for contamination that occurred during its ownership period and subject to cost recovery efforts.
4 An evaluation of the involvement of prior site owners and operators: Like the example of the federal government above, the researcher would also conduct a liability analysis of owners and operators of both the subject properties and adjacent lands to make a potential cost recovery assessment.

In sum, the historical investigation will result in a useful reference guide, complete with documentary evidence, for the future assessment of any of the facilities should they present with environmental liability problems. The information garnered will provide background regarding at least two of the major categories of parties liable under CERCLA: past owners who may have owned the property or facility when the release(s) occurred, and past operators whose activities may have resulted in hazardous materials being released to the environment.

2.3 HYPOTHETICAL SITUATION 2: A multi-party site, such as a regionally spread groundwater plume, where operations that resulted in chemical releases to the environment occurred *in situ*

Problem. In this case, the client is a law firm representing a small city in the southwest that owns a municipal airport that sits atop an extensive groundwater plume contaminated with residual volatile chlorinated hydrocarbon solvents (VOCs), and, in particular, trichloroethylene (TCE). The extent of the plume is not completely defined, but through a program of monitoring wells and sampling, it appears to cover an underground area at least six miles (9.5 km) long and two miles (3.2 km) wide. The airport, which succeeded the Army Air Force's World War II-era use of the facility, has been in operation since about 1954, and there are two repair hangars on site. Even so, there is no record of any use of TCE even in the degreasers there. Instead, the records, which date from about 1970, show that the solvent used in the degreasers has been 1,1,1-trichloroethane (TCA). There are some small industries, particularly automotive related, that reside above the plume as well. Nonetheless, the EPA has assigned responsibility for the cleanup of the plume to the airport, as the current owner of what the agency believes to be the site of origin.

The potential costs for the cleanup of this plume are in the millions of dollars, and there is the added prospect of providing water to, and facing toxic tort liability from, users of a municipal well that has had to be taken out of service. The city has, therefore, hired a law firm to assist in its defense. The law firm, in turn, has hired a consultant specializing in historical research to help clarify the factual issues involving responsible, and potentially responsible, parties. Those issues in particular are:

- Ascertaining whether the airport or any of its lessees used and disposed of TCE at any time in the past.
- Assessing the likelihood of a prior site occupant, such as the Army Air Force, having used and disposed of TCE while on the site.

Workplan. The following discussion illustrates how the historical consultant would approach each of these problem areas to develop the informational basis analyzing the defense potential for the client.

2.3.1 TASK 1: PAST OCCUPANTS AND LOCATIONS OF OPERATIONS

A good place to begin is to determine who the past occupants were, where they operated on the site, and to identify the types of operations they conducted. These entities will most likely be the backbone of the 'target entity' research list. One of the first steps taken to identify the target entities is to perform a chain-of-title search, documenting owners and recorded lessees of the property. Deeds and leases are normally filed at the county recorder's office and are searchable through grantee and grantor indices. For the purpose of this example, the results of the search showed the following information:

Owners	*Lessees*
Agricultural Company A, 1924–1941	
War Department, 1941–1946	
J. Smith, 1946–1953	Company B – auto shop, 1948–1953
	Company C – print shop, 1952–1955
	Company D – warehouse, 1952–1965
Municipal Airport Authority, 1953–present	Company E – storage, 1956–1970
	Company F – sheet metal, 1956–1960
	Company G – auto parts, 1956–1995
	Company H – frozen pizza, 1965–1980
	Company I – dog groomer, 1972–1990

The legal descriptions found in the leases will help to identify where each entity operated at the site, which in turn will help determine their proximity to the plume. In some cases, the legal description of the area may not clearly encompass all of the entities' activities, as operations may have occurred outside of the area designated in the lease. The key is to remember that the chain-of-title search is only a starting point.

Not all leases are recorded, so it would be prudent to check other sources that identify site occupants. For this project, two good sources would be city directories and the airport annual reports. If available, both of these sources may provide some additional names and some context about the types of operations. For instance, a city directory listing from 1952 might include a brief description or an advertisement about the company's services. Airport annual reports may provide an indication about the magnitude of the operations or even discuss site conditions. Other sources that would be of particular use at this point in the research are any historic site tenant maps that the airport authority may have as well as fire insurance maps, which typically identify structures, major equipment, site occupants, and other useful information.

2.3.2 TASK 2: TARGET ENTITY LIST

The title, directory, annual reports, and map research will provide a 'target entity' list of companies, with a documented association or connection to the site, to be used for further research. Review of the list may indicate a two-pronged research approach. On one hand, the target entities are companies that operated at the site in a relatively recent time period. On the other hand, there are industrial activities dating from the 1940s, when the military owned and operated the site. Sources of relevant information for the two categories in some cases may be the same, but those instances will be relatively few.

2.3.3 TASK 3: STATE AND LOCAL AGENCY RESEARCH

The next step is to implement a scope of work in which information about the chemicals used by site owners and occupants is assembled, as well as to document waste handling and disposal practices at the site. Certain federal, state, and local sources that would be consulted and the types of information that they typically contain include:

- Building permits: facility construction and major equipment installation.
- Fire department: inspections, notices of violations, and incident reports.
- Aerial photographs: site features and changes over time.
- Court records: nuisance/pollution complaints.
- Local historical society: contextual site and/or area history, and local business histories.
- Local public library: newspaper clipping collections, local business histories, contextual site and/or area history.
- County Department of Health, Environmental Section: inspections, investigations, notices of violation, releases and/or incidents.
- State Environmental Agency: inspections, investigations, notices of violation, enforcement activities, releases and/or incidents.
- Regional Air and Water Boards or Districts: inspections, investigations, notices of violation, enforcement activities, releases and/or incidents.
- State Archives: contextual site and/or area history, local business histories, state agency investigations.
- EPA: remediation files, investigations, inspections, permit records, compliance documentation, enforcement activities, and the like.

At this point the research results may seem to be disparate pieces of information. For instance, the researchers may have learned that Company B filed an application to install a paint spray booth with the building department in 1952. It is not clear if the permit was issued or spray booth installed. Court records informed the researchers that an employee sued the company in early 1953 for health problems associated with painting cars, but the case was apparently settled out of court and details about the painting activities and any solvents that may have been used in those activities were not in the file. Company B vacated the portion of the site that it occupied in late 1953 and the airport subsequently leased it to Company G, which began operating there in 1956.

Likewise, records from the fire department and the county environmental health department provide limited information about poor housekeeping practices that the two agencies observed at the operations of Companies C,

D, F, and H between 1954 through 1969, but do not indicate the names of chemicals used at the site. The records found at the state archives indicate that the State Department of Health initiated a large study of a nearby river in 1947 which reported numerous pollution problems in the vicinity, but the report only identified operations that were active at the time of the study and did not examine past operations for their potential contribution.

The list of potentially relevant findings could easily continue. Researchers often prefer using different approaches to the questions at hand. Some prefer to identify the exact type of document they want to locate and then set out to find it, while others prefer to see what they can find and then pull the various pieces of data together to tell a story about the operations at the site. Regardless of the method selected, it is essential to periodically examine the information obtained to identify gaps and leads to other relevant information.

2.3.4 TASK 4: CONDUCT FEDERAL ARCHIVE RESEARCH

At this point the most glaring gap is a lack of information about the activities of the military during the World War II era. The necessary research for the purpose of finding out about the daily operations of the Army Air Force is quite different than that described above. The records of the federal government are stored in various places, including the National Archives, Federal Record Centers, and with individual agencies.

The Federal Records Act of 1950 established a federal record management staff and a series of regional Federal Record Centers (FRCs) throughout the country. The FRCs work as a 'temporary' holding center for the records of various federal agencies. Typically, an agency retains its records for a relatively short period of time (e.g., two to seven years) after which it transfers items with potential 'lasting value' to the FRC. Items without lasting value are usually discarded by the agency at that time. Those materials transferred to the FRC are supposed to contain a notation regarding a schedule for their destruction or an indication that they are to be retained. Ownership of the records remains with the originating agency, and the FRC on average holds records for about twenty years. Prior to carrying out the ultimate disposition of the records – which can include return to the agency, transmittal to the National Archives, or destruction – the FRC usually contacts the agency to verify the action, or to request direction about schedules which were not correctly completed. Both the FRC and the originating agency normally hold documentation of the transfer of records in all aspects of the process. Records from the agencies are assigned a 'record group' number and their materials, for the most part, retain that number at both the FRC and the National Archives.

For this project, record groups (RGs) of interest in this matter might include:

- RG 18, Army Air Forces
- RG 51, Office of Management and Budget
- RG 160, Records of the Air Service Forces
- RG 179, Records of the War Production Board
- RG 219, Records of the Office of Defense Transportation
- RG 270, War Assets Administration
- RG 340, Air Force, Office of the Secretary
- RG 341, Air Force, Headquarters
- RG 342, Air Force, Commands

Another key repository for the task at hand is the Air Force Historical Research Agency (AFHRA), which holds, among other things, the records of Army Air Force facilities. Military units were required to prepare reports detailing their unit or section's activities on a regular basis. The reports were usually submitted to a person designated as the base historian, who in turn compiled the information from the various units into a comprehensive history of the base, camp, post, depot, or whatever. The comprehensive histories were usually compiled semi-annually, quarterly, or even monthly.

The level of detail found in the reports varies tremendously and reflects information that the soldiers who prepared the reports found of interest. While such reports may not contain information explicitly about TCE use, they may provide the most detailed information available about the military operations at the site. In addition, the unit histories will also contain the names of personnel stationed at the base. The individuals who worked at the base can often provide first-hand recollections that could provide pivotal or persuasive information about chemical use, waste handling operations, disposal practices, and the like. In the absence of direct documentary evidence of TCE use at the site, the information supplied by the soldiers can provide convincing testimony. World War II veterans, however, are by now quite elderly, and more are lost to us daily.

Once the research at the National Archives, Federal Record Centers, and AFHRA is complete, the historians will most likely be faced with additional disparate information. The research into the military's operations will have yielded, at a minimum, the types of activities that occurred at the base, such as aircraft maintenance, motor vehicle maintenance, gun maintenance, equipment renovation, railroad spur activity, dry cleaning and laundry facilities at the base, and the like. In an ideal world, they would have found explicit documentation that several of the site occupants used TCE and that they dumped, discarded, and disposed of TCE-containing wastes in an unacceptable manner (e.g., poured down drains, used it for weed or dust control, dumped it in a pit that was located in the middle of what is now the highest 'hot spot' at the site, etc.). If these were, in fact, the findings, and if the entities or their successors were active and financially viable, the historians could compile the information, prepare a report, and submit it to the EPA, which would most likely name the additional entities as PRPs. However, it is not the end of the trail if no direct documentary evidence of TCE use is found.

2.3.5 TASK 5: INDUSTRY STANDARDS RESEARCH

The next step would be to identify standard industry practices regarding chemical use and waste handling and disposal practices. Trade literature as well as military technical orders, manuals, directives, and the like can be used to help make connections – albeit inferred – between some of the activities at the site and the TCE contamination. The additional effort of locating and interviewing people who worked at the site as to their first-hand knowledge of solvent use there could provide further persuasive information. For instance, a soldier might recall that he used a special solvent to clean engines right before inspections so that the equipment would pass a particularly grueling white glove test. Another might recall using a vapor degreaser at the site in which only 'Tri' or 'Triad' could be used. The former soldier might not know that 'Tri' and 'Triad' were brand names for TCE, but that particular information can be documented elsewhere. Another worker might remember mixing a solvent concoction to produce a cleaner that removed carbon from engine parts more effectively.

If direct evidence of TCE use is not found, it would be worthwhile to define further the entities for which additional research is required. To do this evaluation, the historians should consider the types of operations and next conduct a preliminary corporate succession tracing to locate likely viable parties. For instance, it is unlikely that the dog groomer used TCE and also unlikely that the company – if it is still in business – has substantial assets. The auto-related lessees at the site may, however, be worth consideration because TCE was a solvent of choice for metal cleaning during the years in which the companies operated at the site. The agricultural company may also be worth further examination because TCE and carbon tetrachloride were used as grain fumigants prior to World War II. The US Government is definitely worth inclusion in any further research because the military was the largest user of TCE during World War II.

In sum, in order to address both of the research goals, the historians would develop a twofold plan: one in which they investigate possible TCE users among the private industry occupants of the area above the plume, and one to investigate the military's possible use and disposal of TCE during its wartime tenure. They would use the information gleaned from one source to enhance their ability to interpret another, rather than discard as useless documents that are not 'stand-alone' or 'smoking guns.' In the end, the historians' ability to construct a convincing narrative from the unearthed information will depend on the interplay among secondary, primary, and oral sources.

2.4 HYPOTHETICAL SITUATION 3: A multi-party site, such as a landfill, to which wastes generated off-site were transported and commingled over a period of time

Problem. In this situation, the client is a municipality that owned and operated a landfill, and only one landfill, for its citizens for a period of more than forty years, but that landfill ceased operations in 1978, a time period effectively predating waste manifest regulations. Now, chemicals have been found leaching from the landfill and migrating through the geological substrata to threaten the aquifers from which the municipal water district draws its supplies. The client city is facing liability not only for cleaning up the landfill and the aquifer, but also for providing alternate water sources to the affected water district clients, and, possibly, for toxic tort claims based on the citizens' suits alleging harm from consuming contaminated drinking water.

The city's attorneys have hired historical investigators to assist in developing evidence relating to two categories of potentially responsible parties under CERCLA: the generators of the contaminating wastes brought to the site over time and the haulers of those wastes.

Workplan. As the city's own consultants, the historians have unusually direct access to the files and records of the city, and need not await the results of a clerk's file search initiated on receipt of a formal letter request.

2.4.1 TASK 1: LANDFILL HISTORY

In order to narrow the list of potentially responsible parties in such a complex site as a landfill serving innumerable parties over a long period of time, the first step is to set the parameters of the research. Examination of the city's Public Works Department, Accounting, and City Council records, among others, should provide answers to some basic historical questions set out below.

When did the landfill open and was it run continuously until it closed? This will help define the period(s) of time relevant to investigate further and it will reveal the regulatory era(s) in which the landfill operated, so as to suggest which current or historic agencies' records, if found, may prove valuable.

Did the municipality own or lease the property? This will identify any co-owners who may be joined into the landfill action, and, if there was a change in ownership or a renewal of leases or the like along the way, it may signal a time when discussion about the landfill occurred in city administration or council. Successful research often depends on finding occasional, rather than regular documentary evidence, and pinpointing decision times can ease the search for such records.

How did the landfill operate? Did the municipality operate it through its own staff, or contract its operations to a private management company? How was income generated? Were customers billed, and, if so, what municipal agency did the billing? Were receipts retained? Answers to these questions will refine the researchers' search for specific types of prospectively helpful documents.

In this case, the landfill history reveals that this single landfill, when in operation, served all of the municipality. Moreover, the city operated its own refuse pickup and disposal service, paid for by citizen taxes, and disposed of all of the wastes picked up in the subject landfill. This seems to exclude finding anyone other than the city to qualify as a PRP by virtue of having been a hauler. It also suggests that anyone who ever lived in or operated a business in the city is a prospective PRP by virtue of having been a generator. Nonetheless, certain generators are sure to have contributed more chemical wastes to the landfill than others, so the next step is to identify who they may have been.

2.4.2 TASK 2: GENERATOR IDENTIFICATION

Primary source research. Because all of the occupants of the city during the time the landfill operated were generators, that is, their wastes all went there, the researchers will inventory the city through screening types of sources (such as city directories or Chamber of Commerce directories and Sanborn fire insurance maps) and, next, set criteria for priority investigation targets. So, perhaps, the researchers will use a series of city maps covering every five years of the landfill's life. Then, they may create a database of businesses listed in the city directories by their street addresses. Next, they may select some of those businesses by their business type, such as dry cleaner or metal shop or automotive repair, if such type has generally been associated historically with chemical usage and waste discharge, and chart those businesses by their locations on the series of maps. These businesses will represent the prospective PRPs, and there may be among them some entity, an aircraft parts manufacturer, for example, that not only particularly suggests chemical waste generation, but also suggests other document leads, such as regulatory or federal agency files. Even so, the nexus between the existence of the PRP in the city and its wastes being in the landfill has yet to be made directly.

Witness interviews. Barring the finding of some highly suggestive PRPs, the utility of documents to identify generators may diminish until there is another breakthrough to follow. It may be useful, then, to locate individuals with first-hand knowledge about the wastes generated by the businesses identified and/or first-hand knowledge about the disposal of those wastes into the landfill. Given that in this case, the city hired its own employees to conduct the business of refuse pickup and disposal, and given that the historians on the matter are consultants to the city, the personnel files may be available for access and review. In them are identified individuals who worked for the Refuse Division of the Public Works Department at the time the landfill operated, and, if appropriate truck drivers or other refuse workers, such as 'swampers,' those who picked up the cans and rode the back of the truck, can be found, they may provide testimonial evidence about the wastes they picked up from the target entities. Sometimes, a ride with such individuals through the city, following the route of their work, regardless of how altered the streetscape now is, can prompt memories that can flesh out the survey documentation in a convincing way.

2.4.3 TASK 3: NARRATIVE COMPLETION

Through the landfill history and the documentary and interview work undertaken to identify generators, the researchers will have compiled disparate data which will need to be organized and accounted for in order to provide the clients with the PRP evidence they requested at the outset. A narrative that explains the work done, the sources consulted, the rationale for the PRP selection, a summary of the evidence regarding each PRP, and appends copies of all cited documentation or interviews, is the final work product of this level of investigation.

2.5 CONCLUSION

One of the operative terms in the conduct of site histories for environmental forensics purposes is 'disparate.' The types of sources available will depend on the city, county, and state in which the site is located, the years of concern, the environmental issues of concern, the liability issues, and, of course, the researchers' ability to navigate among these concerns in the sea of documentation. The examples here are intended to illustrate these interactions. Equally, they are intended to dispel the notion that a 'smoking gun' document is needed to provide good evidence in a PRP case. After all, in environmental cases especially, the historical information will provide only a piece, even if it is sometimes the cornerstone, of the causation puzzle. While the historian may find that the subject property was home to a dry cleaning operation during the 1920s, and that the now-defunct operation used carbon tetrachloride in its process, it will remain for members of the scientific team to assess the relevance of that information to the contamination problem. If the site is contaminated with chromium, then the information about carbon tetrachloride, a chlorinated hydrocarbon, may be interesting, but not related to cleanup costs. On the other hand, the historical information may lead the scientists to conduct additional tests of, say, the place on the site that formerly held the dry cleaning equipment and, perhaps, add tests for chlorinated solvent analytes to the tests for metals currently being taken from the existing sampling locations. Overall, the historical story will provide the scientists with a way to interpret their findings relative to a past user's on-site activities and to build the core of the environmental forensics argument.

REFERENCES

Bookspan, S. (1991) History requited: historians and toxic waste. In *History and Public Policy* (Mock, D., ed.). Krieger Publishing, Malabar, FL.

Bookspan, S. and Corley, J. (2001) The paper trail: how and why to follow it. In *Practical Environmental Forensics: Process and Case Histories* (Sullivan, P.J., Agardy, F.J., and Traub, R.K., eds). John Wiley & Sons, New York.

National Historical Publications and Records Commission (1988) *Directory of Archives and Manuscript Repositories.* Oryx Press, Phoenix, AZ.

FURTHER READING

Barnett, H.C. (1994) *Toxic Debts and the Superfund Dilemma*, University of North Carolina Press, Chapel Hill, NC. [A study of the Superfund program and its legacy elements.]

Brown, M.H. (1979) *Laying Waste: The Poisoning of America by Toxic Chemicals*, Pantheon Books, New York. [This book, together with those by Tarr (1996) and Colten and Skinner (1996), provides background regarding the history of chemical waste disposal and American industry.]

Colten, C.E. and Skinner, P.N. (1996) *The Road to Love Canal: Managing Industrial Waste before EPA*, University of Texas Press, Austin, TX.

Tarr, J.A. (1996) *The Search for the Ultimate Sink: Urban Pollution in Historical Perspective,* University of Akron Press, Akron, OH.

Winks, R.W. (1969) *The Historian As Detective: Essays on Evidence,* Harper & Row, New York. [A classic discussion of methods of historical investigation and, particularly, forensic use of historical documentation.]

CHAPTER 3

PHOTOGRAMMETRY, PHOTOINTERPRETATION, AND DIGITAL IMAGING AND MAPPING IN ENVIRONMENTAL FORENSICS

James I. Ebert

3.1 THE AERIAL PHOTOGRAPHIC RECORD

Aerial photographs comprise a huge international archive of hundreds of millions of images taken over the period of slightly more than a century that document the earth's natural landscape and humankind's interaction with it.

As soon as the process of chemical photography was developed, photographers began experimenting with taking aerial photos, first from balloons and then aircraft. The collection of aerial photographs systematically covering large contiguous areas began in the United States in the early 1930s by government agencies largely for purposes of making topographic, soils, agricultural, intelligence, and other sorts of maps and photointerpretations.

The maps made by the United States Geological Survey (USGS), Department of Agriculture, National Aeronautics and Space Administration (NASA), the United States Army, and other agencies from systematic aerial photographs through time are 'problem oriented,' and each map-making or interpretive effort focused on only a narrow range of the vast amount of information that can be extracted from those aerial images. Although they are conserved and made available at widely varying levels, the aerial photographs remain in government and private archives for others to use in the context of their own problem orientations, background, and expertise. Regardless of colloquialisms that suggest that aerial or any other sort of photographs 'speak for themselves,' they do not. In order to get answers from aerial photographs, questions must be asked of them. And there are as wide a range of kinds of questions to be asked, and ways of asking them, as there are scientific, technical, and other disciplines.

Some of the earliest applications of aerial photographs were by the military, with balloons being used to photograph enemy lines by Union forces during the Civil War. Another early use of balloon platform photography was the imaging of Stonehenge in the United Kingdom in 1906 by archaeologists, who have been pioneers in aerial photographic methods. Geographers base much of their science on information – much of it represented in maps – derived from aerial photos. Aerial photographs also serve as basic data sources for geologists, foresters, range managers, agricultural scientists, hydrologists, biologists, and environmental scientists. Engineers and architects use aerial photographs to interpret, map and measure information important in their work such as topography, soils, and the locations of the built environment. Many other creators and users of aerial photographs do so for historical or artistic purposes, and a search of the aerial photographic literature turns up as many 'hobbyist' or artistic titles as scientific or engineering ones.

Photogrammetrists are scientists and experts specializing in the development and application of methods specifically focused on using aerial photographs and other sorts of imaged data, as well as many associated digital imaging and mapping methods. Photogrammetry is defined by the American Society for Photogrammetry and Remote Sensing (ASPRS) as 'the art, science and technology of obtaining reliable information about physical objects and the environment, through processes of recording, measuring, and interpreting imagery and digital representations of energy patterns derived from noncontact sensor systems' (Colwell, 1997).

Photogrammetry as a practice originated before the advent of aerial photographs, but as a profession emerged by about 1930 when hundreds of photogrammetric engineering firms became established in the United States and around the world, taking photos with sophisticated aerial cameras and using optical-mechanical plotters to make contour maps. Over time these firms, and government agencies have blanketed the landscape with aerial photographs. Since the beginning of systematic coverage of the United States, in most places from the early to late 1930s, it is not uncommon to find up to 30–40 different dates of aerial photographs from then to the present, and in urban areas there can be hundreds.

The shift to digital data and techniques, beginning in the mid-1970s and accelerating through the next two decades, profoundly affected the profession and practice of photogrammetry. In November 1985 the American Society of Photogrammetry changed its name to the American Society for Photogrammetry and Remote Sensing in recognition of the fact that the profession encompasses not only interpreting and measuring from aerial photos, but using space pictures, other digital imagery, digital imaging techniques, and computer mapping technologies including geographic information systems (GIS), computer aided drafting (CAD), and global positioning systems (GPS), which are parts of the field of imaging and mapping today.

3.2 PRINCIPLES OF PHOTOGRAMMETRY

An understanding of the principles of photogrammetry can assist any scientist, expert or other aerial photo user better understand and interpret the information contained in aerial photographs and other sorts of images. The most basic properties of photographic images derive from the fact that photographs are the accommodation of three-dimensional, real-world scenes to a two-dimensional medium.

3.2.1 PHOTOGRAPHIC SCALE

The scale of objects in an image, i.e., the ratio of their size in the image to their actual size, is a function of camera geometry and the distance of objects from the camera. For aerial photographs, this relationship is:

$$\text{scale} = \text{focal length} / \text{flight height} \qquad (3.1)$$

If, for instance, an aerial photo is taken at an altitude of 12,000 feet with a six-inch lens (a common focal length for aerial cameras), then the scale of the photo is 1:24,000. Engineers sometimes express this in terms of inches to feet; in this example 1″ = 2000′.

Scales of about 1:20,000 to 1:30,000 are commonly found in historic aerial photographs taken for mapping purposes until about the 1980s, when improvements in films and plotting devices made it practical to use somewhat smaller scale aerial photos and the USGS began using scales between 1:40,000 and 1:80,000 for producing topographic maps. The scale of aerial photos has a profound impact on what can be seen in them; at 1:24,000 scale, for instance, a 50-gallon drum measures only about a thousandth of an inch in diameter on film or contact prints made from it. Quite clearly, optical or digital magnification of an image is required to recognize such objects in an aerial photo of this scale.

The scales given for aerial photographs by those who took them are nominal scales calculated by using the average height of the aircraft above the ground. Actually, the exact scales of objects with an aerial photo are a bit more complicated than that, varying with the distance of each specific object from the camera. For instance, an object on top of a hill is closer to the camera than the same-sized object in a valley, so their scales vary. This is not much of a problem in relatively flat terrain, or with small-scale aerial photos, but can result in significant variation in the sizes of objects in large scale (low altitude) photos taken over rugged terrain.

3.2.2 RADIAL DISPLACEMENT

The heights of objects has another effect on how they are represented in an aerial photo, which photogrammetrists refer to as radial displacement. Every object in a vertical aerial photograph, except at its exact center, appears to lean outward from the photo's principal point, to a degree that increases with its height and the distance it falls from the photo center. This is why one sees the sides of some objects, for instance buildings, in an aerial photograph and why the exact placement of objects with height differs in two overlapping photos. Radial displacement can be thought of as a photographic 'distortion,' but actually is a property of photographic geometry, and in fact provides the information that allows one to measure from, and see, a three-dimensional view of the landscape, in stereo photographs.

3.2.3 STEREO PHOTOGRAPHS AND STEREO VIEWING

Systematic aerial photo coverage is typically flown in straight flight lines with 60% overlap between consecutive frames, insuring that everything on the ground can be seen in at least two of the photos. When adjacent photos – a stereo pair – are viewed with a stereoscope, a three-dimensional view of the ground and objects in the photo is seen. Stereoscopes range in sophistication from 'pocket' models costing $20 to precision instruments offering high-power zoom magnification costing tens of thousands of dollars. Perhaps the most efficient type of stereoscope for serious use is the mirror stereoscope, which can be fitted with a parallel-tracking mount, allowing the photointerpreter to scan across the stereo image without having to constantly reposition the prints. Binocular eyepieces available for mirror stereoscopes typically offer about 2×–6× magnification. If higher magnifications are necessary, to see very small details in small-scale photos, more expensive instruments are required; some zoom stereoscopes can magnify images as much as 200×. The light-efficiency of such optics quickly decreases as magnification is raised, however, and extremely intense illumination is required. This and the fact that it is hard to find tiny features on photos at high magnification can make using such stereoscopes difficult.

Stereoscopic viewing is always very helpful, and in most cases absolutely necessary, in aerial photointerpretation. Identifying many sorts of features at environmentally relevant sites, for instance drums, structures, piles, pits and trenches, which can be nearly impossible with a monoscopic photo, becomes much easier with stereo. Topographic details, which are of vital importance in many environmental situations, cannot be appreciated without stereoscopic viewing.

3.2.4 RESOLUTION AND PHOTOGRAPHIC MEDIA

There is some debate in the professional 'photointerpretive community' regarding the best photographic medium to obtain from those firms or agencies holding the original photo negatives. Aerial photos are exposed on film, and the original film is referred to as the first generation by photogrammetrists. The original film is archived at firms and agencies which make second-generation copies of aerial photographs for users. Black-and-white negatives are the first generation for almost all black-and-white photo products; color aerial photos can be taken with color negative or color positive transparency film. Second-generation products can be made from negatives as either transparencies or paper prints; second-generation positive transparencies or prints can be made from positive, first-generation film by using reversal film or paper (such as Cibachrome). Positive paper or film products are sometimes also made from first-generation, positive film by making an internegative, in which case the positive copies are third-generation products.

Some photogrammetrists assert that one should obtain positive film transparencies as the basic data for environmental photointerpretation, stating that what one sees in a film transparency is far superior in resolution than a paper print. While it is true in general that paper prints provide less spatial resolution than film duplicates, it is often difficult to tell the difference unless one is inspecting extremely small objects or features in the film. The resolution and contrast performance of duplicating films and papers vary widely, as well (Teng, 1997). The main advantage of a photographic film product over a paper print for most uses is the wide exposure latitude and spectral resolution of film, especially black-and-white panchromatic films. In first-generation panchromatic film negatives, there are rarely any portions of an exposed frame that are totally black or totally white; some level of gray-scale detail appears almost everywhere. Each successive generation of photo products – film or paper – introduces increased contrast and less spectral resolution. If details in areas that appear very dark or bright are important in a specific photo-interpretive effort, exposures and details in those areas can be maximized by under- and over-exposure or dodging. A disadvantage of using film products for photointerpretation is that illumination must be provided by a light table.

A better strategy for evaluating potential differences in resolution between film and paper second-generation aerial photo products is to use enlargements made from original negatives for interpreting details that are small in small-scale aerial photographs. If identical areas are enlarged from the two frames making up a stereo pair, these can be used with a stereoscope just like contact prints. Enlargements can be produced digitally, but even with the high-resolution scanners used by photogrammetric engineering firms (typically about 12.5 microns, or 2032 dpi), tiny details scanned from 10″ aerial film – for instance drums that measure 1/1000″ on 1:24,000 scale photos – are practically unresolvable. If such details are important, obtaining photographic enlargements and then scanning those to produce digital image files allows comparable or better results to be obtained with much more affordable scanners.

3.2.5 DIGITAL IMAGE PROCESSING

Digital imaging and image processing can be advantageous in a number of ways in environmental forensic photointerpretation. When these methods were first becoming widely available in the mid to late 1980s, there was considerable discussion concerning whether they could be used to alter or fabricate information, thereby making them problematic in legal applications. Such concerns have become less frantically voiced as scientists and the public in general become more familiar with digital imaging. Nonetheless, it is true that digital imaging can be used to alter the appearance of images. Therefore, it is necessary when using digital imaging techniques in a legal context to fully understand digital imaging processes and their effects on images.

Of course, traditional photographic techniques 'alter' images as well – two photo prints from the same negative can look quite different. It has always been incumbent upon those using photographs for forensic purposes to exercise ethical behavior and thought processes when interpreting the meaning of images. An example of a common photointerpretive pitfall is the attribution of meaning to 'dark staining' of surface sediments, the appearance of which can be changed or in fact created by many factors, for instance varying contrast caused by photographic or digital processing, or variation among lighting, vegetation and other factors in aerial photos taken on different dates. This theme will feature prominently in a case study to be presented later in this chapter.

One of the most straightforward and immediately valuable ways digital imaging can aid in environmental photointerpretation is as a means of enlarging small details from aerial photos. As discussed above, obtaining photographic enlargements from the original data – i.e., the film negatives – is often a useful first stage in examining important details. The enlargements can then be scanned using relatively affordable flatbed scanners to create digital images.

One way to insure that interpretations are not biased by differences introduced by digital imaging processes is to work interactively with both photographic products and the digital images made from them. A useful practice is to view the stereo photographs with an optical stereoscope, while at the same time zooming in and out to see specific details on a computer screen. Although optical stereoscopes are still the best way to view stereo, they have some disadvantages including that only one person can view the stereo image at a time. Many people can view an image together on a computer screen, and this can often be quite advantageous in 'cooperative photointerpretation' among the photo-grammetrist and other experts in a case, and of course for showing others what the photogrammetrist is seeing.

Stereo can be viewed on a computer screen, and although shuttered or polarized glasses can be used to do so with special software, a quite workable and affordable alternative is anaglyphic viewing, in which one frame of a stereo pair is assigned to the red band, and the other frame to the blue band, of a color (RGB: red-green-blue) image. The image is then viewed with red/blue glasses, which are really just colored filters in a frame and can be purchased from a number of sources for from less than a dollar for cardboard models, to $20+ for comfortable plastic frames. Anaglyphic stereo viewing for extended periods takes a little getting used to, and what one sees is qualitatively somewhat different than viewing through a stereoscope, but this technique can be particularly valuable for seeing small details of an image in stereo, and for showing them to others. Anaglyphs can be made 'manually' in digital imaging software programs such as Adobe Photoshop, or by a number of special software packages.

Once one has a digital image file, image processing techniques can be used to increase the visibility of certain sorts of details in a fairly simple and understandable way. The 'ideal' black-and-white, second-generation aerial photo product is of relatively low contrast to preserve details at both ends of the gray-scale (light-to-dark) range, but digitally changing the brightness and contrast of parts of an image can help clarify details. One way to vary these image properties without losing the dark or light ends of the spectrum is to adjust an image's histogram by moving the 'gamma' point, i.e., the center of the dark-to-light range, without moving either end of the histogram (Figure 3.1). Certain sorts of digital filtering, such as unsharp masking, can increase the contrast of the places in the image where the dark-to-light changes occur most abruptly. This is often called 'edge enhancement', since in fact such changes often do define the edges of objects or patterns in a photo.

There are many types of image processing that can be applied to digital images made from aerial photographs. For instance, it is sometimes possible to remove periodic noise or image blur using certain procedures. An almost unlimited number of digital spatial filters are available or can be designed that have widely varying effects on images. There are many software packages that do image processing at various levels, each doing slightly different things, ranging from freeware through packages costing tens of thousands of dollars. New versions constantly become available. Each original aerial photograph shows things differently, so it is impossible to specify any sort of 'formula' for how to do image processing in any specific instance. Image processing is best viewed as a tool to be used interactively, along with optical stereoscopy, to help make different sorts of details in aerial photos easier to see and interpret.

The spatial arrangements of the picture elements or pixels that make up digital images can be altered in controllable ways to achieve image registration or rectification, or in fact to produce planimetric orthophotos. One of the most efficient and flexible ways to rotate and scale a photo to fit map coordinates or another photo is to use software packages that work with Computer Aided Design (CAD) programs to display and manipulate images within a spatial coordinate system. Many of these programs also perform 'rubber sheeting,' which can be used to match multiple points across an image to map- or field-derived control point coordinates. These methods can be used to register images in relatively flat portions of the landscape – where many sorts of industrial sites are in fact located – quite acceptably for many mapping purposes. The software creates a 'world file' (∗.res or ∗.wrl) which translates the image to the coordinate system again when it is brought into a CAD drawing, or geographic information system (GIS) such as ArcView. Lines and polygons showing photointerpreted objects or areas can then be precisely drawn on top of the image in CAD or GIS, and are then also registered to real-world coordinates.

a

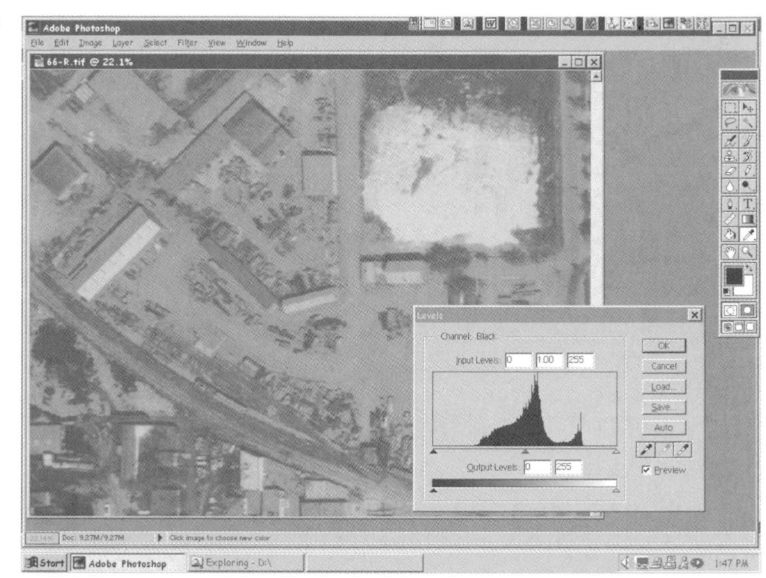

b

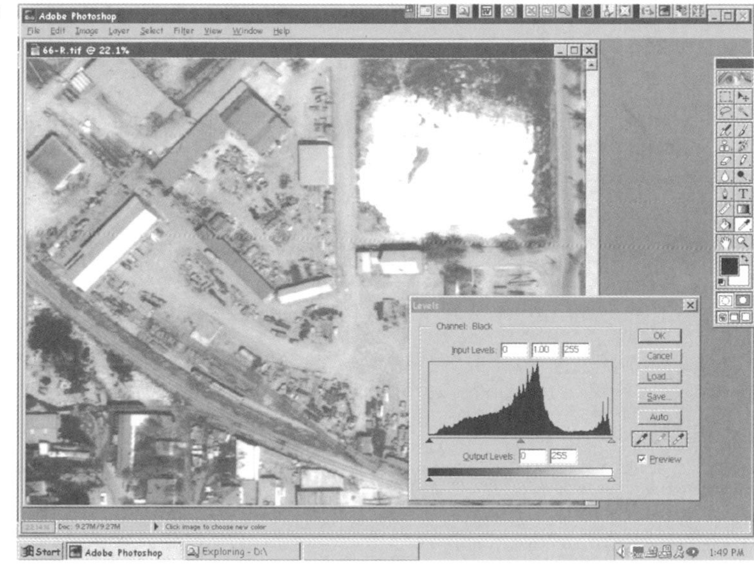

Figure 3.1

Histogram equalization redistributes the range of dark-to-light values in a digital image to increase contrast and thus the visibility of details. In Figure 3.1a, a scanned image exhibits a narrowed range of dark-to-light values (shown in the little graph in the lower right corner). By moving the light and dark ends of the graph to the edges of the histogram, and adjusting the middle of it (the 'gamma') the image is optimized. The new image and its histogram are shown in Figure 3.1b.

3.3 PHOTOINTERPRETATION

Aerial photographs reveal information about natural and cultural features, processes, and interactions through time that cannot be derived from any other source, and for this reason they are important data sources in virtually every science and profession that focuses on spatially distributed phenomena. Aerial photos are used by biologists to map distributions of plants and animals, and by geologists, soil scientists, and engineers to characterize sediments and mineral deposits. Some of the earliest uses of aerial photographs, taken from kites and balloons before the advent of airplanes, were for military purposes, and military funding has usually been responsible for cutting-edge advances in imaging and mapping technology.

Interestingly, another profession that has pioneered many advances in photointerpretive and remote sensing methods is archaeology. Aerial photo-interpretation has been an integral part of both archaeological fieldwork and the archaeological literature since the early 1900s in both the United States and Europe. The first archaeological aerial photographs were taken in 1906 over Stonehenge from a balloon. A number of early military pilots flying over Europe and northern Africa in World War I noted subtle patterning in vegetation, particularly in agricultural fields, that could not be seen from the ground and which revealed the locations of buried Roman forts and other ancient sites. Following the war, some of these aviators became some of the first 'aerial archaeologists' at British and European universities. At about the same time, American archaeologists began using aerial photos, photographing sites such as the Cahokia Mounds and Southwestern pueblo ruins (Avery and Lyons, 1981). Some of the first and still most useful expositions of principles of photointerpretation were written by archaeologists such as O.G.S. Crawford (1923, 1924) and Poidebard (1929, 1930) (Ebert, 1984, 1997).

3.3.1 PHOTOINTERPRETATION VERSUS PHOTO READING

There are interesting parallels between the uses made of aerial photographs and remote sensor data used by archaeologists, and the ways environmental forensic scientists should approach the use of aerial photos as an adjunct to their work. Archaeologists' basic data consist of objects such as flaked stone tools or bits of pottery discarded or lost by past people in the course of their lives and interaction with the environment to reconstruct not only how they were using a scraper or a pot immediately before they left it behind, but to explain things that can never be seen: mobility, settlement systems, religion – the sum of past human behavior. The pottery sherd may be mildly interesting, and in some very rare cases one might even be able to derive a 'story' from it, for instance if a number of sherds with the same pattern are found adjacent to one another: someone dropped the pot and it broke. But anecdotes such as this are of passing interest, illuminating only an instant in time. It is the intangible components of continuous and ongoing behavior and culture that are the goal of archaeological analysis. In fact, archaeologists do this same thing with another data source, aerial photographs, looking for indications of past sites and structures.

In much the same way, one can view objects relevant to environmental forensic studies in aerial photographs and assign names, but the ultimate goal is to elucidate the systemic effects of ongoing human behavior on the environment through time from aerial photos that show only instants. This fact is the basis of a distinction made by Thomas Eugene Avery, one of the US's foremost teachers of photointerpretation, in the fifth edition of his book *Fundamentals of Remote Sensing and Airphoto Interpretation*:

> *Photo interpretation* is defined as the process of identifying objects or conditions in aerial photographs and determining their meaning or significance. This process should not be confused with *photo reading*, which is concerned with only identifications. As such, photo interpretation is both *reading the lines* and *reading between the lines*.
>
> (Avery and Berlin, 1992: 51)

Utilization of aerial photographs in environmental investigations and litigation frequently follow the model of photo reading, rather than photo-interpretation. Experts involved in environmental photointerpretation insist that film transparencies must be used rather than paper prints, or that one kind of scanner is better than another, because it lets you see certain small objects better. Reports contain elaborate overlays with arrows pointing to 'drums,' 'fences,' 'lagoons,' and 'trenches.' Part of the reason for this may be, in a sense, historical. Some of the first photointerpretive products most environmental scientists are exposed to are the reports of the Environmental Protection Agency's Environmental Monitoring Systems Laboratory. The United States Environmental Protection Agency (EPA) and photointerpreters under contract with them prepare *Waste Site Discovery Inventories* and *Detailed Waste Site Analyses* following very specific procedures and formats (Mace *et al.*, 1997). The purpose of the photointerpretive efforts is to identify environmental problem areas in a very general way (waste site inventories), and then to identify more specifically features of properties that give reason to suspect there may be environmental problems at individual locations (detailed waste site analyses). The intent is also clearly *not to* propose activities or events in the past, *not to* 'read between the lines.' The EPA's reports provide an excellent starting place for those involved in environmental forensic studies and environmental litigation – they supply identifications of objects and features, to be interpreted by others, and are *very purposely* the products of 'photo reading.'

3.3.2 PHOTOINTERPRETATION IN ENVIRONMENTAL FORENSICS

What environmental forensic scientists and environmental litigators need from historic aerial photos is information that beyond the identifications of features, to understanding the past *processes* by which environmental damages occurred. This is similar to what archaeologists do with aerial photos. What the archaeologist is looking for is quite clearly not what was going on when the aerial photo was taken, but rather indications of things that happened long ago. What is more, it is indications of things that are often buried beneath the surface of the ground, contained in subsurface sediments, or indications of landscape modification that subtly mirror past interactions between people and their environment. Of course, just about everything the archaeologist is interested in happened long before any aerial photos, historical or not, were taken. Archaeologists use all of the aerial photos available because weather conditions, sun angle, season of the year, and recent landscape modifications change from photo date to photo date, making indications of past landscape use such as the condition of vegetation or ground compaction more or less visible.

Environmental forensic scientists have an easier task in many ways, since activities important at environmental sites really *were* going on when the aerial photo was exposed. Sometimes one can actually see a truck pouring dark liquid onto the ground, or a leaking drum or a burst storage tank. Much more often, however, what is environmentally and legally important is what was happening between the aerial photo exposures. After all, photographic exposures take place in only hundredths of a second. The task of the forensic environmental photointerpreter is to extract information from aerial photographs to elucidate the activities and processes that occurred that are the cause of environmental damages. While this information may be obvious in aerial photos, some of that which is most interesting in environmental litigation is much less so.

It is interesting that a survey of 25 years of literature concerning the use of aerial photographs and other remote sensor data in environmental investigations shows an overwhelming focus on purposeful waste disposal, and dedicated waste disposal sites (see, for instance, Garofalo and Wobber, 1974; Philipson and Sangrey, 1977; Erb *et al.*, 1981; Getz *et al.*, 1983; Nelson *et al.*, 1983; Evans and Mata, 1984; Lyon, 1987; Bagheri and Hordon, 1988; Barnaba *et al.*, 1991; Pope *et al.*, 1996). Waste disposal is an activity that is directly responsible for much environmental damage, and locating waste disposal sites, determining what waste materials were disposed of through time and exactly where, determining ownership and whether disposal methods were proper or improper are important issues in environmental litigation. But *deliberate and organized* waste disposal at sites designated in the past for such purposes is only one sort of activity responsible for environmental damage. Often, waste disposal of a less formal nature occurred at industrial sites, and identifying these locations using aerial photographs requires deductive photointerpretation.

Another contribution of archaeology to a better understanding of what it is environmental forensic scientists and environmental litigators might be most interested in learning from historic aerial photographs and other spatial data sources such as historic maps is the realization that our 'environment' today is the result of a complex overlay of human activities through time. This is not something that archaeologists initially realized; in the mid-1980s, a series of theoretical changes within the field questioned whether one could assume that clusters of artifacts were the result of single episodes of human activity. Archaeology looked to anthropology for the answer. Anthropologists study present-day human behavior, and a century of anthropological observation focusing on hunter-gatherer societies showed that human activities of various sorts often occur repeatedly over long time periods at a place. During such activities, tools and debris are discarded or lost and form a complicated overlay which must be understood with reference to the ways people do things rather than relying on anecdotal evidence (Ebert, 1992).

Human *industrial behavior*, which is responsible for the environmental damages and other characteristics important in environmental forensics and environmental litigation, is a subset of general human behavior. Over approximately the last 200 years in the United States, and longer in some European countries, industrial and manufacturing activities have affected the environment in locations that have shifted through time, in a wide variety of ways as industrial processes and practices evolved. While most environmental engineers and attorneys are quite aware that industrial perceptions and operations have changed radically since the advent of the Comprehensive Environmental Response, Compensation and Liability Act (CERCLA), far less attention is paid to the fact that only a very small percentage of possibly as many as half a million hazardous waste sites in the United States are inventoried by the EPA (Colten, 1990). And most of those sites are not 'disposal' sites per se, but rather the places where industrial activities took place.

At industrial sites, activities are of course organized to get the job done. Raw materials as well as chemicals used in processes arrive at the plant by water, rail, road, or pipeline transport (or sometimes by other means such as aerial tramways). When materials arrive at the plant, they are transferred from whatever means of transport by which they arrive into some form of storage. For instance, solvents are brought to the plant in tank trucks, and are transferred by means of a hose and pump to storage tanks. From the storage tanks, they are carried in glass carboys into the plant to be used to clean circuit boards in open baths. Once the solvents are contaminated, they are transferred by hand into wheeled, open-topped metal carts, taken out of the plant and into a storage area where they are poured into 55-gallon drums. The drums sit in the storage area until they are taken away by trucks to a recycling or disposal facility.

At each juncture when products are transferred from one means of containment, transport or storage to another, a small amount is spilled. In storage, containers leak or chemicals that have been spilled onto containers are washed off by rain. Empty containers are typically rinsed out, often with a hose on a loading dock, before being reused. Although the amounts of chemicals that enter the environment through such *transfer and storage spillage* are small per event, hundreds or even thousands of such events can take place daily, involving dozens of sorts of materials, metals, chemicals and the like at large industrial facilities.

At many industrial sites, a majority of environmental contamination has occurred relatively constantly through time, and instead of being accidental by any definition, was instead part of the normal operation of the plant. When true accidents occur – for instance, a drum falls from a truck and bursts open, the event is obvious, notice is taken, and cleanup efforts ensue. But small amounts of chemicals released as the consequence of routine product transfer and storage spillage or leakage are, in legal parlance, *expected and intended.* Distinguishing among accidental and expected and intended environmental damages is an especially crucial point in insurance defense litigation in the United States, for insurers cover their clients against accidental occurrences, and not routine, intended ones.

Although routine and intended occurrences are individually minor and go largely unnoticed at industrial plants, indications of their locations and extent are often visible in aerial photographs. Of course, the spillage of a few ounces of liquid chemicals that took place during transfer from a drum to a carboy in the yard of a plant cannot be seen on an aerial photo. The most important indicators are contextual: i.e., the identification of facilities or features that indicate areas of activities such as product transfer or spillage, or other means by which contaminants are released. Docks or racks where tank trucks or rail cars are unloaded can be the sites of extensive spillage, often indicated by dark staining, and even tracks of spilled material made by vehicles. Storage areas are another contextual indicator of places where spillage, leakage, and other transfers of contaminants to the environment can take place, and can be of several sorts. Bulk materials such as ore, fuel and other process materials, and slag at metal refineries, are stored in obvious piles that usually dominate the landscape. Other liquids or solid materials are often stored in drums or other uniform containers which are distinguishable even in small-scale aerial photographs by their regular patterning. Particularly important are places where materials are poured from, or into, drums or other containers. This generally takes place at docks or near entrances to plants. It is worthwhile noting that, due to radial displacement of objects, the sides of buildings and entrances on them can sometimes be seen in aerial photographs if the site falls away from the photo's center.

The products of industrial operations are often themselves contaminants, or the source of them; for instance, metal is often heavily oiled before being placed in outdoor storage for transport from a site. Treated wood is another notorious example of products from which contaminants can be washed into the site drainage by rainfall. Waste materials, such as used solvents, are often removed from operational areas and poured into storage containers such as drums outside of a plant. Any information garnered from depositions or knowledge of operating procedures about how and what is transferred can be vitally important in guiding the photointerpretation of such activities.

In some rare instances, when one fortuitously has access to extremely large-scale aerial photographs, it may be possible to actually see 'spills' or liquids leaking from a drum or a tank. It is really the cumulative effects of such occurrences that are important, and for which photographic evidence can usually be seen. 'Staining,' that is, dark patterning noted on the surface of the ground or pavement, is often cited as evidence of spillage or leakage of contaminants. There are in fact many things that can cause dark patterning in an aerial photograph. Most environmental photointerpreters recognize that water, for instance puddles or damp ground from recent rainfall, can cause surfaces to appear dark in places in both color and black-and-white film, and check local precipitation records before pronouncing dark places to be chemical spills or stains. Puddles appearing in places where chemicals would not be expected to be spilled are another indication to look for (Morrison, 2000). Water is copiously used in many industrial operations, too. When staining seems to be caused by water, some care should be taken to attempt to determine whether it is 'just water,' or might contain contaminants, for example water drawn from tank bottoms at petroleum refineries.

Shadows also look dark in aerial photos, and distinguishing shadows from other dark patterning is one very important reason for interpreting aerial photographs stereoscopically so that objects which would project shadows can be detected. This works the other way, too, with shadows sometimes providing important cues in identifying objects that cast them. Shadows can be modeled and illuminated three-dimensionally using rendering software in CAD programs, to determine whether ambiguously appearing shadows are cast by known objects (Ebert, 2000).

The overall tone of aerial photographs is also affected by photographic properties such as exposure settings and the way film and subsequent generations of photo products made from it were exposed and processed. Vignetting at the edges of photo prints, particularly those made in black-and-white with early cameras, can cause some portions of a photo to look darker than others. Differences in sunlight intensity and angle can cause such differences from date to date when comparing photos of a site through time.

3.4 ANALYTICAL PRODUCTS AND PREPARATION OF EXHIBITS

3.4.1 MAPS

Ultimately, images have to be registered to maps. Photointerpretations made from multiple years of aerial photos are not of much use unless they can be considered in spatial relation to one another and to the landscape of today. In some cases, for instance at some industrial sites, large-scale surveyed plant maps are available and should be used to scale and register aerial photos. In many situations, however, engineering maps are not forthcoming and often the only maps that can be found are US Geological Survey (USGS) $7\frac{1}{2}$-minute topographic sheets, which are fortunately available for the entire country. Compiled at 1:24 000 scale, they usually contain enough cultural detail to allow the registration of at least relatively recent aerial photos. The USGS calls the digital versions of their topographic sheets, which have in the last few years become available, digital raster graphics (DRGs). These scanned, georeferenced TIFF files are available for purchase in groups of 36 on CD ROM; they can also be downloaded without charge from a number of web sites as well. DRGs are digitized or rasterized at a resolution of 250 dpi, and since they come with associated world files they can be brought directly into CAD or GIS in their respective UTM (metric) map coordinates, where digital images can be registered to them. They can also be used as a base coverage for display purposes, although at 240 dpi (about 2.4 meters per pixel) they can be a bit difficult to look at closely, and one might want to scan a paper USGS map at higher resolution, and register it to the DRG, to be used as a base map for display purposes.

An orthophoto is a digital image of an aerial photograph that has been systematically corrected so that objects and features across the entire scene appear in their correct planimetric positions; orthophotos fit exactly onto a planimetric map of an area (although some sorts of features with abrupt topography, such as buildings, are difficult to correct in this manner). Orthophotos are created by using a digital elevation model of the area covered by the image to 'rubber sheet' the image to a regular network of closely spaced control points. Orthophotos (called digital ortho quarter-quads or DOQQs and covering 3.75×3.75 minutes of latitude and longitude, one-quarter of a $7\frac{1}{2}$-minute topo quad) are currently available from the US Geological Survey for many areas of the country, with total coverage planned. Where they are available, they provide better registration control source than DRGs since many features absent on maps, such as non-cultural landscape details like trees, drainages and the like, are visible for use as registration points in the orthophotos. USGS DOQQs are based on recent USGS aerial photographs, and conform to National Map Accuracy Standards, which require that objects on the ground be faithful to reality within 1/50″ on the map, or 40 feet on the ground as in a 1:24,000 scale map.

Contemporary aerial photographs can frequently be easily registered to either recent maps or orthophotos. When cultural features such as buildings and road intersections are visible on the maps or orthophotos, enough common points can be defined to allow registration as well as the calculation of the closeness of fit arrived at in the registration process. Historic aerial photos taken decades ago, however, often contain little detail that is the same as that on recent maps or even orthophotos. In many instances, it is necessary to 'work back' through time, registering relatively recent photos to the map, and then using details on those photos which can also be seen in aerial photos a few years older to register *those* photos. The serial registration of historic aerial photos back through time to real-world coordinates can be a difficult process requiring much concentration.

Registration of aerial photographs to maps, or to other photographs, using simple scaling and rotation, or projective transforms, can have varying results in terms of closeness of fit depending on the nature of the terrain and cultural features in the area covered by the photos. Scaling and rotation work well when the area of interest is level. Projective transformation of an image can remove perspective effects caused by a sloping surface, but that surface must be relatively planar to produce good results. If there is considerable non-planarity in the topography of the study area, there will be mismatches of points across the terrain except for those used as control in the matching process.

True parallel axis photogrammetric plotting can produce highly accurate maps from historic aerial photographs. Photogrammetric engineering firms that have such plotting systems can produce planimetric and contour maps from historic aerial photos, but there are often limitations imposed by the scale of the aerial photos and the control points that can be found that mitigate against such maps being much more accurate than USGS $7\frac{1}{2}$-minute quads.

3.4.2 GEOGRAPHIC INFORMATION SYSTEMS

The reason to register aerial photographs in a real-world coordinate system is so that photointerpretations made from successive historic aerial photo dates can be compared spatially through time. Some site features such as property boundaries, roads, and railroads can be defined as lines, but most objects or areas relevant to past activities and present-day problems at a site are more appropriately polygons so that their correspondence or lack thereof through time can be analyzed. Buildings, tanks, pits, trenches, pools, stains, storage areas, loading and unloading areas, and many other sorts of features can and often do change locations through time at sites, and one of the best ways to keep track of such changes and correlate them with other site data *not* derived directly from aerial photos (for instance, sample locations and depictions of chemical plumes derived from them) is by incorporating them all in a geographic information system (GIS) database. Coordinating the efforts and data of the diverse experts involved in any large-scale, serious environmental litigation virtually requires a unified database that can be used by all parties concerned.

Only a few years ago, compiling and using a GIS database required hardware, software and technical expertise and training that was beyond even many large environmental firms. Fortunately, more affordable and user-friendly software coupled with the geometric increase in the capabilities of computers has resulted in GIS being a tool that belongs on everyone's desk. There are dozens of GIS software packages with varying levels of capability and sophistication available today, most of which are affordable enough that multiple licenses can be purchased within the scope of any significant litigation. In a GIS database, information is spatially organized in multiple 'layers' by data type, time, and/or other appropriate classification and can then be compared, combined, and manipulated in a number of ways. Overlapping areas of the same type can be combined, differenced, or viewed additively through time, something which will be discussed in the context of a case study later in this chapter. Phenomena that vary across space, for instance chemical concentrations, can serve as the basis for producing contour maps, or sample locations compared in terms of their proximity to buildings, roads or other spatially organized features. Some GIS software packages have sophisticated graphic and cartographic capabilities which can be used to produce exhibits illustrating locations of features and areas within a plant or other study site, and the results of analyses. A properly compiled GIS database can organize all of the relevant spatial data in a case in 'one place' so it can be used by everyone in many places.

3.4.3 TERRAIN VISUALIZATION

Nonanalytical digital visualization techniques are available for the exposition of information from aerial photographs and other data sources in environmental forensic cases. Terrain visualization is one of these, and can be used to help orient judges, juries, deponents, and other participants in litigation to the topography and the arrangement of site features. Terrain visualization consists of draping a map or aerial photograph over a surface constructed with a digital elevation model (DEM). The resultant rendered surface can then be viewed from any angle, elevation, or distance, or multiple views rendered and chained together to produce an animation that is essentially a 'walk-through' or 'fly-by.' A number of GIS, image processing, and other software packages are available which will create three-dimensional surface/subsurface visualizations.

3.5 CASE STUDIES

Four case studies illustrated by stereo anaglyphic images are included to illustrate some of the principles of environmental forensic photointerpretation and analysis.

3.5.1 CASE STUDY: INFORMAL DUMPING AREA COVERED BY PLANT ADDITION

In older urban areas, the environmental state of industrial sites and the sediments beneath them is the result of a composite overlay of past industrial activities and changing industrial behavior. At a bearing manufacturing plant site in the northeastern United States, subsurface contamination was concentrated beneath the southern part of the factory, and there were questions as to its possible source. Leakage of oil and solvents from process areas within the plant, and through the concrete floor, was suggested but seemed implausible. Plate 1 is a high quality anaglyphic stereo image of the plant in 1975, and around the southern portion of the plant under which contaminant concentrations were highest (marked A), large numbers of drums and containers can be seen to be stored in a haphazard manner in outside corners and along the sides of a service road. Such containers, whether filled with chemicals going into the plant, waste materials, or even 'empty,' are potential sources of leakage and product transfer spillage, but if they were the major source of contamination then concentrations around the perimeter of the buildings would be expected to be as high or higher than under the slab.

The deposition testimony of a machinist who worked at the plant in the 1930s mentioned informal but continuous dumping and pouring of wastes 'on the island,' which was described as a place you would not want to go if you did not have to because it could ruin your shoes. Plate 2 is a stereo anaglyphic image derived digitally from aerial photographs flown by the Soil Conservation Service in 1938. Anaglyphs must be viewed with the axis of the flight line from side to side; the 1975 aerial photos were flown east-to-west, so in them north is up, while the 1938 photos were flown north-to-south, so north is toward the left. Although the resolution of the 1938 photos is low and the contrast very high, it can be seen that in 1938, in the area where the southern portion of the plant had yet to be constructed (A), an abandoned river channel is visible by virtue of reflected sunlight, and between it and the river lies the 'island.'

Dumping there was at least part of the source of hydrocarbon contamination beneath the southern annex to the factory in later times.

3.5.2 CASE STUDY: PRODUCT TRANSFER AND STORAGE SPILLAGE AT CIRCUIT BOARD FABRICATION PLANTS

Circuit board manufacture involves the use of etching acids, fluxes, and solvents for cleaning parts. These chemicals are brought to these industrial sites in bulk, usually in carboys and drums which are unloaded from trucks and often stored outside the plant on or near loading docks. Since they are used inside the plant in relatively small amounts, for instance in tubs or machines in which circuit boards are cleaned, the liquids are transferred from drums or carboys into other containers – in the case of the two plants discussed here, according to deposition testimony, in specially made 'carts' consisting of a vertical tank attached to a two-wheeled hand dolly – using centrifugal pumps (for drums) or by hand pouring (from carboys). Centrifugal pumps operate by turning a crank affixed to a pump with a pipe which screws into the open drum bung; when pumping stops, and the hose is withdrawn from the receptacle, whatever liquid remains in the hose drains out suddenly, and spillage onto the ground and drum is inevitable. Other methods of transferring liquids from drums include rocking cradles and funnels. Liquids in carboys, typically containing 5 or 10 gallons, are often simply poured, using funnels, into other containers. Filled carboys may weigh 100 pounds (45 kg) or more, so pouring carefully is difficult. Because spillage in such transfer operations is expected, these activities often occur outside the plant buildings, in storage areas.

This was the case in 1963 at the circuit board fabrication plant shown in Plate 3, where drums and other containers are stored around the loading docks at A, and in a long row along the side of the parking lot at B. At C in this stereo anaglyphic image are a number of aboveground storage tanks into which other liquid chemicals were transferred from tank trucks as well as drums to be piped into the plant. The parking lot at D shows mottled staining (staining can be light-colored, as well as dark) which may be the result of transfer spillage. The objects stored in the parking lot below D appear to be largely equipment, although some of them may be containers as well.

Some of the containers visible in Plate 3 are undoubtedly filled with used chemicals, to be removed from the plant for disposal or recycling. Spent liquids are poured into drums with more pumps and funnels, with additional spillage – often outside the plant, in the storage yard, because of the mess it would create inside.

Plate 4 shows another circuit board fabrication facility in the western United States. The most obvious 'black patterning' is water flowing from the cooling tower, which probably should not be happening in such volume, but which is also probably environmentally benign. That water is flowing into a drain consisting of a small ditch in the parking lot with a perforated cover.

It is the product transfer area at this plant which is far more interesting in terms of environmental forensics, an area where, according to deposition testimony, you 'wouldn't want to drive.' In the enlargement of this area in Plate 5, a hydrant and hose near the spill-stained loading dock, and a huge plume of water flowing hundreds of feet down the drive to the drain, are apparent. The deponents said the dock manager liked to rinse out carboys and drums in the driveway from which chemical liquids were transferred into the plant so that they would be clean for reuse. Bulk liquid chemicals were also brought in drums and tank trucks to the covered loading dock, and drums unloaded there were stored outside at its right end, as can be seen from the stereo anaglyph. Also visible in the anaglyph is staining on the asphalt from liquid running from this outdoor container storage area into the nearby drain ditch.

3.5.3 CASE STUDY: CHEMICAL DRAINAGE FROM TREATED WOOD AT A CREOSOTE PLANT

While creosote used to treat wood is a natural product and relative involatile, only transmitted in sediments slowly, it is susceptible to transport in water. This can occur when treated wood is stored outdoors (which of course it must be) and subjected to rainfall. Plates 6 and 7 show a wood treatment plant in the southeastern United States in 1959 and 1965. At issue in this case was the question of when wood treatment chemicals from the plant could have entered the new drainage ditch, designated as such on the 1959 aerial photo.

Untreated wood piles are light colored, almost white, in the aerial photos, while treated wood is dark. In 1959 (Plate 6) there is little treated wood near the new drainage ditch, while in 1965 (Plate 7) the practice seems to be to have stored more treated wood along the ditch, from which rainfall-washed chemicals could have drained into it.

3.5.4 CASE STUDY: 'DARK STAINING' AT A MAINTENANCE YARD

An illustration of why it is necessary to exercise caution when citing staining or dark patterning as evidence of site contamination is provided by a case study involving an urban property in the southwestern United States damaged largely by petroleum hydrocarbons and associated chemicals. At issue was the question of the differential responsibility of two consecutive owners of the facility, a construction company which used the property for maintenance and storage of construction equipment for approximately 40 years, versus the subsequent owner and plaintiff, which conducted an oil recycling business there.

The plaintiff engaged the services of an experienced and qualified photogrammetrist who located all of the available aerial photos of the facility, photointerpreted them using mirror stereoscopes as well as using stereo on-screen computer viewing. The photointerpreter counted construction vehicles on-site at various times of the year in a series of 20 sets of aerial photos spanning a period of 26 years, illustrating that over this time many vehicles had potentially been on the site. For each of the aerial photo dates, the photointerpreter also delineated areas of what was defined as dark staining on the ground surface. This being a vehicle maintenance site, it was further reasoned, the dark staining was due to the spillage of lubricating oils and greases.

The historical aerial photos of the plant were registered to a common coordinate system, and the areas of dark surface staining were delineated and digitized and exported to ArcView GIS for analysis. Coverages within a GIS database, for instance the dark surface stains digitized from the air photos for each year, can be compared, combined, measured and analyzed in a number of ways. One of these is a union of all of the coverages through time. Essentially, the polygons or bounded areas in all of the coverages are overlain upon one another, and their internal boundaries are dissolved, resulting in combining overlapping areas into one. This is what the plaintiff's expert did to show the extent of what had been interpreted as dark stained areas at the property which was the subject of litigation, resulting in a map like that shown in Plate 8, left. The implication, and the impression given, by this exhibit is that somewhat more than 40% of the surface of the plant site is covered with lubricating oils and greases.

At least two levels of criticism might be leveled at this analysis and its resulting implications. First, even if all of those dark stains *were* indications of spills of lubricating oils and greases, each ArcView stain coverage records only those areas interpreted as dark when the aerial photos were taken. Representing their union as a single stain 'type' does not take into account that some areas must have been 'stained' only once, whereas other areas would have been stained on more than one aerial photo date. Determining which areas were interpreted as stained more than once might be very informative in terms of determining where the environmental damage would be expected to be most severe.

A related but more severe critique, however, of what the unioned GIS coverages imply is more clearly behavioral, and that is that it is probably very unlikely, if not impossible, that so much of the facility's surface could be stained with oils and greases over even the 40 years that construction equipment maintenance took place there. The configuration of the facility (building

Plate 1

A bearing manufacturing plant in 1875 in the northeastern United States had unexplainably high concentrations of solvents and other chemicals under its slab at A.

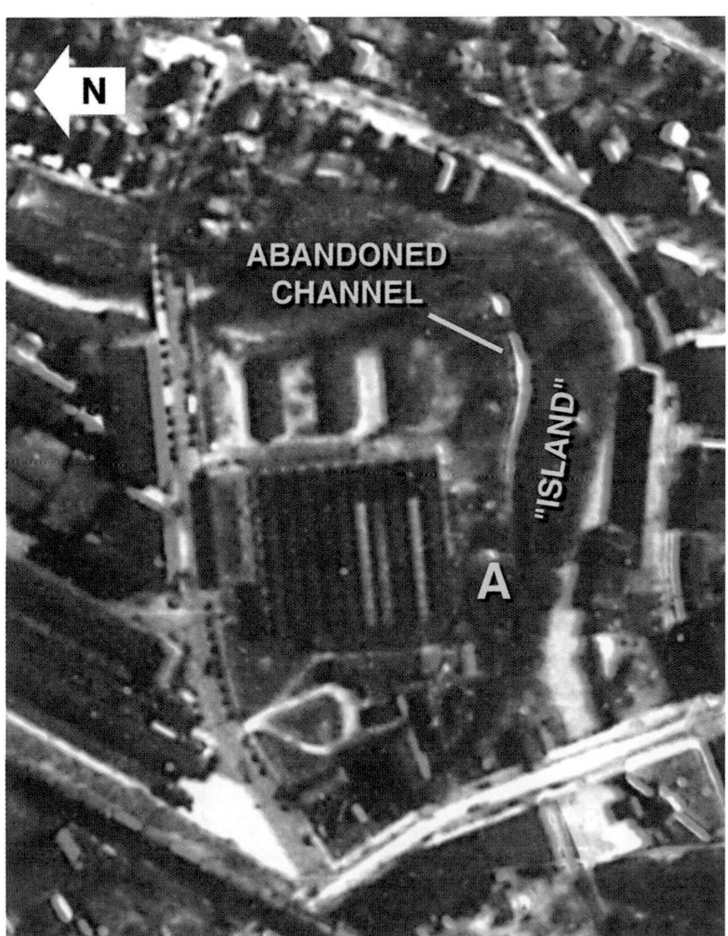

Plate 2

A 1938 aerial photograph of the bearing manufacturing plant in Plate 1 shows an abandoned river channel and an area called 'the island' by a deponent, where large amounts of dumping occurred prior to the building's annex at A being built.

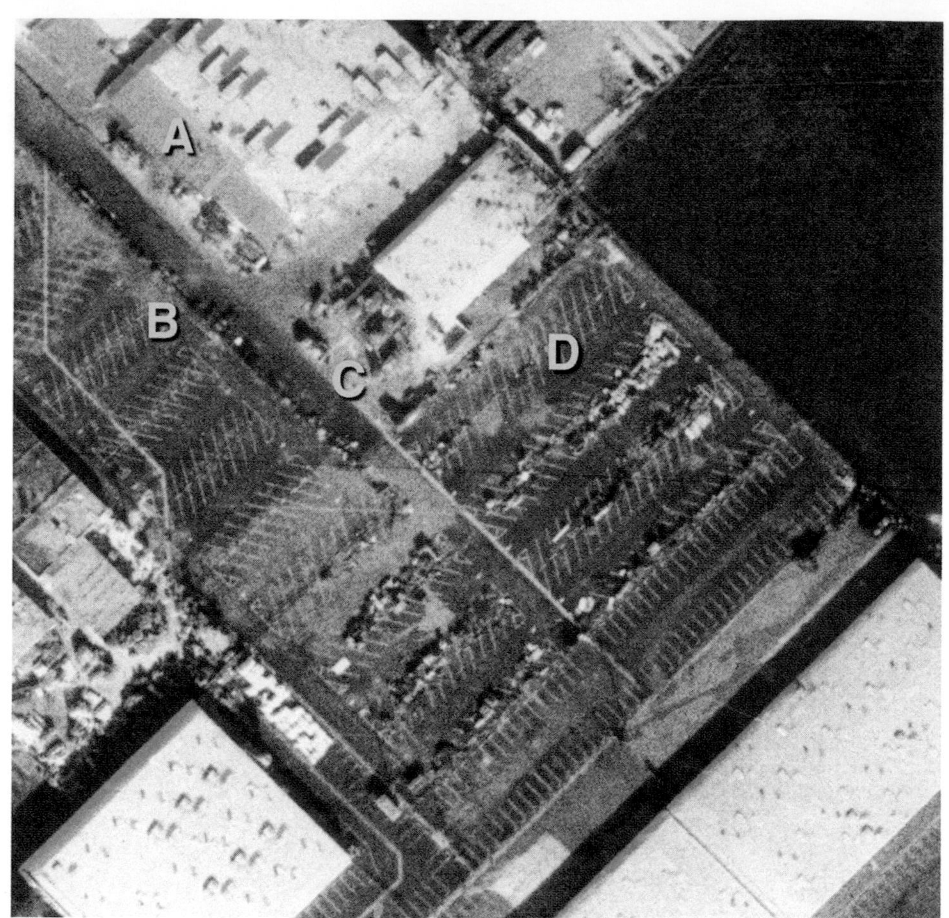

Plate 3

At a semiconductor board fabrication plant in 1963, liquid chemicals were being transferred from drums and carboys at the loading dock (A) and in the container storage area at B. Chemicals were being transferred from tank trucks to aboveground storage tanks at C. Mottled patterning on the asphalt at D may be the result of product transfer spillage as well.

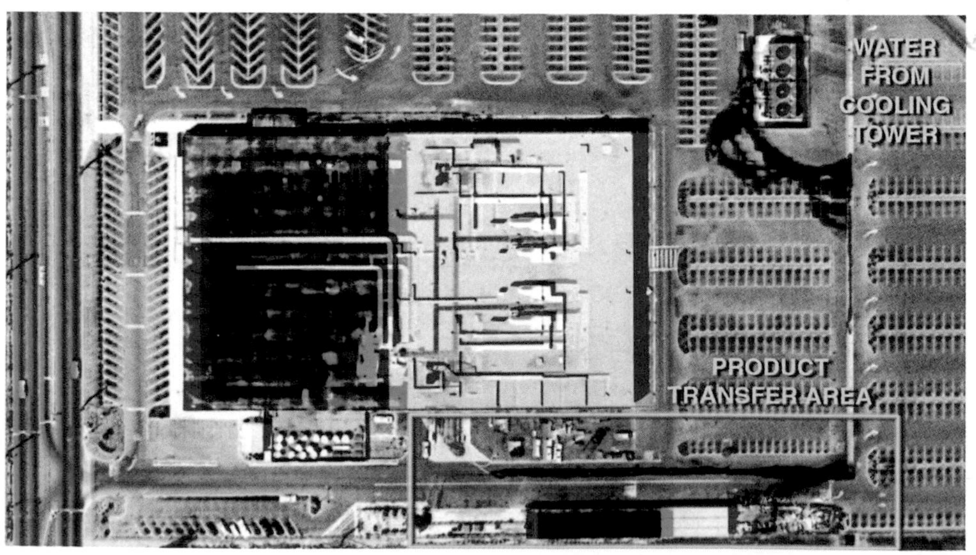

Plate 4

While water from the cooling tower is the most obvious 'dark stain' in this image, the product transfer area is of greater interest from an environmental forensic standpoint. A close-up of the product transfer area is shown in Plate 5.

Plate 5

This close-up of the circuit board fabrication plant product transfer area outlined in Plate 4 shows a hydrant and hose used to wash out drums and carboys. The effluent ran the length of the drive, and emptied into a drain at its end. The covered dock and the adjacent uncovered container storage area to its east were used to store chemical containers, and a dark stain resulting from product transfer spillage is visible issuing from the open storage area into the drain.

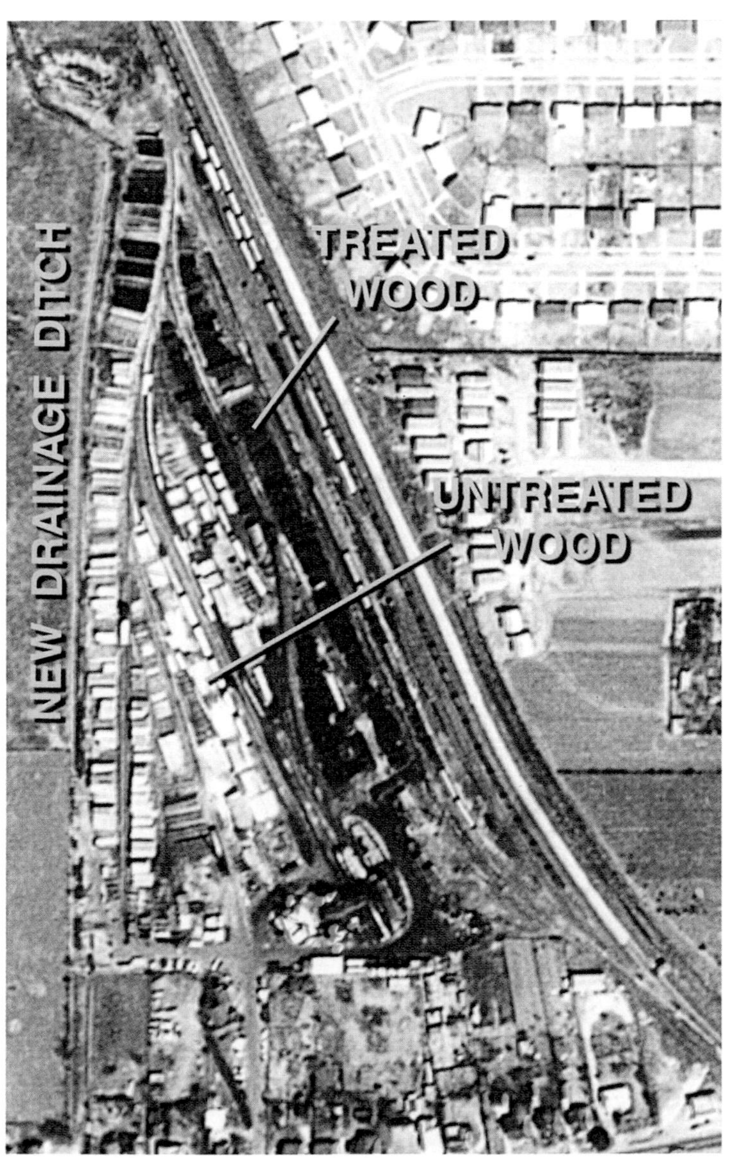

Plate 6

In 1959, the treated (dark colored) wood at this wood treatment plant in the southeastern United States was mostly stored on the east side of the plant. A new drainage ditch had just been dug on the west edge of the facility.

Plate 7

Storage of large piles of treated wood adjacent to the western edge of the wood treatment plant by 1965 allowed creosote laden runoff to enter the drainage ditch there and flow into nearby neighborhoods.

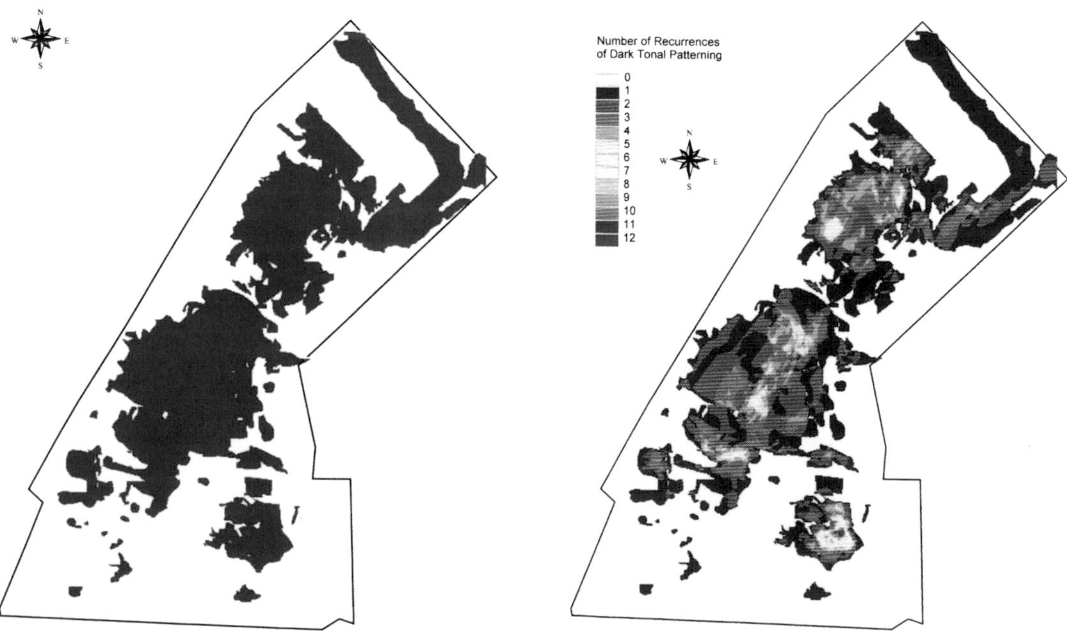

Plate 8

Two maps show alternate analyses of photointerpretation of 'dark staining' in the yard of a construction maintenance plant over 20 sets of aerial photographs spanning a period of 26 years. The map on the left is all of the interpreted dark staining unioned or overlain in by ArcInfo, a geographic information systems (GIS) software package. In the map on the right, the cumulative frequency of interpreted dark staining is shown in colors ranging from dark blue through dark red, showing that only a very small portion of the facility exhibited repetitive dark tones through time, a pattern consistent with human industrial behavior.

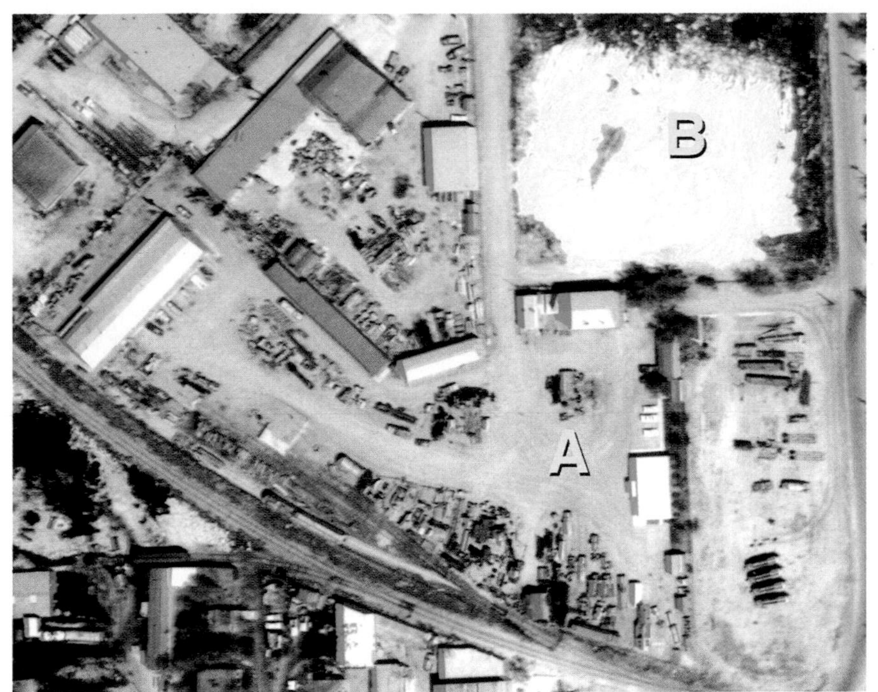

Plate 9

In this stereo anaglyph of the construction vehicle maintenance yard in 1996, there is little contrast difference in the yard indicative of 'dark staining'. The steam cleaning pad is located at A and a very reflective area of bare earth or possibly stockpiled light colored materials is at B.

Plate 10

The construction vehicle maintenance yard as it appears in a 1969 aerial photograph shows considerably more contrast than it does in Plate 9, probably due to the printing of the aerial photograph coupled with grading of the open ways between stored equipment. Differences in photo exposure and sunlight often cause one aerial photo to look dark while another looks much lighter.

Plate 11
A through-type, monorail conveyorized vapor-spray-vapor degreaser.

Plate 12
Thin section of soil impregnated with resin showing a decayed root channel with the remains of the root in the middle of the root channel. The lighter colored soil around the root channel is due to the osmotic stripping of nutrients from the surrounding soil (reproduced with permission from Morrison, R. (1999) Environmental Forensics: Principles and Applications. *CRC Press, Boca Raton, Florida).*

Garofalo, D. and Wobber, F.J. (1974) Solid waste and remote sensing. *Photogrammetric Engineering and Remote Sensing* 15(2), 45–59.

Getz, T.J., Randolph, J.C. and Echelberger, W.F. Jr. (1983) Environmental application of aerial reconnaissance to search for open dumps. *Environmental Management* 7(6), 553–562.

Lyon, J.G. (1987) Use of maps, aerial photographs and other remote sensor data for practical evaluations of hazardous waste sites. *Photogrammetric Engineering and Remote Sensing* 53(5), 515–519.

Mace, T.H., Williams, D.R., Duggan, J.R., Norton, D.J. and Muchoney, D.M. (1997) Environmental monitoring. In *Manual of Photographic Interpretation*, 2nd edn (Philipson, W.R., ed.), pp. 591–612. American Society for Photogrammetry and Remote Sensing, Bethesda, MD.

Morrison, R. (2000) *Environmental Forensics Principles and Applications*. CRC Press, Boca Raton, FL.

Nelson, A.B., Hartshorn, L.A. and Young, R.A. (1983) A methodology to inventory, classify and prioritize uncontrolled waste disposal sites. US EPA, No. 68-03-3049, Las Vegas, NV.

Philipson, W.R. and Sangrey, D.A. (1977) Aerial detection techniques for landfill pollutants. In *Proceedings of the 3rd Annual EPA Research Symposium on Management of Gas & Leachate in Landfills*, p. 11. St Louis. EPA, Washington, DC.

Poidebard, A. (1929) Les révélations archéologiques de la photographie aérienne: une nouvelle méthode de recherches d'observations en région de steppe. *L'Illustration*, May 25, 600–662.

Poidebard, A. (1930) Sur les traces de Rome: exploration archéologique aérienne en Syrie. *L'Illustration*, December 19, 560–563.

Pope, P., Van Eeckhout, E. and Rofer, C. (1996) Waste site characterization through digital analysis of historical aerial photographs. *Photogrammetric Engineering and Remote Sensing* 62(12), 1387–1394.

Teng, W.L. (1997) Fundamental of photographic interpretation. In *Manual of Photographic Interpretation*, 2nd edn (Philipson, W.R., ed.), pp. 49–113. American Society for Photogrammetry and Remote Sensing, Bethesda, MD.

CHAPTER 4

FORENSIC REVIEW OF SOIL GAS, SOIL AND GROUNDWATER DATA

Robert D. Morrison

4.1 INTRODUCTION

The use of forensic evidence for age dating and source identification is premised on the assumption that the underlying information is accurate and reliable. A thorough examination of the underlying environmental information is therefore required, with a focus on sample collection techniques, sample handling, and analytical testing. This chapter presents information on three of the most frequently used field techniques (i.e., soil gas surveys, soil sampling, and groundwater sampling) for collecting subsurface samples as illustrative of the value in forensically reviewing environmental data. While these techniques and matrices differ, the data review and interpretation process is surprisingly similar.

4.2 INTRUSIVE TECHNIQUES

4.2.1 SOIL GAS SURVEYS

Soil gas surveys provide a screening method for detecting volatile compounds so that subsequent investigative activities can be located with the highest probability of contaminant detection. The application of soil gas surveys for source identification assumes that the compounds of interest are susceptible to detection. A general rule is that a compound with a Henry's law constant of at least 0.05 kPa m^3/mol and a vapor pressure of 1.0 millimeters of mercury at 20°C or greater is amenable to a soil gas survey (Marrin, 1988; Erickson and Morrison, 1995). Compounds with a low Henry's law value do not readily partition out of the aqueous and into the vapor phase.

When used for source identification, the sampling probe location, density and depth interval (shallow to depth) must be sufficient to detect the anticipated type of release (e.g., leaking sewer, surface spillage, dumpster leaks, runoff from drain spouts from the roof of a dry cleaner, infiltration from a ditch, etc.). A degree of judgment is exercised to determine what constitutes a sufficient number and density of sampling probes relative to the circumstances of a release. Soil gas probes at five foot (1.5 m) intervals parallel to a suspected leaking sewer may be adequate while multiple depth probes at 20 foot (6 m) centers spaced in a grid pattern may be appropriate for source identification of suspected surface spills on a parking lot.

The selection of soil gas technology (e.g., active, passive and flux chambers) impacts the value of the resulting data for use for source identification purposes. An active soil gas survey consists of the withdrawal of a soil gas sample, typically from a perforated sampling probe, followed by analysis in a stationary or mobile laboratory (Morrison, 1999). Detectors used in various combinations for analysis include electrolytic conductivity detectors (ELCD or Hall detector), flame ionization, mass spectrometer and/or electron capture detectors (California Regional Water Quality Control Board, 1996). Vapor analysis is performed with a gas chromatograph and the results reported in units of parts per million on volume (ppmv). This is the most popular soil gas method due to the availability of companies providing this service and the relatively low cost. Active surveys are appropriate for locations with highly permeable soils (i.e., high gas permeability) and high concentrations of a volatile compound.

Passive soil gas surveys provide for the burial of probes containing adsorbent materials to identify a broad range of volatile and semi-volatile organic compounds (Foley, 1998a,b). The adsorbent is usually a polymeric and/or carbonaceous resin buried two to three feet (0.6–0.9 m) below the ground surface and remains *in situ* for one to two weeks. Passive soil gas survey data are reported as the total adsorbed mass, as the volume of vapor in contact with the absorbent while the absorbent is buried is unknown.

Petrex tubes were a common passive soil gas device used in the United States in the 1980s. Values from Petrex tube surveys are frequently encountered in older environmental investigation reports. Petrex tubes contain a thin ferromagnetic (Curie-point) wire coated with activated charcoal. The wire is housed within a glass tube (Petrex tube) and buried several inches below the ground surface. Upon retrieval, the wire is placed into a vacuum chamber, heated, and the desorbed compounds are analyzed by a Curie-point mass spectrometer and reported in ion counts.

Flux chamber measurements are primarily used in research applications, risk assessments and toxicological studies when direct vapor flux values are required (ASTM, 1997b; Hartman, 1998c). Flux chambers are placed on the ground surface or lowered to a discrete depth within a borehole. Surface flux chambers are equipped with ports for sweep air inlet and outlet, thermocouples for measuring the inside air and surface temperature and pressure outlets. The emission rate is described by

$$\mathrm{ER} = (Q)\,(C_{\mathrm{I}})\,/\,A \tag{4.1}$$

where ER is the emission rate of a species in micrograms per minute per meter squared, Q is the sweep air flow rate in cubic meters per minute, C_{I} is the concentration of the contaminant in micrograms per cubic meter and A is the sampling area in square meters. Sweep air is fed into the chamber at a metered flow rate (e.g., 5 liters per minute or a chamber retention time of about six minutes) and is then withdrawn until steady state conditions are attained within the chamber. A down-hole flux chamber is similar to a surface flux chamber with the exception that a discrete geologic horizon is isolated and the vapor flux rate from this layer is measured (Eklund, 1991; Schmidt and Simon, 1993).

Issues to be considered when evaluating soil gas results obtained from active or passive soil gas surveys are summarized in Table 4.1.

4.2.2 INTERPRETATION OF SOIL GAS DATA

While the interpretation of soil gas data is highly contaminant and site specific, there are common errors frequently encountered in the data analysis.

Table 4.1

Effects of active and passive soil gas sampling equipment on sample results and interpretation (after Hartman, 1999).

Active Soil Gas Survey

Sample spacing. The sampling location selection depends on the objectives of the program, the need for adequate coverage, and budget. Predetermined and widely spaced grid patterns are used for reconnaissance work while closely spaced, irregularly situated locations are used for identifying specific source areas.

Collection depth. Collection depths should maximize the probability of detecting contamination, yet minimize the effects of vapor movement, changes in barometric pressure, surface temperature, or breakthrough of atmospheric air from the surface. To optimize the probability of contaminant detection and minimize biases associated with vapor movement, soil gas samples are collected as close to the suspected contamination source as possible.

Probe seals. For collection systems with large purge volumes, it is often necessary to seal the probe at the ground surface. Surface seals are necessary for small volume systems if the soil is highly porous and the sampling depth is close to the ground surface (≤3 feet (90 cm)). Common sealing techniques include packing the upper contact of the probe and the soil with grout or an inflatable seal. Seal integrity is confirmed with a tracer gas (e.g., propane or butane) that flows around the probe at the contact point with the ground surface. A soil gas sample is then collected and analyzed for the presence of the tracer.

Probe decontamination. All external parts should be wiped clean and washed as necessary to remove soil or contaminants. The internal vapor pathway should be purged with a minimum of five volumes of air or an inert gas, or replaced. Probes fitted with internal tubing offer advantages because the internal tubing can be replaced between uses.

Excessive vacuums applied during sample collection. Soil gas samples collected under high vacuum conditions may reflect contaminants that are desorbed from soil grains created by the collection process, rather than contaminants present in the undisturbed soil gas. For collection systems employing vacuum pumps, the vacuum applied to the probe is measured and recorded.

Systems with vacuum pumps. Soil gas samples from collection systems employing vacuum pumps should be collected on the intake side of the pump to prevent potential contamination from the pump. Because the pressure on the intake side of the pump is below atmospheric pressure, soil gas samples must be collected with appropriate collection devices, such as gas-tight syringes and valves, to ensure that the samples are not diluted by outside air.

Passive Soil Gas Equipment

Collection depth. Passive collectors are buried near the ground surface (6 inches to 3 feet (15–90 cm)). This procedure originates from the convenience in deploying and retrieving the collector. Ideally, collectors are located close to the suspected contamination source to minimize the effects of vapor movement. Collectors buried several feet in the ground may be susceptible to air infiltration due to changes in barometric pressure and surface temperature. If the ambient air is contaminated (e.g., at an operating gasoline station or inside of a dry-cleaning operation), the collector can conceivably adsorb more contamination from infiltration of the surface air than from subsurface contamination. In this situation, it is advisable to bury the collector deeper than 3 feet (90 cm) below the ground surface.

Exposure period. The exposure period for passive collectors is generally selected for convenience rather than for technical reasons. An assumption inherent with the interpretation of passive soil gas data is that each collector is exposed to the same quantity of soil gas. Passive collectors are typically deployed for the same period (several days to three weeks) so that the data are normalized based on the exposure time. The exposure period for a passive collector depends on the concentration of the contaminant of interest and the desired detection levels. In areas of suspected high concentration, collectors may remain in the ground for shorter periods (1–5 days) than in areas of suspected low concentrations (2–3 weeks).

Method blanks. Since passive soil gas methods do not allow the collection of real-time data, the analysis of method blanks is important to verify that detected contamination is not from another source. A *method blank* or *trip blank* should therefore be included as part of the sample batch. If data from method and/or trip blanks are unavailable, values from the collectors can be argued as originating from other sources.

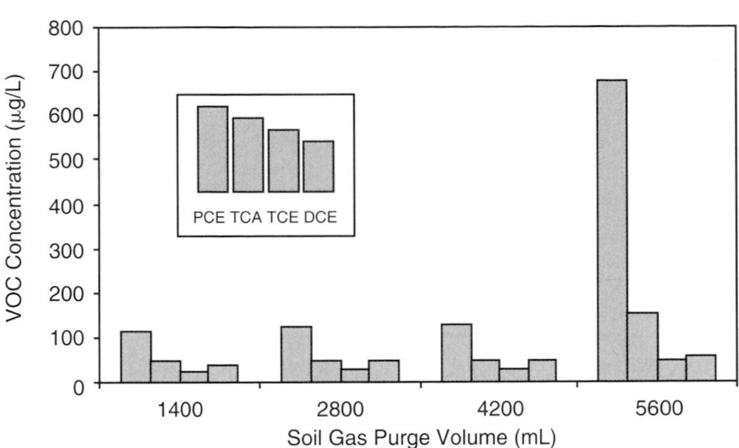

Figure 4.1
Impact of soil gas purge volume on chlorinated solvent concentrations in soil gas.

Interpretation errors include the comparison of soil gas data from multiple surveys with different purge volumes, incorrect unit conversions and excluding the consideration of subsurface barriers or preferential pathways impacting vapor transport.

A greater degree of uncertainty occurs regarding the origin of a vapor sample with increased purge volumes (DiGiulio, 1992). The extraction of a large purge volume, especially for shallow soil gas probes (<5 ft), increases both the uncertainty regarding the location of the contaminant source along with the potential that atmospheric air is drawn from the ground surface and into the sampling probe.

The potential impact of different purging volumes on soil gas concentrations from the same sampling probe is depicted in Figure 4.1. While the use of a purge volume of 5600 mL indicates the presence of perchloroethylene (PCE) at high concentrations (albeit the distance of the soil gas probe from the source is unknown), the use of smaller purge volumes in the immediate vicinity of the soil gas probe results in lower concentrations of PCE. It is therefore important, especially when comparing soil gas results from active surveys performed by different firms, to evaluate the potential impact of the purge volume on the soil gas results and the resulting data interpretation relative to source identification.

Another potential error in the interpretation of soil gas data is the use of different units and/or errors associated with unit conversion (Robbins *et al.*, 1990). Soil gas data are usually reported on a volume per volume (volume of contaminant per volume of soil air) or mass per volume (mass of contaminant per volume of soil air) basis. Soil gas data reported in μg/L are frequently, and incorrectly, assumed to be equivalent to parts per billion. The conversion of vapor data from units of μg/L to ppbv is $C_{soil\ gas}$ in ppbv = ($C_{soil\ gas}$ in

Table 4.2
Soil gas concentration unit conversions.

Units	To Convert to	Multiply by
μg/L	mg/m^3	1
$\mu g/m^3$	mg/m^3	0.001
ppmv	mg/m^3	MW/24 (20°C)
ppbv	mg/m^3	MW/24 000 (20°C)
μg/L	$\mu g/m^3$	1000
μg/L	ppmv	24/MW (20°C)
ppbv	ppm	0.001
ppmv	ppbv	1000

μg/L, micrograms per liter; mg/m^3, milligrams per cubic meter; $\mu g/m^3$, micrograms per cubic meter; ppmv, parts per million by volume; ppbv, parts per billion by volume. MW, molecular weight of the compound.

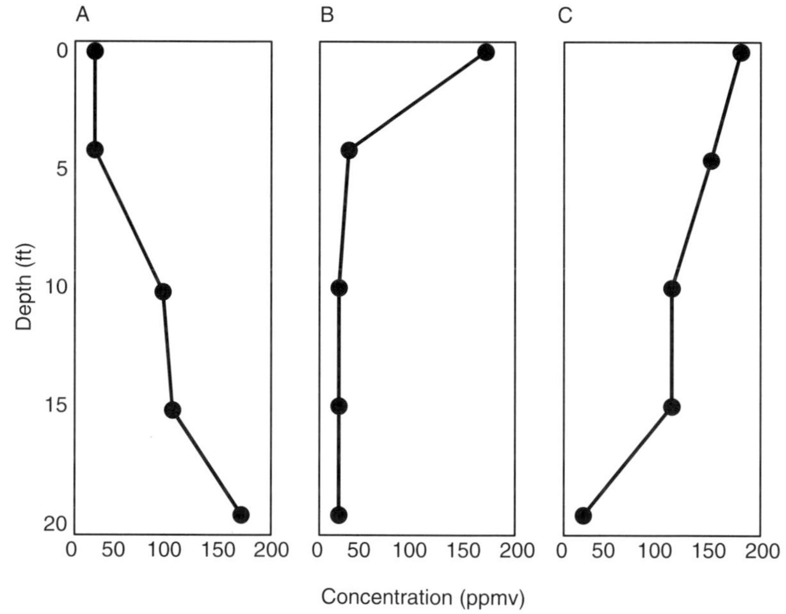

Figure 4.2
Depth versus soil gas concentration profiles for a homogeneous soil and possible source interpretations. (A) deep source, (B) shallow surface source with little depth penetration, and (C) surface source with partial penetration of the soil column.

μg/L × 24 000)/molecular weight of the compound, where the 24 000 is the milliliters per mole at 20°C. Table 4.2 lists common soil gas unit conversions (Hartman, 1998b).

Soil gas samples collected from multiple depths can provide useful qualitative information regarding the origin of a contaminant release. Figure 4.2 depicts three depth/concentration profile scenarios based on five samples along with possible interpretations (after Hartman, 1998b; Morrison, 1999).

As suggested in Figure 4.2, it is important to identify potential natural or artificial barriers as well as conduits that can impact vapor transport. Examples of natural barriers include moist silt or clay layers, low gas permeable soils, caliche horizons, and perched water tables. Artificial barriers include building foundations, basements and utility corridors, if enclosed in concrete trenches. For example, the presence of a soil with a high soil moisture content adjacent to structures such as clarifiers and/or sump can suppress the transport of a contaminant as a vapor cloud, resulting in artificially low soil gas concentration results.

The preferential migration of vapor through soil with different gas permeability values, such as utility trench backfill, may contribute to the rapid diffusion of soil gas relative to indigenous soils (Peargin, 1994). Gas permeability is a function of a soil's intrinsic permeability and liquid content; relative values can be estimated using methods developed by Brooks and Corey (1964) and Van Genuchten (1980). In addition to the identification of potential barriers and/or preferential pathways, the following issues deserve examination when interpreting soil gas data:

- the liquid–gas partitioning coefficient (volatility) of the compound(s) detected;
- the velocity or vapor diffusivity of the compound with time and whether the contaminant source has remained constant;
- the spatial impacts of vapor retardation (e.g., soil matrix adsorption, tortuosity or entrapment in unconnected soil pores) on the geometry of a vapor cloud;
- the potential contribution of localized releases in the ambient air;
- biodegradation; and
- comparison of shallow soil gas data from stationary probes collected at different times with significant ambient air temperature and/or barometric pressure differences.

Figure 4.3 depicts additional examples of soil gas concentration profiles and interpretations when the effects of artificial and lithologic pathways and barriers are considered.

4.2.3 SOIL SAMPLING

When soil sampling data are combined with field screening techniques, historical chemical usage, operational practices, and aerial photography interpretation, convincing causal relationships between contaminants and a source and/or operation often emerge. Given the opportunities for sample bias introduced by the sampling process or subsequent misinterpretation of the chemical data, activities most susceptible to bias require careful examination. Generic categories of sampling bias include the improper selection of sampling equipment relative to the analysis, subsampling, sample transfer, compositing and extended holding times.

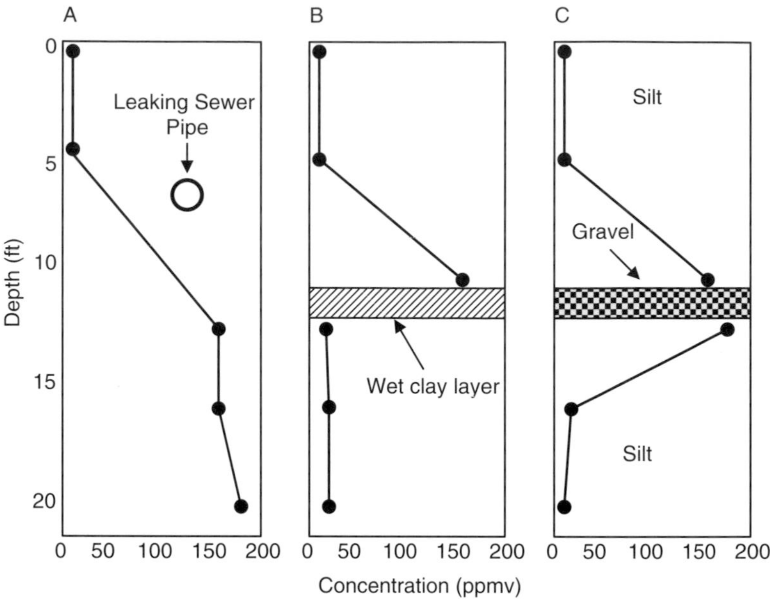

Figure 4.3

Depth versus soil gas concentration profiles and possible source interpretations. (A) Introduction of volatile compounds via a sewer, (B) surface release impeded by a wet clay soil, and (C) a lateral source of volatile organics fluxing above and below a gas permeable horizon.

4.2.4 SOIL SAMPLING EQUIPMENT

A variety of soil sampling equipment is available with different levels of potential chemical bias associated (Lewis *et al.*, 1994; Hewitt *et al.*, 1995; Hewitt and Lukash, 1996; Liikala *et al.*, 1996; ASTM, 1997a,b). Split spoon sampling is the most commonly used method but is not recommended if there is poor sample recovery (e.g., the metal or brass rings are not completely filled with soil) due to the potential losses of compounds by volatili-zation into the headspace of the partially filled brass tubes. Confirmation that the brass tubes are decontaminated prior to use is required. Pre-cleaned rings or tubes can be purchased with a decontamination certification. Recycled tubes can also be cleaned at a laboratory with the requisite number of rinsate samples and testing. In the field, the sampling barrel is attached to the drive rod of the drill rig and is driven into the soil with soil filling the sampling barrel. The sampling barrel is then retrieved at the surface, broken open, and the soil in the brass or steel rings sealed or transferred into another container. The exposed end of the soil in each ring should be quickly covered and sealed in the field using an inert film that is then covered with plastic or threaded metal caps to minimize sample loss or degradation, especially for volatile compounds. The use of electrical tape for sealing the plastic end caps on the brass rings is discouraged as permeation of toluene from the tape adhesives into the soil sample can occur (Morrison, 1999).

Volatilization of organic compounds (VOCs) from open and unsecured containers can result in significant decrease of VOC concentrations from soil samples. Hewitt (1999a,b), examined the loss of trichloroethylene (TCE) from the center of a silty sand held at ambient temperatures (18 ± 2°C) in an uncovered 3.6 cm inner diameter by 5.1 cm long metal core barrel liner stored in a plastic bag (Hewitt and Lukash, 1996). Within 30 and 40 minutes, approximately 80 and 90% of the initial TCE in the soil was lost via volatilization.

4.2.5 SUBSAMPLING AND SAMPLE TRANSFER

Subsampling is the repackaging of a sample into another container. Volatile organic compound losses occur during the soil transfer from a split spoon sampler into a 40 mL glass vial, an 8-ounce glass jar, or plastic bag, primarily through volatilization. The amount lost depends on the vapor pressure of the compound, the volume of headspace in the sample container, the ambient temperature, and the degree of sample disturbance (Hewitt *et al.*, 1992). Soil subsampling can result in volatile organic compound losses of as much as nearly 100% of the actual value (Siegrist and Jenssen, 1990; Siegrist, 1993; Hartman, 1998a).

If additional soil sampling is required due to concerns about the chemical integrity of previously collected samples, the sample transfer steps should be minimized and the sample analyzed as soon as possible, preferably with a mobile laboratory.

4.2.6 SOIL COMPOSITING

A composite soil sample consists of multiple samples collected at various sampling locations and/or points in time. The constituent information for the individual samples is lost although it enters into the composite measurement. Composite sampling reduces concentration variability, thereby narrowing the confidence interval of the population as contrasted with grab samples that maximize concentration variability (Johnson and Patil, 1994; Numbers, 1994).

When soil samples are composited, an averaged value results. Compositing can result in the loss of valuable information via sample dilution, especially at or near detection levels, as well as sample degradation from physiochemical and biological interactions introduced during the sample compositing process (especially for volatile organic compounds) (ASTM, 1997d; Lancaster and Keller-McNulty, 1998a,b). Compositing can also mask information that is useful for age dating the release of a particular compound. Figure 4.4 illustrates the opportunity to age date a release with discrete soil sampling and testing as contrasted with composite sampling. As indicated in Figure 4.4, soil samples collected at discrete depths in the borehole, when correlated with historical water level and groundwater results, confirm that PCE entered the groundwater after 1992. If soils from the 10, 20 and 30 foot depths were composited and trichloroethane (TCA) and PCE detected, however, the approximate date of

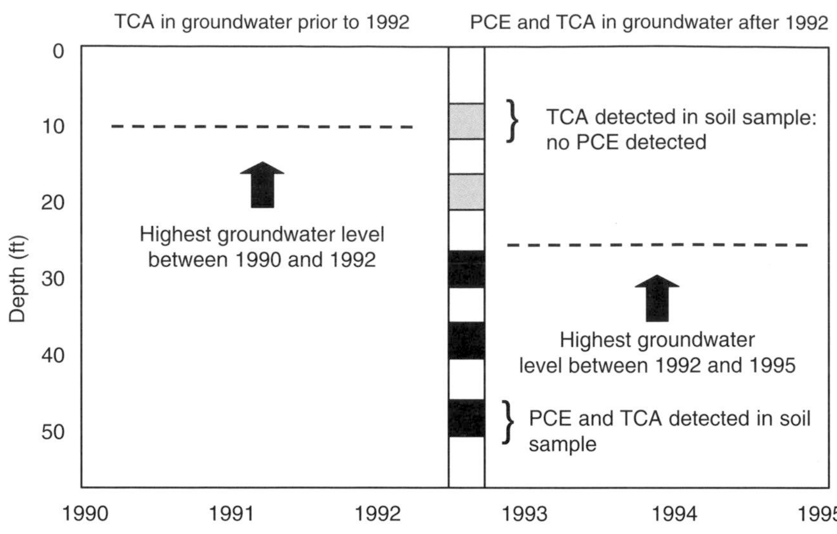

Figure 4.4

Use of discrete soil data with historical groundwater fluctuation data to confirm the release of PCE into the groundwater after 1992.

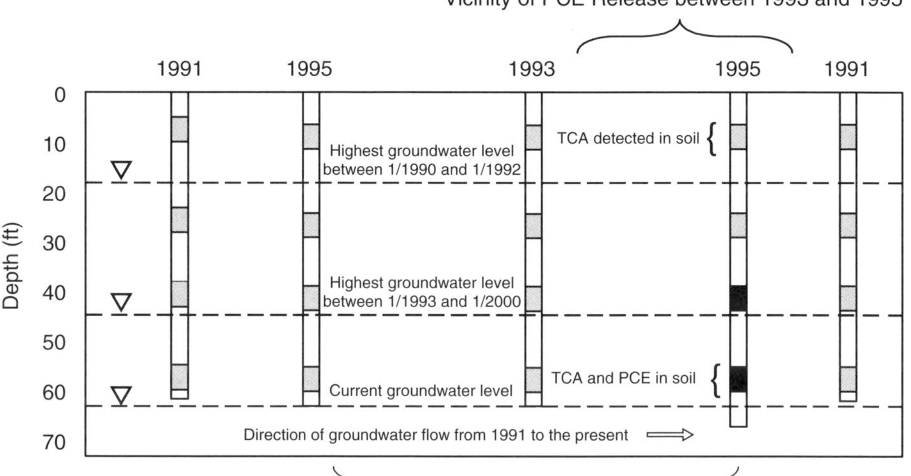

Figure 4.5

Use of discrete soil samples and test results to estimate the age and location of a PCE release.

entry of PCE into the groundwater could not be determined, based exclusively on the soil sample results.

A similar analysis to the Figure 4.4 example is shown in Figure 4.5 to estimate the location and age of a contaminant release. In Figure 4.5, soil samples collected from borings drilled between 1991 and 1995 provide the basis to identify the general location of a PCE release, based exclusively on discrete soil sample test results; the same interpretative methodology is used three-dimensionally and often with multiple compounds to provide greater spatial resolution to identify the source of a contaminant release.

Composite sampling is routinely encountered in soil confirmation sampling. In general, individual samples that are composited should be of a similar mass, although proportional sampling may be appropriate. An example of proportional sampling is the collection of soil cores from contaminated soil overlying an impermeable zone (ASTM, 1997c). Soil cores of different length can provide an averaged contaminant concentration value of the overlying soil, regardless of core length.

4.3 GROUNDWATER SAMPLING

A myriad of considerations arise when interpreting chemical results from groundwater samples for source identification and/or use in groundwater modeling (Morrison, 2000). Examination of chemical results and statistical relationships without examination of other factors can result in the misinterpretation of the data relative to identifying the contaminant source(s). If groundwater samples are collected from a monitoring well, common issues to examine are well design, location, water level measurements, and sample purging/sampling procedures.

4.3.1 MONITORING WELL DESIGN

Contaminant plume geometry is defined, in part, by the monitoring well location, depth and screen interval. If the monitoring well network is suspected of intentionally biasing sample chemistry, the well screen interval relative to the expected behavior of the contaminant in the subsurface and source should be studied. If monitoring wells located nearest to the source are constructed with long screens while those farthest from the source are short-screened, the monitoring well network may be intentionally designed to manipulate plume geometry and sample chemistry. While the potential bias associated with different monitoring well construction can be identified, quantification of this bias is difficult. The qualitative impact of different well construction designs in addition to an evaluation of groundwater purging and sampling techniques can be examined and a judgment made concerning the reliability of the chemical data for identifying or discriminating between contaminant sources.

4.3.2 MONITORING WELL LOCATION

Understanding the impact of the well location and well screen interval on sample chemistry is a necessary step when interpreting chemical data for source identification and age dating. Potential sources of bias include well construction materials, grout chemistry, improperly installed security covers, and cross-contamination via drilling methods (Powell, 1997).

If groundwater entering a well screen is in contact with a bentonite seal, elevated concentrations of sodium and sulfate can result, even after several years (Remenda and van der Kamp, 1997). The longer the contact, the greater the opportunity for detecting elevated sodium and sulfate concentrations in the sample. Bentonite pellets used in monitoring wells in Saskatoon, Saskatchewan, Canada, for example, resulted in sodium and sulfate concentrations in groundwater collected from the well of 3900 and 5800 mg/L, respectively (Wassenaar and Hendry, 1991). These results are consistent with water-soluble extract analyses on bentonite samples (Keller *et al.*, 1991).

Elevated pH values in groundwater from improper grouting procedures or the use of grout materials containing potential contaminants can impact sample chemistry. Cement grout ($CaCO_3$) can raise the pH of the surrounding soil several pH units. Elevated pH values in the vicinity of the well screen may cause precipitation of otherwise soluble metals as they enter a halo of higher pH groundwater. If this phenomenon is suspected, it is advisable to examine whether the pH values are high relative to other wells in the area and to purge the well heavily prior to sampling (e.g., 10–20 casing volumes); the pH should then be measured to observe if it suddenly drops several pH units. If pH values decrease abruptly, this may suggest that the grout material has impacted groundwater pH in the immediate vicinity of the well.

If contamination is present in multiple aquifers and cross-contamination via drilling is suspected, it should be ascertained whether the driller penetrated a confining layer(s) and introduced contamination into a deeper, previously uncontaminated aquifer. A boring log that indicates that the borehole was over-drilled and the lower portion was grouted often suggests penetration through a confining layer. A reverse situation occurs when contamination from a lower aquifer is allowed to mix within a well screened across several aquifers during pumping and then flow into a previously uncontaminated shallower zone when the pump is not operating.

4.3.3 WATER LEVEL MEASUREMENTS

An integral step prior to sampling a monitoring well is to measure the depth to the groundwater surface. Water level measurements from a series of wells are used to approximate groundwater flow directions, and hence velocity, and for estimates of contaminant mass in the case of the presence of an LNAPL (light non-aqueous phase liquid). The thickness of a phase separate liquid on the water surface in a monitoring well is greater than the thickness of LNAPL in the formation (a LNAPL can also exist in the formation but not the monitoring well (Ballestero *et al.*, 1994)). In addition, the approximation of free phase characteristics, especially liquid density, and the corresponding use of this information to construct a groundwater surface can result in variations from the actual groundwater gradient, especially in unconfined aquifers with a low gradient (Hampton and Miller, 1988; Lenhard and Parker, 1990). In cases with monitoring wells with and without LNAPL present, a groundwater contour map with wells without LNAPL can be constructed, followed by a second map that includes the LNAPL corrected water level measurements and the two maps compared.

4.3.4 GROUNDWATER PURGING

A review of groundwater purging procedures is useful as inconsistent purging practices can result in the erroneous identification of intermittent or multiple sources of groundwater contamination (Martin-Hayden and Robbins, 1997). The traditional practice is to purge 5 to 10 times the volume of standing water in the well to remove the stored water in the well casing (ASTM, 1992). Micro-purging a smaller volume of water is currently considered as an alternate as it can provide representative chemical results while minimizing issues regarding disposal of purge waters and/or sample oxygenation. Micro-purging or low flow sampling is generally considered to be between about 100 to 200 mL per minute (Puls *et al.*, 1990). Higher rates are acceptable (i.e., 1 liter per minute) for more transmissive formations (Powell and Puls, 1993). Research indicates that a representative groundwater sample for volatile organic sampling is achievable via micro-purging without pumping 2 to 5 well casing volumes (Kearl *et al.*, 1994; Puls and Paul, 1995; Powell, 1997).

While micro-purging may maintain sample integrity especially for compounds sensitive to redox changes, the small radius of sampling influence provides an opportunity for manipulating sample results. In Figure 4.6, each bar represents the PCE concentration detected in five groundwater monitoring wells on five separate occasions. PCE is present as a DNAPL (dense non-aqueous phase liquid) down-gradient of wells 3 and 4. The sampling radius of influence for wells 3 and 4 and whether the DNAPL is intersected is dependent on the purge rate. For well 3, the selection of a purging rate of less than 0.3 gallons per minute on the first and third sampling round resulted in significantly lower PCE concentrations than for the other rounds where a purging rate of 5 gallons per minute was selected.

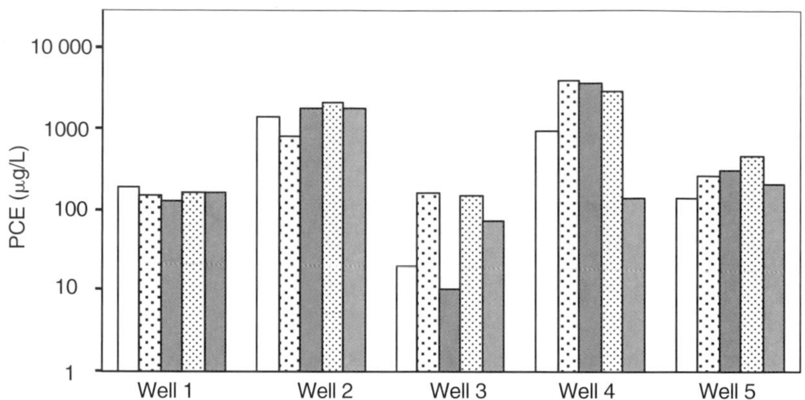

Figure 4.6
Bar graph with PCE concentrations from five monitoring wells sampled on five separate occasions. All wells were purged at a rate of 5 gallons per minute except for the first and third sampling event for well 3, where the purge rate was less than 0.3 gallons per minute.

4.3.5 GROUNDWATER SAMPLING PROCEDURES AND EQUIPMENT

The selection of groundwater sampling equipment and associated procedures impacts the chemistry of the water sample (Stolzenburg and Nichols, 1986; Puls and Barcelona, 1989; Puls *et al.*, 1992). The greatest opportunity for negative bias occurs when sampling groundwater for volatile organic compounds. Because the true value of a volatile organic compound in the groundwater sample is unknown, the bias introduced by the sampling equipment and operation is relative. Comparison of equipment and procedures resulting in higher levels of volatile compounds are therefore considered the most accurate (i.e., unless evidence of a false positive bias is identified).

Volatile compounds are sensitive to losses by degassing at reduced pressures from suction devices such as peristaltic or centrifugal pumps or turbulence created by mechanical sampling devices such as gas lift samplers and bailers. Volatile organic compound losses from improper sampling equipment selection and use are cumulative in nature. Sampling procedures that potentially represent sources of volatile organic compound loss are:

- sampling with a bailer and aerating the sample during transfer from the bailer into the sample container;
- over pressurization of the sampling equipment during sample retrieval; and
- not immediately placing the cap on the sampling container or chilling the sample.

This level of procedural information is usually not included in the environmental report but may be available in field notes or acquired via deposition testimony of the sampler. The forensic review of these data is critical as considerable differences in sample integrity can result, depending on the type of sampler and compound of interest. For concentrations near the detection limit, low-flow groundwater sampling and aqueous diffusion samplers may provide the most representative results (Nimmer and Chase, 2001). An evaluation of the potential temporal impacts (e.g., tidal cycles, irrigation and pumping well cycles, seasonal changes in groundwater direction, etc.) on sample chemistry are also required prior to interpreting the data for source identification and/or age dating. The sampling sequence may also be designed to maximize particular temporal impacts on sample chemistry. If such relationships exist, it should be ascertained whether the sampling schedule results in a systematic impact on sample chemistry, and if so, this information should be incorporated into the analysis.

4.3.6 EQUIPMENT DECONTAMINATION

The chemical results from equipment decontamination samples allow an assessment of the potential presence and nature of cross-contamination originating from the sampling equipment. Federal, state and American Society of Testing Materials (ASTM) standards are available that describe decontamination procedures for non-contact and contact equipment used for soil and groundwater sampling (Mickam *et al.*, 1989; ASTM, 1990; Parker, 1995). Examples of non-contact equipment include drilling rod racks, a backhoe extension arm, drill rods, and the backend of a drilling rig. Non-sample contact equipment can be rinsed with a portable power washer or steam cleaner. In addition, hand washing with a brush and detergent solution may be required followed by rinsing with water of known chemical composition (ASTM, 1990).

Contact equipment includes drill augers, well casing, well screens, backhoe buckets, submersible pumps, spatulas, knives, bailers and split-spoon samplers. In general, contact sampling equipment is washed with a detergent solution (Alconox or Liquinox or similar non-phosphate/ammonia detergent) followed by a series of water, desorbing agents and deionized water rinses (USEPA, 1991; Wilson, 1998). This decontamination sequence, however, may not improve the removal of volatile organic compounds from polymer materials. In one study of methanol and hexane solvent rinses, were not found to significantly improve the removal of TCE, PCE, *m*-nitrotoluene and *p*-dichlorobenzene from polytetrafluoroethylene (Parker and Ranney, 2000). Solvent rinsing, however, was found to improve the removal of lindane, heptachlor, aldrin and dieldrin from low-density polyethylene, although washing followed by over-drying ($\sim$117°C for 24 hours) was more effective.

Barring any other chemical interaction between a contaminant, relatively hydrophilic organic solutes (e.g., compounds with log octanol water partition coefficient values less than 4) are not sorbed by non-permeable materials such as glass and stainless steel (Parker and Ranney, 1994). Hydrophobic contaminants, such as polychlorinated biphenyls (PCBs), chlorinated pesticides, and polyaromatic hydrocarbons can be sorbed by these surfaces (Strachan and Hess, 1982).

Permeable polymeric materials used for well casings and sampling tubing however can adsorb substantial quantities of relatively hydrophilic organic contaminants (Parker and Ranney, 1997; 1998). For example, Gillham and O'Hannesin (1990) found that after 10 minutes, sorption of low parts per million concentrations of benzene by flexible PVC (polyvinyl chloride) was approximately 35% versus 1% for rigid PVC.

4.3.7 SAMPLE CONTAINERS, PRESERVATION, AND HOLDING TIMES

In the United States, Environmental Protection Agency (EPA) standards exist that specify the proper container for soil and water samples for a given test method. If not followed, the containers can introduce a potential bias, especially at near detection limits. Samples with low concentrations may require preservation by refrigeration or the addition of chemical reagents to slow physical, chemical or biological changes. Chemical and biological activities altering the sample chemistry include the formation of metal and organic complexes, adsorption/desorption reactions, acid–base reactions, redox reactions, precipitation/dissolution reactions, and microbiological activities affecting the disposition of metals, anions and organic molecules. Samples analyzed for volatile organic compounds are acidified because hydrochloric acid effectively prevents biodegradation of many volatile organic compounds (Maskarinec *et al.*, 1990). The United States EPA requires that samples analyzed for volatile organic compounds by Standard Method 524.2 for example be preserved by hydrochloric acid (HCl) to a pH of less than 2. The addition of hydrochloric acid does not result in the formation of trihalomethanes (Squillace *et al.*, 1999).

While sample handling may violate various regulatory sampling/handling requirements, the sample results should not be disregarded. In the United States, such data may be relied upon by an expert witness. In *People* v. *Hale* (1976), for example, an element of the case involved the illegal disposal of TCA into waste dumpsters (Simmons, 1999). While sampling procedures (no sampling plan was used), handling (samples were frozen rather than cooled to 4°C), holding times (the specified 14 day holding time exceeded) and analytical methods (Method 8015 with a flame ionization detector was used instead of accepted EPA Methods 8010 or 8240 using a mass selective detector) deviated from EPA SW-846 standards, the court found that these deviations did not preclude the introduction of this information as evidence.

Another example of the value of chemical testing information that violated recommending sampling and/or handling procedures is illustrated in the matter of *People* v. *Sangani* (1994), which involved the illegal disposal of hazardous waste into a sewer system. The defendant was convicted, but appealed because the laboratory performing the analysis was not certified. The court found that 'Failure to follow precise regulatory or statutory requirements for laboratory tests generally does not render the test results inadmissible, provided the foundation requirements for establishing the reliability of the tests are met.' The necessary foundation requirements are: (1) the testing apparatus is in proper working order, (2) the test was properly administered, and (3) the operator was competent and qualified. The axiom that all data have value is important to consider in the forensic interpretation of any analytical data and reliance that a court will find the analytical data inadmissible may be misplaced.

4.4 ANALYTICAL METHODS

When examining analytical methods, compare the procedures used with published procedures (USEPA, 1982, 1983, 1986; American Public Health Association, 1985): Areas of inquiry include the following (Mishalanie, 1995):

- misidentification of compounds (false and negative bias);
- chemical interference (false and negative bias);
- incomplete recovery of analytes from the sample matrix (negative bias);
- matrix effects (false and positive bias);
- instrumentation calibration (false and positive bias); and
- cross contamination (positive bias).

An example of the impact of analytical methods on sample results is the analysis of methyl tertiary butyl ether (MTBE). For MTBE analysis using USEPA Standard Method 8020, shortened run times can result in the coelution of compounds (e.g., 2-methylpentane, 3-methylpentane, cyclopentane, 2-methyl-2-pentene, 2,3-dimethylbutane) that elute similarly to MTBE and are reported as MTBE. A review of EPA Method 8020 conducted by the Lawrence Livermore National Laboratory in Livermore California noted that for the detection of low MTBE concentrations (<100 μg/L) in the presence of high gasoline or BTEX (benzene, toluene, ethylbenzene, xylene) concentrations overestimates of MTBE concentrations by greater than 40% are common. Another analytical option when testing for MTBE is to use a chromatographic column that separates methyl pentanes, which in some cases, can be used with EPA Method 8020 (Jacobs *et al.*, 2000).

EPA Standard Methods 8240 and 8260 are gas chromatography methods using a mass selective detector that can quantify MTBE and other coeluting compounds (Hartman and Hitzig, 1998). Another reported technique is a modified ASTM Method D4815 specifically designed to test for MTBE, ethyl tertiary butyl ether (ETBE), tertiary amyl methyl ether (TAME), as well as methanol, ethanol and tertiary butyl alcohol (TBA) in soil and groundwater.

4.4.1 MISIDENTIFICATION OF ANALYTES

The misidentification and laboratory misrepresentation of compounds is common. A report performed by the Advancement of Sound Science Coalition for the United States Environmental Protection Agency found that 11% of 2000 laboratory reports showed 'serious deficiencies' with pesticides testing (Meyer, 1999). In a few cases, fraud is the culprit such as in the case of the *United States* v. *Hess Environmental Laboratories* and *United States* v. *Klusaritz*, where the director of a testing laboratory accepted payment for fabricated data from measurements that were not performed. Numerous unintentional reasons for misidentification are possible, including the coelution of compounds (e.g., dibromochloromethane, dichloropropene and 1,1,2-trichloroethane). Laboratories may automatically assign a name to a product type or a material that elutes within a given carbon range or retention range, for example, that is reported as in the 'gasoline range' but may not be the same as gasoline (Rhodes, 1999).

Another example is shown in Table 4.3 for TPH (total petroleum hydrocarbons) analysis for gasoline and diesel using a gas chromatograph equipped with a flame ionization detector (Bruya, 1994, 1998). These tests were performed according to USEPA mandated total petroleum hydrocarbon testing protocols. These data demonstrate that common plants contain organic compounds that boil in the same boiling ranges associated with gasoline and diesel.

The selected extraction method can bias sample results. For example, for TPH analyses, an acetone/methylene chloride extractant is more rigorous than methanol; the former extractant extracts more hydrocarbons from the sample resulting in a higher sample concentration.

The use of multiple laboratories in a project can produce dramatically different analytical and interpretative results. A 10–20% variation in reported concentrations for the same sample (spiked concentration) between laboratories is not uncommon.

The presence of field and laboratory contaminants should be examined when reviewing test results. Common field contaminants include

Table 4.3
Total petroleum hydrocarbon values in various materials.

Matrix	TPH as gasoline (ppm)	TPH as diesel (ppm)
Spinach	<10	60
Carrots	<10	10
Orange juice	300	<10
Cedar tree	1400	2200
Pine tree	450	400
Dandelion	<10	140
Daisy	40	40
Moss	<10	<10

bis-(2-ethylhexyl)phthalate, di-butylphthalate and adipates from sampling tubing, toluene from electrical tape, 2-butanone from Duct tape, *N*-nitroso-diphenyl amine from rubber sealants or gloves, and trihalomethanes from domestic water. Laboratory contaminants include the laboratory solvents methylene chloride, Freon 113, chloroform, and carbon tetrachloride (Erickson and Morrison, 1995; Mabey, 1995). Other laboratory contaminants include laboratory internal standards such as fluorobenzene, chlorobenzene d_5 and 1,4-dichlorobenzene d_4, which are typically analytes of interest that are spiked into a blank prior to analysis for the purpose of measuring the accuracy of the measurement.

System monitoring compounds are potential contaminants that are used by laboratories in volatile compound analyses and can include dibromo-fluoromethane, 1,2-dichloroethane d_4, toluene d_8 and 4-bromofluorobenzene. Detection of these chemicals is usually only considered positive when their concentrations are greater than ten times the amount detected in any blank samples (State of California, 1994; USEPA, 1994). If the concentration of a common laboratory contaminant is less than ten times its concentration in the blank, then one may conclude that the chemical was not detected in the soil, soil gas or air sample. If the blank contains detectable levels of one or more organic or inorganic chemicals that are not common laboratory contaminants, the sample results are considered positive only if the concentration exceeds five times the maximum amount detected in the blank. If all of the samples contain concentrations less than five times the level of contamination detected in the blank, the chemical should be eliminated from the set of sample results (State of California, 1994).

BTEX (benzene, toluene, ethylbenzene and xylene) are common laboratory artifacts, often due to the large volume samples with high concentrations that are processed through many laboratories. If anomalous BTEX readings are suspected, the laboratory throughput of these analyses during the time period for which these samplers were tested should be examined to assess whether this potential for cross-contamination is a reasonable explanation.

4.5 CONCLUSIONS

The use of chemical data acquired from soil gas, soil and/or groundwater sampling for source identification and age dating requires a careful analysis of the underlying equipment selection, sample handling procedures, and analytical methodology. When performing the forensic analysis of this information, it is important to acquire the original copies of this information, including the analytical laboratory results, chain of custody forms, laboratory bench notes and sampling field notes rather than relying upon summaries of information in an environmental report and/or electronic spreadsheet. It is also important to verify that unit conversions were performed correctly.

REFERENCES

American Public Health Association (1985) *Standard Methods for the Examination of Water and Waste-water*, 16th edn. Washington, DC.

American Society for Testing Materials (ASTM) (1990) Standard Practice for Decontamination of Field Equipment Use at Nonradioactive Waste Sites. D 5088-90. In *ASTM Standards on Environmental Sampling*, 2nd edn, pp. 538–540. ASTM Publication Code PCN 03-418097-38. ASTM, West Conshohocken, PA.

American Society for Testing Materials (ASTM) (1992) Standard Guide for Sampling Groundwater Monitoring Wells. D 4448-85a. In *ASTM Standards on Environmental Sampling*, 2nd edn, pp. 430–443. ASTM Publication Code PCN 03-418097-38. West Conshohocken, PA.

American Society for Testing Materials (ASTM) (1997a) Section 4.2. Specific Waste Sampling Procedures. In *ASTM Standards on Environmental Sampling*, 2nd edn, pp. 558–658. ASTM Publication Code PCN 03-418097-38. ASTM, West Conshohocken, PA.

American Society for Testing Materials (ASTM) (1997b) Standard Guide for Soil Gas Monitoring in the Vadose Zone. ASTM Designation D5314-92. In *ASTM Standards on Environmental Sampling*, 2nd edn, pp. 185–215. ASTM Publication Code PCN 03-418097-38. ASTM, West Conshohocken, PA.

American Society for Testing Materials (ASTM) (1997c) Standard Guide for Direct-Push Water Sampling for Geoenvironmental Investigations. ASTM

Designation D 6001-96. In *ASTM Standards on Environmental Sampling*, 2nd edn, pp. 444–457. ASTM Publication Code PCN 03-418097-38. ASTM, West Conshohocken, PA.

American Society for Testing Materials (ASTM) (1997d) Composite Sampling and Field Subsampling for Environmental Waste Management Activities. ASTM Designation D 6051-96. In *ASTM Standards on Environmental Sampling*, 2nd edn, pp. 514–520. ASTM Publication Code PCN 03-418097-38. ASTM, West Conshohocken, PA.

Ballestero, T., Fiedler, F., and Kinner, N. (1994) An investigation of the relationship between actual and apparent gasoline thickness in a uniform sand aquifer. *Groundwater* 32(5), 708–718.

Brooks, R. and Corey, A. (1964) Hydraulic properties of porous media. Colorado State University. Fort Collins, CO. Hydrology Paper No. 3. 27.

Bruya, J. (1994) Interpretation of laboratory analysis of chlorinated solvents. In *Proving the Technical Case: Soil and Groundwater Contamination Litigation with Emphasis on Chlorinated Solvent Contamination*. November 9–10, 1994. San Jose, CA. Department of Engineering Professional Development, College of Engineering. University of Wisconsin, Madison, WI.

Bruya, J. (1998) Review of Analytical Data. *Proceedings of the National Environmental Forensic Conference: Chlorinated Solvents and Petroleum Hydrocarbons*. College of Engineering and Engineering Professional Development. University of Wisconsin, Madison, WI. August 27–28, 1998. Tucson, AZ.

California Regional Water Quality Control Board (1996) *Interim Site Assessment & Cleanup Guidebook*. California Environmental Protection Agency, Los Angeles and Ventura Counties, Region 4. May 1996.

DiGiulio, D. (1992) Evaluation of soil venting application. United States Environmental Protection Agency. Groundwater Issue. EPA/540/2-92/004.

Eklund, B. (1991) User's guide for the measurement of gaseous emissions from subsurface wastes using a downhole flux chamber. United States Environmental Protection Agency, Office of Research and Development. RCN 228-069-03-01. 26.

Erickson, R. and Morrison, R. (1995) *Environmental Reports and Remediation Plans: Forensic and Legal Review*. John Wiley & Sons, New York. Environmental Law Library.

Foley, G. (1998a) Passive soil gas sampler. Gore-Sorber® screening survey passive soil gas sampling system. Environmental Technology Verification Program Verification

Statement. EPA-VS-SCM-19. United States Environmental Protection Agency, Washington, DC.

Foley, G. (1998b) Passive soil gas sampler. Emflux® soil gas investigation system. Environmental Technology Verification Program Verification Statement. EPA-VS-SCM-22. United States Environmental Protection Agency, Washington, DC.

Gillham, R. and O'Hannesin, S. (1990) Sorption of aromatic hydrocarbons by materials used in construction of groundwater sampling wells. *Groundwater and Vadose Zone Monitoring*, pp. 108–122. American Society for Testing and Materials. Special Testing Publication 1053. Philadelphia, PA.

Hampton, D. and Miller, P. (1988) Laboratory investigation of the relationship between actual and apparent product thickness in sands. In *Proceedings of the Conference on Petroleum Hydrocarbons and Organic Chemicals in Groundwater*, pp. 157–181. National Water Well Association. November 5–7, 1998. Houston, TX.

Hartman, B. (1998a) To methanol preserve or not to methanol preserve? *LUSTline Bulletin 28*, pp. 17–18. New England Interstate Water Pollution Control Commission, Wilmington, MA.

Hartman, B. (1998b) MTBE – Beware the false positive. *LUSTline Bulletin 26*, p. 18. New England Interstate Water Pollution Control Commission. Wilmington, MA.

Hartman, B. (1998c) Applications and interpretation of soil vapor data. *Petroleum Hydrocarbon Contamination: Legal and Technical Issues*, pp. 81–110. Argent Communications Group, Forresthill, CA.

Hartman, B. (1999) Soil vapor analysis and interpretation. *Chlorinated Solvent Contamination: Legal and Technical Issues*, pp. 84–123. Argent Communications Group, Forresthill, CA.

Hartman, B. and Hitzig, R. (1998) A layman's guide to the new EPA methods for VOC analysis. *LUSTline Bulletin 30*, pp. 21–23. New England Interstate Water Pollution Control Commission, Wilmington, MA.

Hewitt, A. (1999a) Frozen storage of soil samples for VOC analysis. *Environmental Testing and Analysis* Sept/Oct. 1999. 18.

Hewitt, A. (1999b) Storage and preservation of soil samples for volatile compound analysis. United States Army Corps of Engineers. Cold Regions Research & Engineering Laboratory, Hanover, NH. Special Report 99-5.

Hewitt, A. and Lukash, N. (1996) Obtaining and transferring soils for in-vial analysis of volatile organic compounds. United States Army Corps of Engineers. United

States Cold Regions Research and Engineering Laboratory. Hanover, NH Special Report 96-5.

Hewitt, A., Miyares, P., Leggett, D., and Jenkins, T. (1992) Comparison of analytical methods for determination of volatile organic compounds in soils. *Environmental Science and Technology* 28(10), 1932–1938.

Hewitt, A., Jenkins, T., and Grant, C. (1995) Collection, handling, and storage: Keys to improved data quality for volatile organic compounds in soil. *American Environmental Laboratory* 7, 25–28.

Jacobs, J., Guertin, J., and Herron, C. (2000) *MTBE: Effects on Soil and Groundwater Resource*. Lewis Publishers, Boca Raton, FL.

Johnson, G. and Patil, G. (1994) Observational economy for hazardous waste site characterization and evaluation using methods of composite sampling. In *Proceedings of the International Specialty Conference on Cost Efficient Acquisition and Utilization of Data in the Management of Hazardous Waste Sites*, (Lewis, R., ed.), pp. 83–99. Air and Waste Management Association, Pittsburgh, PA.

Kearl, P., Korte, N., Stites, M., and Baker, J. (1994) Field comparison of micropurging vs. traditional groundwater sampling. *Groundwater Monitoring Review* 183–190.

Keller, C., van der Kamp, G., and Cherry, J. (1991) Hydrogeochemistry of a clayey till. Spatial variability. *Water Resources Research* 27(10), 2543–2554.

Lancaster, V. and Keller-McNulty, S. (1998a) Composite sampling, Part I. *Environmental Testing and Analysis* 7(4), 15–18 and 32.

Lancaster, V. and Keller-McNulty, S. (1998b) Composite sampling. Part II. *Environmental Testing and Analysis* 7(5), 14–15.

Lenhard, R. and Parker, C. (1990) Estimation of free hydrocarbon volume from fluid levels in monitoring wells. *Groundwater* 28(5), 57–67.

Lewis, T., Crockett, A., Siegrist, R., and Zarrabi, K. (1994) Soil sampling and analysis for volatile organic compounds. *Environmental Monitoring and Assessment* 30, 213–246.

Liikala, T., Olsen, K., Teel, S., and Lanigan, D. (1996) Volatile organic compounds: Comparison of two sample collection and preservation methods. *Environmental Science and Technology* 30(12), 3441–3447.

Mabey, W. (1995) Verifying data quality-quality assurance. Section 7. In *Environmental Chemistry for Investigating and Remediating soil and Groundwater Contamination.* Department of Engineering Professional Development College of Engineering, University of Wisconsin at Madison. September 18–20, 1995. Madison, WI.

Marrin, D. (1988) Soil-gas sampling and misinterpretation. *Groundwater Monitoring Review* 8(2), 54–57.

Martin-Hayden, J. and Robbins, G. (1997) Plume distortion and apparent attenuation due to concentration averaging in monitoring wells. *Groundwater* 35(2), 339–347.

Maskarinec, M., Johnson, L., Holladay, S., Moody, R., Bayne, C., and Jenkins, R. (1990) Stability of volatile organic compounds in environmental water samples during transport and storage. *Environmental Science and Technology* 24(11), 1664–1670.

Meyer, C. (1999) Distinguishing good science, bad science and junk science. *Expert Witnessing: Explaining and Understanding Science* pp. 99–120. CRC Press, Boca Raton, FL.

Mickam J., Bellandi, R., and Tifft, E. (1989) Equipment decontamination procedures for groundwater and vadose zone monitoring programs. Status and prospects. *Groundwater Monitoring Review* 9(2), 100–121.

Mishalanie, E. (1995) Testing biases associated with chlorinated solvents and hydrocarbon analyses. Section 3. In *Environmental Litigation: Hydrocarbon, Chlorinated Solvents and Visual Display of Evidence*. University of Wisconsin at Madison, College of Engineering, Department of Engineering Professional Development. Kahuku, Oahu, Hawaii, December 1–2, 1995.

Morrison, R. (1999) *Environmental Forensics. Principles and Applications*. CRC Press, Boca Raton, FL.

Morrison, R. (2000) Application of forensic techniques for age dating and source identification in environmental litigation. *Environmental Forensics* 1(3), 131–153.

Nimmer, P. and Chase, S. (2001) Aqueous diffusion samplers – a low-cost alternative to low-flow sampling. Submitted Abstracts. *11th Annual West Coast Conference on Contaminated Soils, Sediments and Water*. Association for the Environmental Health of Soils. March 19–21, 2001. San Diego, CA.

Numbers, R. (1994) PCB sampling BCM Engineers, Inc. *The National Environmental Journal* November/December, 24–27.

Parker, L. (1995) Decontamination of organic contaminants from groundwater sampling devices: A literature Review. United States Army Cold Regions Research and Engineering Laboratory. CRREL Special Report 95-14. Hanover, New Hampshire.

Parker, L. and Ranney, T. (1994) Effect of concentration of dissolved organic on sorption by PVC, PTFE, and stainless steel casings. *Groundwater Monitoring and Remediation* 14(3), 139–149.

Parker, L. and Ranney, T. (1997) Decontaminating materials used in groundwater sampling devices. United States Army Cold Regions Research and Engineering Laboratory. CRREL Special Report 97-24. Hanover, New Hampshire.

Parker, L. and Ranney, T. (1998) Sampling trace-level organic solutes with polymeric tubing. Part 2: Dynamic studies. *Groundwater Monitoring and Remediation* 18(1), 148–155.

Parker, L. and Ranney, T. (2000) Decontaminating materials used in groundwater sampling devices: organic contaminants. *Groundwater Monitoring and Remediation* 20(1), 56–68.

Peargin, T. (1994) Unsaturated zone air flow and soil vapor extraction theory. *Designing Air-based In Situ Soil and Groundwater Remediation Systems*. University of Wisconsin at Madison, College of Engineering and Professional Development. Cheyenne, WY. August 21–23, 1996.

Powell, R. (1997) Hitting the bull's eye in groundwater sampling. *Pollution Engineering* June, 51–54.

Powell, R. and Puls, R. (1993) Passive sampling of groundwater monitoring wells without purging: multilevel well chemistry and tracer disappearance. *Journal of Contaminant Hydrology* 12, 51–77.

Puls, R. and Barcelona, M. (1989) Filtration of groundwater samples for metals analysis. *Hazardous Waste and Hazardous Materials* 6(4), 385–393.

Puls, R. and Paul, C. (1995) Low-flow purging and sampling of groundwater monitoring wells with dedicated systems. *Groundwater Monitoring Review* Winter, 116–123.

Puls, R., Eychaner, J., and Powell, R. (1990) Colloidal-facilitated transport of inorganic contaminants in groundwater, Part I: Sampling considerations. Robert S. Kerr Environmental Research Laboratory, Ada, OK. United States Environmental Protection Agency, Environmental Research Brief. EPA/600/M-90/023.

Puls, R., Clark, D., Bledsoe, B., Powell, R., and Paul, C. (1992) Metals in groundwater: sampling artifacts and reproducibility. *Hazardous Waste and Hazardous Materials* 9(2), 149–162.

Remenda, V. and van der Kamp, G. (1997) Contamination from sand–bentonite seal in monitoring wells installed in aquitards. *Groundwater* 35(1), 39–46.

Rhodes, I. (1999) Pitfalls using conventional TPH methods for source identification. In *Proceedings of Environmental Forensics: Integrating Advanced Scientific Techniques for Unraveling Site Liability*. International Business Communications. June 24–25, 1999. Washington, DC.

Robbins, G., Deyo, B., Temple, M., Stuart, J., and Lacy, M. (1990) Soil-gas surveying for subsurface gasoline contamination using total organic vapor detection instruments. Part II. Field experimentation. *Groundwater Monitoring Review*, Fall, 110–117.

Schmidt, C. and Simon, M. (1993) Application of a direct emission assessment technology for conducting site investigations for subsurface contamination. 88th Annual Meeting & Exhibit of the Air and Waste Management Association. June 13–18, 1993. Denver, CO.

Siegrist, R. (1993) VOC measurement in soils: The nature and validity of the process. In *National Symposium on Measuring and Interpreting VOCs in Soils: State of the Art Research Needs*, Las Vegas, NV. January 12–14, 1993.

Siegrist, R. and Jenssen, P. (1990) Evaluation of sampling method effects on volatile organic compound measurements in contaminated soils. *Environmental Science and Technology* 24, 1387–1392.

Simmons, B. (1999) Are the data legally defensible. *Hydrovisions* (Groundwater Resources Association of California) 8(2), 8–9.

Squillace, P., Pankow, J., Barbash, J., Price, C., and Zogorski, J. (1999) Preserving groundwater samples with hydrochloric acid does not result in the formation of chloroform. *Groundwater Monitoring and Review* 19(1), 67–74.

State of California (1994) *Preliminary Endangerment Assessment Guidance Manual*. Department of Toxic Substances Control, Environmental Protection Agency.

Stolzenburg, T. and Nichols, D. (1986) Effects of filtration method and sampling devices on inorganic chemistry and sampled well water. *Proceedings of the Sixth National Symposium and Exposition on Aquifer Restoration and Groundwater Monitoring*. May 19–22, 1986. National Water Well Association, Dublin, OH.

Strachan, S. and Hess, F. (1982) Dinitroaniline herbicides adsorb to glass. *Journal of Agricultural Food Chemistry* 30(2), 389–391.

United States Environmental Protection Agency (1982) Methods for organic chemical analysis of municipal and industrial wastewater. EPA-600/4-82-057.

United States Environmental Protection Agency (1983) Methods for chemical analysis of water and wastes. EPA-600/4-79-020.

United States Environmental Protection Agency (1986) Test methods for evaluating solid waste: physical/chemical methods, 3rd edn. SW-846.

United States Environmental Protection Agency (1991) Final Draft of Chapter 11 of SW-846, Groundwater-Monitoring. Washington, DC.

United States Environmental Protection Agency (1994) Risk assessment guidance for Superfund sites (Part A) Final. United States Environmental Protection Agency, Office of Emergency and Remedial Response, Washington, DC.

Van Genuchten, M. (1980) A closed-form equation for predicting the hydraulic conductivity of unsaturated soils. *Soil Science Society of America Journal* 44, 982–998.

Wassenaar, L. and Hendry, J. (1991) Improved piezometer construction and sampling techniques to determine pore water chemistry in aquitards. *Groundwater* 37(4), 564–571.

Wilson, N. (1998) *Soil Water and Groundwater Sampling*. CRC Press, Boca Raton, FL.

CHAPTER 5

APPLICATION OF STABLE ISOTOPES AND RADIOISOTOPES IN ENVIRONMENTAL FORENSICS

R. Paul Philp

5.1 INTRODUCTION

In recent years there have been many published studies concerned with the origin and fate of pollutants in aquatic environments such as rivers, lakes, groundwater, and coastal waters (Macko *et al.*, 1981; Farran *et al.*, 1987; Galt *et al.*, 1991; Eganhouse *et al.*, 1993; Oros *et al.*, 1996). Sources of contaminants are many and varied and include: natural seepage of crude oils, ship traffic, supertankers spilling crude oils, leaking storage tanks, pipelines, polychlorinated biphenyls (PCBs), pesticides, dioxins, and many other types of chemical spills from a wide variety of sources. Identification, quantification, and monitoring the fate of these pollutants in the environment are important in providing a better response to such spills and for the purposes of determining responsibility for the spill and liability for the cleanup.

Common approaches for characterization of such spills, comprised of organic compounds, and identification of potential sources, generally rely upon analyses by gas chromatography (GC) and gas chromatography–mass spectrometry (GCMS). Correlations are made on the basis of the molecular distribution of saturate or aromatic hydrocarbons and more specifically, in the case of crude oil spills, biomarker fingerprints (Wang *et al.*, 1994). In certain cases GC and GCMS data may be ambiguous, and inconclusive, as a result of weathering processes such as evaporation, photooxidation, water washing, and biodegradation changing the distribution of components in the spilled material vs. the original sample (Milner *et al.*, 1977; Lafargue and Barker, 1988; Palmer, 1993). Depending on the nature of the compounds involved, evaporation can occur in the first few hours after a spill and remove the more volatile components. If the accident occurs in an aquatic environment water-washing will remove the more water-soluble components, such as hydrocarbons below C_{15}, and some of the C_{15+} aromatic compounds which are more water-soluble than paraffins (Lafargue and Barker, 1988; Fuentes *et al.*, 1996; Palmer, 1998). At the same time, biodegradation will also start to affect the distribution of individual compounds, although the rate will depend upon the nature of the spill and environment into which it spilled.

Correlation of spilled material with its suspected source, whether it is hydrocarbon-based or of some other chemical origin, requires discriminative parameters that are relatively insensitive to weathering processes. GC and GCMS are not always the ultimate answer to such a problem. One alternative approach is the use of stable isotopic compositions, typically carbon or hydrogen isotopic values (Hartman and Hammond, 1981; Macko *et al.*, 1981; Macko and Parker, 1983; Farran *et al.*, 1987) but in selected cases chlorine, oxygen or sulfur isotope compositions. The most common parameter would be the carbon isotopes $^{13}C/^{12}C$, but in relatively recent years there has been an increasing number of papers reporting the use of $^{35}Cl/^{34}Cl$; D (deuterium)/H, $^{18}O/^{16}O$, in a variety of geochemical applications, including spill/source correlations and rate of biodegradation of individual contaminants. These isotopic compositions may be determined as either the bulk isotopic composition of the total spill, or by the use of combined gas chromatography–isotope ratio mass spectrometry (GCIRMS; Hayes *et al.*, 1990; O'Malley *et al.*, 1994; Boreham *et al.*, 1995; Dowling *et al.*, 1995; O'Malley *et al.*, 1996) as the isotopic composition of individual compounds in a mixture. Recent developments in GCIRMS systems permit determination of both carbon and hydrogen isotopic values for the individual compounds.

Uses of isotopes in environmental forensics can be divided into two main areas, those associated with stable isotopes and those associated with radioisotopes. The division is actually more specific than that since radioisotope applications are generally associated with age dating applications whereas stable isotopes are commonly associated with correlations or source determination. Radioisotopes commonly used for dating of sediments and groundwater in the context of environmental forensics include caesium-137 (^{137}Cs), lead-210 (^{210}Pb), and tritium (^{3}H) which will be described in more detail below. Whilst there are other radioisotopes that may have applications in various geochemical exploration topics, it should be remembered that the current chapter is concerned with environmental applications. Stable isotopes have found a much greater range of applications in environmental studies than have radioisotopes and such a bias is naturally reflected in this chapter.

The chapter is divided into five sections covering these general areas. Section 5.2 will deal with radioisotope age dating; Section 5.3 with the use of stable isotopes in the identification of contaminant sources; Section 5.4 with other uses of stable isotopes in environmental forensics; Section 5.5 with biodegradation; and finally Section 5.6 will examine the combined use of isotopes and data from other techniques in environmental forensics. It is the intent of this chapter to provide the interested reader with a comprehensive insight into the use of isotopes in environmental forensics rather than an exhaustive review of all the published literature.

5.2 RADIOISOTOPE AGE DATING OF CONTAMINANTS IN SEDIMENTS AND GROUNDWATER

The need or desire to age date a contaminant in a sediment or groundwater like most environmental matters is basically driven by money and the need to assign blame or determine the responsible party. In many cases a contaminated site will have a long list of previous owners and the ability to provide an indication as to when any contaminant spill occurred will enable some of the previous owners to be absolved of cleanup costs and other liability. Hence the main uses of radioisotopes in environmental forensics are for dating sediments and groundwater. However, it should be noted that unlike other techniques described below, this particular application of radioisotopes is still in its infancy and at the present time is probably more commonly used for research purposes.

5.2.1 SEDIMENT DATING

Whilst sediment cores can be dated to determine historical inputs to aquatic systems, it needs to be emphasized there are limitations to the methods.

Disturbance of the sediment cores in any way by storms or animals can preclude accurate dating of the cores or require adjustments in interpretation. The two main isotopes used for sediment dating are ^{137}Cs and ^{210}Pb. ^{137}Cs is derived from atmospheric nuclear weapons testing and when removed from the atmosphere is incorporated into the sediments. Once deposited it is bound into the sediment by sorbing to fine clay particles and organic material and is essentially non-exchangeable (Tamura and Jacobs, 1960; Lieser and Steinkopff, 1989). The first observation of ^{137}Cs in sediments dates back to about 1954 and reaches a well-documented maximum in 1963, followed by minor increases in 1971 and 1974 (Ritchie and McHenry, 1990). The global deposition of ^{137}Cs can be estimated based on the better documented deposition of strontium-90 (^{90}Sr) (Hermanson, 1990). However, for many purposes having the absolute deposition rate is unnecessary since dating can be accomplished by noting the 1963 ^{137}Cs peak and measuring its depth in the sediment (Pennington and Cambray, 1973). ^{137}Cs sediment dating has been used for a number of purposes, including reconstructing a history of arsenic inputs along a tidal waterway in Tacoma, Washington (Davis *et al.*, 1997), historical trends in PCBs and pesticides in reservoir sediments from central and southeastern United States (van Metre *et al.*, 1997), and the identification of dioxins and dibenzofurans in remote lakes (Czuczwa and Hites, 1984; Juttner *et al.*, 1997).

^{210}Pb has a half-life of 22.6 years and is useful for dating sediments deposited during the past one hundred years particularly when incorporated with ^{137}Cs dating. Uranium-238 (^{238}U) decays to radon-222 and then through a series of isotopes to ^{210}Pb. Radon is a gas which seeps up into the atmosphere from ^{238}U containing rocks. Once it enters the atmosphere it soon decays to ^{210}Pb, which falls out and can be deposited in sediments where it decays. The decrease in ^{210}Pb with sediment depth indicates the age of a sediment level and of any undisturbed sediments at that level. If the sediments contain ^{238}U a correction for ^{210}Pb formed in the sediments may be necessary. ^{210}Pb measurements have been used to date Lake Ontario sediments and historical input of PCBs, pesticides, and other hydrophobic organic compounds (Wong *et al.*, 1995) and metal accumulations in Lake Zurich (von Gunten *et al.*, 1997).

5.2.2 GROUNDWATER DATING

In addition to sediment dating, isotopes have been used for dating groundwater for use in determination of hydrogeological parameters. Such parameters are used for purposes such as reverse plume modeling to determine source location and release date, modeling of leachate and infiltration to link groundwater and surface contamination. Determination of the age of the groundwater can also indicate where active local recharge is occurring. Age determinations of the groundwater at two different locations along a streamline can be used to give an estimate of groundwater velocity.

Tritium is a radioactive isotope of hydrogen (^3H) with a half-life of 12.3 years, which makes it particularly useful for dating modern groundwaters (Clark and Fritz, 1997). The presence of significant amounts of tritium in the groundwater is evidence for active recharge. Tritium is formed in the atmosphere by high energy cosmic radiation bombarding the earth, the decay of radioactive materials, in nuclear reactors and from the atmospheric testing of thermonuclear devices. Tritium is deposited on the earth's surface with precipitation and makes its way to the groundwater. The amount of tritium in the atmosphere increased after 1953 due to thermonuclear testing and reached a peak in 1963, after which it started to decline. In areas with small net infiltration and large overburden, the 1963 peak may still occur in the unsaturated or vadose zone and can be reflected in groundwater samples (Dincer *et al.*, 1974). If the 1963 peak is preserved at a known distance from a recharge zone, flow velocity can be estimated, as demonstrated at a waste disposal site in Glouster, Ontario (Michel *et al.*, 1984).

Groundwater can also be dated by measuring the decay product of tritium, ^3He. The problem is complicated by the fact that the concentration of ^3He normally in the atmosphere that was dissolved in the groundwater during recharge must be estimated and subtracted. In addition, a prerequisite for using this approach is a rapid vertical flow of the groundwater such that losses by gas diffusion from the top of the water table are limited. Comparison between the use of this method and the tritium 1963 bomb peak method resulted in dates that differed by only 15% (Schlosser *et al.*, 1989).

Whilst not a radioisotope method, chlorofluorocarbons (CFCs) provide another age marker in sediments and groundwater. Since the input functions of CFCs have been well established since the 1940s, both their presence and distribution provide a means of estimating sediment ages (Clark and Fritz, 1997). Detection of CFCs in the deeper parts of large lakes such as Lake Balkail, Russia, provides an indication for the rate of turnover in these lakes.

Finally two decay products of ^{238}U also offer the possibility of yielding age-related information on groundwater samples, in addition to the information available for sediments. Radium-226 has a half-life of 1620 years, whereas radon-222 has a half-life of 3.8 days. With accurate information on dissolution rates, there is a good possibility of establishing fairly accurate dates over the past 1000–5000 year period (Hillaire-Marcel and Ghaleb, 1997). Whilst of significant interest to hydrogeologists and of direct relevance to obtaining information on recharge rates, such information is probably of limited use in terms of the forensic chemistry being discussed in this monograph.

5.3 USE OF STABLE ISOTOPES TO IDENTIFY CONTAMINANT SOURCES

For many years, as described in other chapters of this book, techniques such as gas chromatography (GC) and gas chromatography–mass spectrometry (GCMS) have been used extensively for the purpose of identifying contaminant sources. Identification in this sense can mean unambiguous identification of each individual compound in the sample or it might mean the ability to correlate a spilled product with its suspected source thus identifying the source of the spill. The isotopic composition of a specific compound, or the spilled material, will play little, if any, role in the task of actually identifying individual components in the contaminant. The power in the use of stable isotopes is their ability to permit correlations between source samples and spilled samples even if the samples are weathered. This surge in the use of isotopes for correlation purposes has been due in large part to the development of combined gas chromatography–isotope ratio mass spectrometry (GCIRMS). Before discussing this development let us consider some of the basic concepts of stable isotope geochemistry.

5.3.1 ORIGIN OF ISOTOPIC CARBON DIFFERENCES

Carbon exists as a mixture of two stable isotopes, ^{12}C and ^{13}C, with the approximate natural abundance of $^{12}C/^{13}C$ ratio being 99:1. The carbon isotopic composition of living organic matter in part depends on the species but is also determined by a number of environmental properties. Atmospheric carbon dioxide is assimilated by living plants during photosynthesis. The nature of the plants and whether they assimilate CO_2 via a C_3 or C_4 photosynthetic cycle determines the extent of preferential assimilation of the lighter ^{12}C isotope. This results in a 5 to 25‰ depletion in ^{13}C. The amount of fractionation depends on the pathway followed, with the so-called C_4 plants typically being associated with warmer and more arid climates and in general having isotopic values in the −10‰ to −18‰ range. C_3 plants on the other hand are more typically associated with cooler and wetter climates, and generally have lighter isotopic values in the −22‰ to −30‰ range. The C_3 pathway operates in about 85% of plant species and dominates in most terrestrial ecosystems. The natural vegetation in temperate and high latitude regions, as well as tropical forests, is almost exclusively C_3. Major crops such as wheat, rye, barley, legumes, cotton, tobacco, tubers (including sugar beets) and fallow grasses are C_3 plants. The C_4 pathway evolved as atmospheric CO_2 concentrations began to drop in the early Tertiary. C_4 plants represent less than 5% of floral species but dominate in hot open ecosystems such as tropical and temperate grasslands and include common crops such as sugar cane, corn and sorghum. As a point of interest it should be noted that the isotopic composition of food products marked as 100% natural provides a useful tool for customs and excise departments to monitor and check the origin of these and other imports (Hillaire-Marcel, 1986). Desert plants such as cacti operate under yet another mechanism of photosynthesis referred to as the CAM (crassalacean acid metabolism) photosynthesis cycle and can operate under both the C_3 and C_4 cycles. Photosynthesis will have an effect on the composition of the residual CO_2 in the atmosphere and lead to an enrichment of the heavier ^{13}C isotope. The composition of atmospheric CO_2 also varies depending upon whether it is over land or over ocean and upon surrounding temperatures.

Fossil fuels, coal, crude and natural gas are formed as a result of a series of very complex and long-term reactions where the living organic matter dies and is deposited in the sedimentary environment. A significant amount of this material will be degraded and recycled as CO_2 and water. However, a small proportion is buried and ultimately converted to fossil fuels of one type or other depending on the nature of the depositional environment. During this formation and generation stage, organic matter from many different sources is deposited in a sedimentary sink and becomes a heterogeneous mixture of organic material. At the same time the unique isotopic signatures associated with specific sources of organic matter are lost as a result of this mixing process. Whilst hydrocarbons generated in this process result in the formation of crude oils, or other fossil fuels, which possess isotopic signatures, these signatures can no longer be related to specific sources of organic material. From an environmental forensic point of view the fact that it is not possible to relate the isotopic numbers to a specific source of organic material is not critical since the more important application is the ability to use these isotopic values for correlation of the spilled product with its suspected source(s). As will be seen below, isotope values are even more valuable in this application when the spilled product is a single component or a mixture of relatively light hydrocarbons that may not contain any of the components commonly used for correlation purposes following analysis by GCMS. Correlations on the basis of isotopic compositions utilize either the bulk isotopic compositions or the isotopic composition of individual compounds as determined by GCIRMS, as discussed below in more detail. In complex mixtures, the bulk isotopic composition of the mixture will actually reflect the isotopic compositions of all the components in the mixture and their relative proportions.

So in summary, the carbon isotopic composition of any compound or material is a reflection of the relative proportions of the two stable isotopes and a measure of how depleted the compound or material is in the heavier isotope relative to a standard. The next question therefore is how is this depletion measured?

5.3.2 REPORTING ISOTOPE MEASUREMENTS

There are two common methods for the determination of the carbon isotopic composition of a sample. It is possible to measure either the *bulk carbon isotope value* or the *isotopic composition of each individual compound* in the mixture. The principle behind the determination of either of these values is very similar. Each method requires complete combustion of the sample, regardless of its origin, and conversion to CO_2 and water. For bulk determinations, the combustion is undertaken in a sealed tube in the presence of CuO. The CO_2 is then transferred and analyzed in a stable isotope ratio mass spectrometer and the isotopic composition measured relative to that of a standard material (Pee Dee belemnite or PDB) whose isotopic composition has been assigned a value of 0. Virtually all environmental samples will contain less ^{13}C than the standard, and thus being depleted in ^{13}C will have negative $\delta^{13}C$ values.

Stable carbon isotope ratios ($R = {}^{13}C/{}^{12}C$) are expressed relative to a standard and typically in 'delta' notation, where: $\delta^{13}C = (R_{sample}/R_{standard} - 1) \times 1000$ (units are ‰ or per mil). Although this refers specifically to carbon isotopes, the ratios for the other elements such as O, H, Cl, S or N are expressed in the same way relative to their specific standard. Figure 5.1 shows the typical isotopic compositions for a range of organic material of different geological ages. Note both the differences between various types of organic material and also the relatively narrow spread of values for the bulk isotopic composition of crude oils.

5.3.3 BULK ISOTOPE VALUES

Whilst the bulk isotope numbers represent weighted averages of all components in a mixture they have still been used successfully in many environmental applications. For example, in the case of crude oils, correlations can be made using the bulk isotopic compositions of the saturate and aromatic fractions rather than the whole oil itself. In order to do this the oils are fractionated by a simple column chromatography approach using alumina or silica as the solid support and solvents of varying polarity to separate the required fractions. (It should be noted that in this application, the saturate and aromatic fractions are typically comprised of the C_{15+} fraction since the lighter components are lost during fractionation.) It is a very simple application, since the isotopic values for the saturate and aromatic fractions to be correlated are plotted against each other (Figure 5.2).

Samples that are related will plot very close to each other, those that are not related will plot in totally different areas. The use of bulk isotopes in this manner is a preliminary screening tool and simply because two samples plot close

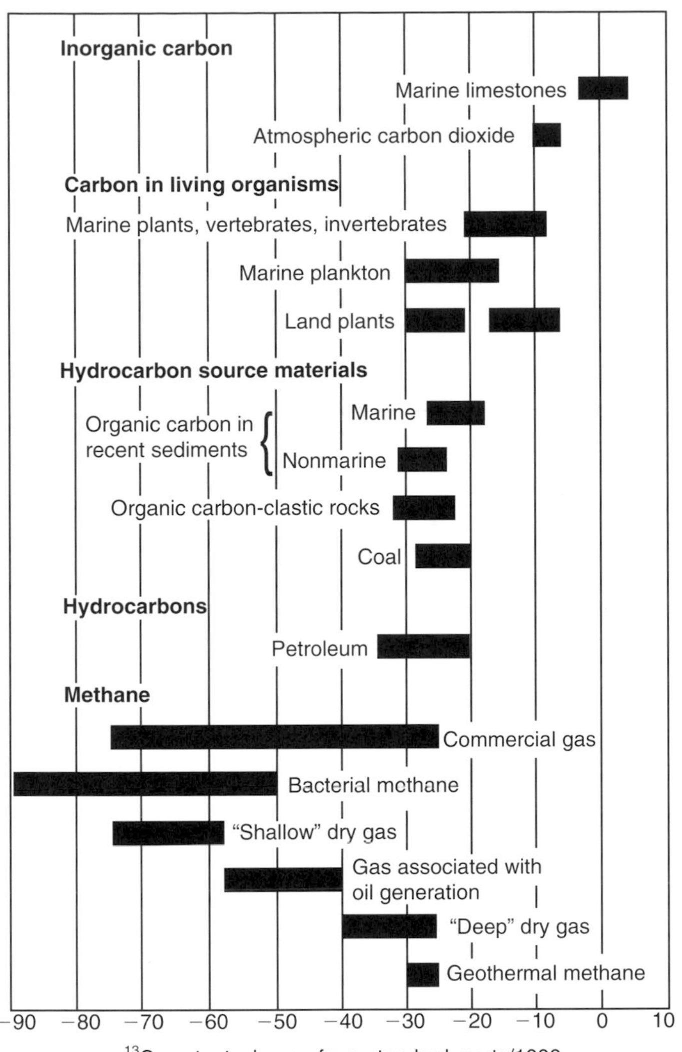

Figure 5.1

Variations in the isotopic composition of different types of organic matter. Note both the differences between various types of organic material and the relatively narrow spread of values for crude oils. (Reprinted from A.N. Fuex (1977) The use of stable carbon isotopes in hydrocarbon exploration. J. Geochem. Expl. *7(2), 155–188, with permission from Elsevier Science.)*

to each other does not necessarily mean that they are unequivocally related to each other. Once a preliminary relationship between samples has been established more definitive correlations should be undertaken using techniques such as GC and GCMS. Correlation of degraded and non-degraded samples is a little more complicated, depending on the degree of degradation. The isotopic composition of a sample can change as a result of biodegradation, particularly in the case of crude oils due to the removal of the predominant *n*-alkanes. In general the saturate fraction of the biodegraded crude oil sample will become isotopically heavier as a result of preferential removal of the isotopically lighter compounds.

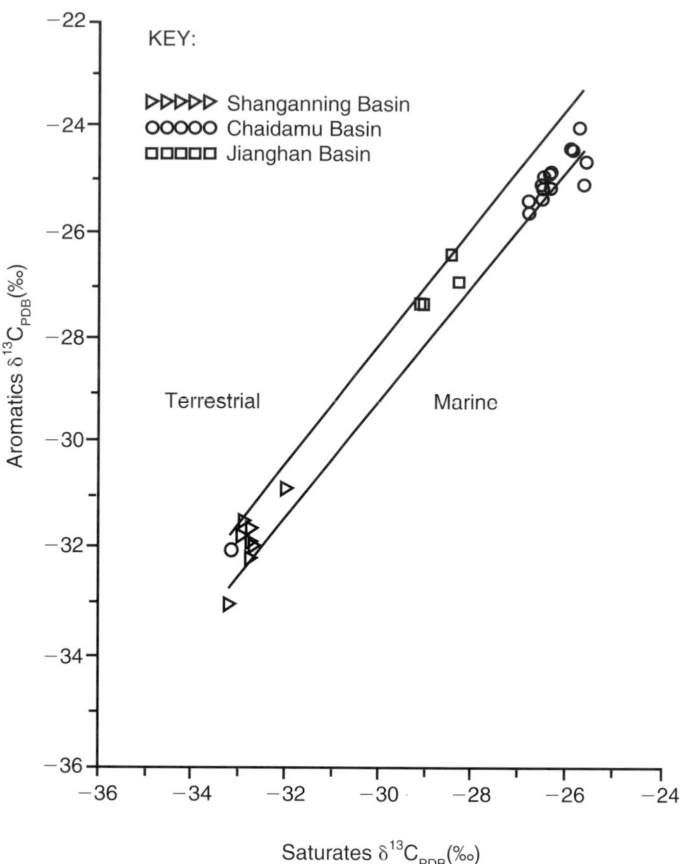

Figure 5.2

A plot of the $\delta^{13}C$ values for the saturates vs. aromatics provides a means of correlating samples of similar origin as shown in this figure using China crude oils from different locations. (Reprinted from Kaplan et al.*, 1997 with permission from Elsevier Science.)*

It should be noted that although utilization of the bulk isotopic values works well for crude oils and heavier refined products, it becomes a little more difficult for the lighter refined products. Separation of the lighter refined products into saturate and aromatic fractions by column chromatography in the manner described above will lead to loss of significant proportions of the individual fractions due to evaporation. These losses may have an effect on the bulk isotopic composition of the individual fractions, making it difficult to make accurate correlations. Once again though the bulk isotopic composition of the non-fractionated samples may be used as a preliminary screening tool to determine similarities and differences between selected samples.

A classic example of the use of bulk carbon isotopes in a forensic type study would be the work of Kvenvolden *et al.* (1995). Tar ball residues from the beaches of Prince William Sound were collected several years after the *Exxon Valdez* accident and characterized on the basis of their bulk carbon isotope ratios. A summary of the bulk isotope numbers for a number of these residues is shown in Table 5.1. The important point to notice here is the presence of two

Location	$\delta^{13}C$ (‰)	Description
Bligh Reef (EV Oil)	−29.2	North Slope Crude EV tanker
EV Residues		
Naked I.	−29.4	Oiled cobbles, McPherson Bay
Eleanor I.	−29.1	Solid oil on rock, Northwest Bay
Evans I.	−29.3	Solid oil mat, Shelter Bay
Non-EV Residues		
Old Valdez	−23.6	Asphalt mats, old asphalt plants
Naked I.	−23.9	Tar on metashale, Bass Harbor
Evans I.	−23.7	Tar on rock, Shelter Bay
Other		
Latouche I.	−24.8	Tar in mine dump
Knight I.	−25.3	Solid oil on cobble, Herring Bay
Pavements		
Whittier	−24.8	Street pavement > 20 years old
Valdez	−27.2	New airport runway
Seward	−28.1	New airport runway
Cold Bay	−22.7	End of old airport runway
Amchitka	−23.4	World War II C airport runway

Table 5.1

Examples of bulk isotope numbers for a number of residues collected from Prince William Sound several years after the Exxon Valdez *(EV) accident indicated the presence of two distinct families of tar ball residues. GCMS was used to confirm that one family was from* Exxon Valdez *crude whereas the other was from California crude imported to Alaska many years earlier (modified from Kvenvolden* et al., *1995).*

distinct families of tar ball residues. One family corresponds closely to the *Exxon Valdez* oil on the basis of its isotopic composition, and the second and isotopically heavier group was ultimately related to a California crude oil on the basis of its biomarker fingerprints. How did the California crude end up in Prince William Sound? In retrospect it is very simple – prior to the discovery of crude oil in Alaska it had to be imported from California. Much of this oil was temporarily stored in tanks on the edge of Prince William Sound. In 1964 a massive earthquake struck the region of Valdez, the tanks ruptured, oil spilled into the Sound and in view of the magnitude of the disaster, the oil was left in the Sound to degrade naturally with very little cleanup. Crude oils contain many compounds that are relatively resistant to biodegradation and hence 40 years later these can still be detected in these tar residues. However, the point of this example is to demonstrate that it was the bulk carbon isotopes that originally permitted the distinction to be made between the two groups of oils. Furthermore, the same approach was extended to tar residues on building, airport runways and other locations showing that most of the bitumens used there prior to the discovery of the Alaska reserves could be correlated with an origin from California crudes on the basis of their isotopic compositions.

Whilst most emphasis has been placed on the use of bulk carbon isotopes for correlation purposes, there are a number of forensic applications that have used other stable isotopes. For example, Kaplan *et al.* (1997) clearly showed that a plot of deuterium values against carbon isotope values for groundwater extracts and suspected gasoline sources was a viable correlation tool

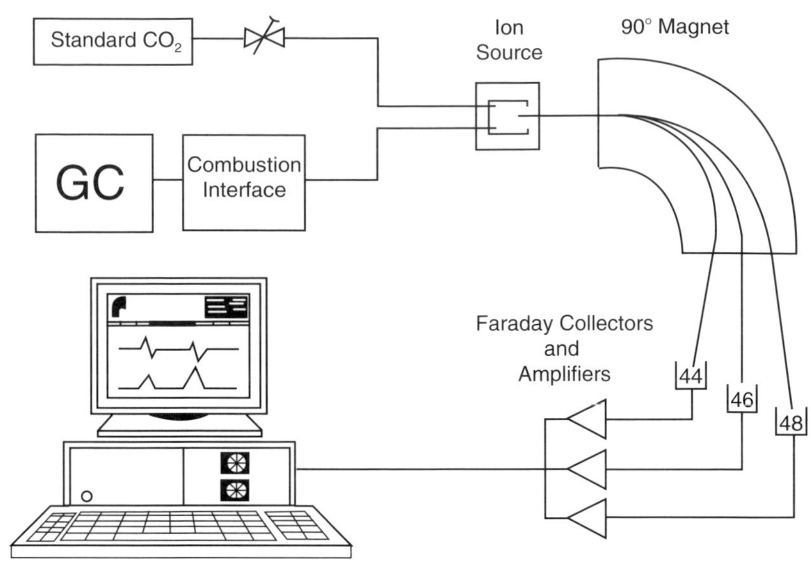

Figure 5.4

In a GCIRMS system separated compounds elute from the GC column and pass through a combustion tube; the CO_2 and water generated in this manner pass through a separator to remove the water, and the CO_2 continues into the mass spectrometer, where the $\delta^{13}C$ values are determined relative to the standard Pee Dee belemnite.

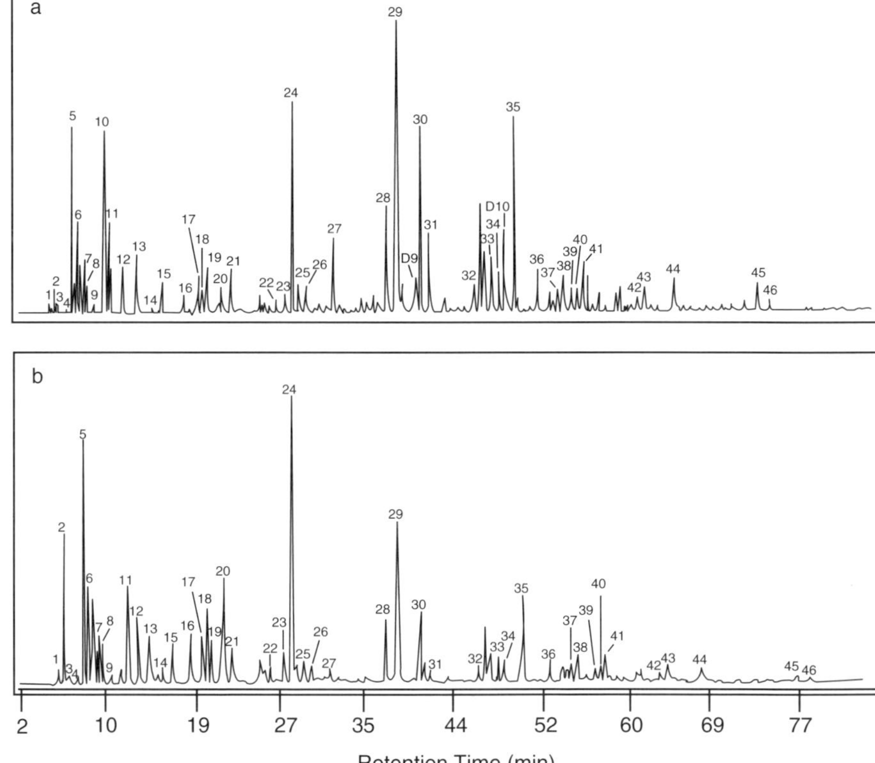

Figure 5.5

Chromatograms for two different gasoline samples. The initial output from the GCIRMS analysis is a m/z 44 chromatogram which appears to be very similar to a conventional GC trace but actually depicts the CO_2 produced from the various components in the mixture. Data for each individual component are then converted into $\delta^{13}C$ values relative to the PDB standard.

Location	$\delta^{13}C$ (‰)	Description
Bligh Reef (EV Oil)	−29.2	North Slope Crude EV tanker
EV Residues		
Naked I.	−29.4	Oiled cobbles, McPherson Bay
Eleanor I.	−29.1	Solid oil on rock, Northwest Bay
Evans I.	−29.3	Solid oil mat, Shelter Bay
Non-EV Residues		
Old Valdez	−23.6	Asphalt mats, old asphalt plants
Naked I.	−23.9	Tar on metashale, Bass Harbor
Evans I.	−23.7	Tar on rock, Shelter Bay
Other		
Latouche I.	−24.8	Tar in mine dump
Knight I.	−25.3	Solid oil on cobble, Herring Bay
Pavements		
Whittier	−24.8	Street pavement > 20 years old
Valdez	−27.2	New airport runway
Seward	−28.1	New airport runway
Cold Bay	−22.7	End of old airport runway
Amchitka	−23.4	World War II C airport runway

Table 5.1

Examples of bulk isotope numbers for a number of residues collected from Prince William Sound several years after the Exxon Valdez *(EV) accident indicated the presence of two distinct families of tar ball residues. GCMS was used to confirm that one family was from* Exxon Valdez *crude whereas the other was from California crude imported to Alaska many years earlier (modified from Kvenvolden* et al., *1995).*

distinct families of tar ball residues. One family corresponds closely to the *Exxon Valdez* oil on the basis of its isotopic composition, and the second and isotopically heavier group was ultimately related to a California crude oil on the basis of its biomarker fingerprints. How did the California crude end up in Prince William Sound? In retrospect it is very simple – prior to the discovery of crude oil in Alaska it had to be imported from California. Much of this oil was temporarily stored in tanks on the edge of Prince William Sound. In 1964 a massive earthquake struck the region of Valdez, the tanks ruptured, oil spilled into the Sound and in view of the magnitude of the disaster, the oil was left in the Sound to degrade naturally with very little cleanup. Crude oils contain many compounds that are relatively resistant to biodegradation and hence 40 years later these can still be detected in these tar residues. However, the point of this example is to demonstrate that it was the bulk carbon isotopes that originally permitted the distinction to be made between the two groups of oils. Furthermore, the same approach was extended to tar residues on building, airport runways and other locations showing that most of the bitumens used there prior to the discovery of the Alaska reserves could be correlated with an origin from California crudes on the basis of their isotopic compositions.

Whilst most emphasis has been placed on the use of bulk carbon isotopes for correlation purposes, there are a number of forensic applications that have used other stable isotopes. For example, Kaplan *et al.* (1997) clearly showed that a plot of deuterium values against carbon isotope values for groundwater extracts and suspected gasoline sources was a viable correlation tool

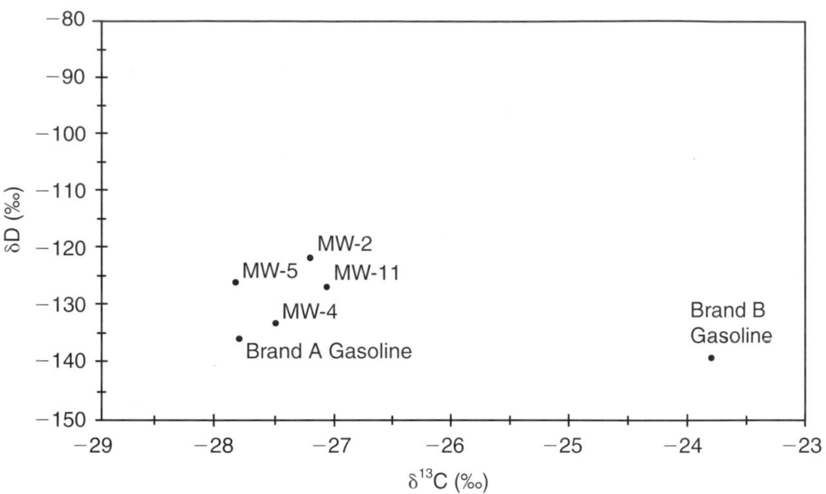

Figure 5.3

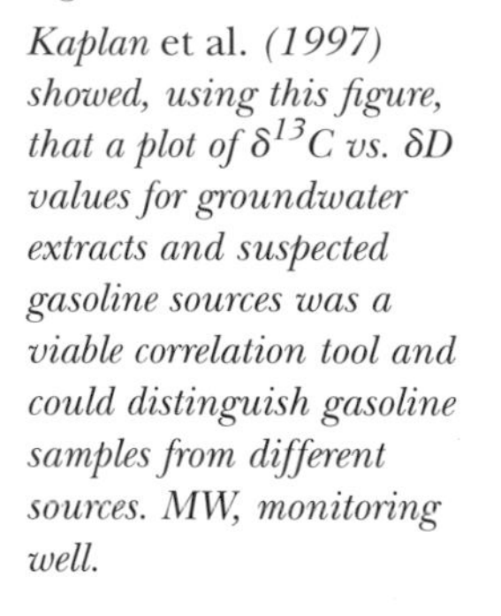

Kaplan et al. *(1997) showed, using this figure, that a plot of $\delta^{13}C$ vs. δD values for groundwater extracts and suspected gasoline sources was a viable correlation tool and could distinguish gasoline samples from different sources. MW, monitoring well.*

(Figure 5.3). Determination of δD values is a more time-consuming process than determination of δC values and not so commonly used. However, the capability to determine δD values for individual compounds could change this situation dramatically in the next few years (Sessions *et al.*, 1999). In addition, problems such as the possibility of H exchange reactions with the groundwater have to be minimized or eliminated to prevent the possibility of erroneous values being obtained for the H/D ratio.

Whilst the above discussion has centered on the use of bulk carbon and hydrogen isotopes, variations in the bulk isotopic composition of other isotopes such as hydrogen, oxygen, sulfur and chlorine have also found limited application in a number of environmental studies. For example, a study of chlorinated solvents used a plot of $^{13}C/^{12}C$ against $^{37}Cl/^{35}Cl$ to illustrate a relationship between isotopic content and manufacturer for a variety of solvent types such as perchloroethylene (PCE), trichloroethylene (TCE) and trichloroethane (TCA) (Beneteau *et al.*, 1999).

Sulfur has four naturally occurring isotopes: ^{32}S, ^{33}S, ^{34}S, and ^{36}S. The ^{32}S and ^{34}S isotopes are most commonly used for ratio analysis. Whilst reports have been presented on the use of sulfur isotopes to differentiate anthropogenic and natural sources of sulfur dioxide, it would not appear that the sulfur isotopes have been used extensively for the purposes of identifying contaminants in the environment. One recent application published by Becker and Herner (1998) examined the combined use of bulk carbon and sulfur isotopes as a means of discriminating crude oils from various sources. It was noted that since many carbon isotope values are very similar, the additional use of the sulfur values permits separation of oils from different sources.

5.3.4 CARBON ISOTOPE GCIRMS FOR HYDROCARBONS

Bulk isotopic values have been readily available for many years but one of the most significant analytical advances in geochemistry in the past few years has undoubtedly been the development of combined gas chromatography–isotope ratio mass spectrometry (GCIRMS). This technique permits acquisition of $\delta^{13}C$ values for individual components in complex mixtures in real time and without the need to physically isolate each individual compound (Freeman *et al.*, 1990; Hayes *et al.*, 1990). This technique permits continuous-flow acquisition of carbon isotope ($\delta^{13}C$) values for individual components in complex mixtures (Mansuy *et al.*, 1997; Abrajano and Sherwood Lollar, 1999). A comprehensive review of the subject was presented recently by Meier-Augenstein (1999) and specific details of the actual technique will not be presented again in this chapter.

The most recent development in the area of the combined GCIRMS is the capability to determine the H/D ratio of individual compounds as well as C isotopes of individual compounds. To date the number of applications has been fairly limited and these have been simply designed to show that the technique actually works (Hilkert *et al.*, 1999; Sessions *et al.*, 1999). However, the combination of the carbon and hydrogen isotope measurements of individual compounds, determined by GCIRMS, will provide a powerful tool, in the future, to discriminate compounds from different sources. A recent publication demonstrated that biodegradation of toluene produced a significant change in the H/D isotopic composition of the toluene (Ward *et al.*, 2000). This in turn gives rise to the possibility that such isotopic changes may be useful indicators of the extent of biodegradation of compound such as toluene present as groundwater contaminants.

The basic concept for the determination of the isotopic composition of individual compounds is the same as for the bulk isotopic values in that the components are completely combusted to CO_2 and water and the isotopic composition of the resulting CO_2 determined. However, the big difference is that these values are determined in real time as the individual compounds elute from the GC column. The separated compounds pass through a combustion tube where they are combusted, the CO_2 and water pass through a separator to remove the water, and the CO_2 continues into the mass spectrometer (Figure 5.4).

With the GCIRMS system, the combustion has to be complete in the time it takes for the sample to pass through the reactor, which is typically 12–15 s. However, the preliminary output from the analysis is a m/z 44 chromatogram, which appears to be very similar to a conventional GC trace but actually depicts the CO_2 produced from the various components in the mixture (Figure 5.5).

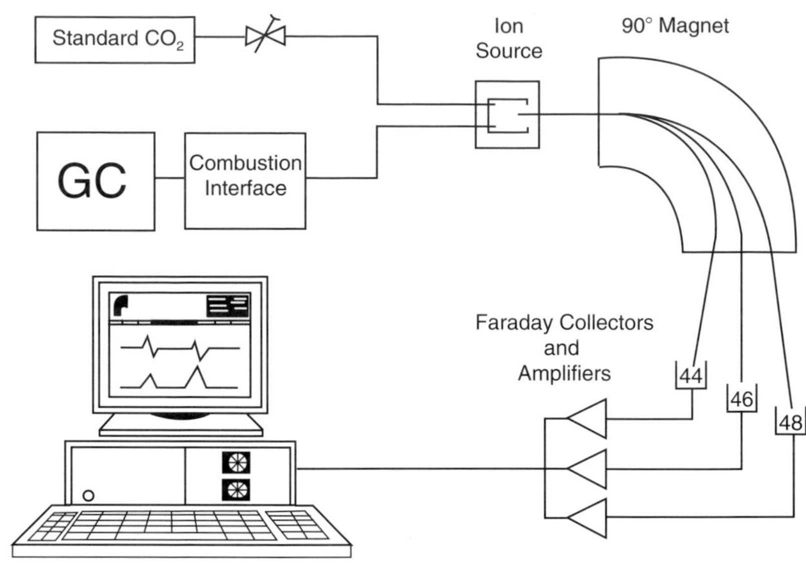

Figure 5.4

In a GCIRMS system separated compounds elute from the GC column and pass through a combustion tube; the CO_2 and water generated in this manner pass through a separator to remove the water, and the CO_2 continues into the mass spectrometer, where the $\delta^{13}C$ values are determined relative to the standard Pee Dee belemnite.

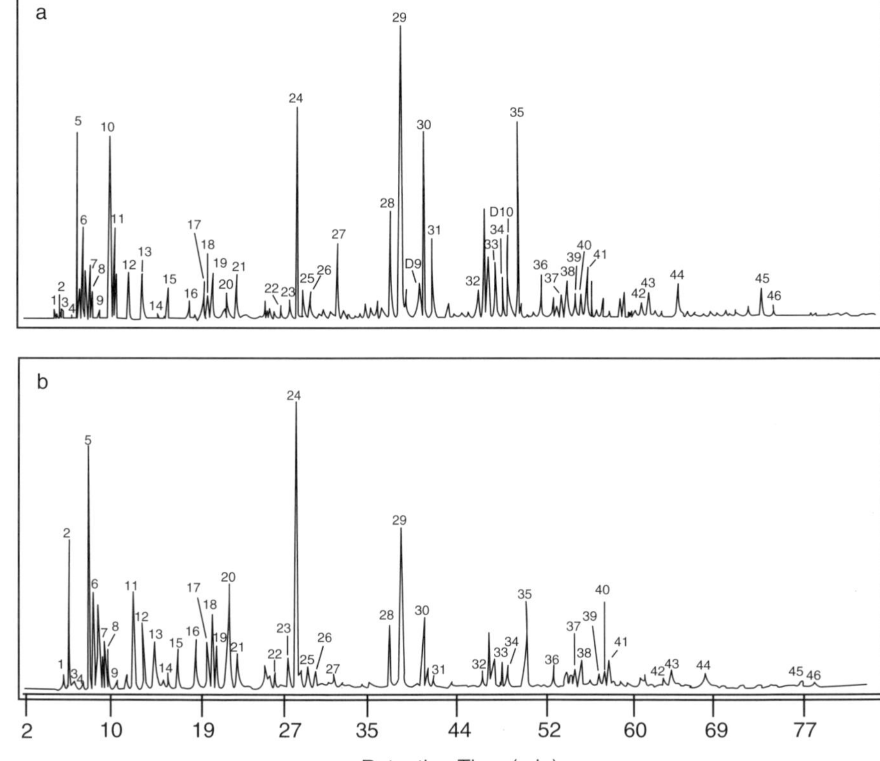

Figure 5.5

Chromatograms for two different gasoline samples. The initial output from the GCIRMS analysis is a m/z *44 chromatogram which appears to be very similar to a conventional GC trace but actually depicts the CO_2 produced from the various components in the mixture. Data for each individual component are then converted into $\delta^{13}C$ values relative to the PDB standard.*

Data for each individual component are then converted into $\delta^{13}C$ values relative to the PDB standard. It should be noted that although it sounds simple, there are a number of problems attached to the technique, particularly that of coelution.

Any coelution between two peaks can have a significant effect on the isotopic composition of the other component. This requires manual manipulation and processing of the data in order to minimize and correct for this interference. Despite all of these problems, the data obtained from this approach can be very useful for correlation purposes as will be demonstrated below. To illustrate the nature of the results obtained in this way, Figure 5.6 shows the results from the analyses of several oils from various geographic origins. The isotopic compositions for each of the major components have been determined and are plotted in this diagram. Differences between these samples are immediately apparent and reflect their different origins. In a second example, results from the analyses of four gasoline samples are shown in Figure 5.7. These results

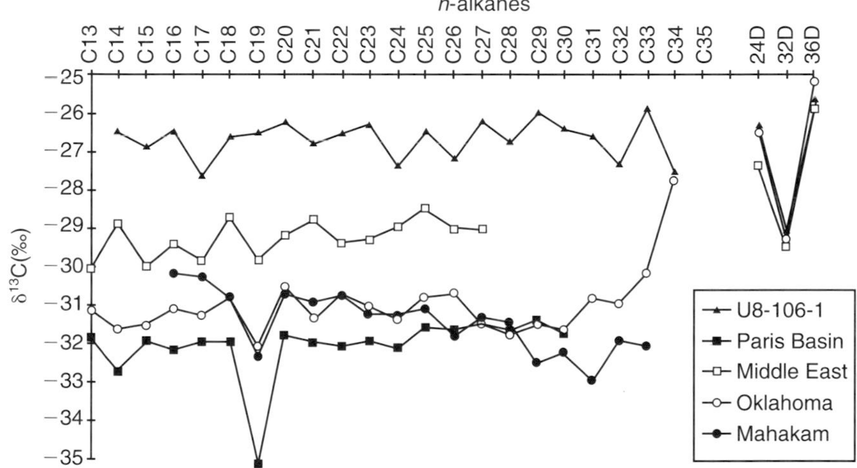

Figure 5.6

Characterization of a number of oils from different geographical regions by GCIRMS showed that this technique was a viable approach for differentiating these oils. Again this information should be used in conjunction with the GC and GCMS data. (Reprinted with permission from Mansuy et al., *1997, American Chemical Society.)*

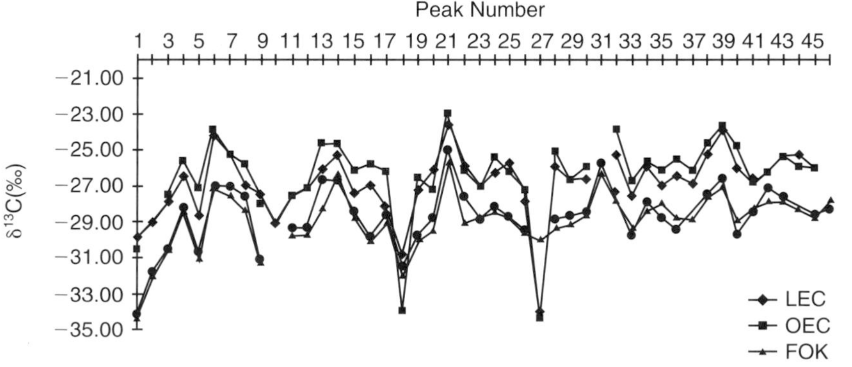

Figure 5.7

A similar example to Figure 5.6 but instead of crude oils, GCIRMS has been used to differentiate gasolines from Oklahoma (FOK, GOK) and the East Coast (LEC, OEC). (Reprinted with permission from Mansuy et al., *1997, American Chemical Society.)*

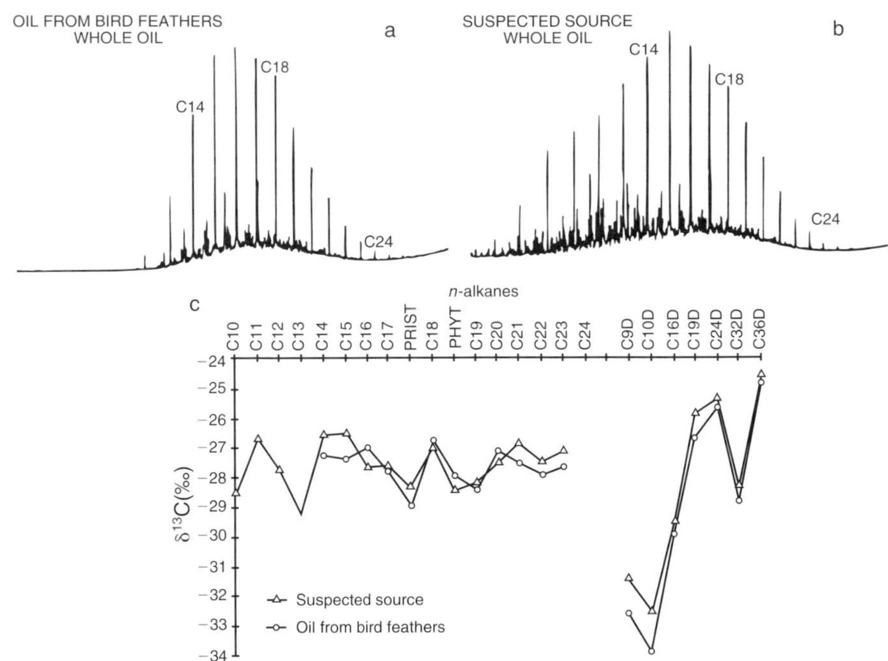

Figure 5.10

(a) and (b) Bird feathers coated with a light fuel oil were not readily correlated by conventional techniques such as GC and GCMS due to the absence of the biomarkers. (c) GCIRMS successfully showed a relationship between the samples in Figures 5.10a and 5.10b. (Reprinted with permission from Mansuy et al., *1997, American Chemical Society.)*

water-washing. The suspect oil was analyzed by GCIRMS without any fractionation step prior to the analysis. Comparison of the isotopic values for the individual components in the two samples suggested the suspect oil as the possible source for the oil on the bird feathers (Figure 5.10c). The standard deviation calculated between the two samples (0.36‰) is close enough to the range of the reproducibility (0.21–0.26‰) to consider that these two light fuel oils are derived from the same source. These results provided a high level of confidence suggesting similar origins for the oils found on the bird feathers and the suspected sources. The correlations could be supported by the conventional GC and GCMS fingerprinting but the absence of the conventional biomarkers make this a more difficult task.

Determination of the origin of polyaromatic hydrocarbons (PAHs) provides another challenging environmental problem. PAHs in the environment may be derived from sources such as coal tar, crude oils, or creosotes, for example. In many cases the chromatographic distributions of these compounds for different sources may be very similar and be of little use in terms of distinguishing one source from another. In addition, weathering can often result in additional ambiguities. However, with the development of GCIRMS, the possibility of determining the isotopic composition of the individual components provides a potential method for differentiating these sources of the PAHs. For example, Hammer *et al.* (1998) studied the PAHs from two creosote-contaminated sites and showed that at these two sites there was a suite of compounds that had very similar isotopic values which agreed within 1‰ of each other and which may be useful for identification of PAHs of creosote origin.

5.3.5 USE OF GCIRMS FOR MTBE

Methyl *tert*-butyl ether (MTBE) is rapidly becoming a major environmental problem but one that may be ideally suited to study using both carbon and hydrogen isotope compositions. The recent controversy over addition of MTBE to gasoline to boost oxygen content and decrease carbon monoxide emissions to the atmosphere has led to a proposed phase out of this compound by 2002. MTBE has been blended with some gasolines at low levels (2%) to improve octane performance. It was known to be a carcinogen since 1979 but was thought to be less of a health risk than many of the other constituents of gasoline (Hanson, 1999). MTBE has become a major contaminant of water (i.e., when gasoline leaks from underground storage tanks (UST) and makes contact with groundwater) and, to a lesser extent, air (i.e., when dispensed at a gasoline pump although it would be washed out relatively rapidly following the first rain). Furthermore in air, MTBE is relatively rapidly photodegraded. Examination of MTBE-contaminated groundwater samples shows that in many cases they are dominated by MTBE with only minor concentrations of other components present in the chromatograms (Figure 5.11). Once again the problem is to establish the origin of the MTBE.

This can be an extremely difficult question since many sites may have had multiple sources of MTBE from different gas stations in the area. The recalcitrant

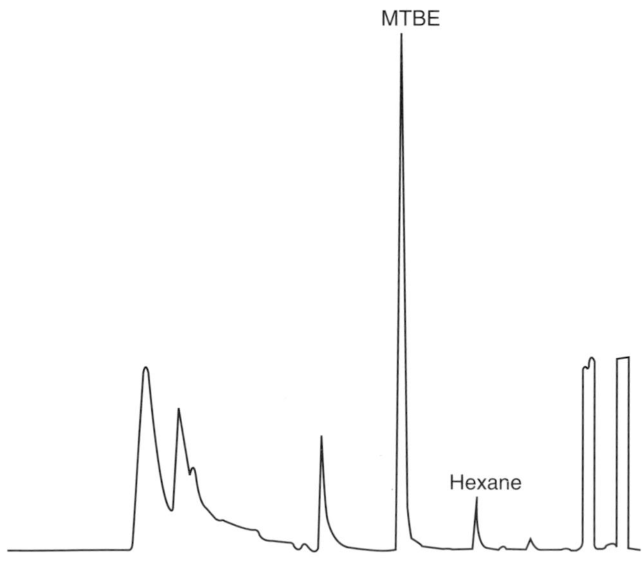

Figure 5.11
Examination of MTBE-contaminated groundwater samples shows that in many cases they are dominated by MTBE with only minor concentrations of other components present in the chromatograms.

nature of MTBE provides a potential opportunity for the source/spill correlation of gasoline or MTBE using isotope-ratio mass spectrometry (Smallwood *et al.*, 2001). Whilst not the ultimate solution, GCIRMS is probably the only viable technique which has any chance to discriminate various sources of MTBE in the groundwater samples. Since these samples are typically dominated by MTBE, GC and GCMS will be of little use in correlating or distinguishing samples of different origins. In a number of studies undertaken in our laboratory we have shown that a combination of carbon and hydrogen isotope values GCIRMS can be used to distinguish MTBE from different sources, and these values do not appear to be affected by evaporation or water washing. Four pure MTBE samples showed little variation in their δ^{13}C values but these samples could be discriminated on the basis of their δD values (Sessions *et al.*, 1999; Figure 5.12). When a larger number of gasoline-containing MTBE samples were examined, a much greater variation in the carbon isotope compositions was observed, as a result of the MTBE coming from different suppliers.

The carbon isotope composition of MTBE can be determined reproducibly, to concentrations as low as 15 parts per billion (ppb) using a purge and trap system in conjunction with the GCIRMS technique. This is very significant since concentrations of 15–20 ppb of MTBE in drinking water are the threshold for taste and odor and also the limit set by most states within the United States for the maximum concentration of MTBE allowed before remediation is enforced or drinking water wells are shut down.

In summary, the results discussed above provide a number of illustrations concerned with the application of both bulk isotopic compositions and isotopic compositions of individual compounds for the purposes of correlating and identifying samples of various origin. It is essential to stress that whilst these data are extremely useful, particularly in undertaking correlations that may not be possible by other approaches, the approach will be most powerful when used in conjunction with currently existing techniques, such as GC and

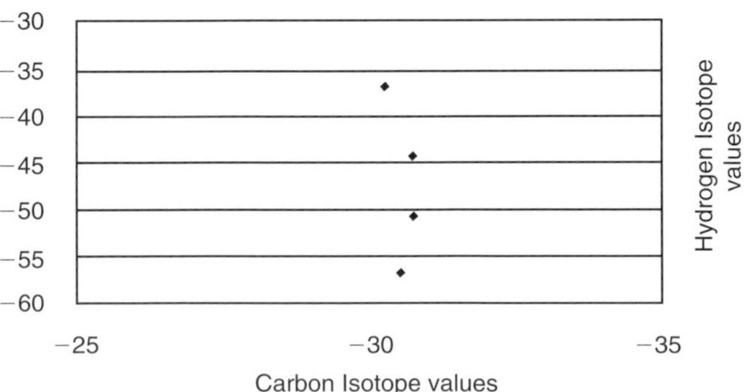

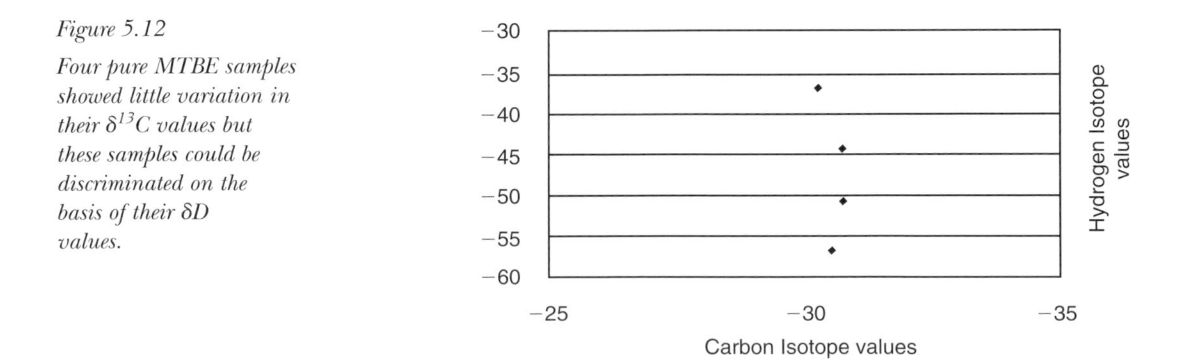

Figure 5.12
Four pure MTBE samples showed little variation in their $\delta^{13}C$ values but these samples could be discriminated on the basis of their δD values.

GCMS. The isotope values can provide that additional piece of information, which may be used to confirm conclusions based on these other data sets. In the situation where the contaminant is a single component, determination of the isotopic composition, preferably carbon and hydrogen, is probably the only approach that has any chance of providing a successful answer.

5.4 OTHER ISOTOPES

Lead was added to gasolines in the United States for many decades in the form of tetraethyl lead (TEL) and other forms of alkylate derivatives. In recent times, as health-related effects of lead in gasoline became a major issue, the use of TEL was phased out and was completely eliminated in all US gasolines by 1996. The concentrations of lead were greatly reduced in the early 1980s. If the lead content of a groundwater sample containing gasoline is measured and found to be non-existent or negligible, the normal conclusion is that the sample must have been an unleaded gasoline. If lead is present then typically the assumption is made that the spill must have been of a leaded gasoline, or a mixture of leaded and unleaded gasolines. A number of combustion methods can be pursued to measure lead content or alternatively GC methods are available to determine the distribution of alkyl lead derivatives in the extracts.

An alternative approach is to measure the lead isotope content of the extract from the groundwater. There are two reasons for doing this; first the ability to measure the lead isotope ratio will simply indicate the presence of lead in the sample. Second is the possibility of obtaining an approximate age for the sample release. There are four naturally occurring lead isotopes: ^{204}Pb is the stable isotope and ^{206}Pb, ^{207}Pb and ^{208}Pb are stable radiogenic isotopes. (Whilst the latter three are actually radiogenic, they have extremely long half-lives and can be thought of as being stable in the timeframe under consideration in the current context.) The lead isotope ratios are measured relative to ^{204}Pb or ^{206}Pb and similar isotope ratios should imply a common source (Hurst *et al.*, 1996). The age dating application here is not directly related to the decomposition of any of these isotopes but rather due to the fact that the lead isotope ratios reflect the source of the lead used in the preparation of the TEL. Furthermore, there were basically only two major manufacturers of TEL from 1960 through 1990, namely DuPont and Ethyl Corporation. In a plot of δ^{206}Pb (^{206}Pb/^{207}Pb) in gasoline vs. time it can be seen that the isotopic value for the lead becomes progressively heavier as it gets closer to the present time (Figure 5.13). However, the number of sources of lead was somewhat limited and divided between Mississippi Valley type ores and imported ore, largely from Australia. The ^{206}Pb/^{207}Pb ratio from the former was approximately 1.30 whereas for

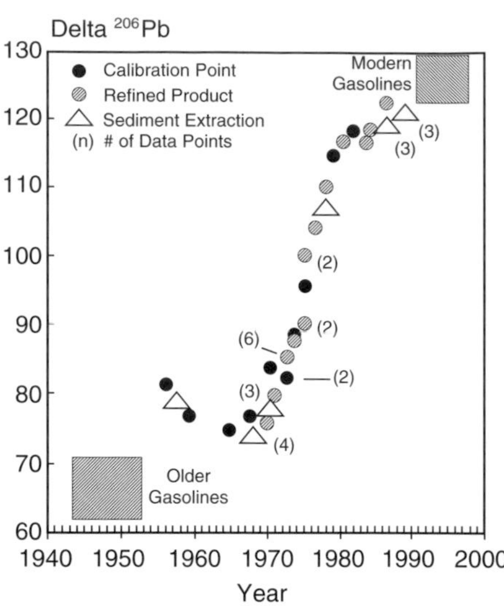

Figure 5.13

ALAS model calibration curve. In a plot of $\delta^{206}Pb$ in gasoline vs. time it can be seen that the isotopic value for the lead becomes progressively heavier as it gets closer to the present. (Reprinted with permission from Hurst et al., *1996, Environ. Sci. and Technol., 30(7), p. 306A.)*

imported ore, the $^{206}Pb/^{207}Pb$ ratio was less than 1.1. The plot of delta $^{206}Pb/^{207}Pb$ against time shows a steady increase with time for lead in gasolines from a number of different sources (Hurst *et al.*, 1996). From this plot came the Anthropogenic (i.e., gasoline derived) Lead Archeostratigraphy Model (ALAS) to distinguish between multiple sources of gasoline. The model plots one lead isotope ratio against the other or one isotope ratio against concentration and it is proposed by Hurst that patterns emerge which can distinguish multiple sources of gasoline plus time of formulation within a five-year time period (Hurst, 2000). If correct, the lack of scatter within the data over a forty-year period clearly demonstrates the potential viability of using this approach to provide an approximate indication of the age, or date of manufacture, of the TEL added to the gasoline. However, the gradual shift in isotopic ratios for individual spills is puzzling since this implies lead from Mississippi Valley (MV) was being mixed with lead from Australia, or elsewhere, prior to making the tetraethyl lead and the relative proportions were changing systematically with time. In all probability, as with any commodity, lead was being purchased on the spot market at the best possible price and the variations in the amounts of lead being used from different sources in all probability varied significantly from month to month. Secondly, if we assume that MV lead has an isotopic value of −1.32 and the other source is −1.04, on the basis of the ALAS model it can be computed that MV lead constituted 46% in 1965 and 64% in 1987, where Hurst claims that over the same period MV ore use rose from 9% to 82%. Other problems include the potential for additional sources, such as lead-based paints or pipeline contributions to the groundwater; this could be significant and therefore one must exercise some caution whilst applying this method at the current level of knowledge. The problem is also exacerbated by the fact that mixing of gasolines of several vintages at the same site could greatly distort the age obtained on the basis of the lead isotope ratio. It may be possible to unravel the situation but this could be a complex problem if end-members (pure samples of the individual components that comprise the mixture) are unavailable.

As far as can be ascertained the bulk of isotope studies to date that are concerned with environmental problems have been based on carbon isotope values. H and Cl isotopes are starting to see more widespread use in specific areas. N isotopes, whilst widely used in a number of geochemical studies not related to environmental issues, have yet to see any significant use in this area. Once again it should be noted that with the capability of determining the N isotope composition of individual compounds, in the near future one could expect to see an increase in the use of the N isotopes to study the fate of N-containing pesticides and other compounds in the environment. Similarly O isotopes have yet to see any significant applications in forensic environmental studies. It should be noted of course that O, N, Cl, S and certain radioisotopes have been used extensively in many hydrogeology studies for such purposes as dating groundwater, characterizing groundwater quality, examining changes in hydrogeological cycles, and studying precipitation (Clark and Fritz, 1997a). However, the majority of these applications are not strictly forensic as discussed in the context of the current monograph. As mentioned above, lead isotopes are used in environmental forensics but for a very specific application.

5.4.1 CHARACTERIZATION OF CHLORINATED COMPOUNDS

The widespread occurrence of chlorinated contaminants in the environment, PCBs, dioxins, DDT and a wide variety of other compounds, has led to an increased interest in the possible use of the chlorine isotope ratios as a means of studying the transport and fate of these compounds in the environment. A recent paper by Reddy *et al.* (2000) studied a number of PCBs from a variety of sources and noted that the δ^{37}Cl values fell within a narrow range from -3.4 to -2.1‰ but found no correlation between the mass percent of chlorine and δ^{37}Cl values. It should be noted that the samples used in this particular study were obtained from environmental standard suppliers who in turn had obtained the materials from Bayer (Germany) and Caffaro (Italy). In the US there was only one manufacturer of PCBs, namely Monsanto. The variability observed in the standards was suggested to arise from differences in the starting materials, synthesis methods, purification, or postproduction storage and handling of these products or some combination of all these factors. PCBs extracted from sediments ranged from -4.54 to -2.25‰. These isotopic variations suggest that $\delta^{37}Cl$ could be a useful tool for tracing the sources and fate of PCBs. In a limited number of experiments there did not appear to be any significant differences between the $\delta^{37}Cl$ values for individual congeners. Heraty *et al.* (1999) noted that in a laboratory study monitoring the aerobic degradation of dichloromethane (DCM), the $\delta^{37}Cl$ values of the residual DCM increased as a function of biodegradation.

In another slightly different approach with PCBs, Jarman *et al.* (1998) examined the carbon isotope variations on individual congeners in a variety of commercial PCBs. The bulk carbon isotope values for individual PCBs varied from −22.34 to −26.95‰, whereas individual congeners showed values ranging from −18.65 to −27.98‰. Large differences were observed for individual congeners between mixtures and the $\delta^{13}C$ values for individual congeners showed a depletion of ^{13}C with increasing chlorine content probably as result of a kinetic isotope effect caused by the position of the chlorine atom on the biphenyl molecule. Such a wide diversity of carbon isotope values could prove to be very powerful in source determination for PCBs in the environment. Recent developments in the GCIRMS capabilities plus the results from the paper by Heraty *et al.* (1999) would suggest that a combination of both the carbon and chlorine isotope compositions could prove to be a powerful tool in distinguishing PCBs, and other chlorinated compounds, from different sources. Difficulties in obtaining authentic standards of PCBs manufactured in the 1960–1970 period could make it difficult to assign absolute sources of the PCBs in sediments. However, variations in the carbon and chlorine isotopic compositions should be sufficient to determine whether the presence of PCBs in the sediments resulted from one or multiple inputs to a particular site.

5.5 BIODEGRADATION

Whilst carbon, hydrogen, and chlorine isotopes are now being used in a fairly regular and systematic manner for evaluating the origin of contaminants and the correlation of these contaminants with their suspected sources, there are a number of other potential applications in environmental chemistry based on changes in isotopic compositions, particularly carbon values. The most important application for these values, in addition to correlations, would be monitoring the rate and/or extent of biodegradation of individual compounds in the environment. The potential of this approach is enormous, particularly in the case of groundwater samples and specifically for compounds that are very soluble in water such as MTBE. One of the problems with compounds such as MTBE is that they are mobile and concentration changes at the monitoring wells do not necessarily reflect changes in concentrations resulting from biodegradation. The changes in concentration at a specific water well may simply reflect movement of the samples within the plume.

However, the potential of this approach has been demonstrated in a number of field and laboratory studies, both published and unpublished. For example, in some unpublished studies in our laboratory, we have shown that with increasing biodegradation MTBE becomes progressively isotopically heavier. This effect has been observed at more than one site and appears to be reproducible. If such changes can be quantified it could provide an excellent tool to demonstrate progressive mineralization of the MTBE. There are some potential limitations since biodegradation anywhere depends upon a number of factors such as oxicity, presence of nutrients, the nature of the microbial community, and the salinity of groundwater. Similar isotopic changes have also been observed for other compounds undergoing biodegradation, particularly in laboratory studies. All processes that work well in the laboratory may not work as well or in such a predictable manner in the field. Ahad *et al.* (2000) noticed a small but reproducible isotopic enrichment of approximately 2‰ during the degradation of toluene in the laboratory. As with the MTBE this introduces the possibility of using the approach in the field although such a relatively small fractionation could be quite difficult to detect at a complex field site. However, with the current generation of GCIRMS systems precision of 0.5‰ is easily attainable.

Another problem that must be taken into consideration is the fact that rates and changes in these isotopic compositions may not be the same at different sites, meaning that each site would need to be calibrated independently. Microbial degradation of toluene by Meckenstock *et al.* (1999) produced a shift in the $^{13}C/^{12}C$ values of the residual fraction under different environmental conditions. Many biological reactions are known to produce a $^{13}C/^{12}C$ isotopic fractionation of the substrate. In general the ^{12}C isotopes are used preferentially and the naturally occurring ^{13}C is discriminated against. As a result of the preferential consumption of the lighter isotope, the residual substrate will show a corresponding increase in its ^{13}C content. In their paper Meckenstock *et al.* (1999) demonstrated, using toluene as a substrate, that shifts of up to 10‰ in favor of the heavier isotope could be observed depending on the particular strain of bacterium being utilized. Similar experiments in a non-sterile soil column induced changes of about 2–3‰ in the toluene. In a study of the degradation of C_1–C_5 alkylbenzenes in crude oils, Wilkes *et al.* (2000) observed that for *o*-xylene a shift in the $\delta^{13}C$ of 4‰ was observed after loss of 70% of the original material whereas a shift of 6‰ was observed for the *o*-ethyltoluene.

Clearly a great deal of work remains to be done in terms of the isotopic changes which may be expected during biodegradation since such changes depend on a variety of factors including the compound itself as well as all the associated environmental factors and strains of bacteria responsible for the degradation reactions. At this stage of development it certainly does not appear that any universal set of observations or guidelines have started to appear for individual compounds that can be used in a wide variety of environments. To complicate the matter further, not all compounds show any degree of isotopic fractionation upon biodegradation. O'Malley *et al.* (1994) noted that naphthalene showed no isotopic fractionation as a result of biodegradation and in similar studies, Kelly *et al.* (1995) and Trust *et al.* (1995) showed that there was no change in the isotopic composition of acenaphthene, fluorene, phenanthrene, and fluoranthene. Clearly in addition to environmental conditions, the extent of changes in the isotopic composition appears to be dependent upon the compound under consideration.

An alternative approach to this problem was described several years ago, when Aggarwal and Hinchee (1991) proposed measuring the CO_2 produced by hydrocarbon degradation. It was anticipated that the carbon isotopic composition of the CO_2 produced in this way from petroleum hydrocarbons could be distinguished from that produced from other sources and other mechanisms. In the analysis of CO_2 from three sites contaminated with jet fuel it was noted that the CO_2 was about 5‰ lower than the values for CO_2 at the uncontaminated locations. Hence it would appear that in areas undergoing aerobic biodegradation this could provide viable evidence for *in situ* biodegradation of the hydrocarbons. Extensive levels of biodegradation could lead to anaerobic conditions for the biodegradation and under these conditions degradation of the CO_2 could affect the $\delta^{13}C$ values for the residual CO_2; clearly this is an area where additional studies are required.

The isotopic composition of the dissolved inorganic carbon (DIC) can also be used as a means to monitor mineralization of contaminants in the environment. For example, at a site where a groundwater aquifer was contaminated with jet fuel, the $\delta^{13}C$ values for the DIC ranged from −28 to +11.9‰. The $\delta^{13}C$ value for the jet fuel was −27‰ and following degradation under aerobic conditions produced DIC with values of −26‰ or −18‰ under sulfate-reducing conditions. In periods of low rainfall or lack of oxygen when methanogenesis was the major terminal electron acceptor process, the DIC values ranged up to +11.9‰. The results obtained in this study demonstrated that the $\delta^{13}C$ values of the DIC could be used to indicate zonation of biodegradation processes under the influence of hydrologically controlled electron acceptor availability (Landmeyer *et al.*, 1996).

In a similar approach, stable isotope ratios of C and H of dissolved CH_4 and the DIC in the groundwater in the vicinity of a crude oil spill near Bemidji, Minnesota, were used to support the concept of CH_4 production by acetate fermentation (Revesz *et al.*, 1995). Two biogenic CH_4 production pathways are commonly recognized – the reduction of CO_2 by H_2 and the fermentation of acetate. Acetate fermentation refers to methanogenesis that involves transfer of a CH_3^- group from a substrate, e.g. acetate, methylamines etc. The fractionation factor between coexisting CO_2 and CH_4 can be expressed as:

$$\alpha_{CO_2-CH_4} = \delta^{13}C_{CO_2} / \delta^{13}C_{CH_4} \quad (5.1)$$

For reduction of CO_2 this value is in the range 1.05–1.09 and for fermentation in the range 1.04–1.06. If the pathways cannot be distinguished solely by use of the carbon values, it is possible to utilize the H/D values to confirm the suspected pathway. In freshwater environments methanogenesis usually occurs by both acetate fermentation and CO_2 reduction pathways, with fermentation being the source of about 70% of the CH_4 in freshwater environments, and the remainder coming from CO_2 reduction. In the case of the Bemidji samples, the CH_4 production by acetate fermentation caused an increase in concentration and $\delta^{13}C$ values for the dissolved inorganic carbon in the saturated zone. In the unsaturated zone, the CH_4 is oxidized at shallower depths as indicated by an increase in $\delta^{13}C$ values and decrease in DIC concentration. The $\delta^{13}C$ values for the CO_2 increase with depth in accordance with oxidation of CH_4.

Biological degradation of organic refuse is another prolific source of both CO_2 and CH_4. Landfill methane or biogenic methane will typically have $\delta^{13}C$ values ranging from −42‰ to −61‰. Non-landfill environmental values of $\delta^{13}C$ in methane range from −60‰ to −95‰ for biogenic reduction from CO_2 and from −47‰ to −63‰ via acetate formation (Whiticar *et al.*, 1986; Schoell, 1988). A very practical application for these variations in the isotopic composition of the methane is to distinguish natural gas sources of methane from biogenic methane (Kaplan *et al.*, 1997). There have been numerous reported cases of anomalously high concentrations of methane in populated areas. The first reaction is to place responsibility with the operators of any natural gas pipeline operators in the area. In many cases this may indeed be the source of the gas. However, there have also been cases where these anomalous concentrations simply result from a buildup of methane coming from degradation of organic refuse, typically from landfills. Source identity can be established very quickly on the basis of the isotopic composition of the methane. Natural gas methane will be relatively heavy with $\delta^{13}C$ values in the −35 to −45‰ range. Natural gas will also typically contain C_{2+} components, which will also be isotopically heavy. Biogenic methane as noted above will be isotopically much lighter than the natural gas methane.

Hence in summary it would appear that in addition to the various correlation approaches mentioned above, the most significant additional use of isotope values would have to be monitoring the effects, rate, and possible mechanisms of biodegradation. The approach should be particularly useful in the case of single components which are very water soluble and whose rates of degradation are difficult to measure by other methods.

5.6 COMBINING ISOTOPE METHODS WITH OTHER METHODS

Whilst stable isotopes have many powerful applications in environmental forensics, in most cases they should be used in conjunction with other analytical techniques, such as GC and GCMS. In almost all applications the isotope values should be used as a preliminary tool to provide guidance on possible relationships between samples. This is true of both bulk isotope numbers and values for individual components, although for single components, isotope values may be the only tool available for correlation purposes.

The techniques that will be most commonly used in conjunction with the carbon isotopes, either the bulk isotopes or those obtained via GCIRMS, will be GC and GCMS or GC combined with tandem mass spectrometry (GCMSMS). It is not proposed to discuss these techniques in detail since they are well documented in the literature and elsewhere in this monograph (see Chapter 6). However, to illustrate the value of such combinations, two examples will be provided. The first of these involves a number of tar balls collected from the Indian Ocean washing on to the beaches of the Seychelle Islands. The question of interest here was whether or not there was any relationship between various samples and secondly to determine whether these samples were derived from passing tanker traffic or from underwater seeps. The chromatograms showed the samples had been weathered to varying degrees in the ocean as a result of evaporation, water-washing, and biodegradation (Figure 5.14a). Varying amounts of the *n*-alkanes had been removed from the samples, hence the bulk numbers for the samples would not be very useful since these values will change as a result of the *n*-alkane removal. Since these samples were crude oils, the sterane and terpane biomarker distributions could be used to demonstrate the relationships between the four samples. The biomarker data showed the two samples on the left hand side of the diagram were clearly related to each other and different from the two samples on the right hand side of the diagram. Detailed interpretation reveals that the two samples on the left hand side were derived from carbonate source rocks and had many characteristics similar to oils from the Middle East. Hence it is suspected that these samples were indeed derived from tanker traffic. The other two samples were not derived from carbonate source rocks but had characteristics similar to other samples known to be derived from lacustrine source rocks in the Seychelles area. A cursory examination of the two chromatograms on the right hand side could easily suggest that these two samples are virtually identical to each other.

However, carbon isotope values for the individual *n*-alkanes of the two samples showed the samples to be isotopically distinct (Figure 5.14b). This finding prompted closer examination of the biomarker chromatograms, which subsequently revealed the presence of a rather unique biomarker, oleanane, in one

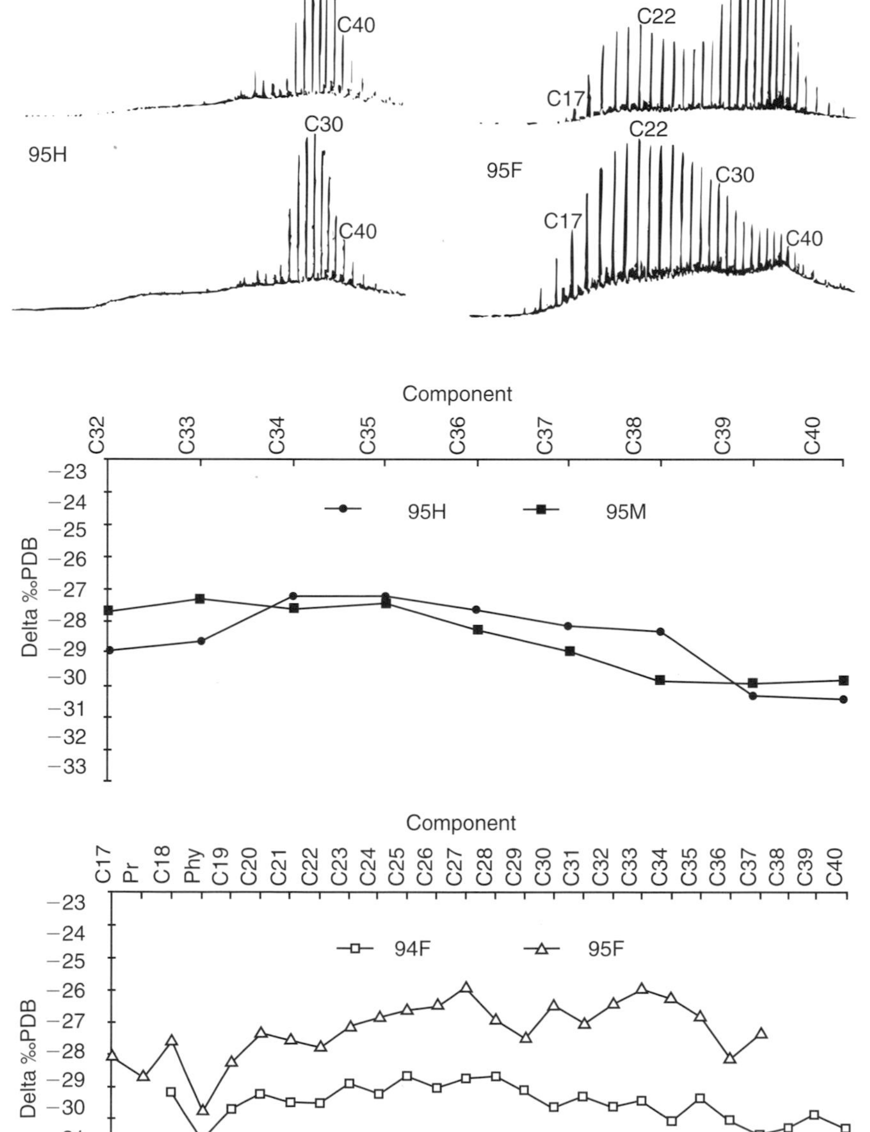

Figure 5.14a

Gas chromatograms of the heavily biodegraded Seychelles Islands tar ball samples. Samples could not be correlated on the basis of the bulk isotope numbers due to removal of varying amounts of n*-alkanes.*

Figure 5.14b

GCIRMS could be used to establish a relationship between the Seychelles Islands tar ball samples on the basis of their sterane and terpane distributions.

sample but not the other. The presence of this compound plus the isotopic differences permitted the conclusion to be made that these two seep samples originated from different geological facies of the same source rock. Hence this is an excellent example of where GC, GCMS and GCIRMS were combined to differentiate origins of these samples. Whilst this example illustrates differences between two naturally occurring seeps, such small differences could also be utilized in the differentiation of two sources of crude oil responsible for contamination of the environment.

Another classic example of the isotope approach is the Kvenvolden *et al.* (1995) paper, which described a bulk carbon isotope study on oil residues collected from the beaches of Prince William Sound after the *Exxon Valdez* spill. It was noted that basically there were two populations of residues. One with an isotopic composition similar to the *Exxon Valdez* oil of around −29‰. The second group had a value of around −23‰ – quite different from any Alaskan oil. So the carbon isotope data suggested two populations of crude oil residues. Once this information became available, more detailed analyses of the various residues by GC and GCMS revealed that the heavier residues were actually derived from California crudes sourced from Monterey Shale.

Despite increased specificity, the GCIRMS technique still suffers from a number of problems and limitations related to the correlation of moderate to severely biodegraded oils. In order to be successful, GCIRMS requires the presence of well-resolved components in the chromatogram. Severely weathered, particularly biodegraded, samples typically have lost their *n*-alkanes and the resulting chromatogram will be dominated by a hump of unresolved components. However, it has been shown that isolation and pyrolysis of the asphaltenes from these types of samples provides an alternative method for undertaking correlations (Behar *et al.*, 1984). Previous studies have demonstrated the use of bulk isotopic composition of asphaltenes as a correlation parameter for severely weathered oils (Hartman and Hammond, 1981; Macko *et al.*, 1981). Hartman and Hammond (1981) did not see any change in the isotopic composition of these compounds although Stahl (1977) observed a slight ^{12}C enrichment with increasing artificial biodegradation. In general the asphaltene pyrolysis is performed off-line (Eglinton, 1994) and the pyrolysates collected, fractionated into saturates + unsaturates, aromatics and polar fractions. The saturate + unsaturate fraction containing the *n*-alkene/*n*-alkane doublets can be subsequently analyzed by GCIRMS. This approach was applied to the asphaltenes from the oils that were artificially biodegraded for two and four months described above (Fig. 5.8). The pyrograms of the biodegraded oil samples showed a good correlation with the original oil and had a predominance of the lighter hydrocarbons, maximizing at C_{15}–C_{17} (Figure 5.15). However, the biodegraded samples did have a higher contribution of *n*-alkanes around C_{28}

and C_{38} compared to the original oil. Analyses of the fractions by GCIRMS showed the isotopic compositions of individual *n*-alkanes of these fractions from the original and degraded samples to be quite similar (Figure 5.16). Standard deviations calculated between the original and biodegraded oils were relatively low considering the complexity of the mixtures analyzed. The

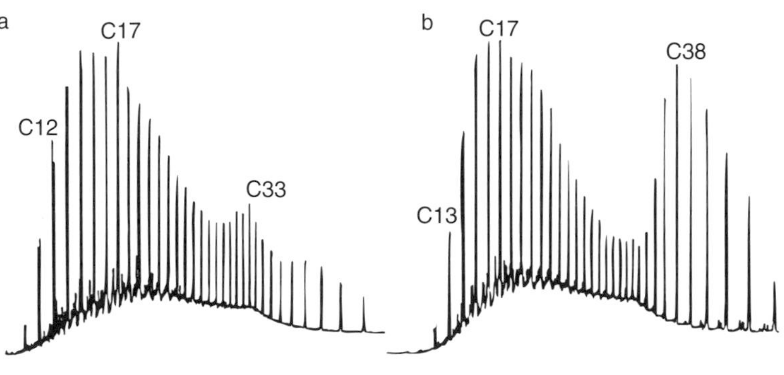

Figure 5.15

The pyrograms of the asphaltenes from the two biodegraded oils (a, after two months; b, after four months) are similar to that of the initial oil with a predominance of the light ends, maximizing at C_{15}–C_{17}. These pyrolysis fractions were used to correlate the degraded and non-degraded samples.

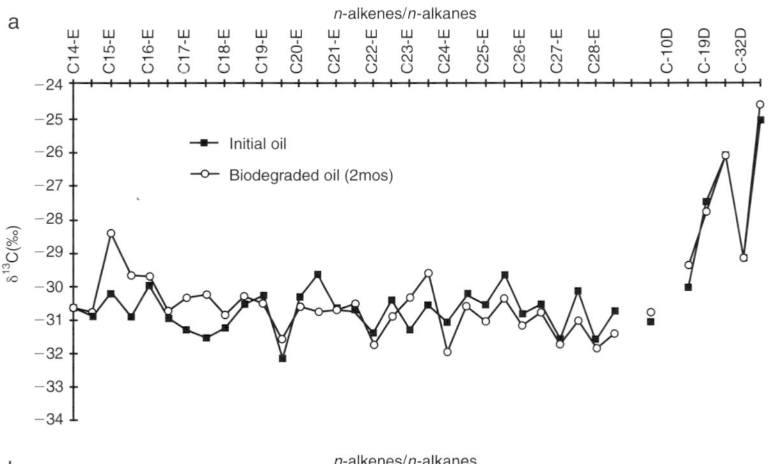

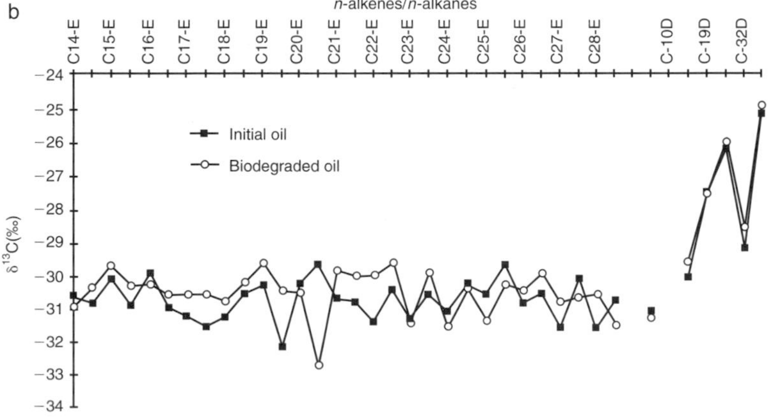

Figure 5.16

Analyses of the asphaltene pyrolysis fractions by GCIRMS showed the isotopic compositions of individual n*-alkenes/* n*-alkanes from these two fractions (Figure 5.15) to be very similar and capable of correlating degraded and non-degraded samples. (Reprinted with permission from Mansuy* et al.*, 1997, American Chemical Society.)*

n-alkene/*n*-alkane doublets and the high background (Figure 5.15) increase the chance of coelutions and affect the reproducibility of the analyses, which range between 0.33 and 0.40‰. The isotopic composition of the *n*-alkanes generated by the asphaltene pyrolysis is very close to the isotopic composition of the *n*-alkanes of their respective oils and of the initial oil.

These results confirm the ability of GCIRMS and the asphaltene pyrolysates to correlate severely biodegraded oils with their unweathered counterparts with an analytical error still acceptable despite the chromatographic problems encountered during the analyses of such complex mixtures. Whilst it may be argued the number of steps involved in this process may render it ineffective as a correlation tool it should be remembered that there are many situations where oil samples have been recovered from old storage tanks or tar balls in the oceans that have been weathered over many years severely altering the biomarker fingerprints. In this situation, isolation of the asphaltenes, off-line pyrolysis, fractionation of the pyrolysates, followed by GCIRMS may prove to be extremely valuable when other techniques provide ambiguous or inconclusive results.

SUMMARY

This chapter has attempted to discuss and summarize some of the significant developments in the application of isotopic geochemistry to environmental forensic problems. Applications are many and varied and involve a wide range of isotopes. Developments and new ideas continue to occur with the availability of the combined gas chromatograph–isotope ratio mass spectrometer systems being the most recent example of the available new technology. Applications include age dating, correlations between contaminants and suspected sources, measuring rate of biodegradation, and evaluating fate of contaminants in the environment. The successful applications of these isotopic approaches to environmental forensics will undoubtedly assure a continued future for this approach in combination with other available techniques such as GC and GCMS for many years to come.

REFERENCES

Abrajano, J. and Lollar, B.S. (1999) Compound-specific isotope analysis: tracing organic contaminant sources and processes in geochemical systems. *Organic Geochemistry* 30(8B), v–vii.

Aggarwal, P.K. and Hinchee, R.E. (1991) Monitoring in-situ biodegradation of hydrocarbons by using stable carbon isotopes. *Environmental Science and Technology* 25, 1176–1180.

Ahad, J., Lollar, B.S., Edwards, E., Slater, G., and Sleep, B. (2000) Carbon isotope fractionation during anaerobic biodegradation of toluene: Implications for intrinsic bioremediation. *Environmental Science and Technology* 34, 892–896.

Becker, S. and Herner, A.V. (1998) Characterization of crude oils by carbon and sulphur isotope ratio measurements as a tool for pollution control. *Isotopes in Environmental Health Studies* 34, 255–264.

Behar, F., Pelet, R., and Roucache, J. (1984) Geochemistry of asphaltenes. In *Advances in Organic Geochemistry 1983* (Schenck, P.A., de Leeuw, J.W., and Lijmbach, G.W.M., eds), pp. 587–595. Pergamon Press, Oxford.

Beneteau, K., Aravena, R., and Frape, S. (1999) Isotopic characterization of chlorinated solvents-laboratory and field results. *Organic Geochemistry* 30, 739–753.

Boreham, C., Dowling, L.M., and Murray, A.P. (1995) Biodegradation and maturity influences on n-alkane isotopic profiles in terrigenous sequences. *Abstracts 17th International Meeting EAOG*, pp. 539–541. AIGOA Donostic, San Sebastian.

Clark, I. and Fritz, P. (1997) *Environmental Isotopes in Hydrogeology*. Lewis Publishers/CRC Press, Boca Raton, FL.

Czuczwa, J.M. and Hites, R.A. (1984) Environmental fate of combustion-generated polychlorinated dioxins and furans. *Environmental Science and Technology* 31, 2193–2197.

Davis, A., Curnou, P.D., and Eary, L.E. (1997) Discriminating between sources of arsenic in the sediments of a tidal waterway in Tacoma, Washington. *Environmental Science and Technology* 31, 1985–1991.

Dincer, T., Al-Mugrin, A., and Zimmermann, U. (1974) Study of the infiltration and recharge through the sand dunes in arid zones with special reference to the stable isotopes and thermonuclear tritium. *Journal of Hydrology* 23, 79–109.

Dowling, L.M., Boreham, C.J., Hope, J.M., Murray, A.P., and Summons, R.E. (1995) Carbon isotope composition of hydrocarbons in ocean transported bitumens from the coastline of Australia. *Organic Geochemistry* 23, 729–737.

Eganhouse, R.P., Baedecker, M.J., Cozzarelli, I.M., Aiken, G.R., Thorn, K.A., and Dorsey, T.F. (1993) Crude oil in a shallow sand and gravel aquifer-II. *Organic Geochemistry. Applied Geochemistry* 8, 551–567.

Eglinton, T.I. (1994) Carbon isotopic evidence for the origin of macromolecular aliphatic structures in kerogen. *Organic Geochemistry* 21, 721–735.

Farran, A., Grimalt, J., Albaiges, J., Botello, A.V., and Macko, S.E. (1987) Assessment of petroleum pollution in a Mexican river by molecular markers and carbon isotope ratios. *Marine Pollution Bulletin*, 18, 284–288.

Freeman, K.H., Hayes, J.M., Trendel, J.M., and Albrecht, P. (1990) Evidence from carbon isotope measurements for diverse origins of sedimentary hydrocarbons. *Nature* 343, 254–256.

Fuentes H.R., Jaffe, R., and Shrotriya R.V. (1996) Weathering of crude oil spills in seawater under simulated laboratory conditions. *5th Latin American Congress on Organic Geochemistry*, Cancun, Mexico, Abstracts, Mexican Petroleum Institute, Mexico City, Mexico. pp. 310–312.

Galt, J.A., Lehr, W.J., and Payton, D.L. (1991) Fate and transport of the Exxon Valdez oil spill. *Environmental Science and Technology* 25, 202–209.

Hammer, B.T., Kelly., C.A., Coffin, R.B., Cifuentes, L.A., and Mueller, J. (1998) $\delta^{13}C$ values of polycyclic aromatic hydrocarbons collected from two creosote-contaminated sites. *Organic Geochemistry* 152, 43–58.

Hanson, D. (1999) MTBE – villain or victim. *Chemical and Engineering News* October 19th, p. 49.

Hartman, B. and Hammond, D. (1981) The use of carbon and sulphur isotopes as a correlation parameter for source identification of beach tars in southern California. *Geochimica Cosmochimica Acta* 45, 309–319.

Hayes, J.M., Freeman, K.H., Popp, B.N., and Hoham, C.H. (1990) Compound specific isotope analysis: A novel tool for reconstruction of ancient biogeochemical processes. *Organic Geochemistry* 16, 1115–1128.

Heraty, L., Fuller, M.E., Huang, L., Abrajano, Jr, T., and Sturchio, N. (1999) Isotopic fractionation of carbon and chlorine by microbial degradation of dichloromethane. *Organic Geochemistry* 30, 739–753.

Hermanson, M.H. (1990) ^{210}Pb and ^{137}Cs chronology of sediments from small, shallow Arctic lakes. *Geochimica Cosmochimica Acta* 54, 1443–1451.

Hilkert, A., Douthitt, C.B., Schluter, H.J., and Brand, W.A. (1999) Isotope ratio monitoring gas chromatography/mass spectrometry of D/H by high temperature conversion isotope ratio mass spectrometry. *Rapid Communications in Mass Spectrometry* 13, 1226–1230.

Hillaire-Marcel, C. (1986) Isotopes and food. In *Handbook of Environmental Isotope Geochemistry*, vol. 2, *The Terrestrial Environment*, B, (Fritz, P. and Fontes, J. Ch., eds), pp. 507–548. Elsevier, Amsterdam.

Hillaire-Marcel, C. and Ghaleb, B. (1997) Thermal ionization mass spectrometry measurements of ^{226}Ra and U-isotopes in surface and groundwaters-porewater/matrix interactions revisited and potential dating implications. *Isotope Techniques in the Study of Past and Current Environmental Changes in the Hydrosphere and Atmosphere*, IAEA Symposium 349, April 1997. International Atomic Energy Administration, Vienna.

Hurst, R.W. (2000) Applications of anthropogenic lead archaeostratigraphy (ALAS model) to hydrocarbon remediation. *Environmental Forensics* 1(1), 11–24.

Hurst, R.W., Davis, T.E., and Chinn, B.D. (1996) Lead fingerprints of gasoline contamination. *Environmental Science and Technology* 30, 304A–307A.

Jarman, W.M., Hilkert, A., Bacon, C.E., Collister, J.W., Ballschmitter, P., and Risebrough, R.W. (1998) Compound specific carbon isotope analysis of Aroclors, Clophens, Kaneclors, and Phenoclors. *Environmental Science and Technology* 32(6), 833–836.

Juttner, I., Henkelmann, B., Schramm, K., Steinberg, C.E.W., Winkler, R., and Kettrup, A. (1997) Occurrence of PCDD/F in dated lake sediments of the Black Forest, Southwestern Germany. *Environmental Science and Technology* 31, 806–812.

Kaplan, I.R., Galperin, Y., Lu, S., and Lee, R-P. (1997) Forensic environmental geochemistry: differentiation of fuel types, their sources and release time. *Organic Geochemistry* 27, 289–317.

Kelly, C.A., Coffin, R.B., Cifuentes, L.A., Lantz, S.E., and Mueller, J. (1995) The use of GC/C/IRMS coupled with GC/MS to monitor biodegradation of petroleum hydrocarbons. Platform Abstracts of the Third International Symposium on In Situ and On Situ Bioreclamation, San Diego, CA. April 24–27, 1995.

Kelly, C.A., Hammar, B.T., and Coffin, R.B. (1997) Concentrations and stable isotope values of BTEX in gasoline contaminated groundwater. *Environmental Science and Technology* 31, 2469–2472.

Kvenvolden, K.A., Hostettler, F.D., Carlson, P.R., Rapp, J.B., Threlkeld, C.N., and Warden, A. (1995) Ubiquitous tar balls with a California-source signature on shorelines of Prince William Sound, Alaska. *Environmental Science and Technology* 29, 2684–2694.

Lafargue, E. and Barker, C. (1988) Effect of water washing on crude oil compositions. *American Association of Petroleum Geologists Bulletin* 72, 263–276.

Landmeyer, J.E., Vroblesky, D.A., and Chapelle, F. (1996) Stable carbon isotope evidence of biodegradation zonation in a shallow jet fuel contaminated aquifer. *Environmental Science and Technology* 30, 1120–1128.

Lieser, K.H. and Steinkopff, T. (1989) Chemistry of radioactive cesium in the hydrosphere and in the geosphere. *Radiochimica Acta* 46, 39–47.

Macko, S.A. and Parker, L. (1983) Stable nitrogen and carbon isotope ratios of beach tars on South Texas barrier islands. *Marine Environmental Research* 10, 93–103.

Macko, S.A., Parker, L., and Botello, A.V. (1981) Persistence of spilled oil in a Texas salt marsh. *Environmental Pollution Bulletin* 2, 119–128.

Mansuy, L., Philp, R.P., and Allen, J. (1997) Source identification of oil spills based on the isotopic compositin of individual components in weathered oil samples. *Environmental Science and Technology* 31, 3417–3425.

Meckenstock, R.U., Morasch, B., Warthmann, R.W., Schink, B., Annweller, E., Michaelis, W., and Richnow, H.H. (1999) $^{13}C/^{12}C$ Isotope fractionation of aromatic hydrocarbons during microbial degradation. *Environmental Microbiology* 1(5), 409–414.

Meier-Augenstein, W. (1999) Applied gas chromatography coupled to isotope ratio mass spectrometry. *Journal of Chromatography* 842, 351–371.

Michel, F.A., Kubasiewicz, R.J., Patterson, R.J., and Brown, R.M. (1984) Ground water flow velocity derived from tritium measurements at the Glouster landfill site, Glouster, Ontario. *Water Pollution Research Journal of Canada* 19, 13–22.

Milner, C.W.D., Rogers, M.A., and Evans, C.R. (1977) Petroleum transformations in reservoirs. *Journal of Geochemical Exploration* 7, 101–153.

O'Malley, V.P., Abrajano, T., and Hellou, J. (1994) Determination of the $^{13}C/^{12}C$ ratios of individual PAH from environmental samples: can PAH sources be apportioned. *Organic Geochemistry* 21, 809–822.

O'Malley, V.P., Abrajano, T., and Hellou, J. (1996) Stable isotopic apportionment of individual polycyclic aromatic hydrocarbons in St. Johns Harbour, Newfoundland. *Environmental Science and Technology* 30, 634–639.

Oros, D., Aboul-Kassim, T.A., Simoneit, B.R.T., and Collier, R.C. (1996) *5th Latin American Congress on Organic Geochemistry*, Cancun, Mexico, Abstracts, pp. 330–332.

Palmer, S.E. (1993) Effect of biodegradation and water washing on crude oil compositon. In *Organic Geochemistry* (Engel, M.H. and Macko, S.E., eds), pp. 511–534. Plenum Press, New York.

Pennington, W. and Cambray, R.S. (1973) Observations on lake sediments using fallout ^{137}Cs as a tracer. *Nature* 242, 324–326.

Reddy, C.M., Heraty, L., Holt, B., Sturchio, N., Eglinton, T.I., Drenzek, N.J., Xu, L., Lake, J.L., and Maruya, K.A. (2000) Stable carbon isotope compositions of Aroclors and Aroclor contaminated sediments. *Environmental Science and Technology* 34(13), 2866–2870.

Revesz, K., Coplen, T.B., Baedecker, M.J., and Glynn, P. (1995) Methane production and consumption monitored by stable C and H ratios at a crude oil spill site, Bemidji, Minnesota. *Applied Geochemistry* 10, 505–516.

Ritchie, J.C. and McHenry, J.R. (1990) Application of radioactive fallout caesium-137 for measuring soil-erosion and sediment accumulation rates and patterns – a review. *Journal of Environmental Quality* 19, 215–233.

Schlosser, P., Stute, M., Sonntag, C., and Munnich, K.O. (1989) Tritogenic ^{3}He in shallow ground water. *Earth and Planetary Science Letters* 94, 245–254.

Schoell, M. (1988) Multiple origins of methane in the earth. *Chemistry and Geology* 71, 1–10.

Sessions, A.L., Burgoyne, T.W., Schimmelmann, A., and Hayes, J.M. (1999) Fractionation of hydrogen isotopes in lipid biosynthesis. *Organic Geochemistry* 30(9), 1193–1200.

Smallwood, B.J., Philp, R.P., Burgoyne, T.W., and Allen, J. (2001) The use of stable isotopes to differentiate specific source markers for MTBE. *Journal of Environmental Forensics* (in press).

Stahl, W.J. (1977) Carbon and nitrogen isotopes in hydrocarbon research and exploration. *Chemistry and Geology* 20, 121–149.

Tamura, T. and Jacobs, D.G. (1960) Structural implications in cesium sorption. *Health Physics* 2, 391–398.

Trust, B.A., Coffin, R.B., Cifuentes, L.A., and Mueller, J. (1995) *Monitoring and Verification of Bioremediation*, (Hinchee, R.E., Douglas, G.S., and Ong, S.K., eds). pp. 233–239. Battelle Press, Colombia, OH.

van Metre, P.C., Callendar, E., and Fuller, C.C. (1997) Historical trends in organochlorine compounds in river basins identified using sediment cores from reservoirs. *Environmental Science and Technology* 31, 2339–2344.

Von Gunten, H.R., Strum, M., and Moser, R.N. (1997) 200-Year record of metals in lake sediments and natural background concentrations. *Environmental Science and Technology* 31, 2193–2197.

Wang, Z., Fingas, M., and Sergy, G. (1994) Study of 22-year old Arrow oil samples using biomarker compounds by GCMS. *Environmental Science and Technology* 28, 1733–1746.

Ward, J., Ahad, J., LaCrampe-Couloume, G., Slater, G., Edwards, E., and Lollar, B.S. (2000) Hydrogen isotope fractionation during methanogenic degradation of toluene: Potential for direct verification of bioremediation. *Environmental Science and Technology* 34(21), 4577–4581.

Whiticar, M.J., Faber, E., and Schoell, M. (1986) Biogenic methane formation in marine and freshwater environments: CO_2 reduction vs. acetate fermentation-isotopic evidence. *Geochimica Cosmochimica Acta* 50, 693–709.

Wilkes, H., Boreham, C., Harms, G., Zengler, K., and Rabus, R. (2000) Anaerobic degradation and carbon isotopic fractionation of alkylbenzenes in crude oil by sulphate-reducing bacteria. *Organic Geochemistry* 31, 101–115.

Wong, C.S., Sanders, G., Engstrom, D.R., Long, D.T., Swackhamer, D.L., and Eisenreich, S.J. (1995) Accumulation, inventory, and diagenesis of chlorinated hydrocarbons in Lake Ontario sediments. *Environmental Science and Technology* 29, 2661–2672.

CHAPTER 6

CHEMICAL FINGERPRINTING OF HYDROCARBONS

Scott A. Stout, Allen D. Uhler, Kevin J. McCarthy, and Stephen Emsbo-Mattingly

6.1 INTRODUCTION

Petroleum-, coal-, and combustion-derived hydrocarbons are the most frequently discovered chemicals of concern at contaminated sites. Careful forensic investigation of these hydrocarbons in free phase products, soils and sediments, tissues, groundwater and surface water, and atmospheric particulates can yield a wealth of data necessary to determine the precise chemical nature of the contamination. Armed with this chemical 'fingerprinting' data, and drawing upon whatever historic (both regulatory and operational), geologic, hydrologic data that may be available, the forensic investigator is in a strong position to determine the origin(s) or source(s) of the hydrocarbon contamination. In some instances, this same information can help to constrain the most likely duration of time that has passed since the hydrocarbons were released into the environment.

This chapter focuses on the two critical aspects of hydrocarbon fingerprinting for environmental forensic applications. The first critical aspect lies in the *type and quality* of the chemical data to be utilized. In addressing this aspect we present a thorough review of the types of laboratory analytical methods required to adequately and defensibly fingerprint hydrocarbon contamination in environmental media. This first aspect also must include the quality control (QC) and quality assurance (QA) requirements for the chemical measurements, especially considering the rigors to which these data must be defended in the course of (potential or existing) litigation.

The second critical aspect of hydrocarbon fingerprinting addressed in this chapter is a review of those *controlling factors* that influence the interpretation of hydrocarbon fingerprinting data. In this regard, we detail three controlling factors that we have classified as *primary*, *secondary*, and *tertiary* (Table 6.1). Primary controls are those features of hydrocarbon fingerprinting data that are inherited directly from the parent form of the contamination, i.e., these features are 'genetic.' For instance, the chemical compositions of crude oils around the world are highly variable due to the conditions under which they have formed, migrated, and accumulated prior to their discovery and production (Tissot and Welte, 1984). These 'genetic' features allow forensic investigators to distinguish or correlate one crude oil from or with another crude oil (Page *et al.*, 1995; Wang *et al.*, 1999), or among petroleum products derived from different crude oils (Sauer and Uhler, 1994–1995; Kaplan *et al.*, 1997).

Secondary controls on hydrocarbon fingerprinting are those features that are introduced in the course of producing usable products from the raw crude oil or coal feedstocks. In other words, the anthropogenic effects of petroleum

Table 6.1

Critical aspects affecting hydrocarbon fingerprinting.

Type and Quality of Fingerprinting Data		
Primary Controls	**Secondary Controls**	**Tertiary Controls**
Petroleum composition Coal composition	Petroleum refining Manufactured gas production Coal carbonization Tar processing Combustion processes	Evaporation Dissolution in water Biodegradation Sorption to particles Uptake by Biota Photo-oxidation Commingling with background[a] Other fractionation processes

[a]Naturally occurring hydrocarbons.

refining or coal/oil processing (gasification, coking, liquifaction) can modify, overprint, or in some cases retain various primary features of the 'parent' hydrocarbon mixtures. For example, considering the array of refining processes and blending efforts that are necessary to yield modern automotive gasolines, few if any primary features are retained in gasoline that are inherited from the parent crude oil feedstock. By contrast, heavy fuel oil (bunker C) often retains many of the primary features of the parent crude oil feedstock. Thus, understanding the effects of refining or other anthropogenic processing (e.g., combustion or pyrolysis of fuels) can have on hydrocarbon fingerprinting is critical in most forensic investigations.

Finally, the tertiary controls on hydrocarbon fingerprinting are those features that are introduced or affected after the hydrocarbons are released into the environment, i.e., the effects of 'weathering' (Table 6.1). Evaporation, dissolution into water, sorption to soils, photo-oxidation, and biological degradation can each affect the distribution of hydrocarbons once released to the environment (Atlas *et al.*, 1981; Douglas *et al.*, 1996; McCarthy *et al.*, 1998; Rodgers *et al.*, 2000), and thereby alter its hydrocarbon fingerprint. It is the responsibility of the forensic investigator to recognize these effects and account for them in their investigation.

Thus, environmental forensic investigators must have knowledge of all three controls' effects on hydrocarbon fingerprinting. Later in this chapter each of these controls on hydrocarbon fingerprinting are reviewed. Our emphasis throughout this chapter lies in the hydrocarbons derived from petroleum, although coal and other forms of hydrocarbon contamination are introduced wherever possible. Finally, while the focus herein lies on hydrocarbon fingerprinting, it is important to remember that the other components of environmental forensics (as discussed throughout this book) can each contribute to a thorough and defensible forensic investigation.

6.2 ANALYTICAL ASPECTS OF HYDROCARBON FINGERPRINTING

Petroleum and other complex mixtures of hydrocarbons such as coal tars and coal distillates or combustion- and pyrolysis-related by-products are the most frequently discovered chemicals of concern at contaminated sites. In conventional remedial investigation and feasibility studies (RI/FS) carried out at hydrocarbon-contaminated sites, one of the primary objectives is to determine the 'nature and extent' of the contamination. In strictly regulatory driven assessments, e.g., Comprehensive Environmental Response, Compensation, and Liability Act of 1980 (CERCLA) or Resource Conservation and Recovery Act of 1976 (RCRA) type investigations, only a limited number of parameters and chemicals of concern are typically measured and used to determine the nature and extent of hydrocarbon contamination (Uhler *et al.*, 1998–1999). Examples of these measurements include total petroleum hydrocarbon (TPH), concentrations of water-soluble benzene, toluene, ethyl-benzene, and *o*-, *m*-, and *p*-xylenes (BTEX), and the 16 Priority Pollutant polycyclic aromatic hydrocarbons (PAHs). As described below, these compliance-driven measurements – while adequate for gross descriptions of the types of contaminants found at a site (e.g., TPH or BTEX above or below some regulatory action limit) – are largely insufficient to address the fundamental questions raised in an environmental forensics investigation. As such, one of the primary goals of an environmental forensic investigation at a hydrocarbon-contaminated site or following a spill/release incident is a detailed understanding of the precise nature of in-place hydrocarbon contamination. Achieving this goal requires that the appropriate chemical fingerprinting data be obtained, in a defensible manner.

6.2.1 OVERVIEW OF SELECTED STANDARD EPA ANALYSES

In the last 20 years, chromatographic techniques – particularly gas chromatography (GC) using detectors as simple as flame ionization (GC/FID) and as sophisticated as mass spectrometers (GCMS) – have evolved for the measurement of purgable (e.g., volatile) and extractable (e.g., semi-volatile) organic compounds in environmental media. In the regulatory environment, the US Environmental Protection Agency (EPA) has developed and codified a series of gas chromatographic techniques to measure industrial chemicals of concern using GC methods. Those most frequently cited are the EPA's so-called SW-846 Methods (EPA, 1997). Those of greatest interest to the environmental forensic chemist include

- *Method 8015B, Non-Halogenated Organics Using GC/FID.* This method is used to measure, among other things, total petroleum hydrocarbons (TPH), particularly TPH as diesel (TPH_d). This gas chromatography with flame ionization detection (GC/FID) technique often is used for gross petroleum product identification.
- *Method 8260, Volatile Organic Compounds by Gas Chromatography/Mass Spectrometry (GC/MS).* This purge-and-trap GCMS method is designed to measure a limited number of volatile-range hydrocarbons including the petroleum-related BTEX; TPH as gasoline (TPH_g) can be determined using this technique.
- *Method 8270, Semivolatile Organic Compounds by Gas Chromatography/Mass Spectrometry (GC/MS).* This powerful GCMS method was designed as a multiresidue analytical technique for the quantification of a broad range of industrial chemicals. The target compound list for the standard method includes over

100 well-known chemicals of potential concern. Of interest to the forensic chemist is the inclusion of the 16 so-called Priority Pollutant PAHs as target compounds.

- *Method 8310, Polynuclear Aromatic Hydrocarbons.* This high performance liquid chromatography (HPLC) method was designed for the low level measurement of the 16 Priority Pollutant PAHs. The method is used less frequently than Method 8270, because of the greater selectivity of the mass spectrometric method.

In addition to these standard chromatographic techniques, there have been two historic wet chemical methods used for estimating total hydrocarbons in water and soil/sediment that the environmental forensics investigator might encounter: EPA Method 413.1, *Total Recoverable Oil and Grease* and EPA Method 418.1, *Total Recoverable Petroleum Hydrocarbons by Infra-Red Spectroscopy* (EPA, 1983). In May 1999, these two methods were effectively replaced by EPA Method 1664, *N-Hexane Extractable Material (HEM; Oil and Grease) and Silica Gel Treated N-Hexane Extractable Material (SGT-HEM; Non-polar Material) by Extraction and Gravimetry* (EPA, 1999). In these simple methods, EPA uses the term 'total petroleum hydrocarbons' (TPHs) to designate the substances that remain after the *n*-hexane extractable material is exposed to silica gel. These screening methods are rarely, if ever, used in a detailed forensic site assessment. The limitations of these standard methods are discussed in the following section.

The preponderance of hydrocarbons that constitute crude oil, refined petroleum products and coal-derived liquids are largely non-polar volatile and semi-volatile compounds. From a practical standpoint, petroleum- and coal-derived distillates and wastes are ideally suited for detailed chemical characterization using modern high resolution capillary gas chromatography (Nordtest, 1991; Douglas and Uhler, 1993; Boehm *et al.*, 1998; Wang *et al.*, 1999). The standard EPA chromatographic methods described above are, indeed, excellent base techniques for the analysis of petroleum residues in environmental media. However, as discussed in the next section, there are fundamental restrictions to the standard EPA methods that limit their usefulness in forensic investigations. The solution to these impediments (a rational adaptation of the standard methods for detailed, forensic petroleum analysis) is described in a subsequent section.

6.2.2 LIMITATIONS OF STANDARD EPA METHODS OF ANALYSIS IN FORENSIC INVESTIGATIONS

Conventional RI/FS site investigations or related due diligence investigations, carried out under CERCLA or RCRA, require the use of standard EPA Methods of chemical analysis for the measurement of chemicals of concern (McHugh, 1997). Herein lies a fundamental barrier for environmental forensic investigators: standard EPA methods, exemplified by the popular SW-846 methodologies, are useful for measuring a wide variety of discrete industrial chemicals. However, with the exception of measurement of extractable hydrocarbons (e.g., EPA Method 1664) or total petroleum hydrocarbons (e.g., TPH_g by EPA Method 8260 or TPH_d by EPA Method 8015B), none of the standard EPA methods were developed for the detailed measurement of petroleum or petroleum-derived constituents (Douglas and Uhler, 1993). Of the more than 160 EPA Priority Pollutant volatile and semi-volatile organic compounds, only 20 are petroleum-type hydrocarbons that would be useful in hydrocarbon-contaminated site assessments. This list includes benzene, toluene, ethyl-benzene, xylenes, naphthalene, acenaphthylene, fluorene, anthracene, phenanthrene, fluoranthene, chrysene, pyrene, benzo[*a*]anthracene, benzo[*b*]fluoranthene, benzo[*a*]pyrene, indeno[1,2,3-*c,d*]pyrene, dibenzo[*a,h*]anthracene, and benzo[*g,h,i*]perylene (Sauer and Boehm, 1991). Only half of these compounds are found in significant concentrations in petroleum- or coal-derived products and wastes. Even together, these chemicals alone are of virtually no forensic benefit because they reveal virtually nothing regarding the primary, secondary, or tertiary aspects (Table 6.1) of the hydrocarbon contamination, which is essential to the environmental forensics investigator.

A listing of the standard EPA analytical methods often used in conventional assessments for measuring petroleum-, coal-derived, or combustion-related hydrocarbons is presented in Table 6.2. This table identifies the potential use of the method, and summarizes some of the limitations of these measurements in environmental forensic investigations.

The fundamental shortcoming with virtually every conventional EPA method of analysis when used for measuring hydrocarbons in environmental media is a lack of detailed measurement of those chemicals known to comprise these complex mixtures. However, using the knowledge of the composition of hydrocarbons of environmental concern – petroleum- and coal-derived liquids, wastes, and combustion products – environmental forensics chemists have adapted the standard EPA methods of analysis to yield the necessary data to support detailed forensic investigations. These adaptations, employed in a logical, tiered fashion, are the key to a rational environmental forensic chemistry investigation. As an important sidebar, EPA SW-846 guidelines allow flexibility in the deployment of the 'standard' analytical methods. While most commercial laboratories are not interested in modifying 'production line' chemistry methods, more flexible and experienced laboratories have experience in altering standard methods to meet project goals without violating the standard method guidelines. This means that, properly planned, most of the modified methods described below can be used to support both forensic and regulatory monitoring.

Table 6.2

Standard EPA methods of analysis, applications, and limitations in environmental forensic investigations.

Standard Method	**Potential Application**	**Limitations for Environmental Forensics**
EPA 1664 n-*Hexane Extractable Material (HEM: Oil and Grease) and Silica Gel Treated* n-*Hexane Extractable Material (SGT-HEM; Non-Polar Material) by Extraction and Gravimetry*	■ Screening tool for extractable hydrocarbons	■ Measures only total extractable material ■ Subject to interferences from naturally occurring hydrocarbons and particulates ■ High detection limits ■ No quantitative information
EPA 413.1 *Total Recoverable Oil and Grease*		
EPA 418.1 *Total Recoverable Petroleum Hydrocarbons by Infra-Red Spectroscopy*	■ Measurement of total petroleum hydrocarbons (TPH)	■ Subject to multiple interference from co-extracted natural organic compounds or other anthropogenic contaminants ■ False positive and negatives common ■ High detection limits ■ No quantitative information
EPA 8015B *Nonhalogenated Organics using GC/FID*	■ Measurement of TPHs ■ Measurement of selected industrial chemicals ■ Product identification	■ Sub-optimal GC conditions lead to poor chromatography; many petroleum products appear similar when analyzed under such conditions ■ False positives common due to presence of biogenic material, other contaminants ■ Suffers from mass discrimination unless stringent chromatographic conditions are established
EPA 8260 *Volatile Organic Compounds by Gas Chromatography/Mass Spectrometry (GC/MS)*	■ Measurement of benzene, toluene, ethylbenzene, xylenes (BTEX) hydrocarbons ■ Measurement of selected industrial chemicals	■ BTEX only; misses 100 + important gasoline range hydrocarbons and automotive fuel additives ■ Provides little/no diagnostic source information, e.g., identification of gasoline, aviation gas, kerosene often confused
EPA 8270 *Semivolatile Organics by Gas Chromatography/Mass Spectrometry (GC/MS)*	■ Measurement of 16 Priority Pollutant PAHs ■ Measurement of selected industrial chemicals	■ Ignores the most important petrogenic and pyrogenic PAHs and biomarkers ■ Provides little/no diagnostic source information ■ Detection limit limitation for low-concentration but diagnostically important target compounds

6.2.3 A TIERED ANALYTICAL APPROACH TO HYDROCARBON FINGERPRINTING

The identification of petroleum-, coal-, or combustion-derived hydrocarbons, whether as free products or as contaminants in soil and sediment, water, tissue, or air-borne particulates, can best be carried out in a tiered fashion (Uhler *et al.*, 1998–1999; Wang *et al.*, 1999). The objective of a tiered analytical approach is to gather sufficient detail concerning the composition of the contaminant(s) under investigation so that the forensic questions of concern pertinent to the site under investigation can be answered. The tiered approach, summarized in Figure 6.1, gives the environmental forensic investigator the flexibility to gather as little, or as much, information as necessary to address site- or incident-specific questions about the nature and extent, and ultimately source(s), of hydrocarbon contamination. The utility of the methods recommended in the tiered approach is summarized in Table 6.3.

The development of the tiered approach presented in this chapter rests on understanding the composition of petroleum-, coal-, and combustion-derived hydrocarbons, and the chromatographic behavior of the major and minor constituents of each of these assemblages. The progression of analytical techniques used at each step of the tiered approach focuses on the detailed analysis of particular hydrocarbon boiling point ranges (e.g., volatile-range hydrocarbons that comprise light distillates, or semi-volatile range hydrocarbons that comprise middle and residual distillates of petroleum or coal liquids) and particular classes of chemical compounds that comprise those fractions. A typical tiered approach for assessments involving unknown hydrocarbon mixtures will differ from one in which the nature of the contamination is well known. For example, a site survey for unknown hydrocarbon contamination will require a Tier 1 screening method that provides a look at the whole range of potential hydrocarbon products, utilizing most of the methods depicted in Figure 6.1. Sometimes, even when a site's contamination is 'known', a Tier 1 analysis is still prudent as it can either confirm, reveal, or refute the presence of different types of contamination.

6.2.3.1 Tier 1 – Total Hydrocarbon Characterization

Tier 1 analysis involves a GC/FID screening method that is based upon a modified EPA Method 8015B method. The results of this include measurement of the C_8+ TPH content of any matrix and a screening level 'fingerprint,' i.e., a chromatogram that shows the boiling point range and, hence, the hydrocarbon product makeup. Chromatographic 'fingerprints' can be especially useful to the forensic investigator because almost all hydrocarbon assemblages – crude oil, petroleum distillates, coal-derived liquids, and their combustion and

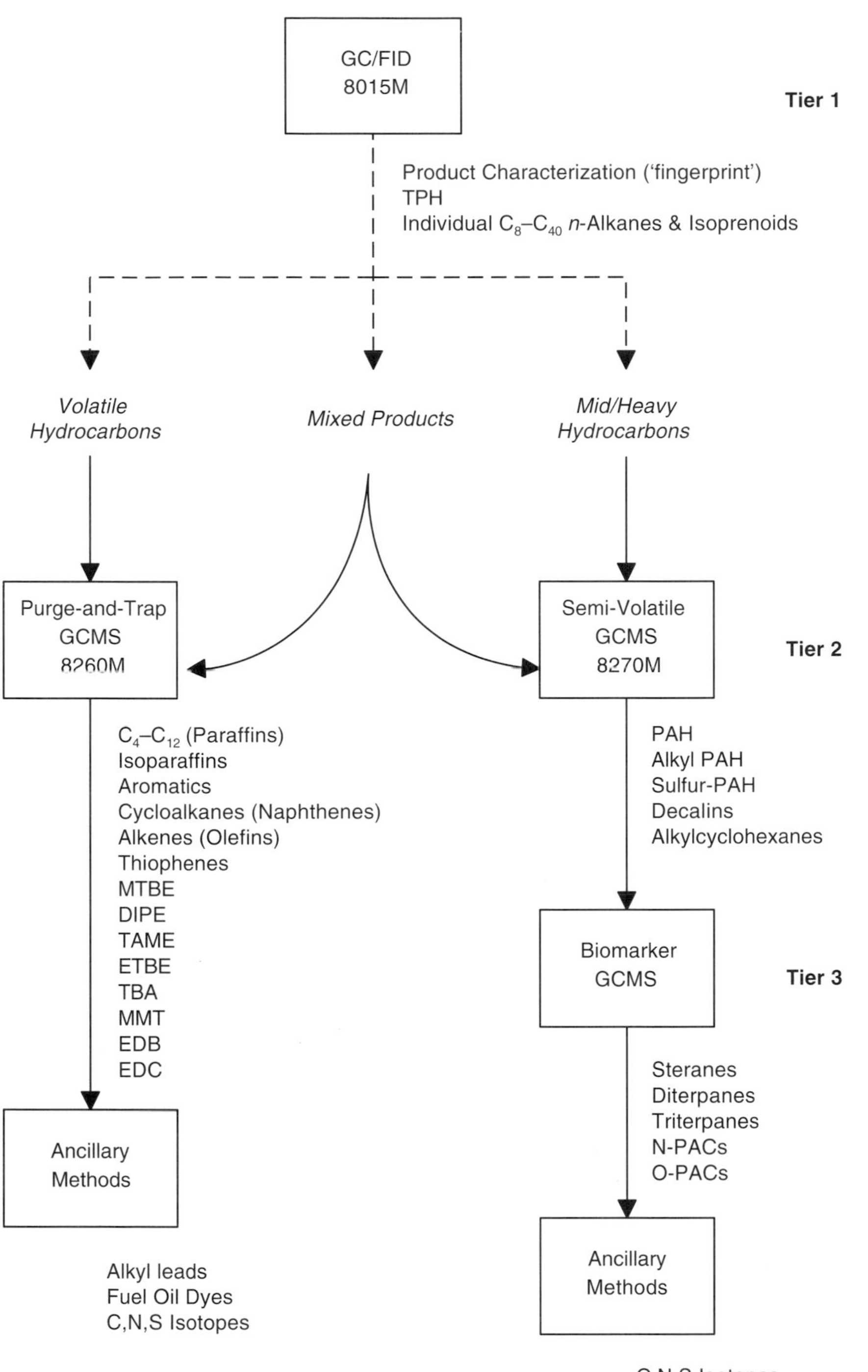

Figure 6.1

Tiered analytical approach to hydrocarbon fingerprinting.

Table 6.3

Useful analytical methods to support environmental forensic investigations.

Measurement Method	Target Compounds	Utility for Environmental Forensics
Modified EPA Method 8015: Measurement of total petroleum hydrocarbon (TPH) product identification screen and measurement of individual saturated alkanes by high resolution gas chromatography with flame ionization detection (GC/FID)	■ Total extractable hydrocarbons ■ Total petroleum hydrocarbons ■ C_8 to C_{44} normal and branched-chain hydrocarbons ■ High resolution product fingerprints	■ Accurate determination of total extractable hydrocarbons and petroleum in the light distillate to residual petroleum product range ■ Development of diagnostic indices (e.g., pristine, phytane) ■ Accurate product identification in the light distillate to residual petroleum product range ■ Evaluation of weathering state of petroleum product(s) identification in the light distillate to residual petroleum product range

Modified EPA Method 8260:
 	
Measurement of a broad range of volatile petroleum hydrocarbons by purge-and-trap GCMS
	■ Measurement of C_5 to C_{12} hydrocarbons and related chemicals
	  ■ Paraffins
	  ■ Isoparaffins
	  ■ Aromatics
	  ■ Naphthenes
	  ■ Olefins
	  ■ Oxygenates
	  ■ Selected additives
		■ Product identification and differentiation in the light distillate range
		■ Evaluation of weathering
		  ■ Evaporation
		  ■ Water-washing
		  ■ Biodegradation
		■ Presence/absence of important automotive gasoline blending agents and additives

Modified EPA 8270:
 	
Measurement cyclic hydrocarbons, polycyclic aromatic hydrocarbons (PAHs), heteroatomic PAHs, and biomarker compounds by high resolution GCMS by selected ion monitoring (SIM)
	■ Measurement of petroleum, coal, and combustion-derived PAH
	  ■ Alkyl homologues of parent PAH
	  ■ S-, N- containing PAHs
	  ■ Sterane, triterpane and other selected biomarkers
		■ Detailed chemical indices used for evaluation of
		  ■ Product identification and differentiation
		  ■ Product mixing and allocation
		  ■ Weathering
		  ■ Evaporation
		  ■ Water washing
		  ■ Biodegradation
		  ■ Petroleum vs. coal vs. combustion sources

Modified EPA 8270: High resolution GCMS fingerprint of non-aqueous phase liquid (NAPL) in the C_4 to C_{44} hydrocarbon range	■ Broad boiling point range GCMS analysis of petroleum, coal-derived, or other NAPL	■ 'Snapshot' view of entire environmentally important hydrocarbon boiling point range ■ Allows investigator to visualize mixing of different hydrocarbon products ■ GCMS allows for the identification of specific peaks in the chromatogram
Ancillary measurements	■ C, H, S stable isotopes ■ Lead speciation by GCMS ■ Fuel dyes ■ Density, viscosity ■ New methods (GCIRMS, GC × GC, etc.)	■ Source/manufacturer differentiation ■ Age/production period classification ■ Data for fate and transport modeling ■ Useful data for mix modeling

waste products – have distinctive chromatographic signatures (Figure 6.2). Although weathering processes (described later in this chapter) can greatly influence the appearance of the chromatographic signature of hydrocarbon contaminants, more often than not, distinctive features of the contaminants of concern can be seen in the GC/FID trace. In addition to producing chromatographic 'fingerprints', Tier 1 analysis can involve quantitation of any individual compounds that can be chromatographically separated and measured from the GC/FID chromatographic trace, e.g., the normal alkanes and selected acyclic isoprenoids (e.g., pristane and phytane; Table 6.4).

The Tier 1 GC/FID methodology is a straightforward adaptation of EPA Method 8015B, *Non-Halogenated Organics Using GC/FID*. The most appropriate chromatography column for forensics purposes is narrow bore (0.25 mm internal diameter (i.d.)) fused silica coated with a 0.25 μm 100% methylsilicone cross-linked stationary phase. Minimum column length for forensic purposes is 30 m, although longer columns, e.g., 60 m, will afford even better resolution and 'fingerprints'. The key to the successful application of this

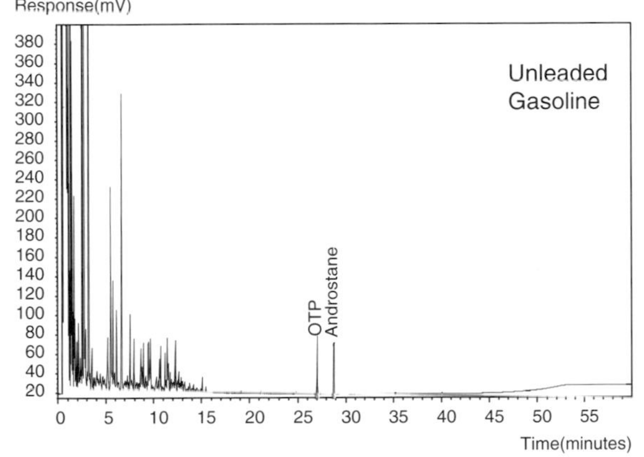

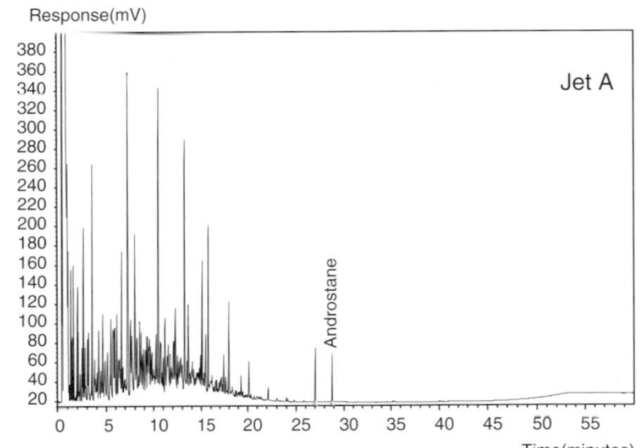

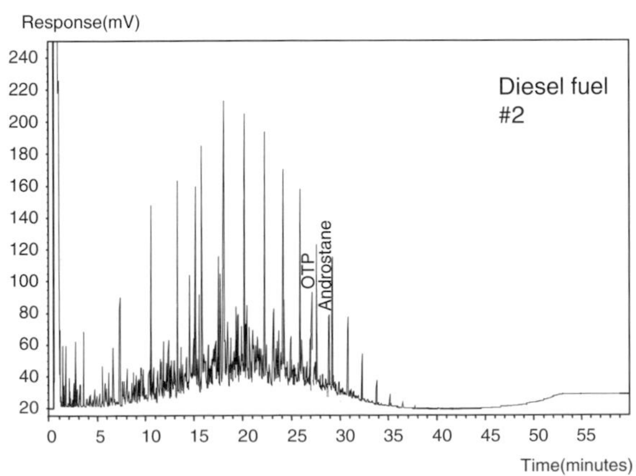

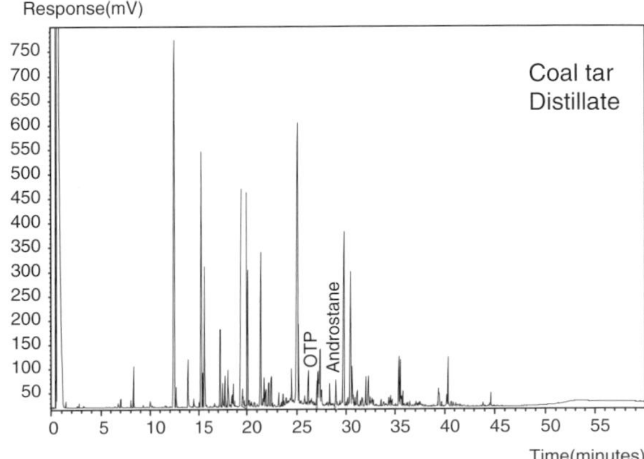

Figure 6.2
Sample gas chromatograms (FID) for a crude oil, selected petroleum distillates and a coal-derived liquid.

Table 6.4

Normal and acyclic isoprenoid hydrocarbons measured by high resolution GC/FID in environmental forensic investigations.

Target Compound	Abbreviation
Octane	n-C_8
Nonane	n-C_9
Decane	n-C_{10}
Undecane	n-C_{11}
Dodecane	n-C_{12}
2,6-Dimethylundecane	IP13
2,6,10-Trimethylundecane	IP14
Tridecane	n-C_{13}
2,6,10-Trimethyldodecane	IP15
Tetradecane	n-C_{14}
2,6,10-Trimethyltridecane	IP16
Pentadecane	n-C_{15}
Hexadecane	n-C_{16}
Norpristane (2,6,10-Trimethylpentadecane)	NP
Heptadecane	n-C_{17}
Pristane (2,6,10,14-Tetramethylpentadecane)	Pr
Octadecane	n-C_{18}
Phytane (2,6,10,14-Tetramethylhexadecane)	Ph
Nonadecane	n-C_{19}
Eicosane	n-C_{20}
Heneicosane	n-C_{21}
Docosane	n-C_{22}
Tricosane	n-C_{23}
Tetracosane	n-C_{24}
Pentacosane	n-C_{25}
Hexacosane	n-C_{26}
Heptacosane	n-C_{27}
Octacosane	n-C_{28}
Nonacosane	n-C_{29}
Triacontane	n-C_{30}
Hentriacontane	n-C_{31}
Dotriacontane	n-C_{32}
Tritriacontane	n-C_{33}
Tetratriacontane	n-C_{34}
Pentatriacontane	n-C_{35}
Hexatriacontane	n-C_{36}
Heptatriacontane	n-C_{37}
Octatriacontane	n-C_{38}
Nonatriacontane	n-C_{39}
Tetracontane	n-C_{40}

method is (1) optimal operation of the GC inlet (McCarthy and Uhler, 1994), which will minimize mass discrimination of heavier hydrocarbons (e.g., $> C_{30}$), and (2) utilization of a very slow GC oven temperature program, to facilitate optimal resolution of close-eluting compounds. GC oven profiles that can provide forensic-quality gas chromatographic 'fingerprints' typically have initial temperatures of 30–35°C, and oven ramp rates of 6°C per minute or less. Total GC run times of about 1 hour are typical for these analyses. Appropriate calibration standard concentrations range from approximately 1 μg/mL to 100 μg/mL per individual component. Using this methodology, method detection limits (MDLs) for individual components are approximately 1 μg/L and 0.1 mg/kg (dry weight) for water and soils, respectively. MDLs for C_8+ TPH are approximately 100 μg/L and 5 mg/kg (dry weight) for water and soil/sediment, respectively.

Strategically, the results of the Tier 1 analysis guide the forensics investigator toward logical next steps in an analytical program. If the Tier 1 analysis reveals only the presence of volatile-range hydrocarbons such as gasoline, then only detailed purge-and-trap GCMS is necessary as part of a Tier 2 measurement plan (Figure 6.1). Similarly, if the Tier 1 analysis reveals only the presence of middle or residual distillates (e.g., fuel oils) or coal-derived liquids, the next steps in the analytical program would focus on semi-volatile GCMS characterization of the samples.

6.2.3.2 Tier 2 – Volatile Hydrocarbon Fingerprinting

Light refined petroleum products such as automotive gasoline, aviation gas, various refinery intermediates (e.g., straight run gasoline or naphthas), and water-soluble fractions of many hydrocarbon mixtures are comprised largely of hydrocarbons in the C_4 to C_{12} range. These products can be composed of a wide range of compounds, the most common being those defined by the so-called 'PIANO' groupings: *p*araffins, *i*soparaffins, *a*romatics, *n*aphthenes, and *o*lefins. Fortunately, chemical standards containing the most abundant PIANO chemicals of concern to the forensic investigator are available commercially, and can be analyzed by Tier 2 purge-and-trap GCMS (Table 6.5). Importantly, this extended list of target compounds includes volatile compounds of regulatory concern – benzene, toluene, ethyl-benzene, and *o*-, *m*-, and *p*-xylenes. Thus, using methods and analyte lists such as those described here allows the forensic site investigator to gather information for both regulatory and forensic applications.

Automotive gasolines can or have in the past contained various additives or blending agents, e.g., ether or alcohol oxygenates such as methyl-*tert*-butyl ether, ethyl-*tert*-butyl ether, *tert*-amyl-methyl ether, and *tert*-butyl alcohol (MTBE, ETBE, TAME, and TBA, respectively), ethylene dibromide (EDB) and ethylene dichloride (EDC), methylcyclopentadienyl manganese tricarbonyl (MMT; Gibbs, 1990). These chemicals can also be measured by purge-and-trap GCMS. In addition, with the evolving regulations on sulfur in gasolines, determination of volatile sulfur compounds (e.g., thiophenes) by purge-and-trap GCMS is often appropriate and useful from the forensic standpoint.

Implementation of a Tier 2 GCMS analysis typically involves the use of purge-and-trap concentration methods, e.g., straightforward adaptation of EPA Method 5035, *Closed-System Purge-and-Trap and Extraction for Volatile Organics in Soil and Waste Samples* and EPA Method 5030, *Purge-and-Trap for Aqueous Sample*

Table 6.5

Diagnostic volatile-range hydrocarbons useful for characterizing light refined products and water-soluble fractions of petroleum- and coal-derived materials.

Paraffins	Aromatics	Olefins
n-C_5 (Pentane)	Benzene	1-Pentene
n-C_6 (Hexane)	Toluene	2-Methyl-1-butene
n-C_7 (Heptane)	Ethylbenzene	*trans*-2-Pentene
n-C_8 (Octane)	*m*-Xylene	*cis*-2-Pentene
n-C_9 (Nonane)	*p*-Xylene	1-Hexene
n-C_{10} (Decane)	*o*-Xylene	*trans*-2-Hexene
n-C_{11} (Undecane)	Isopropylbenzene	2-Methylpentene-2
n-C_{12} (Dodecane)	Propylbenzene	*cis*-2-Hexene
	1-Methyl-3-ethylbenzene	2-Methyl-1-hexene
Isoparaffins	1-Methyl-4-ethylbenzene	*trans*-3-Heptene
Isopentane	1,3,5-Trimethylbenzene	*cis*-3-Heptene
2,2-Dimethylbutane	1-Methyl-2-ethylbenzene	*trans*-2-Heptene
2-Methylpentane	1,2,4-Trimethylbenzene	*cis*-2-Heptene
3-Methylpentane	*sec*-Butylbenzene	1-Octene
2,2-Dimethylpentane	1,2,3-Trimethylbenzene	*trans*-2-Octene
2,4-Dimethylpentane	2,3-Dihydroindene (Indane)	*cis*-2-Octene
2-Methylhexane	1,3-Diethylbenzene	1-Nonene
2,3-Dimethylpentane	1,4-Diethylbenzene	*trans*-3-Nonene
3-Methylhexane	1,3-Dimethyl-5-ethylbenzene	*cis*-3-Nonene
2,2,4-Trimethylpentane	1,2-Diethylbenzene	*trans*-2-Nonene
2,4-Dimethylhexane	1-Methyl-2-*n*-propylbenzene	*cis*-2-Nonene
2,3,4-Trimethylpentane	1,2-Dimethyl-4-ethylbenzene	1-Decene
2,3,3-Trimethylpentane	1,3-Dimethyl-2-ethylbenzene	
2,3-Dimethylhexane	1,2,4,5-tetramethylbenzene	**Chlorinated Compounds**
2-Methylheptane	1,2,3,5-tetramethylbenzene	Trichloroethane (TCE)
2,2-Dimethylheptane	1,2,3,4-tetramethylbenzene	Tetrachloroethylene (PCE)
2,2,4-Trimethylhexane	*n*-Pentylbenzene	
2,4-Dimethylheptane		**Sulfur Compounds**
2,6-Dimethylheptane	**Naphthenes**	Thiophene
2,3-Dimethylheptane	Cyclopentane	2-Methylthiophene
4-Methyloctane	Methylcyclopentane	3-Methylthiophene
2-Methyloctane	Cyclohexane	2-Ethylthiophene
2,2-Dimethyloctane	Methylcyclohexane	Benzothiophene
3,3-Dimethyloctane	*ctc*-1,2,4-Trimethylcyclopentane	
3-Methylnonane	*ctc*-1,2,3-Trimethylcyclopentane	**Gasoline Oxygenates**
	Isopropylcyclopentane	Tertiary butyl alcohol (TBA)
	1,1,4-Trimethylcyclohexane	Methyl tertiary butyl ether (MTBE)
	Ethylcyclohexane	Di-isopropyl ether (DIPE)
	ctt-1,2,4-Trimethylcyclohexane	Ethyl tertiary butyl ether (ETBE)
	ctc-1,2,4-Trimethylcyclohexane	Tertiary-amyl methyl ether (TAME)
	1,1,2-Trimethylcyclohexane	
	Isopropylcyclohexane	**Gasoline Additives**
	n-Butylcyclopentane	1,2-Dichloroethane (EDC)
		1,2-Dibromoethane (EDB)
		Methylcyclopentadienyl manganese tricarbonyl (MMT)

for soils/sediments and waters, respectively. For forensics purposes, the GCMS system should be operated per Method 8260 guidelines. Volatile hydrocarbon analysis is best accomplished using a 50 m or longer 0.32 mm i.d. Rtx-1 PONA® 60 m capillary column (Restek, Inc.) or equivalent, and a slow GC oven temperature program to facilitate optimal resolution of close-eluting compounds suggested. GC oven programs that can provide the resolution necessary for the analysis of the target compound list presented in Table 6.5 typically have initial temperatures of 30°C or lower, initial hold times of 10–15 minutes, oven ramp rates on the order of 5°C/min or less, and total run times of approximately 1 hour. GC run times should extend beyond the retention times of the C_5-substituted benzenes (Figure 6.3). This is beyond that where most standard gasoline range organics (GRO) analyses end. Using this methodology, MDLs for individual analytes are approximately 0.5 µg/L and 5 µg/kg (dry weight) for water and soil/sediment, respectively.

6.2.3.3 Tier 2 and Tier 3 – Semi-volatile Hydrocarbon Fingerprinting

The majority of petroleum-, coal-, and combustion-derived products encountered in the environment contain important, diagnostic hydrocarbons in the C_8 to C_{44} carbon range. The most notable classes of compounds in this range are the homo- and hetero-atomic polycyclic aromatic hydrocarbons (PAH; Tier 2 in Figure 6.1), and so-called 'biomarker' compounds (Tier 3 in Figure 6.1;

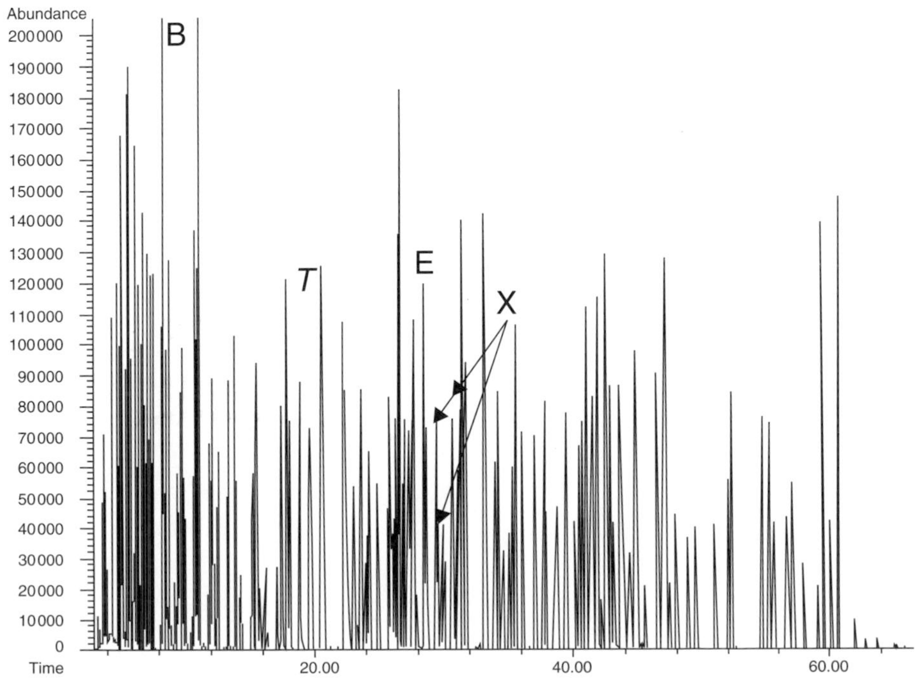

Figure 6.3
Chromatographic separation of over 100 volatile constituents in a PIANO standard as measured by GCMS (TIC).

which are described later in this chapter). In crude oil and most refined petroleum, these compounds are present at much lower concentrations than the more abundant normal and branched alkanes that can easily be detected by GC/FID techniques. Once entrained in environmental media, the concentrations of PAHs and biomarkers are often in the parts-per-million to low parts-per-billion range. In order to detect and measure these diagnostic, semivolatile hydrocarbons in environmental samples, the forensic investigator is best served using GCMS techniques, with the mass spectrometer operated in the selected ion monitoring (SIM) mode.

Table 6.6 presents a list of target polycyclic aliphatic and aromatic compounds that should be considered in Tier 2 analysis of hydrocarbons. This compilation includes parent and alkyl-substituted PAHs with demonstrated diagnostic utility in forensic investigations (Boehm and Farrington, 1984; Colombo *et al.*, 1989; Sauer and Uhler, 1994–1995). Other compounds of diagnostic value

Table 6.6

Diagnostic polycyclic aliphatic and aromatic hydrocarbons useful in advanced chemical characterization of petroleum-, coal-, and combustion-derived hydrocarbons.

PAHs and Alkyl PAHs		Individual PAH Isomers
Decalin	Fluoranthene	1-Methyldibenzothiophene
C_1-Decalins	Pyrene	3-Methyldibenzothiophene
C_2-Decalins	C_1-Fluoranthenes/Pyrenes	2-Methyldibenzothiophene
C_3-Decalins	C_2-Fluoranthenes/Pyrenes	4-Methyldibenzothiophene
C_4-Decalins	C_3-Fluoranthenes/Pyrenes	1-Methylphenanthrene
Naphthalene	Benzo[a]anthracene	2-Methylphenanthrene
C_1-Naphthalenes	Chrysene	3-Methylphenanthrene
C_2-Naphthalenes	C_1-Chrysenes	4-Methylphenanthrene
C_3-Naphthalenes	C_2-Chrysenes	9-Methylphenanthrene
C_4-Naphthalenes	C_3-Chrysenes	2-Methylanthracene
Acenaphthene	C_4-Chrysenes	Retene
Acenaphthylene	Benzo[*b*]fluoranthene	Cadalene
Biphenyl	Benzo[*k*]fluoranthene	5-Methylchrysene
Dibenzofuran	Benzo[*a*]pyrene	
Fluorene	Benzo[*e*]pyrene	**Surrogate compounds**
C_1-Fluorenes	Perylene	Naphthalene-d_8
C_2-Fluorenes	Indeno[1,2,3-c,d]pyrene	Phenanthrene-d_{10}
C_3-Fluorenes	Dibenzo[*a*,*h*]anthracene	Chrysene-d_{12}
Dibenzothiophene	Benzo[*g*,*h*,*i*]perylene	
C_1-Dibenzothiophenes	Benzo[*j*,*k*]fluoranthene	**Recovery standards**
C_2-Dibenzothiophenes	Benzo[*a*]fluoranthene	Fluorene-d_{10}
C_3-Dibenzothiophenes	Benzo[*b*]fluoranthene	Benzo[a]pyrene-d_{12}
C_4-Dibenzothiophenes		Acenaphthene-d_{10}
Phenanthrene		5β-cholane
Anthracene		
C_1-Phenanthrenes/Anthracenes		
C_2-Phenanthrenes/Anthracenes		
C_3-Phenanthrenes/Anthracenes		
C_4-Phenanthrenes/Anthracenes		

discussed later in this chapter, e.g., diamondoids, alkyl cyclohexanes, etc., can, of course, always be added to make a more robust or project-specific target analyte list.

The operating procedures and GCMS conditions for optimal detection, separation, and measurement of these semi-volatile target compounds are somewhat different from those presented in EPA Method 8270, *Semivolatile Organic Compounds by Gas Chromatography/Mass Spectrometry (GC/MS).* First, GC and MS operating conditions need to be optimized for separation and sensitivity. At a minimum, 30 m, 0.25 mm i.d., 0.25 μm film thickness 5% phenyl–95% methyl-silicone (or equivalent) capillary column should be used for separation. Optimal operation of the GC should be established to minimize mass discrimination of heavier hydrocarbons (e.g., $> C_{30}$) during sample injection (McCarthy and Uhler, 1994). A slow GC oven temperature program should be used to ensure adequate resolution of close-eluting compounds and to clarify alkyl PAH homologue patterns. Appropriate GC oven profiles have initial temperatures of 30–35°C, and oven ramp rates of 6°C per minute or less. Total GC run times of 1 to 2 hours are typical for these analyses. Using this methodology, MDLs for individual components are approximately 1 ng/L and 0.05 μg/kg (dry weight) for water and soil/sediment, respectively.

Next, chromatographic retention times for each target compound must be established under the appropriate GCMS operating conditions. The diagnostic and confirmatory ions for each target compound must be verified from GCMS analyses of authentic standards. Since only a handful of authentic standards exist for the literally hundreds of C_1 to C_4 alkylated isomers of PAH, characteristic ions, retention time windows, and homologue group patterns must be constructed based on carefully documented PAH and alkyl PAH chromatographic retention indices (Lee *et al.*, 1979), and then re-documented by the analysis of a well-characterized reference petroleum product such as Alaska North Slope Crude Oil (Wang *et al.*, 1994a). Alkyl homologues of PAH are quantified using the straight baseline integration method, versus response factors assigned from the parent PAH compound (Sauer and Boehm, 1995).

In Tier 3 GCMS analysis, the presence and relative distribution of biomarkers – e.g., acyclic isoprenoids, steranes, tricyclic, tetracyclic, and pentacyclic terpanes – can be measured in free phase products and environmental media by taking advantage of the knowledge of characteristic ion fragments and well-documented retention indices for each class of compounds (Stout *et al.*, 1999c). With this technique, both biomarker pattern groups and concentrations of important individual biomarkers can be measured (e.g., the prominent pentacyclic terpane 17α(H),21β(H)-hopane). Table 6.7 presents a representative list of biomarkers that can be measured by GCMS for forensic purposes.

Table 6.7

Useful biomarker compounds analyzed by GCMS in environmental forensic investigations.

Biomarker Name	Quantitation Ion
Terpanes	
C_{23} Tricyclic terpane	191
C_{24} Tricyclic terpane	191
C_{25} Tricyclic terpane	191
C_{26} Tricyclic terpanes	191
C_{28} Tricyclic terpane 1	191
C_{28} Tricyclic terpane 2	191
C_{29} Tricyclic terpane 1	191
C_{29} Tricyclic terpane 2	191
18α(H)-22,29,30-trisnorhopane (T_s)	191
17α(H)-22,29,30-trisnorhopane (T_m)	191
C_{30} Tricyclic terpane 1	191
C_{30} Tricyclic terpane 2	191
17α(H),21(β)H-28,30-bisnorhopane	191
17α(H),21(β)H-30-norhopane	191
18α(H)-30-norneohopane ($C_{29}T_s$)	191
17β(H),21α(H)-normoretane	191
18α(H) and 18β(H) oleanane	191
17α(H),21β(H)-hopane	191
17β(H),21α(H)-moretane	191
22S-17α(H),21β(H)-30-homohopane	191
22R-17α(H),21β(H)-30-homohopane	191
Gammacerane	191
22S-17α(H),21β(H)-30-bishomohopane	191
22R-17α(H),21β(H)-30-bishomohopane	191
22S-17α(H),21β(H)-30-trishomohopane	191
22R-17α(H),21β(H)-30-trishomohopane	191
22S-17α(H),21β(H)-30-tetrakishomohopane	191
22R-17α(H),21β(H)-30-tetrakishomohopane	191
22S-17α(H),21β(H)-30-pentakishomohopane	191
22R-17α(H),21β(H)-30-pentakishomohopane	191
Steranes	
13β,17α-diacholestane (20S)	217
13β,17α-diacholestane (20R)	217
5α,14β,17β-cholestane (20R)	218
5α,14β,17β-cholestane (20S)	218
5α,14α,17α-cholestane (20R)	217
5α,14β,17β,24-methylcholestane (20R)	218
5α,14β,17β,24-methylcholestane (20S)	218
5α,14α,17α,24-methylcholestane (20R)	217
5α,14α,17α,24-ethylcholestane (20S)	217
5α,14β,17β,24-ethylcholestane (20R)	218
5α,14β,17β,24-ethylcholestane (20S)	218
5α,14α,17α,24-ethylcholestane (20R)	217

The GCMS analysis of biomarkers can be carried out simultaneously with the analysis of PAHs described above. GCMS data are acquired by including appropriate diagnostic ions (Table 6.7) into the GCMS SIM method. Two types of calibration standards are appropriate for determination of biomarker concentrations. The first are retention time marker compounds (e.g., 5β(H)-cholane), against which measured biomarker retention times can be compared

(Sauer and Boehm, 1995). The second are discrete biomarker chemicals, available commercially, that can be used to generate response factors for individual compounds and compounds of similar structure.

6.2.3.4 Ancillary Fingerprinting Techniques

Both intrinsic bulk hydrocarbon properties (e.g., stable carbon isotope composition, relative proportions of chemical classes) and additives and blending agents introduced into petroleum distillates can prove to be useful chemical markers in some environmental forensic investigations. Several of the common and emerging ancillary measurement techniques are briefly discussed below.

6.2.3.4.1 Alkyl Lead Fuel Additives

Some petroleum distillates, particularly automotive and aviation gasoline, contain additives that can be of forensic value. In the US, pre-1996 (pre-1992 in California) automotive gasoline could contain varying amounts of alkyl lead compounds (Gibbs, 1990). The measurement of these compounds can be exploited to estimate the total lead burden in a fuel and hence, potentially estimate its age (Johnson and Morrison, 1996; Kaplan *et al.*, 1997; Stout *et al.*, 1999a) or deduce the source of the gasoline based on the distribution of the individual alkyl lead compounds (Kaplan and Galperin, 1996). The analysis of alkyl lead compounds is currently restricted to free products, since laboratory and field experiments have demonstrated that, in the absence of free phase gasoline, alkyl leads are strongly adsorbed to soils, precluding their efficient extraction and analysis (Mulroy and Ou, 1998).

The five individual alkyl lead compounds that were contained in alkyl lead additive packages (Gibbs, 1990) – tetramethyl lead (TML; m/z 253, 223), trimethylethyl lead (TMEL; m/z 253, 223), dimethyldiethyl lead (DMDL; m/z 267, 223), methyltriethyl lead (MTEL; m/z 281, 223), and tetraethyl lead (TEL; m/z 295, 237) – can be measured by GCMS in the selected ion monitoring mode following adaptations of EPA Method 8270. In the analysis of these compounds, free product can be diluted to 10 mg/mL in solvent, fortified with conventional Method 8270 internal standards, and analyzed by GCMS using standard splitless injection techniques. A multi-point calibration curve containing authentic standards of each of the alkyl lead target compounds should be developed prior to the analysis of field samples covering the concentration range of 5–500 μg/mL. The same gas chromatographic operating conditions, described above in Section 6.2.3.3, can be used to separate and measure these target compounds (Figure 6.4). The results of such an analysis – typically reported in units of micrograms of individual alkyl lead compound per milliliter of gasoline – can be easily converted to units of grams lead per gallon (glpg).

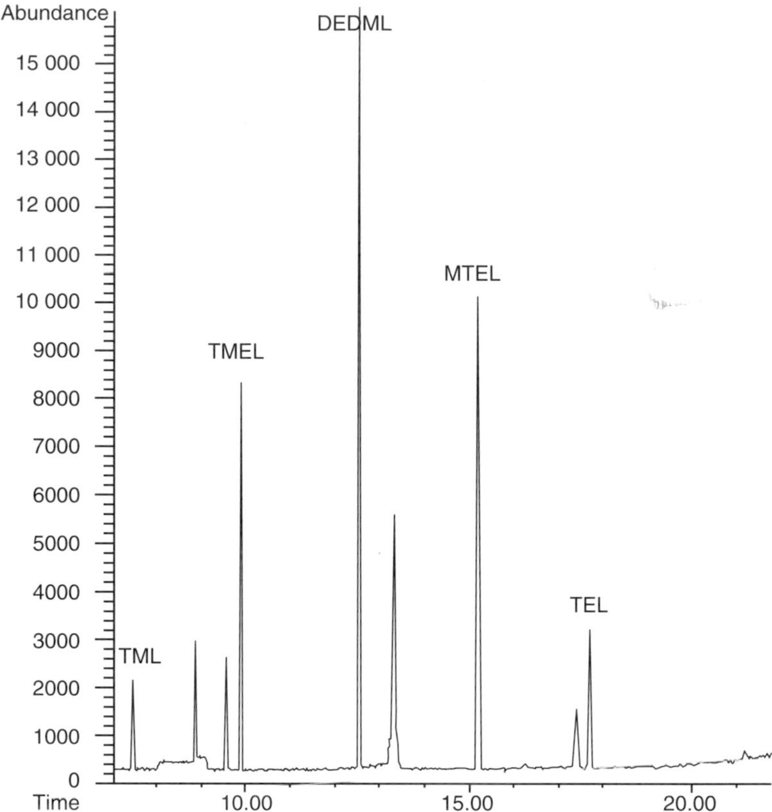

Figure 6.4
Distribution of five alkylated lead compounds (and internal standards) measured by GCMS (m/z *117 + 223 + 253 + 267 + 281 + 295).*

6.2.3.4.2 Stable Isotopes of Bulk Hydrocarbons

Hydrocarbons – be they petroleum-derived, coal-derived, or from recent biogenic material – have intrinsic ratios of the stable isotopes of the carbon and hydrogen atoms that comprise each molecule. As a bulk property, the stable isotope ratio for a given element in a hydrocarbon product or assemblage reflects the many different geochemical and biological processes to which the hydrocarbons were exposed during their formation, refining or processing, and environmental degradation/fractionation (Whittaker *et al.*, 1995). Measurement of stable isotopes has conventionally been applied to 'bulk' hydrocarbon fractions, e.g., aliphatic or aromatic hydrocarbons separately, and this bulk level of fingerprinting has proven useful in some investigations (Stout and Lundegard, 1998). The methodologies and applications of stable isotope measurements, on both bulk and individual hydrocarbons (see below), are presented in Chapter 5.

6.2.3.4.3 Compound-specific Stable Isotope Measurements

Bulk isotope measurements are most useful for situations in which the material measured is relatively pure, insensitive to alteration, and devoid of interferences. This is rarely the case in petroleum and other hydrocarbon mixtures. Instead petroleum and related contaminant contain hundreds or thousands of compounds, each of which may possess a unique isotopic signature and different origin. For this reason, chemists devised two strategies for measuring stable isotope ratios on individual compounds. The first strategy involves the physical isolation of target analyte during the sample preparation procedure, which is followed by a bulk measurement on the isolated compounds. This strategy is difficult to achieve and can require significant sample volumes. The more viable strategy relies on the coupling of gas chromatograph with an isotope ratio mass spectrometer (GCIRMS), which greatly increases the analytical specificity of the isolate at the instrument level using a capillary GC column.

The benefits of compound-specific isotope measurements are compelling. When coupled with sample preparation methods that isolate the aliphatic, aromatic, polar, and asphaltene constituents of an organic extract, individual hydrocarbon compounds can be isotopically measured (Mansuy *et al.*, 1997). This technique accurately identified the source of petroleum collected from bird feathers based on the carbon isotopic signature of *n*-alkanes produced during pyrolysis of the asphaltene fraction. Other workers have compared uncombusted and combusted plant and fuel materials using GCIRMS and concluded that the 2- and 3-ring PAHs exhibited greater source identification potential than heavier PAH, because 5- and 6-ring PAHs were more prone to isotopic alteration during pyrolysis (McRae *et al.*, 1997). O'Malley *et al.* (1996) used GCIRMS measurements of 4- and 5-ring PAHs to apportion the contribution of fire, car soot, and crankcase oil in coastal sediments. This study was not compromised by pyrolytic bias, because the source material was not secondarily combusted prior to release.

The GCIRMS strategy has also been used for volatile hydrocarbons. A classic application of the technique involved the isotopic composition of carbon in benzene, toluene, ethyl-benzene and xylenes (Kelley *et al.*, 1997). In this study, groundwater samples were used to identify one source by the depletion of ^{13}C and the other by both an enrichment in ^{13}C and the presence of an oxygenate. The mechanism of isotopic fractionation of these hydrocarbons was primarily threefold. With respect to the residual material, volatilization resulted in a slight depletion of ^{13}C, aerobic microbial degradation resulted in an enrichment of ^{13}C, and adsorption resulted in no significant change of ^{13}C (Aggarwal and Hinchee, 1991; Whittaker *et al.*, 1995; Harrington *et al.*, 1999).

The hydrogen isotope ratio of $^{2}H/^{1}H$ is interesting, because it exhibits the largest relative mass difference compared to all other elements. Like carbon, the use of GCIRMS has been used to identify the hydrogen isotopic ratio of specific compounds. Different sources of vanillin, ethanol, fatty acid methyl esters, hydrocarbons and sterols have been identified (Jarman *et al.*, 1998; Sessions *et al.*, 1999) demonstrated hydrogen isotope fractionation in

80 individual lipids by algae, bacteria, and higher plants, which revealed significant compound-specific differences among distinct species. To our knowledge, this technique has not yet been used for forensic purposes.

6.2.3.4.4 Fuel Dyes

Many finished automotive gasolines and diesel fuels have been augmented by retailers with pigmented dyes to differentiate among fuel grades or to identify fuels for intended end use (e.g., home heating fuel versus automotive diesel). There are commercially used dyes – red (azobenzen-4-azo-2-napthol), orange (benzene-azo-2-napthol), yellow (*p*-diethyl aminoazobenzene), and blue (1,4-diisopropyl aminoanthraquinone). In free products, these dyes can be quickly identified by thin layer chromatography (Kaplan and Galperin, 1996; Kaplan *et al.*, 1997). Identifying the presence, absence, or mixture of these dyes can be of great forensic utility in cases where one needs to differentiate between or among potentially similar fugitive petroleum products. One limitation of reliance on fuel dye data for forensic purposes is the fact that these dye compounds are environmentally unstable. Thus, the absence of dyes in a fugitive fuel does not necessarily mean that the compound(s) were not initially present in the fresh fuel.

6.2.3.4.5 Stable Lead Isotopes

Recent advances in the understanding of the industrial use of lead and its impacts on the stable lead isotope ratios of organo-lead compounds has led to the development of a model used to estimate the time of production of organo-lead compounds found as additives in automotive gasoline. Hurst (2000) argues that the change in the stable isotope ratios of lead found in environmental samples (e.g., free product and soil) is a function of the sources of lead-containing ore that were mined for commercial production of organo-lead compounds. This author has developed a model, termed ALAS (Anthropogenic Lead Archaeo-Stratigraphy), that relies upon a predictable temporal change in the proportion of two of lead's four naturally occurring isotopes – ^{206}Pb and ^{207}Pb – over time as US production of commercial lead ore shifted from mid-Precambrian/Tertiarty Cenozoic Era ores (with $^{206}Pb/^{207}Pb$ ranging from 1.02 to 1.22) to increased reliance on Mississippi Valley type ore deposits, with $^{206}Pb/^{207}Pb$ of about 1.3 from the late 1950s to the 1990s (Hurst, 1996, 1998, 2000). Using high precision thermal ionization mass spectroscopy, Hurst measured $^{206}Pb/^{207}Pb$ ratios in carefully collected and preserved environmental samples and petroleum, and derived his $^{206}Pb/^{207}Pb$ ALAS calibration curve. The curve has a reported accuracy of plus or minus 2 years between 1965 and 1990. In practice, measurement of the $^{206}Pb/^{207}Pb$ ratio in a free product hydrocarbon such as gasoline allows the investigator to date the age of production of the lead additive package in the fuel. Certain caveats to this methodology must be kept in mind; for example, this dating technique may be most useful for estimating the age of a gasoline discharged in a catastrophic, one-time release, but inappropriate for dating free product that arises from chronic leaks or mixing of different vintage leaded gasolines, because the resulting lead isotope ratios in a mixed product would be a composite of the organo-lead compounds from the different aged gasolines. The ALAS model is discussed further in Chapter 5.

Another useful forensic tool for sediment studies is the use of ^{210}Pb age-dating. This technique relies upon the comparison of the amounts of ^{210}Pb derived from natural decay of ^{238}U versus the excess ^{210}Pb derived from atmospheric deposition and water column (Bloom and Crecelius, 1987). Under constant sedimentation rates the amount of ^{210}Pb can be used to determine the age of the sediments, which can prove valuable in forensic investigations relying upon sediment cores (e.g., Prince *et al.*, 1995).

6.2.3.4.6 *Two-dimensional Gas Chromatography*

The final ancillary method to be discussed was recently described by Gaines *et al.* (1999), where its application was demonstrated for oil spill identifications. Two-dimensional GC (GC × GC) with FID detection provides another dimension to conventional chromatographic separation, greatly improving the resolution attainable. In this method, each analyte in the petroleum mixture is subject to two different separations achieved using two GC columns connected serially by a thermal modulator, the latter of which is heated and rotated in such a way that it desorbs, spatially compresses, and injects a portion of the first GC column eluent into the second GC column. The promising aspect of this technique lies in the potential to 'fingerprint' compounds that could not previously be resolved using GC or GCMS.

6.2.4 QUALITY ASSURANCE AND QUALITY CONTROL

Like any high quality laboratory work, analytical chemistry measurements carried out in support of environmental forensic investigations must be conducted within the framework of a defensible, robust quality assurance (QA) and quality control (QC) program. Beyond the obvious, practical reasons a laboratory should operate under a well-defined QA/QC program is the need for the utmost confidence in the quality of the data because of the likelihood that the data will be used in litigious inquiries. An environmental forensics investigator must be aware that, in a litigious setting, undocumented errors, omissions, or deviations are all potential fodder for undermining the otherwise defensible quality of an environmental data set. The best defense against this kind of attack is to ensure that the environmental forensic laboratory operates under strict QA/QC guidelines.

6.2.4.1 Quality Control

The quality of hydrocarbon fingerprinting data is an essential component of any environmental forensic investigation. In most litigious situations the data quality must be defended (long) before any forensic interpretations. No standard protocols exist for forensic applications (though they do for regulatory purposes) and therefore different laboratories employ different degrees of quality control (QC) in the course of sample collection, analysis, and reporting of forensic data. The practicalities of running a commercial laboratory often compete with QC of the data. Fingerprinting data to be used in a given forensic investigation should be analyzed in exclusive analytical batches, on the same analytical instrument and, to the extent possible, within the same analytical sequence. Whenever practical the data analyses (peak integrations) for a given batch of samples should be conducted by a single GCMS analyst with experience in hydrocarbon (volatiles, PAHs and biomarkers) pattern recognition. These steps, combined with the calibrations (initial and continuing), appropriate QC samples (procedural blanks, laboratory control samples), and replicate (duplicates or triplicates), each contribute to overall data quality.

6.2.4.2 Quality Assurance

A QA program assures laboratory management and project investigators that documented standards for the quality for facilities, equipment, personnel training, and work performance are being attained, and if not, to identify and report the areas that need improvement to meet those standards. A laboratory's QA program should be described in the laboratory's Quality Management Plan (QMP). The QMP should describe the laboratory's policy for management system reviews, quality control and data quality objectives, QA project plans, standard operating procedures, training, procurement of items and services, documentation, computer hardware and software, planning and implementation of project work, assessment and response, and corrective action and continuous improvement. A QA system should consist of a minimum of six components, namely:

1. A formal Work/QA Project Plan that describes all work, QA, and quality control activities associated with a project is developed for each study.
2. Up-to-date standard operating procedures (SOPs) that describe all technical activities conducted by the laboratory.
3. A program to ensure and document that all project personnel are fully trained and qualified to perform project activities before independent activities may begin.

Personnel training records should be maintained by the QA Unit and include records of qualifications, prior experience, professional training, and internal training procedures.

4. A documentation and records system that facilitates full sample and data tracking.
5. A quality assessment program for all projects, conducted through management system reviews, technical system audits, performance evaluation samples, data validation, laboratory inspections, and independent data audits. An independent QA Unit within a laboratory should conduct the latter two activities.
6. A continuous improvement program, facilitated through quality assurance audits, a formal corrective action program, and routine, laboratory-wide performance assessments and reviews.

In addition, all laboratory deliverables should receive an independent review that includes a quality assurance (QA) and technical component. The QA review should ensure that the data are (1) complete, (2) in compliance with the procedures defined in the QAPP, (3) in compliance with regulatory statutes, and (4) accurate and technically sound. The technical review should ensure that the results and interpretations are (1) objective, (2) defensible among peers, and (3) presented in a manner that conveys the conclusions clearly, often to non-expert decision makers.

6.3 PRIMARY CONTROLS ON THE COMPOSITION OF PETROLEUM

6.3.1 EFFECTS OF PETROLEUM'S GENESIS ON HYDROCARBON FINGERPRINTING

Once the appropriate and defensible chemical fingerprinting data are obtained (Section 6.2), the forensic investigator must next interpret those data, to the best of his/her ability, in an attempt to yield a technically sound conclusion that addresses questions raised in the investigation. A prerequisite to forming this conclusion is a thorough understanding of the composition of petroleum and petroleum-derived products. In this section, the primary controls on petroleum composition are discussed in order to provide a basis for understanding their utility in environmental forensic studies.

Petroleum occurs in sedimentary basins around the world as a gas (natural gas), a liquid (crude oil and condensate), and a solid (solid bitumen; Tissot and Welte, 1984). The focus in this chapter is on the primary controls on the composition of liquid petroleum, since it is most often encountered in environmental forensic investigations. The chemical composition of any particular crude oil is determined by the depositional and geologic history of the strata within that particular sedimentary basin. This history imparts certain primary or 'genetic' features to the petroleum that can be used to compare and contrast the chemical composition of (1) petroleum sources from different basins or fields and (2) petroleum products from different 'parent' petroleum.

Most crude oil contains tens of thousands of compounds that on a 'bulk' level can be broadly grouped as either hydrocarbons or non-hydrocarbons. The former consists of both aliphatic and aromatic hydrocarbons while the latter consists of various nitrogen, sulfur, and oxygen (NSO) or metal (e.g., V and Ni) containing compounds. The non-hydrocarbons include smaller molecules known as polars as well as larger, *n*-pentane insoluble asphaltenes. Crude oils from around the world vary widely in the relative abundance of aliphatic hydrocarbons, aromatic hydrocarbons, and non-hydrocarbons (Tissot and Welte, 1984; Figure 6.5). Much of this variation is the result of variable geologic histories of crude oils (described below). Unfortunately, 'bulk' hydrocarbon fraction data are not particularly useful for forensic investigations. The relative abundance of total aliphatic or aromatic hydrocarbons is simply not diagnostic or specific enough for forensic applications. Instead, the forensic investigation involving petroleum and petroleum products typically must rely on chemical fingerprinting petroleum at a 'molecular' level.

On a molecular level, the compositions of crude oils are infinite and therefore not easily depicted in a ternary or other simple diagram (as in Figure 6.5). In fact, it can be safely stated that no two crude oils are identical, i.e., on some level there is a degree of molecular variability. (It is usually the limitations of analytical precision that prohibit distinction among 'identical' crude oils.) The molecular composition of crude oil is largely dependent upon the conditions under which it was formed and existed prior to its discovery and production.

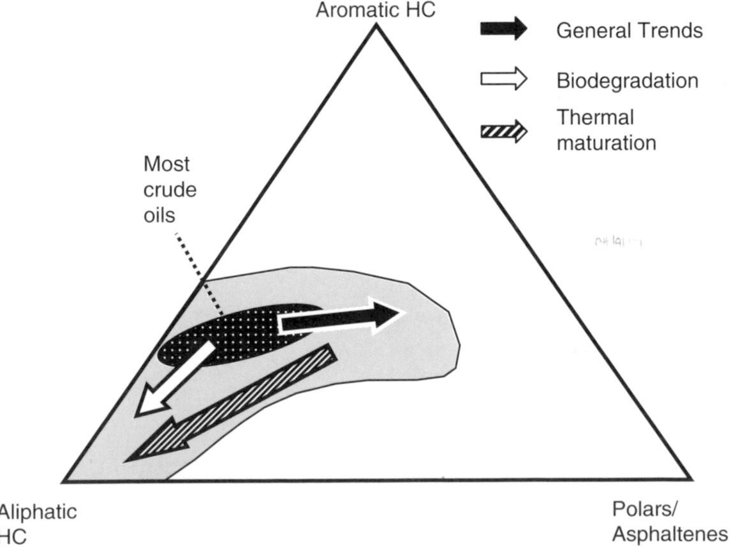

Figure 6.5

Ternary diagram showing gross compound class composition of crude oil.

In the terminology introduced in Table 6.1, these conditions can be broadly classified as the primary or 'genetic' features of the crude oil.

One important control on a crude oil's primary features is the nature of the oil's source rock organic matter type(s) (Tissot and Welte, 1984). Crude oil is generated from naturally occurring organic matter preserved in certain ancient rocks, i.e., *source rocks*. The types of ancient organic matter vary widely with geologic age, lithology, and the particular depositional environment (e.g., lake, ocean, delta, etc.) of the ancient sediments. Exploration geochemists have long understood the effect that the source rock organic matter type has on the molecular (Seifert and Moldowan, 1978a) and isotopic (Sofer, 1984) composition of crude oil (Mackenzie, 1984; Moldowan *et al.*, 1985). Environmental forensic investigators are in a position to use this knowledge in characterizing and comparing crude oil and petroleum products in environmental media to their suspected source(s) or to one another.

An example of a common primary feature of crude oil is the ratio of pristane to phytane (Pr/Ph), which has been linked to (1) the degree of anoxia (Didyk *et al.*, 1978) and (2) the organic matter type (Moldowan *et al.*, 1985) that existed in the ancient source rock sediments. While the secondary effects of refining and blending certainly must be considered (Table 6.1), it is not hard to envision distinct Pr/Ph ratios for, for example, diesel fuels distilled

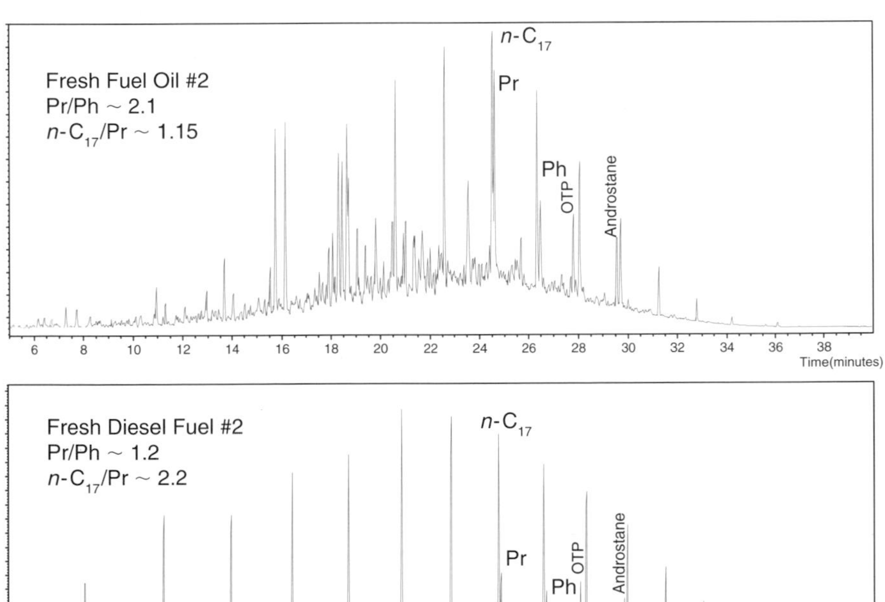

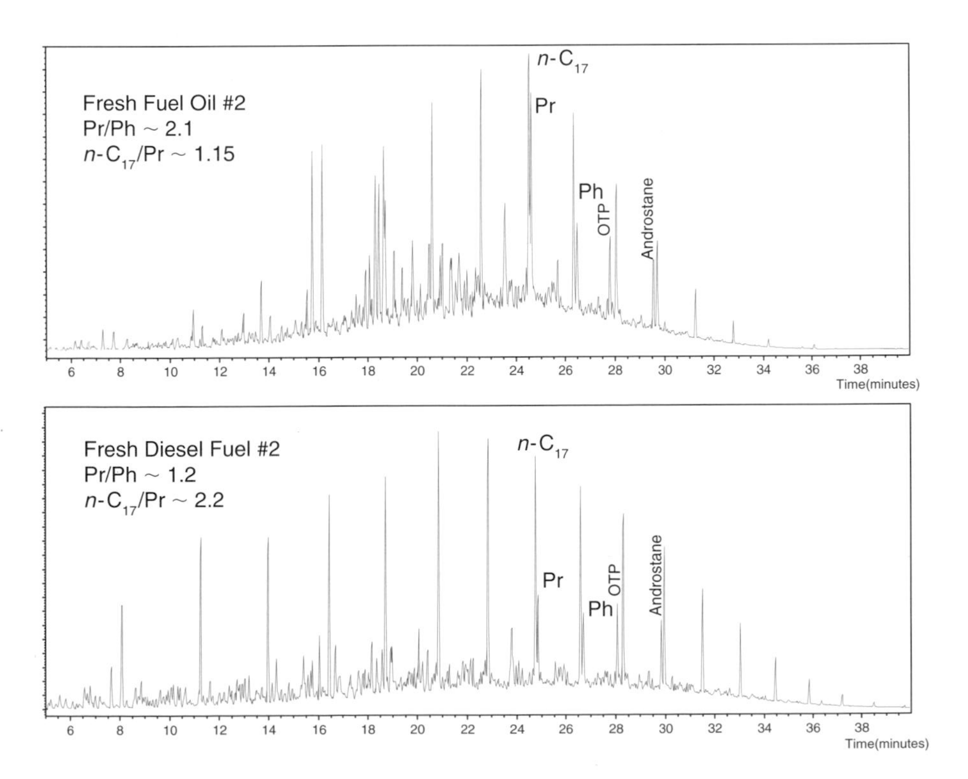

Figure 6.6
GC/FID of middle distillate fuels exhibiting characteristic features of different crude oil origin.

from different crude oils that had been generated from distinct source rock strata. Figure 6.6 demonstrates this for two unweathered middle distillate fuel samples that exhibit markedly different Pr/Ph ratios. The different Pr/Ph ratios probably stem from the primary differences between the parent crude oil stock(s) that gave rise to each.

Other important primary controls on a crude oil's molecular composition include the

1 thermal maturity of the source rock at the time the oil was expelled,
2 compositional changes during crude oil migration to a reservoir, and
3 changes upon reaching the reservoir.

In the first case, increased thermal maturity (heating) of the source rock generally yields increasing proportions of aliphatic hydrocarbons (Figure 6.6), which are normally of decreasing molecular size (this is why natural gas is often produced from very mature source rocks). Some hydrocarbons in crude oil are particularly sensitive to thermal rearrangement reactions (e.g., selected biomarkers; see below), which can impart highly diagnostic (primary) features to crude oils that were generated at different temperatures in the subsurface (Seifert and Moldowan, 1978a; Farrimond *et al.*, 1998). In the second case, migration of petroleum within geologic basins through certain *carrier beds* can induce a form of 'geochromatography,' a process in which there is a selective loss of higher boiling compounds or other fractionation processes that alter its molecular composition (Larter *et al.*, 1996). Finally, in the third case, after reaching a reservoir (but prior to its discovery and production) crude oil composition can further change due to, water-washing by meteoric water, deasphaltening, or biodegradation (Seifert *et al.*, 1984; Cassani and Eglinton, 1991). Thus, collectively the primary controls on crude oil invoke tremendous molecular variability that, when analyzed for and interpreted properly, can provide valuable information for the environmental forensic investigator.

6.3.2 PRIMARY MOLECULAR FEATURES OF PETROLEUM

Many excellent reviews of crude oil composition are already available (Kinghorn, 1983; Tissot and Welte, 1984; Speight, 1991), which allows this section to focus on those primary molecular features of crude oils that are most useful in forensic applications. Many environmental forensic investigations require careful comparison between samples in an effort to determine a 'positive' or 'negative' correlation between the samples. Because many of the primary molecular features of petroleum are relatively stable over environmental time-scales (years and decades), when properly interpreted, diagnostic chemical features can help establish a positive or negative correlation among samples (Douglas *et al.*, 1996). At the same time, this information may yield information allowing for the determination of a spilled crude oil's most likely source/origin (e.g., Eastern Gulf of Mexico versus West Texas). More importantly, while the secondary effects of refining and tertiary effect of 'post-release' weathering (Table 6.1; both of which are addressed later in this chapter) certainly must be considered, this same information often can be extended to assess and compare spilled petroleum products, which are at the center of many environmental forensic investigations. In the following sections, various primary molecular features of crude oil are discussed for each of the compound classes that occur in crude oils and petroleum products.

6.3.2.1 Straight and Branched Aliphatic Hydrocarbons

6.3.2.1.1 n-Alkanes

Aliphatic hydrocarbons comprise the bulk of most non-degraded crude oils (Tissot and Welte, 1984; Figure 6.5). The most prominent aliphatics in crude oil are the normal (straight-chained) paraffins. These compounds can contain between one (methane) and 60+ carbon atoms; measurements beyond about n-C_{45} are only available with special high temperature gas chromatographic columns. Petroleum fingerprinting typically relies upon *n*-alkanes in the C_5 to C_{45} range obtained using GC/FID or GCMS (m/z 57) techniques as described above (Table 6.4). In situations where weathering has not altered the *n*-alkane pattern significantly, certain diagnostic information can be revealed by the shape of the *n*-alkane 'envelope' or profile. Most crude oils exhibit relatively smooth uni- or bi-modal *n*-alkane profiles. The maximum (or most abundant) *n*-alkane within the profile is variable, e.g., varying with thermal maturity or organic matter type of the source rock (Murray *et al.*, 1994). Though certainly many exceptions exist, most crude oils exhibit an *n*-alkane profile with decreasing abundance with increasing carbon number. In refined petroleum products the *n*-alkane maximum is often a function of the distillation conditions or specifications of the product (e.g., diesel fuel #2 will exhibit an *n*-alkane maximum at a higher carbon number than kerosene).

The smoothness of the *n*-alkane profile in crude oils and petroleum products can also be diagnostic. It can be monitored by simple ratios (n-C_{15}/n-C_{16}) or by more sophisticated parameters, e.g., the carbon preference index (CPI) or odd–even preference (OEP). The latter focus on higher carbon number *n*-alkanes and are measured as:

$$\begin{aligned}\mathrm{CPI} = [&(C_{25}+C_{27}+C_{29}+C_{31}+C_{33})/(C_{26}+C_{28}+C_{30}+C_{32}+C_{34})\\ &+(C_{25}+C_{27}+C_{29}+C_{31}+C_{33})/(C_{24}+C_{26}+C_{28}+C_{30}+C_{32})]/2\end{aligned}$$

$$\mathrm{OEP} = (C_{21}+6C_{23}+C_{25})/(4C_{22}+4C_{24})$$

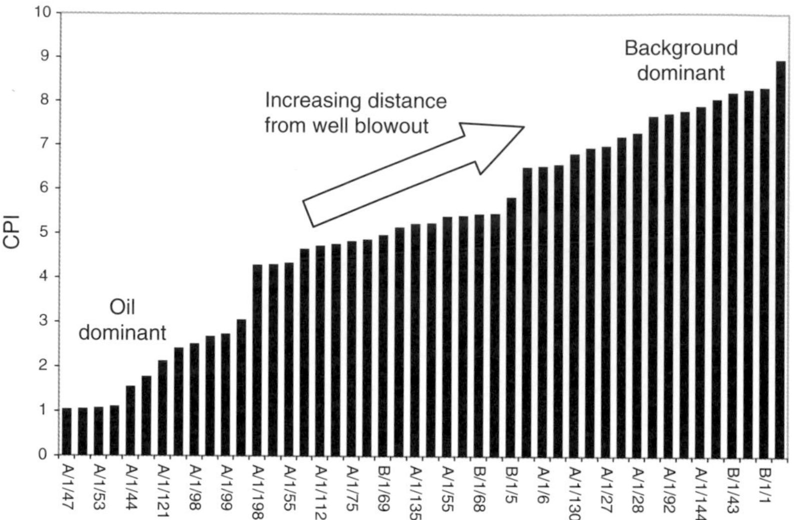

Figure 6.7
CPI as an indicator of oil contamination versus natural plant hydrocarbons in soil.

The CPI or OEP of an oil with a 'smooth' *n*-alkane envelope is ~1 whereas an oil with a strong odd-carbon dominance has a CPI > 1 (Tissot and Welte, 1984). CPI or OEP values ~1 occur in oils that have been generated from source rocks enriched in marine organic matter or of a high thermal maturity. Oils with CPI > 1 are often derived from source rock strata that contained an abundance of land plant (rather than algal) organic debris, including leaf waxes. (The predominance of odd-carbon *n*-alkanes results from the decarboxylation of even carbon *n*-alkyl fatty acids that are present in plant waxes.)

In some environmental forensic investigations, the CPI can be useful in recognizing the contribution of modern hydrocarbons derived from modern leaf debris in soils and sediments. The presence of modern leaf waxes can impart a strong odd-carbon dominance (CPI > 2) that is unrelated to the petroleum contamination (Stout *et al.*, 2000b). An example of this situation is shown in Figure 6.7. In this study, the soils over a wide area were impacted following the blowout of an oil production well. Fingerprints of the surface soil samples from the impacted area showed that the spilled crude oil could be distinguished from naturally occurring background hydrocarbons derived from extant leaf waxes in the soil using the CPI index. Soil extracts exhibiting CPI values of approximately nine were comprised exclusively of background hydrocarbons. Changes in the CPI across the study area were used to show that the proportion of crude oil found in the soils decreased with increasing distance from the blowout site (and vice versa) allowing rapid assessment of the impacted soils.

6.3.2.1.2 Acyclic Isoprenoids

Acyclic isoprenoids are another group of aliphatic hydrocarbons that provide diagnostic value in assessing spilled crude oils and petroleum products. The

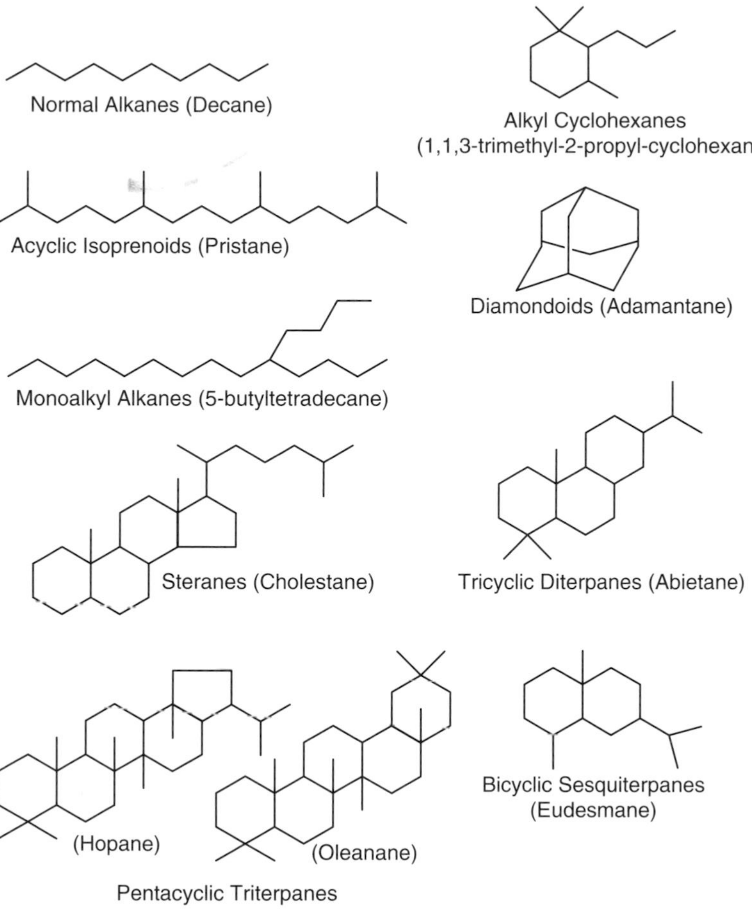

Figure 6.8
Selected aliphatic hydrocarbon structures.

most abundant and frequently measured isoprenoids in crude oil include pristane (Pr) and phytane (Ph) (Table 6.4; Figure 6.8). These compounds are derived primarily from the phytol side-chain in the chlorophyll molecule during early diagenesis. The relative proportion between Pr and Ph is related to the depositional conditions of the source rock (Didyk *et al.*, 1978). Since chlorophyll is pervasive in all natural environments, Pr and Ph are initially present in all crude oils, though they can be depleted or removed by biodegradation (either in the reservoir or following their release into the environment). These and related acyclic isoprenoids can be easily quantified with GC/FID or GCMS (*m/z* 113; see above). Higher carbon number acyclic isoprenoids (up to C_{45}) also occur in crude oils, but their analysis (even by GCMS SIM) often requires removal of the more dominant *n*-alkanes by molecular sieving techniques.

Thus, the Pr/Ph ratio is a primary molecular feature of crude oil and, by extension, petroleum products refined from crude oil. High Pr/Ph values (> 3.0) often indicate suboxic depositional conditions, as is typical in terrestrial depositional environments, while low values (<0.6) often indicate anoxic deposition, as is typical in some marine depositional environments (Peters and Moldowan, 1993). Because these two isoprenoids are similar in structure, boiling point, and solubility, the ratio between them remains stable following a release to the environment. Thus, the Pr/Ph ratio is often useful in forensic investigations where weathering has not affected the ratio. However, it is important to note that while distinct Pr/Ph ratios can be useful in establishing a 'negative' correlation between two similar crude oils or petroleum products, similar ratios (alone) do not warrant a defensible 'positive' correlation.

From the discussions above it follows that the ratios between selected *n*-alkanes and isoprenoids must also have a primary genetic control. For example, the ratios of n-C_{17}/Pr and n-C_{18}/Ph are commonly and easily determined in GC/FID chromatograms (which have achieved adequate peak separation). However, these ratios reflect the combined influences of source rock organic matter type, thermal maturity, and biodegradation. In forensic investigations, the latter's influence can be a reflection of biodegradation (1) in the geologic past (i.e., in the reservoir prior to production) or (2) since the petroleum was released to the environment. n-C_{17}/Pr ratios < 1 are not uncommon among (non-biodegraded) oils sourced from source rocks that were deposited in relatively suboxic environments that were enriched in land plant organic debris, e.g., as are common in deltaic depositional environments (e.g., many Tertiary-sourced Indonesian oils). Furthermore, both ratios are prone to increase with thermal maturity of the source rock (Peters and Moldowan, 1993). Figure 6.6 demonstrated the variability in n-C_{17}/Pr ratio for two unweathered middle distillate fuels. Therefore, the application of these simple and easily measured ratios for diagnostic/correlation purposes in forensic investigations requires considerable caution.

6.3.2.1.3 Mono-alkylalkanes

The final group of alkanes discussed herein are those with a single side branch, i.e., mono-alkylalkanes (also known as T-branched alkanes). These compounds have only recently been recognized in crude oils (Warton *et al.*, 1997) and their application in forensic studies is evolving. Analysis of mono-alkylalkanes requires removal of the interfering *n*-alkanes by molecular sieving techniques. While the origin of these compounds is not yet well established, their potential to provide an additional means of fingerprinting crude oils and petroleum products is under investigation in our laboratory. Most T-branched alkanes elute within the diesel range, which indicates their application in forensic problems involving middle distillate fuels may be appropriate.

6.3.2.2 Cyclic Aliphatic Hydrocarbons

The cyclic aliphatic hydrocarbons in crude oils and refined products provide a wide array of diagnostic features suitable for fingerprinting. This group of compounds includes the alkylcyclohexanes, decalins (tetrahydronaphthalenes), diamondoids, and aliphatic biomarkers, all of which require analysis by GCMS (see above; Figure 6.8). While alkylcyclohexanes, diamondoids, and decalins show considerable promise in forensic applications, it is the aliphatic biomarkers that have already proved to be highly valuable in forensic studies (Douglas *et al.*, 1993; Wang *et al.*, 1994b, 1999; Kvenvolden *et al.*, 1995; Kaplan *et al.*, 1997; Stout *et al.*, 1999c; Wang and Fingas, 1999).

6.3.2.3 Biomarkers

Biomarkers (i.e., molecular fossils, biological markers) are organic compounds in petroleum whose chemical structure can be unequivocally linked to a naturally occurring biochemical (Mackenzie, 1984; Philp, 1985; Waples and Machihara, 1990). The degree of this similarity is exemplified in Figure 6.9, which shows the chemical structure of a common biomarker, 5α(H)-cholestane. The skeletal structure of 5α(H)-cholestane is akin to its well-known precursor, cholesterol (cholest-5-en-3β-ol), a biochemical found in many modern and ancient organisms. Figure 6.9 also demonstrates a similar biochemical-to-biomarker relationship that exists between abietic acid (a constituent of certain plant resins) and retene. The conversion of biochemicals to biomarkers occurs during the early diagenesis of sediments and during the subsequent thermal maturation of the oil source rocks (Simoneit, 1986). The conversion of Nature's vast array of biochemicals into biomarkers creates a vast suite of compounds in crude oils that have distinct chemical structures, often including more than one stereoisomer. Through careful studies of modern biota, modern sediments, crude oils and their source rocks in basins

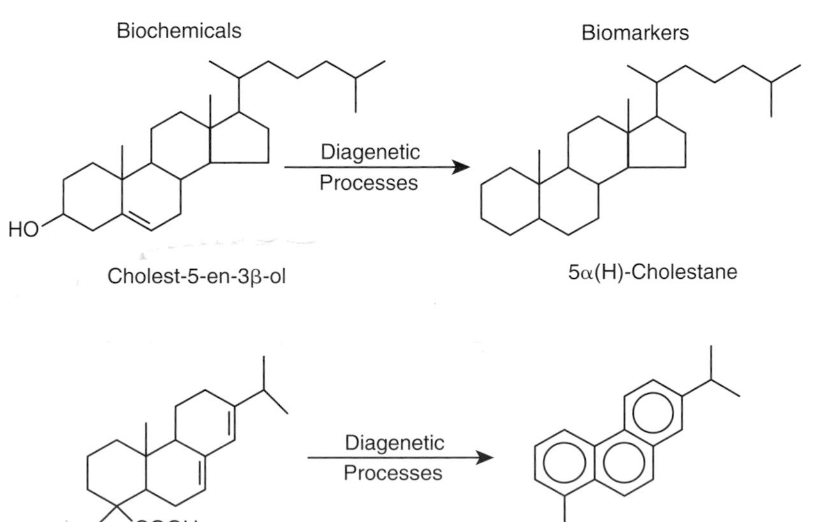

Figure 6.9
Genetic relationship between selected petroleum biomarkers and their biochemical precursors.

around the world, petroleum geochemists can interpret the biomarker distributions in crude oils to determine the type(s) of precursor organic matter and the thermal maturity of the source rock strata of that oil (Seifert and Moldowan, 1978b; Seifert *et al.*, 1979; Seifert and Moldowan, 1986; Peters and Moldowan, 1993). By extension, the biomarker distributions in environmental samples contaminated with petroleum can be highly diagnostic in environmental forensic studies (Wang *et al.*, 1994b; Stout *et al.*, 2000a). Some of the established primary biomarker parameters that are potentially useful in petroleum fingerprinting investigations are listed in Table 6.8.

The wide variety of conditions under which crude oil has formed in the geologic past has imparted an essentially unique biomarker 'fingerprint' on every crude oil. Crude oils from within a given geologic basin that have been sourced from source rocks of the same geologic age normally vary less than crude oils from different basins or from source rocks with a different geologic age. On a gross scale, for example, as might be determined from a GC/FID analysis, two crude oils may exhibit nearly identical *n*-alkane profiles or isoprenoid distributions. However, their biomarkers, as determined from GCMS analysis, may be

Table 6.8

Inventory of selected primary biomarker and PAH parameters and their controlling factors.

Diagnostic Parameter	Basis	Controlling Factor
Terpanes		
Homohopane profile (C_{31}–C_{35})	*m/z* 191	Source rock type
Homohopane 22S/(22S + 22R)	*m/z* 191	Thermal maturity
25-Norhopanes/Hopane	*m/z* 177, 191	Biodegradation
28,30-bisnorhopane/Hopane	*m/z* 191	Source rock type
Oleanane/Hopane	*m/z* 191	Source rock type
Lupane/Hopane	*m/z* 191	Source rock type
Gammacerane/Hopane	*m/z* 191	Source rock type
C_{28}–C_{29} Extended tricyclics/Hopane	*m/z* 191	Source rock type
T_s/T_m	*m/z* 191	Thermal maturity
Steranes		
Regular sterane distribution	*m/z* 217, 218	Source rock type
Diasteranes	*m/z* 217	Source rock type
Mono/Tri aromatic sterane ratios	*m/z* 253, 231	Thermal maturity
Ethylcholestane 20S/(20S + 20R)	*m/z* 217	Thermal maturity
PAH		
Methylphenanthrene indices	*m/z* 192	Thermal maturity
Dimethylphenanthrene indices	*m/z* 206	Thermal maturity
Cadalene/C_4-naphthalenes	*m/z* 184	Source rock type
Retene/C_4-phenanthrenes	*m/z* 234	Source rock type
Methyldibenzothiophene ratio	*m/z* 198	Thermal maturity
C_2-Dibenzothiophene/C_2-phenanthrene ratio	*m/z* 212, 206	Source rock type
C_3-Dibenzothiophene/C_3-phenanthrene ratio	*m/z* 226, 220	Source rock type
C_1–C_3 Methylnaphthalene ratios	*m/z* 142, 156, 170	Thermal maturity

markedly different. Thus, the successful forensic investigator will quickly grasp the utility of biomarkers in environmental forensic problems requiring detailed comparisons among similar product types (Wang *et al.*, 1994b; Kvenvolden *et al.*, 1995; Stout *et al.*, 2000a). Excellent reviews on the applications of biomarkers in petroleum characterization already exist (Peters and Moldowan, 1993); therefore only those aspects relevant to forensic applications are discussed below.

6.3.2.3.1 Terpanes

The terpanes in petroleum include mono- (C_{10}), sesqui- (C_{15}), di- (C_{20}), sester- (C_{25}), and tri-terpanes (C_{30}) (Figure 6.8; Seifert and Moldowan, 1978b; Gallegos, 1981). The chemistry and formation of terpanes has been thoroughly reviewed (Simoneit, 1986). Analysis of these compounds requires GCMS, typically operated in the SIM mode seeking a suite of characteristic ions (Tables 6.7 and 6.8). Various terpanes can occur throughout the diesel and residual boiling ranges of most crude oils and petroleum products (Figure 6.10). As such, forensic analyses can be tailored to look for appropriate suites of biomarker compounds, depending on the nature of the contamination. The relative resistance of the terpanes to weathering on environmental time-scales

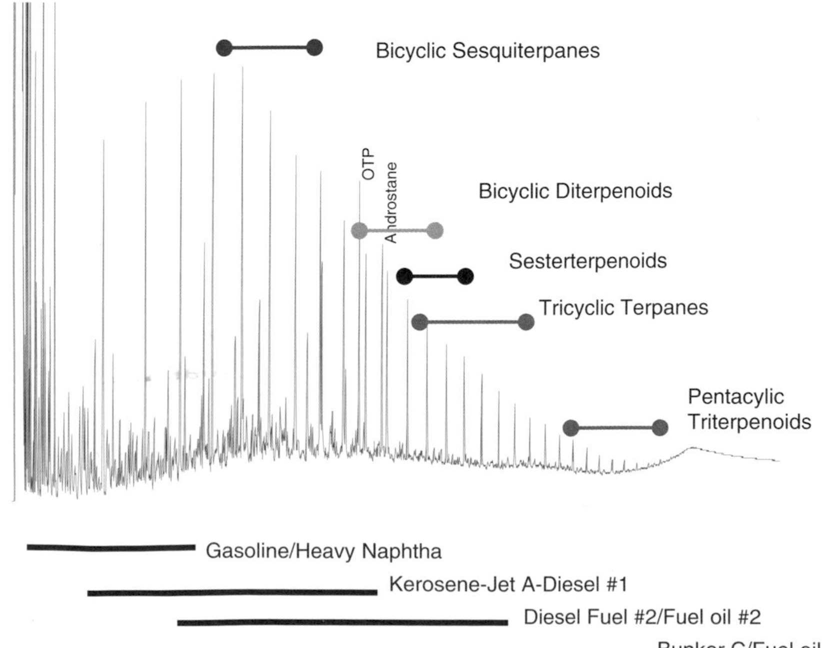

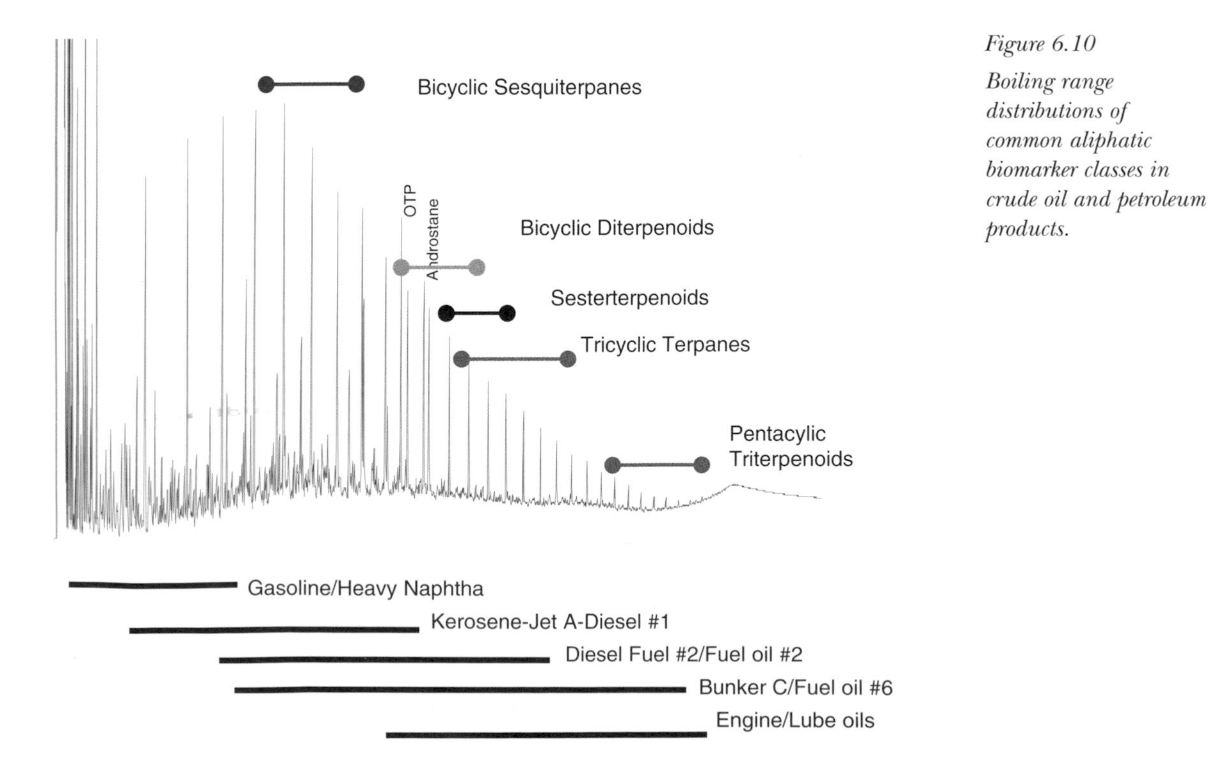

Figure 6.10
Boiling range distributions of common aliphatic biomarker classes in crude oil and petroleum products.

(e.g., Prince *et al.*, 1994) and their specificity makes them particularly useful in petroleum fingerprinting investigations.

Among the most common and useful terpanes found in petroleum is a homologous suite of pentacyclic triterpanes (C_{27} to C_{35}, except for C_{28}) known as *hopanes.* The hopanes are derived from bacterial cell membranes (Ries-Kautt and Albrecht, 1989), and therefore are ubiquitous in sediments, source rocks, and ultimately crude oils. These compounds boil in the n-C_{26} to n-C_{33} range (Figure 6.10) and, therefore, are only useful in forensic investigations involving residual range hydrocarbons. However, they are thermally stable and relatively resistant to biodegradation and thereby can prove useful in many forensic investigations. An example is given in Figure 6.11, which shows a partial m/z 191 trace for a fugitive oil slick and the suspected source oil. While the C_{27} to C_{35} hopane distributions are similar in this example, several primary molecular differences clearly indicate a 'negative' correlation between these two oils. For example, the fugitive oil slick contains a higher proportion of C_{35} homohopanes (both 22S and 22R isomers) than the suspected source oil. In addition, the fugitive oil exhibits a distinct T_s/T_m ratio (see Table 6.7 for compound identifications). In this case, these 'primary' differences clearly indicated that the suspected source was not the source of the fugitive slick.

In addition to the hopanes, the presence/absence and relative abundance of other pentacyclic triterpanes in oils can be extremely useful as they can provide additional diagnostic information on the type(s) of organic matter that gave rise to the crude oil or the parent crude oil of some petroleum products. Among these are compounds such as gammacerane (ten Haven *et al.*, 1989), moretanes, 18α(H)-oleanane (Ekweozor and Udo, 1988), lupanes (Brooks, 1986), bicadinanes (Stout, 1995), bisnorhopanes (Curiale and Odermatt,

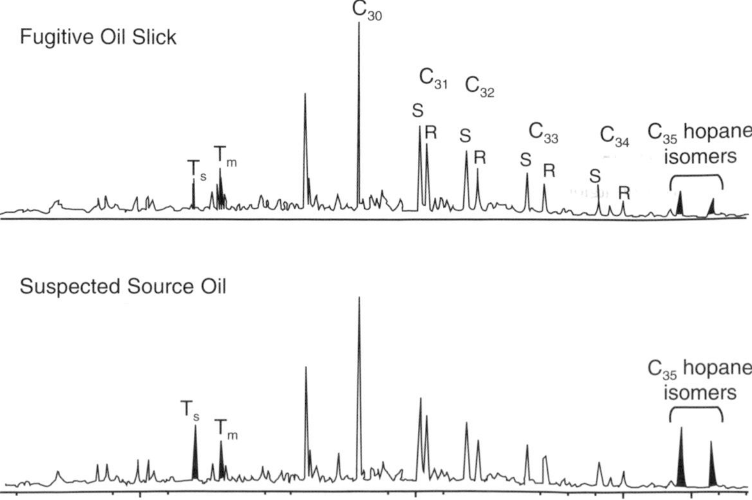

Figure 6.11
Comparison of pentacyclic triterpane biomarkers (m/z *191*) *in fugitive oil slick and suspected source oil.*

1989), and hopane degradation products (Peters *et al.*, 1996). Most of these can be easily monitored using m/z 191 or m/z 177 fingerprints generated by GCMS (see above). Identification of these compounds requires authentic standards, full scan spectral interpretation or careful comparison to published literature.

6.3.2.3.2 *Steranes*

The steranes are a suite of biomarkers containing 26 to 30 carbons that are derived from sterols (e.g., Figure 6.9; Huang and Meinschein, 1979; Volkman, 1986). They boil in the residual range of crude oils and petroleum products (Figure 6.10) and are thermally stable and relatively resistant to degradation, properties that enhance their utility in environmental forensic investigations. Diagenesis converts sterols into steranes and the increases in temperature during burial result in additional rearrangement reactions and aromatization reactions, both of which produce a wide variety of (1) regular steranes, (2) rearranged steranes (i.e., diasteranes), and (3) aromatic (mono- and tri-) steranes (Mackenzie *et al.*, 1982). Each of these three groups of steranes has potential application in environmental forensic investigations.

Perhaps the most straightforward application of sterane biomarkers in forensic studies is the proportion of C_{27} to C_{29} regular steranes, as measured by the proportion of 5α(H),14β(H),17β(H)-steranes in m/z 218 mass chromatograms. (These compounds are described using 'shorthand' as ββ-steranes.) The relative abundance of the C_{27} versus C_{29} ββ-steranes in crude oils reflects the relative proportions of algal versus land plant organic debris in the crude oil's source rock. This difference arises from the nature of the precursor sterols present in different plant types (Huang and Meinschein, 1979). Figure 6.12 shows a ternary diagram of the ββ-sterane distributions for a number of tar balls (washed up on beaches) and two candidate source oils (Stout *et al.*, 1999c). The distribution of ββ-steranes demonstrated that the (highly weathered) tar balls, which had washed up on a beach following an oil spill, were from two distinct sources, mostly from candidate source oil #2. In this case, source oil #2 was recognized as being derived from local, natural oil seeps, a chronic regional oil source unrelated to the spilled oil (source oil #1).

6.3.2.3.3 *Low-boiling Biomarkers*

As described and demonstrated above, the pentacyclic triterpanes (m/z 191) and steranes (m/z 217 and 218) provide considerable ability to describe/compare crude oils and certain petroleum products. However, in petroleum products where the 'high-boiling' pentacyclic triterpanes and steranes are absent (e.g., most diesel fuels; Figure 6.10), several 'low-boiling' terpanes can sometimes

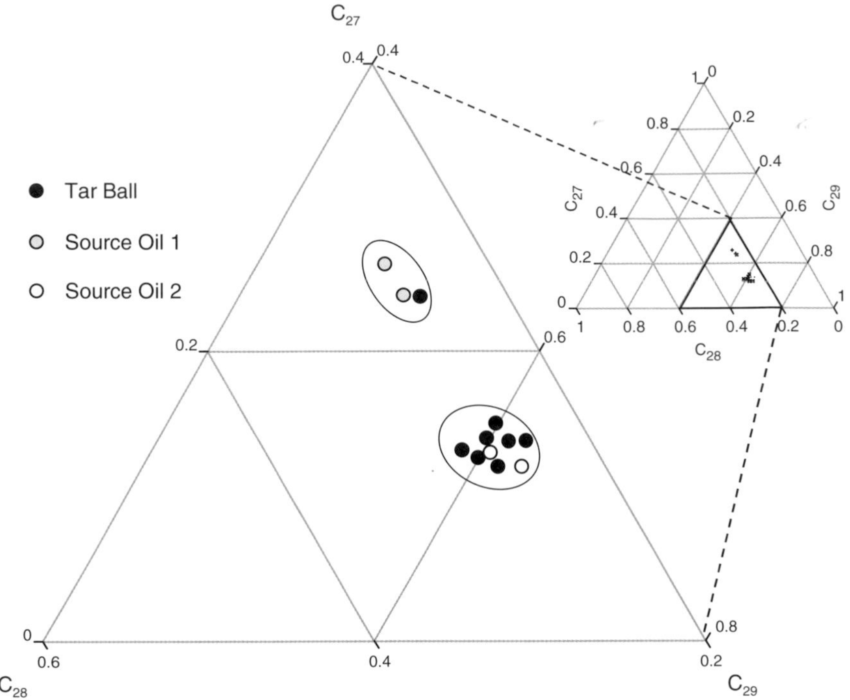

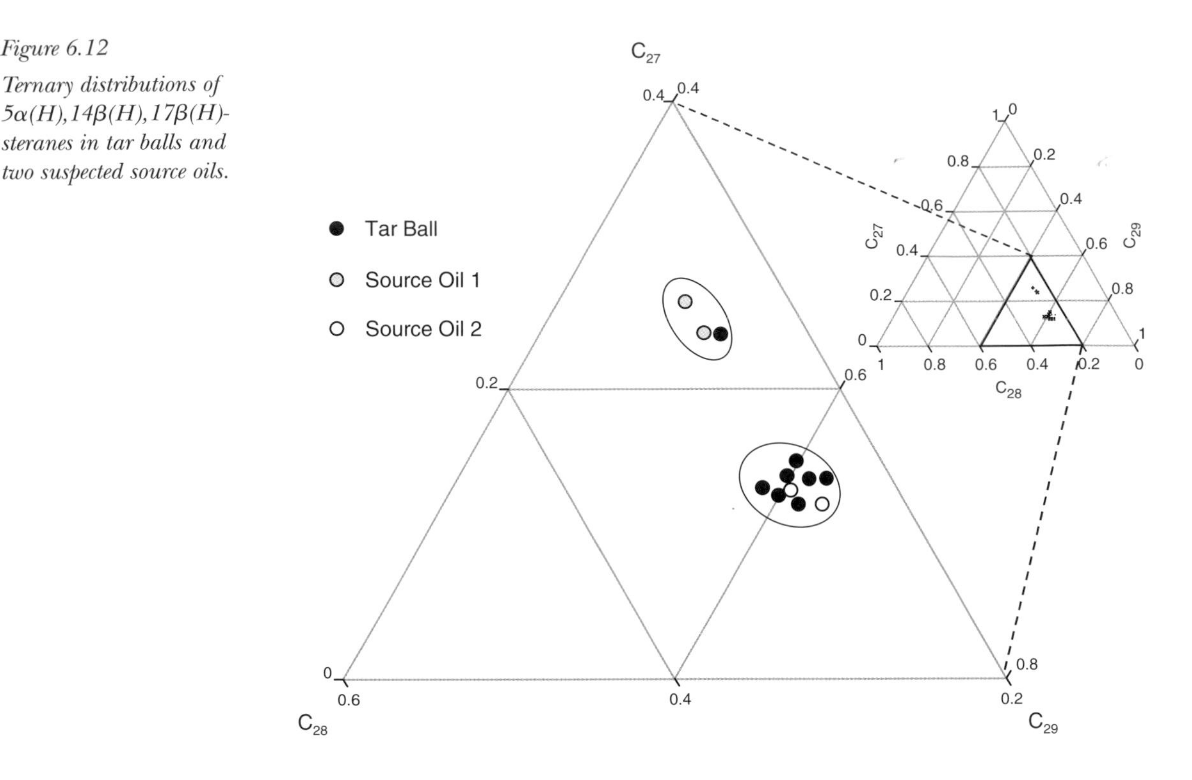

Figure 6.12
Ternary distributions of 5α(H),14β(H),17β(H)-steranes in tar balls and two suspected source oils.

provide a comparable and highly diagnostic means of comparison (Stout *et al.*, 1999b).

Various bicyclic sesquiterpanes and diterpanes boil within the diesel range (Figure 6.10). The presence of these 'low-boiling' biomarkers is not uncommon in oils generated from Tertiary (i.e., geologically 'young') source rocks containing terrestrial plant debris, particularly plant resins (Grantham and Douglas, 1980; Noble *et al.*, 1986; Philp and Gilbert, 1986). The variety and specificity of these compounds sometimes allows for detailed comparisons between middle distillate fuels. For example, Figure 6.13 shows the partial *m/z* 123 mass chromatograms for two weathered diesel fuel #2 (free phase oil) samples from two adjacent petroleum terminal properties. *n*-Alkanes were completely removed from each by biodegradation and the two samples exhibited similar isoprenoid patterns (e.g., Pr/Ph ratio). However, GCMS analysis of the *m/z* 123 ions showed a distribution of (compounds tentatively identified as) bicyclic sesquiterpanes demonstrating distinct differences between the diesel fuels. These differences argued strongly that two distinct sources/releases of diesel fuel existed in the study area.

Finally, a promising group of 'low-boiling' compounds of interest are the adamantanes (C_{10}) and diamantanes (C_{14}), which are collectively termed diamondoids due to their 'cage-like' structures. These are a group of cyclic

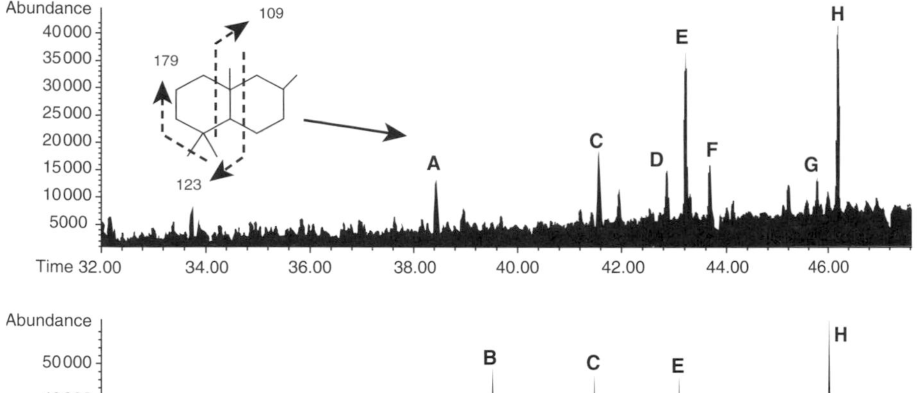

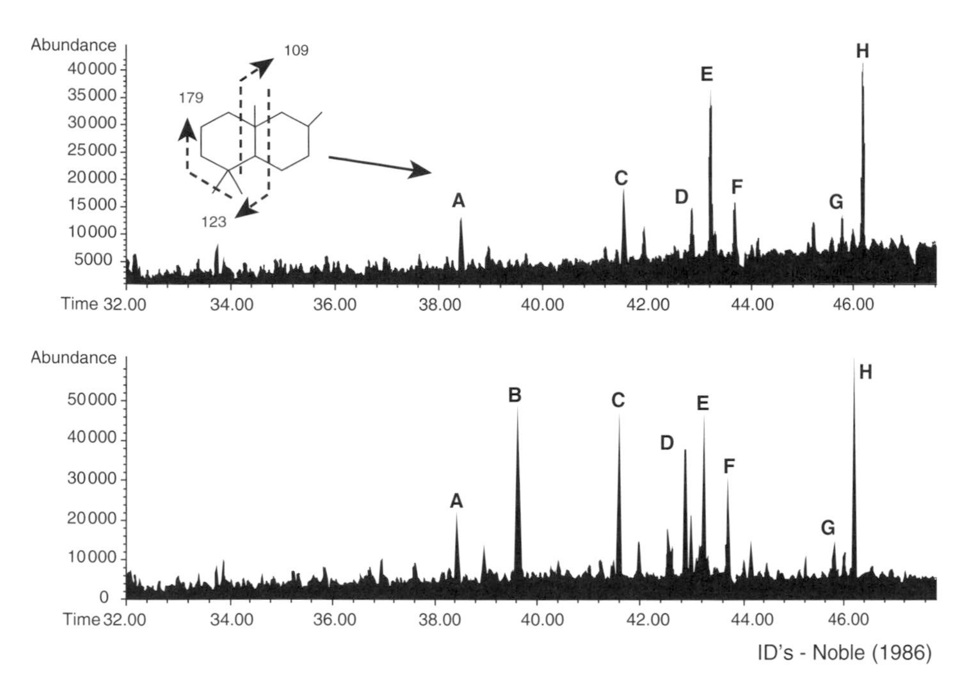

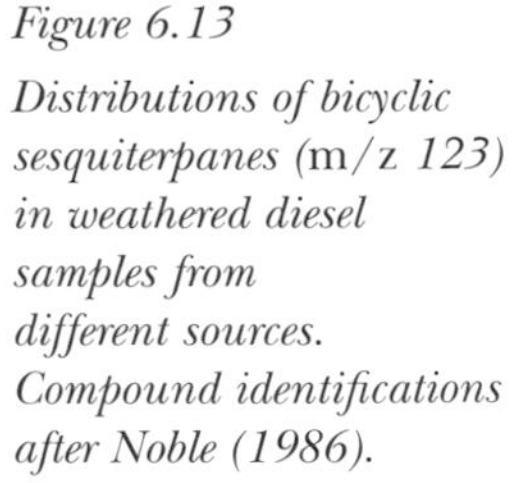

Figure 6.13
Distributions of bicyclic sesquiterpanes (m/z *123*) *in weathered diesel samples from different sources. Compound identifications after Noble (1986).*

aliphatic hydrocarbons with three or five fused rings (e.g., adamantane; Figure 6.8). They are thermally stable and highly resistant to biodegradation, although their biochemical precursor(s) is not yet well known (Grice *et al.*, 2000). Their presence in the early boiling diesel range of liquid petroleum ($< C_{15}$), and even in condensates, may make them useful in characterizing petroleum products that may not contain the higher boiling biomarkers (steranes and triterpanes) described above. These compounds show considerable promise in their ability to assess lower boiling, distillate petroleum products (kerosene, Jet A, etc.) in the environment and are currently being evaluated in our laboratory for application in forensic issues.

6.3.2.4 Aromatic Hydrocarbons

6.3.2.4.1 Mono-aromatic Hydrocarbons

Aromatic hydrocarbons occur in all crude oils (Tissot and Welte, 1984; Figure 6.5). Simple mono-aromatic compounds include the various BTEX compounds (i.e., benzene, toluene, ethyl benzene, and *m*-, *p*-, and *o*-xylenes) and other alkyl-substituted aromatic compounds. The utility of these compounds in forensic applications is limited to instances where weathering has not significantly impacted the distribution of these compounds. However when appropriate, the large number of isomers that exist among, for example, the C_4 alkyl-benzenes (m/z 134) can prove useful in comparing samples (e.g., moderately-to-highly-weathered gasolines). Though not widely reported, many crude oils contain homologous series of *n*-alkylbenzenes ranging up to at least C_{39} (Albaiges *et al.*, 1986) and alkyl-tetralines (1,2,3,4-tetrahydronaphthalenes) (Tissot and Welte, 1984). The potential application of using these compounds in environmental forensics is warranted but, to our knowledge, has not been pursued.

6.3.2.4.2 Polycyclic Aromatic Hydrocarbons

Clearly, the most studied group of aromatic hydrocarbons in crude oils is the polycyclic aromatic hydrocarbons (PAH). Limited only by the ability of conventional gas chromatography, the commonly analyzed PAH compounds range from 2-ring PAHs (naphthalene) up through 6-ring PAHs (e.g., benzo[*g,h,i*]perylene; see Table 6.6). In unweathered crude oils the most abundant PAHs are commonly naphthalene and various alkyl-naphthalenes (Sauer and Uhler, 1994–1995). The concentrations of PAH tend to decrease with increasing PAH molecular weight/boiling point in most crude oils. Figure 6.14 shows the PAH distribution for a composite North Slope crude oil. The total PAH (54 analytes) concentration is almost 14 000 ppm and the dominance of the C_0–C_4 alkyl-naphthalene is evident. Since these 2-ring PAHs are among the most susceptible to weathering (see below), their concentrations can be markedly reduced in environmental samples. A notable feature of this and most crude oils is the relative absence of selected PAHs, e.g., anthracene, fluoranthene, and pyrene, compounds which are typically enriched in non-petroleum sources of PAH (e.g., coal-derived liquids and combustion-derived PAHs; Sauer and Uhler, 1994–1995).

The PAHs in crude oils and petroleum products typically exhibit a characteristic 'bell-shaped' profile within each alkyl-homologue series (Sauer and

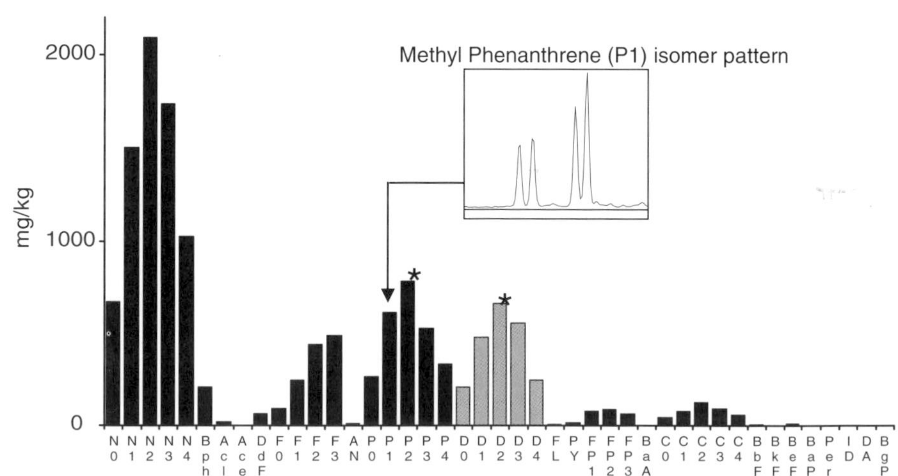

Figure 6.14
PAH distribution for a composite North Slope crude oil. Sulfur-containing PAH shown as hatched. Starred peaks refer to compound groups used in diagnostic ratio D2/P2. TPAH = 14 000 mg/kg.

Uhler, 1994–1995). This is evident in the naphthalenes, phenanthrenes, dibenzothiophenes, fluoranthene/pyrenes, and chrysenes in the North Slope crude oil example shown in Figure 6.14. This 'bell-shaped' profile is another means by which petroleum-derived PAHs are distinguished from non-petroleum sources (which instead tend to be dominated by the parent PAH, and reduced abundance with increasing degree of alkylation). This is discussed further later in this chapter.

Ratios among individual PAH isomers have been demonstrated to be related to the thermal history of a crude oil and its source strata (Radke *et al.*, 1986; Strachan *et al.*, 1988; Radke *et al.*, 1990; Requejo *et al.*, 1996; van Aarssen *et al.*, 1999). For example, the ratios among methyl-phenanthrenes, dimethyl-naphthalenes, and methyl-dibenzothiophenes have been thoroughly studied and, in each case, it was shown that the more thermally stable isomers (β-type) tend to increase in abundance with increasing maturity of the oil's source rocks. Ratios among the four prominent methyl-phenanthrene isomers (3-, 2-, 9- and 1-MP) can be monitored using numerous methyl-phenanthrene indices (MPI 1, MPI 2, and MPR) defined by (Radke, 1988) as follows:

$$\text{MPI } 1 = 1.5\,(2-\text{MP}+3-\text{MP})\,/\,(\text{P}+1-\text{MP}+9-\text{MP})$$
$$\text{MPI } 2 = 3\,(2-\text{MP})\,/\,(\text{P}+1-\text{MP}+9-\text{MP})$$
$$\text{MPR} = 2-\text{MP}\,/\,1-\text{MP}$$

Figure 6.14 demonstrates the distribution of methyl-phenanthrene isomers for the North Slope crude oil. Methyl-phenanthrene ratios can prove valuable in forensic investigations, particularly in situations involving middle distillate fuels, in which the methyl-phenanthrenes are common constituents. Other potentially useful PAH-based thermal maturity parameters also require measurements of individual isomers (rather than isomer groups). This includes the C_1–C_3 methyl-naphthalenes (MNR, ENR, numerous DNRs and TNRs i.e., methyl-, ethyl-, numerous dimethyl-, and trimethyl-naphthalene ratios), dimethyl-phenanthrenes (DPI) and methyl-dibenzothiophenes (MDR; Radke *et al.*, 1984, 1991). Some caution is necessary in applying all of these ratios as slight differences in the apparent susceptibility to degradation among the isomers (on environmental time scales) can influence these aromatic ratio/indices values (Wang and Fingas, 1995).

On the other hand, the considerable amount of research following the 1989 *Exxon Valdez* oil spill has demonstrated that some PAH parameters are largely independent of weathering (Douglas *et al.*, 1996). For example, the ratios of alkyl-phenanthrenes to alkyl-dibenzothiophenes (D2/P2 and D3/P3) have been shown to remain stable over a wide range in the degree of weathering (i.e., these PAH groups tend to weather at comparable rates (Wang and Fingas,

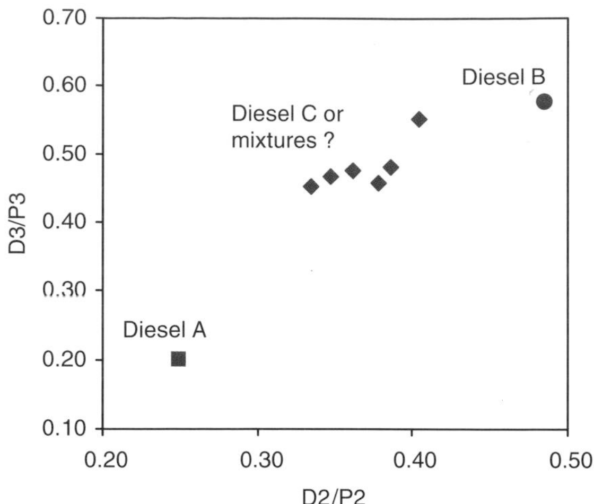

Figure 6.15
Cross-plot of D2/P2 and D3/P3 for soil extracts from a site demonstrating the occurrence of at least two different middle distillate fuels.

1995; Douglas *et al.*, 1996)). The relative abundance of sulfur-containing PAH, however, is a function of the degree of anoxia in the sediments that gave rise to an oil's source rock (higher sulfur indicating greater sulfate reduction). Thus, the ratios of D2/P2 and D3/P3 can provide primary genetic information on the type of oil source rock. This proved particularly useful in monitoring sediments in Prince William Sound where the more sulfur-rich *Valdez* cargo oil could be distinguished from the sulfur-depleted Tertiary oil seeps in the region (Page *et al.*, 1995; Boehm *et al.*, 1998, 2000).

Since the C_2 and C_3 phenanthrenes and dibenzothiophenes boil within the diesel range, diesel fuels distilled from different crude oils (or refined to different levels of desulfurization) will also exhibit different D2/P2 and D3/P3 ratios. This approach has found numerous applications in environmental forensic investigations. Figure 6.15 shows a cross-plot of these indices for a suite of soils from a terminal site impacted by distinct middle distillate fuels. The diesel fuels are sufficiently weathered that distinctions using GC/FID patterns were inconclusive. However, the PAH data revealed that at least two distinct types of diesel fuel could be recognized in the site's soils. A third type (C) may also exist, or this could be evidence for the mixing of types A and B.

6.3.2.4.3 Aromatic Biomarkers

The final group of aromatic hydrocarbons of interest to the forensic investigator is the aromatic biomarkers. Like the aliphatic biomarkers described above, these compounds can provide useful diagnostic information on the nature of the precursor organic matter type and thermal history of a crude oil and, thereby, petroleum products refined from it. It is not uncommon for numerous aromatization reactions to alter biochemicals during diagenesis in sediments (Wakeham *et al.*, 1980b). For example, Figure 6.9 showed the

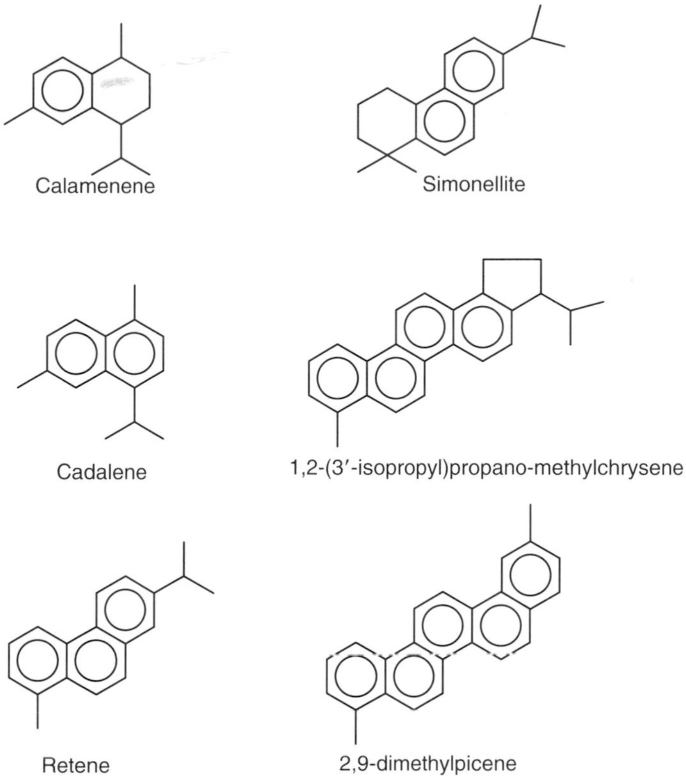

Figure 6.16
Selected aromatic biomarker structures.

conversion of the biochemical abietic acid (a component of certain plant resins) into retene (1-methyl-7-(1-methylethyl)-phenanthrene). Aromatization reactions produce a spectrum of aromatic biomarkers in crude oils and petroleum products refined from them. While aromatization typically results in biomarkers with reduced structural specificity (compared to the aliphatic biomarkers), these compounds can nonetheless, prove useful in assessing environmental forensic issues. Some useful groups of aromatic biomarkers include the aromatic sesquiterpanes (e.g., calamenene and cadalene), aromatic diterpanes (e.g., retene, simonellite), aromatic hopanoids (1,2-(3′-isopropyl)-propano-methyl chrysene), and aromatic oleanoids (e.g., 2,9-dimethylpicene) (Bastow *et al.*, 1999; van Aarssen *et al.*, 1999, 2000; Figure 6.16). Identification of these compounds in environmental samples often requires analysis by full scan GCMS followed by spectral interpretation or, when available, analysis of authentic standards. Therefore, the application of aromatic biomarkers is usually limited to 'tough' forensic questions that justify specialized analysis.

6.3.2.5 NSOs and Metals

Some non-hydrocarbon (N-, S-, or O-containing) compounds in crude oils can also prove useful in environmental forensic applications. The sulfur-containing aromatic compounds, benzo- and dibenzothiophenes, are the most common non-hydrocarbons in most oils. The forensic utility of these compounds was discussed above along with the PAHs. Other NSOs (polars) in crude oil and certain refined products whose forensic applications have not been adequately explored include various phenols, furans, mercaptans, thiophenes, pyrroles, pyridines, quinolines, carbazoles, benzocarbazoles, acridines, and benzacridines. One can imagine, however, that since the concentration of polar compounds tends to increase with increasing biodegradation (e.g., Figure 6.5), their greatest potential use in forensic studies may be in instances of severely weathered petroleum and petroleum products.

Finally, it is necessary to mention the presence of metal-containing organic moieties in crude oils. Among these the best known are nickel and vanadium metallo-porphyrins. Metallo-porphyrins are high boiling biomarkers derived from the tetrapyrroles component of chlorophyll (Sundararaman, 1985). (Recall chlorophyll's phytol side-chain is the precursor to Pr and Ph; see above.) Two major groups of metallo-porphyrins occur in crude oils (deoxyophylloerythro-etioporphyrins and etioporphyrins, or DPEP and etio, respectively). Both of these commonly have Ni or V substituted for original Mg ion in the center of the tetrapyrrole, a phenomenon that occurs relatively quickly during diagenesis (Filby and Van Berkel, 1987). The relative abundance of Ni and V in crude oil, which occurs almost entirely within the metalloporphyrins and is expressed by the Ni/V ratio, is thought to be a function of the source rock depositional environment (Lewan, 1984).

To our knowledge, the metallo-porphyrins have not been utilized in environmental forensic applications. Part of this is due to the difficulty of analyzing for these high molecular weight compounds, which are not amenable to conventional gas chromatographic analyses. Most analyses on petroleums have employed ultraviolet spectroscopy, liquid chromatography, high temperature GC analysis, or supercritical fluid chromatography. Atomic absorption measurements to determine the proportion of Ni and V in petroleum products may be applicable in some environmental forensic investigations.

6.4 SECONDARY CONTROLS ON THE COMPOSITION OF PETROLEUM

6.4.1 *EFFECTS OF PETROLEUM REFINING ON HYDROCARBON FINGERPRINTING*

Knowledge of the primary controls on crude oil composition and its implications (Section 6.3) is critical to defensible forensic interpretations. However, equally important is an understanding of the effects that petroleum refining, a secondary control, has on the composition of petroleum products. Several excellent reviews on general refining practices/processes already exist (Gary and Handwerk, 1984; Speight, 1991), which warrants only a general review of the major refining processes herein. Following the review we discuss the impact refining has on the chemistry of major petroleum fuel products, i.e., information most often needed by the forensic investigator.

6.4.1.1 Distillation (Atmospheric and Vacuum)

Perhaps the most important process in refining is the initial process, distillation. In distillation, a desalted and pre-heated crude oil feedstock (sometimes a blend of different crude oils) is fed into a distillation tower operated at atmospheric pressure (atmospheric distillation). Light and heavy virgin naphthas, various middle distillates, atmospheric gas oil, and atmospheric residuum are separated based upon volatility/boiling point. More often than not in a modern refinery the atmospheric residuum feeds another distillation tower operated at reduced pressure (vacuum distillation), in which the atmospheric residuum is further fractionated into vacuum gas oil(s) and vacuum residuum. Thus, atmospheric and vacuum distillation produces a variety of 'straight-run' boiling fractions that act as feedstocks to other refinery units, or in some instances, yield essentially 'finished' petroleum products that achieve the required product specifications (e.g., kerosene or Jet A can be distilled directly from paraffinic, low sulfur crude feedstocks).

The most common straight-run fractions produced via distillation are summarized in Table 6.9. It is important to consider that the conditions for distillation are subject to a large number of variables and are closely controlled depending on the refinery stream needs, which can alter the character (end

Table 6.9

Inventory of straight-run distillate products.

Straight-Run Products	BP Low (°C)	BP High (°C)	Common Use
Light straight-run gasoline	−1	150	Gasoline blending or feedstock for isomerization unit
Heavy straight-run gasoline (naphtha)	150	205	Feedstock for catalytic hydrotreater and/or reformers
Straight-run Jet (kerosene)	205	260	Feedstock for distillate fuel production (kerosene and jet fuel)
Straight-run diesel[a]	260	315	Feedstock for distillate fuel production (fuel oil and diesel fuel #2)
Heavy atmospheric gas oil	315	425	Feedstock for vacuum distillation
Vacuum gas oil	425	600	Feedstock for fluid catalytic cracker, hydrocracker or lube oil stock
Vacuum bottoms/pitch	600		Feedstock to visbreaker, coker, lube/residual fuel, asphalt

[a]Light atmospheric gas oil.

cuts) of the straight-run products. Thus, the boiling ranges listed in Table 6.9 should be considered as being approximate rather than steadfast cut-offs.

The importance of distillation in forensic investigations rests largely on the effect that it has on the boiling ranges of the petroleum products. Distillation tends to alter the profiles of compounds depending upon their relative volatility (Peters *et al.*, 1992). For example, the progressive loss of tricyclic triterpanes due to distillation is exemplified in Figure 6.17, which compares their distributions in both the 'parent' crude oil feedstock and in the 'daughter' diesel fuel #2. The higher boiling pentacyclic triterpanes (and steranes) are completely absent from the diesel fuel #2. Some distillation cuts are so 'sharp' that even compounds with very close boiling points can be impacted differently. Sometimes this type of partitioning alters the 'primary' molecular features of the parent crude oil, e.g., norhopane/hopane ratio (Peters *et al.*, 1992).

Distillation will also alter the shape of the *n*-alkane envelope proportional to volatility. Distillation generally produces a relatively smooth *n*-alkane envelope due to the regular volatility differences among *n*-alkanes of increasing carbon number. In fact, the 'sharpness' of the distillation cut can be evaluated by the slope of the *n*-alkane envelope, i.e., sharp distillation cuts producing higher slopes. Therefore, a useful diagnostic parameter(s) in some investigations is one that relies on the ratios among consecutive *n*-alkanes (e.g,. $n\text{-}C_{15}/n\text{-}C_{16}$, etc.), particularly the slope among *n*-alkanes less prone to evaporative weathering.

Comparison of refined products to crude oil feedstocks, and particularly interpretations surrounding the primary molecular features of the 'parent' crude oil, requires considerable caution. Certain primary features that rely upon compounds with similar boiling points, and which do not rely upon compounds near the distillation endpoints, should not be significantly affected by distillation. For example, most diesel fuel #2 contains hydrocarbons ranging

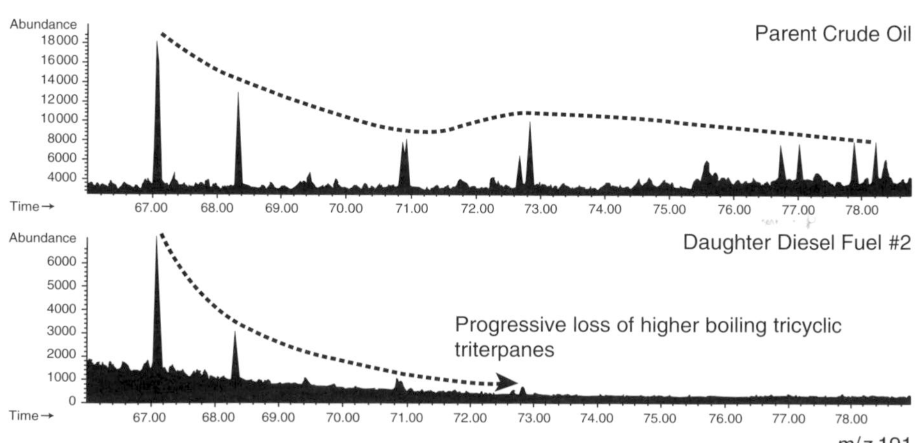

Figure 6.17
Comparison of parent crude oil and daughter diesel fuel #2 demonstrating the loss of tricyclic triterpanes due to distillation (m/z 191).

from about n-C_8 to n-C_{24}, which reach a maximum around n-C_{16}. As such, one would not expect primary parameters within the 'middle' of this range (e.g., the Pr/Ph or n-C_{17}/Pr ratios) to change during distillation of the 'parent' crude oil. However, considerable caution is necessary when evaluating any parameter that is dependent upon (1) compounds that occur near the boiling point maximum or minimum of a given product or (2) compounds of widely varying boiling points. In the first case, consider that Pr and Ph boil at the lowest end of most lubricating oils. As such, one can anticipate a distillation effect such that a 'daughter' lube oil would have a lower Pr/Ph ratio than the 'parent' crude oil stock (since Pr has a slightly higher volatility than Ph).

While the effects of distillation can affect the interpretation of certain primary features of petroleum, and potentially hinder the ability to correlate a petroleum product to the parent crude oil, these effects do not reduce the utility for comparing one petroleum product in environmental media to another. Of course, the tertiary effects of 'natural distillation' (i.e., evaporation) or 'anthropogenic distillation' (e.g., soil vapor extraction) must always be considered.

6.4.1.2 Hydrocarbon Cracking Processes

6.4.1.2.1 Thermal Cracking

As the demand for producing automotive gasoline exceeded that of kerosene, the need to 'crack' larger hydrocarbons in crude oil into smaller ones rose. The history of refining has seen the development of several cracking processes. Thermal cracking was first introduced in 1913 (Speight, 1991). In this process hydrocarbon feedstocks were heated in a reactor so that C–C bonds were broken to yield smaller hydrocarbon fragments, and thereby increase the proportion of gasoline range hydrocarbons (as compared to the feedstock). As these bonds are broken, the fragments produced typically include unsaturated hydrocarbons, i.e., olefins (hydrocarbons containing at least one double bond). The production of olefins served to increase the octane rating of the cracked gasoline over that of the feedstock.

6.4.1.2.2 Catalytic Cracking

The efficiency of hydrocarbon cracking was improved with the development of catalytic cracking in 1936, a process in which an inert catalyst(s) was added to the reactor to further promote the cracking of hydrocarbons. Today, modern fluid catalytic cracking (FCC) is the heart of gasoline production at most refineries. FCC also produces olefins, but generally in lower proportions relative to thermal cracking. In addition, the olefins produced during FCC typically contain three or more carbons, versus the one or two carbon fragments normally produced by thermal cracking. Also formed during FCC are iso-alkanes (branched alkanes) and aromatic hydrocarbons, both of which further improve the octane rating of the cracked gasoline. Catalyst and reactor improvements, along with the increasing demand for octane in the gasoline pool, has made thermal cracking essentially obsolete in the modern refinery (except as it is used to reduce the viscosity of heavy feedstock, i.e., visbreaking).

The catalysts used in FCC prior to about 1965 were synthetic alumina-silicates. Since then catalysts have progressively shifted toward catalysts containing zeolites (Pines, 1981). This shift had reduced the proportion of olefins formed (in favor of alkanes) but the rate of cracking and catalyst lifetime is vastly increased. Thus, one might suppose that historical (pre-1965) gasolines might have contained even higher olefin content than more modern gasolines.

6.4.1.2.3 Hydrocracking

A third cracking process, hydrocracking, was introduced in 1958. During this process the breaking of C–C bonds was conducted in the presence of catalysts in a pressurized hydrogen atmosphere, a combination which immediately saturates any double bonds leading to the formation of an 'olefin-free' cracked product. Thus, a gasoline blended using a hydrocracked product will not contain olefins (whereas an FCC gasoline will). This distinction presents the forensic investigator with an opportunity. Since all refiners that produce automotive gasoline employ some form of cracking process, an automotive gasoline that contains notable concentrations of olefins must have been blended using either a thermally- or FCC-cracked blending product. Nowadays, that is most likely an FCC. Oppositely, the absence of olefins argues for a gasoline source produced via hydrocracking. (Of course, weathering must be considered since olefins are subject to biodegradation; see below.)

Figure 6.18 shows the weight percent of hydrocarbons $<C_6$ determined for two regular unleaded gasolines produced at two different refineries, one using an FCC and the other using a hydrocracker. The gasolines can be easily distinguished depending on the type of cracking process employed in their production. Furthermore, olefins are highly reactive compounds and are thus rare (or at least rarely reported to occur; Hoering, 1977) in liquid petroleum (crude oil and condensates) and straight-run gasoline. Thus, another potential application for the forensic investigator is in the ability to distinguish automotive gasoline from gas condensate. Gas condensate, being a natural petroleum product, is anticipated to contain no appreciable olefins, whereas the gasoline could. This difference is also demonstrated in Figure 6.18.

US refining charge capacity as of January 1, 2000 indicates that FCC capacity is about 4 times higher than hydrocracking capacity (5.5 versus 1.4 million barrels per day; Radler, 1999). This suggests that most modern gasolines will still contain olefins, although modern reformulated gasolines (RFGs) must

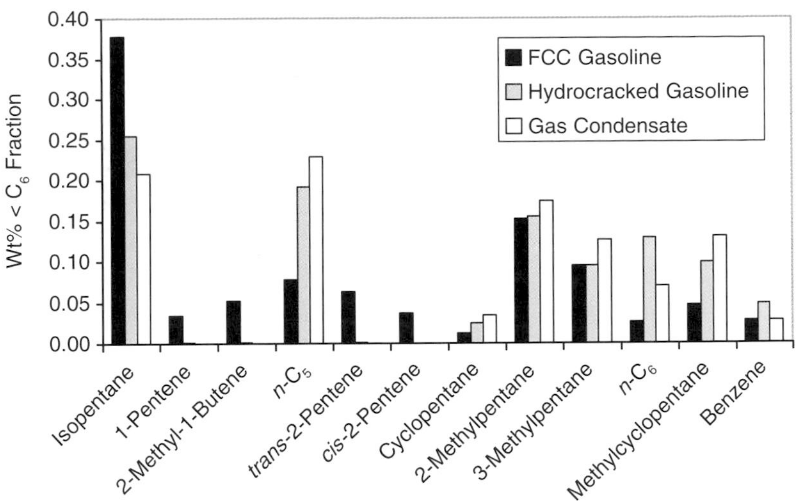

Figure 6.18
Weight percentage of selected hydrocarbons eluting before hexane for regular unleaded automotive gasolines blended with an FCC gasoline and hydrocracked gasoline. A natural gas condensate is shown for comparison purposes.

contain no more than 10% olefins (by volume). In some investigations, the presence or absence of olefins may help to unravel the likely source (or even vintage) of a fugitive gasoline. This is particularly true in markets where branded stations received gasoline exclusively from a local refinery, which employs only one type of cracking unit.

Finally, both FCC and hydrocracking produce a higher boiling 'residual' gas oil that is often blended into residual fuels. It has been reported that hydrocracking 'destroys' any aromatic steranes in the feedstock (Peters *et al.*, 1992). This means that a hydrocracked gas oil will not contain these (and likely other) aromatic biomarkers. Similarly, the temperatures employed in FCC tend to 'crack' pentacyclic triterpanes into novel tricyclic triterpanes and to preferentially destroy compounds that are less thermally stable (e.g., T_m versus T_s; Peters *et al.*, 1992). These types of changes in biomarkers can be important in some forensic studies where weathered crude oil must be distinguished from residual fuels.

6.4.1.3 Catalytic Reforming

The need for improved gasoline octane in the early 1900s also led to the development of thermal reforming processes in which rearrangement ('reforming'), rather than cracking, reactions are dominant. This technology was replaced starting in around 1940 with the development of catalyzed reformation. In this process, reactions occur in the presence of a mixed-catalyst system and hydrogen that promotes the rearrangement of lower octane compounds into higher octane compounds without a significant reduction in carbon number (Speight, 1991). Because of the mixed catalysts used, some compounds tend to lose hydrogen (e.g., forming aromatics from naphthenes) while others tend to gain hydrogen (e.g., saturated alkanes from olefins). The product of catalytic reforming, *reformate*, is a major blending stock for most automotive gasolines.

Dehydrogenation is the primary reaction sought during reformation since it yields the higher octane aromatics (and hydrogen that can be used elsewhere in the refinery, e.g., hydrocracking, desulfurization, or hydrofining). Consequently, most reformates are enriched in the monoaromatic (e.g., BTEX) compounds. Toluene is often the most abundant compound, being formed primarily from the dehydrogenation of methyl-cyclohexane in the feedstock. Since reformates are typically blended directly into gasoline the proportions between various mono-aromatics can provide some diagnostic information potentially related to the reforming conditions (temperature, catalyst type and age, and feedstock). For example, chemical equilibrium during reforming typically results in a stable relationship among the three xylene isomers such that *m*-xylene comprises ~50% of the total xylenes. Substantially higher values are observed in some gasolines and this may be linked to (1) peculiar reforming conditions or (2) the use of *p*- or *o*-xylenes as chemical feedstocks to phthalic acid and anhydride production, respectively. Similarly, the percentage of ethyl-benzene to total C_2-aromatics is normally ~10–20% in fresh gasolines. This percentage can vary as a function of catalyst and conditions and therefore could prove diagnostic in some forensic investigations; e.g., very low percentages of ethyl-benzene versus total C_2-aromatics (<5%) could indicate ethyl-benzene's removal for specialty chemical production (e.g., styrene). Thus, relationships among various aromatics, particularly those with comparable environmental fates, could prove useful in distinguishing gasolines containing reformate blends formed under different catalytic reforming conditions.

6.4.1.4 Isomerization

Isomerization of low boiling (<C_6) normal hydrocarbons to saturated branched hydrocarbons was another refining development triggered by the need for octane, particularly during World War II (Speight, 1991). In this process various catalysts and reactor conditions are used to convert C_4 to C_6 feed streams (butane, pentane, and/or pentane-hexane) into their various isomeric equivalents, i.e., isobutane, isopentane, 2,2-dimethlybutane, 2,3-dimethylbutane, 2-methylpentane, and 3-methylpentane. The specific reaction conditions during isomerization (type and age of catalyst, feed stream and rate, and temperature) will yield rather specific distributions of isomers, collectively known as *isomerate.* Since isomerate is a common blending component in gasoline the distribution among these C_4–C_6 isomers can be used to distinguish among gasolines containing different isomerate blends. For example, studies have shown that the ratio between isopentane and pentane or between 2-methylpentane and 3-methylpentane can vary significantly depending upon the isomerization reaction conditions (Pines, 1981). Monitoring such ratios in environmental samples containing gasoline can yield information about the nature of the isomerate blending stock that may have been used. Of course, the high volatility of these $<C_6$ compounds warrants caution when dealing with environmental media.

6.4.1.5 Alkylation

The acid-catalyzed reactions of olefins with normal hydrocarbons to yield higher boiling, and higher octane, gasoline range branched hydrocarbons (iso-alkanes) was also driven by the need for octane during World War II. Alkylation methods employing hydrogen sulfate (H_2SO_4) and hydrogen fluoride (HF) are still in use today at refineries that utilize alkylation, and not all refineries do. The acids promote the reaction of iso-butane and various C_3 or C_4 olefin streams (e.g., ethylene, propylene, butylenes). The type of olefin stream, combined with the reaction conditions (acid type, temperature, iso-butane/olefin feed ratio, and olefin charge rate), will determine the mixture of iso-alkanes produced (Gary and Handwerk, 1984). While it is difficult to generalize, the reaction of iso-butane with ethylene normally yields large proportions of 2,3-dimethylbutane whereas the reaction(s) of iso-butane with butylenes yields mostly 2,2,4-trimethylpentane (Pines, 1981). Because of its octane and volatility, iso-octane (2,2,4-trimethylpentane) is the primary alkylation product sought by most refiners. In fact, since iso-octane is not present in straight-run gasoline or other blending stocks, its presence in gasoline indicates that the refiner utilized an alkylation unit in the production of the gasoline. For example, a ratio of iso-octane to methylcyclohexane greater than 0.5, and sometimes as high as 20, clearly indicates the presence of an alkylate component in a gasoline. This can sometimes be sufficient to distinguish among gasolines in forensic investigation and, under the right circumstances, eliminate some candidate sources (refiners) which did not utilize an alkylation unit in gasoline production.

Naturally, the high octane of the iso-alkanes makes alkylation an important process at most (but not all) modern refineries. It is the key process in the production of RFG, where octane is achieved largely through high-octane aliphatic hydrocarbons (rather than oxygenates, olefins, or aromatics, which are limited in RFG). For the forensic investigator the variety of compounds formed during alkylation at any one refinery versus those formed at another can provide a useful tool in comparing/correlating gasolines. For example, Figure 6.19 shows a cross-plot of selected iso-alkane ratio for fresh gasolines (regular and premium) from five different refineries that utilize the alkylation

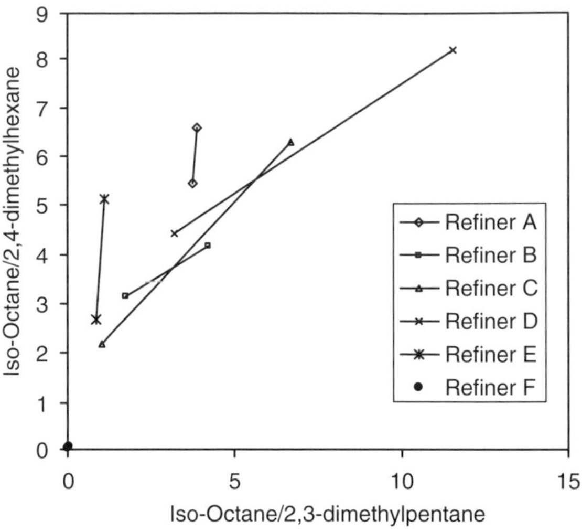

Figure 6.19

Cross-plot of selected iso-alkane ratios for regular and premium gasolines from different refineries. Premium gasolines exhibit higher ratios; refiner F did not use an alkylate blend in the production of their gasoline.

process, and one refinery that does not (refiner F). Note that the gasoline from refiner F contains no iso-octane (and both ratios plotted are zero). Each of the five other refiners' (A–E) regular and premium gasolines exhibit distinct trends in the relative proportion of 2,4-dimethylhexane and 2,3-dimethylpentane normalized to iso-octane. As might be expected, in each case, the premium gasoline contains higher amounts of iso-octane. What is notable is that all five of the trends tend to exhibit only two distinct slopes, which result from consistent proportions among the three compounds represented in the plot. These trends may reflect the distinct alkylation reaction conditions used by each refiner. Recognizing that the proportions among various iso-alkanes can vary among gasolines produced by different refiners could prove useful in certain forensic investigations involving gasolines containing an alkylate component.

6.4.2 FINGERPRINTING CHARACTERISTICS OF COMMON PETROLEUM PRODUCTS

6.4.2.1 Automotive Gasoline

The complicated and dynamic history of automotive gasoline over the past 100 years is not an easily summarized topic. The evolving requirements for gasoline's combustibility, antiknock quality (octane), chemical stability, volatility, gum content, and combustion emission yields, combined with the refiners' varying approaches to meet these requirements, have led to a wide variety in gasoline composition over time. Today, the blending of gasoline at any given refinery depends upon their internal inventory of blending stocks, the operating status of the various refining units, the regulatory requirements for the intended marketing location, and, of course, any economic requirements. Consequently, generalizations regarding the chemical composition of 'typical' automotive gasoline are often oversimplified and dangerous for the forensic investigator.

Even when gasolines of a similar grade (octane), vintage, and from a single region of the country are compared, significant compositional differences can still be observed. This is exemplified in Figure 6.20, which shows the normalized PIANO distribution for two premium RFGs sold in the mid-Atlantic region (an ozone non-attainment area) during the winter of 1999. Both gasolines (presumably) met federal RFG requirements yet each exhibits distinct hydrocarbon distributions. It is apparent that the RFG from refiner A achieved octane primarily from the blending of MTBE and toluene whereas refiner B achieved octane from MTBE and iso-octane. This probably reflects a difference in refining capabilities, e.g., refiner A does not employ an alkylation unit and must rely upon aromatics (reformate) to achieve the necessary octane.

The type of differences exemplified in Figure 6.20 can provide the forensic investigator with an ability to distinguish gasolines from different sources. However, as the regulations governing gasoline formulation/emissions have become stricter, and the gasoline distribution system has become more complicated (e.g., fungible pipelines, unbranded or exchanged products, etc.), the ability to defensibly distinguish a gasoline's particular source on the basis of chemical composition alone has become more difficult. Therefore, the forensic investigation must often incorporate chemistry with local geologic and hydrologic controls, as well as operational and regulatory history, to unravel gasoline issues.

In fact, the regulatory history of gasoline, in particular the additives found in gasoline, is an important consideration for the forensic investigator (Dorn *et al.*, 1983; Gibbs, 1990, 1993, 1996; Frank, 1999; Morrison, 2000). In the

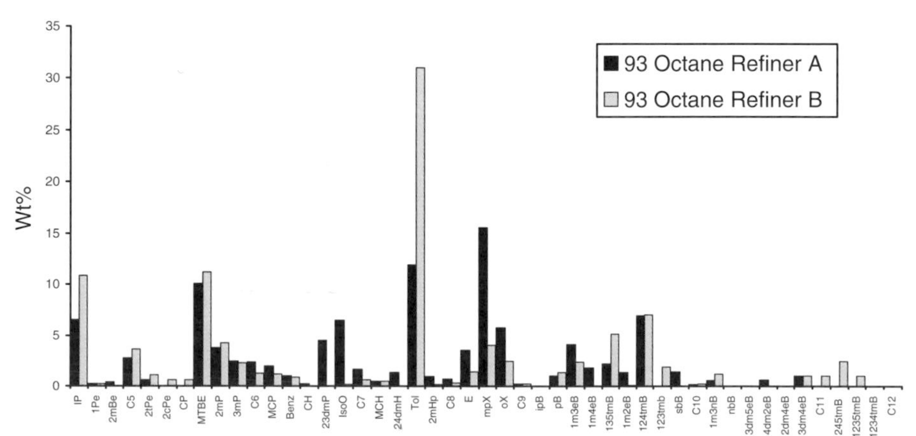

Figure 6.20
Histogram of selected PIANO analytes showing a comparison of premium reformulated gasoline from two refiners.

sections below, the regulatory history of the most common and useful gasoline additives is presented.

6.4.2.1.1 Antiknock Additives

Additives were continually introduced into gasoline in the attempt to improve engine efficiency, principally the antiknock properties or 'octane rating'. Among the most important additives in the early evolution of automotive gasoline was organic lead, specifically tetraethyl lead (TEL; $Pb(C_2H_5)_4$). TEL was first introduced into automotive gasolines in 1923 as an antiknock agent (Gibbs, 1990). After some initial controversy surrounding safety concerns (prompted by the deaths of five Standard Oil of New Jersey refinery workers in 1924; Robert, 1983; Needleman, 1998), the use of TEL increased steadily in the following decades (except during World War II) as the demand for increased engine performance and fuel economy grew. Ultimately, the average levels of lead in the premium gasoline pool reached a high of approximately 3.0 grams lead per gallon (glpg) around 1970; individual gasolines could legally contain up to 4.23 glpg, though they rarely did. Until 1970, with one exception, all grades of gasoline sold in the US contained alkyl lead. After 1970, improvements in refining practices and new federally mandated restrictions (1970 Clean Air Act and 1990 Clean Air Act Amendment) led to a systematic reduction in maximum allowable lead levels in the US gasoline pool (Table 6.10) and the introduction of low-lead and unleaded gasolines. This reduction culminated with a complete elimination of organic lead additives in automotive gasoline in the US in 1996 (1992 in California).

The federal rollback schedule (Table 6.10) can sometimes be useful to the forensic investigator for bracketing the age of spilled gasoline in the subsurface (Johnson and Morrison, 1996; Kaplan *et al.*, 1997). Sometimes the maximum allowable concentrations (Table 6.10) can accomplish this; however, a significant

Table 6.10

Summary of the federal rollback of lead in automotive gasoline.

Date	Legal Maximum Pb (grams/gallon) in Gasoline	
	Leaded Gasoline	Unleaded Gasoline
1926	3.17	
1959	4.23	
Jul. 1974		0.013
Oct. 1982	1.1[a]	
Jul. 1985	0.5[a]	
Jan. 1986	0.1[a]	
Jan. 1992	Banned in California	0.05[b]
Jan. 1996	Banned throughout US	0.05[b]

[a]Average quarterly leaded gasoline production.
[b]Incidental lead in unleaded gasoline.

number of variables can affect gasoline lead levels in terms of both pre-spill (e.g., lead credits, lead banking, whole and leaded gasoline pool averaging, summer vs. winter gasoline) and post-spill (e.g., mixing of multiple gasolines, partitioning, and fractionation) conditions. Thus, extreme caution is necessary in dating spilled gasolines solely based on the concentration of lead in free phase gasoline. Under most circumstances, the best that the concentration of organic lead in a free phase gasoline sample may accomplish is to bracket a time frame of (at least) one component in the spilled gasoline. For example, if a free phase gasoline sample is determined to contain >0.1 g/gal it is reasonable and defensible to conclude that it (at least) contains a pre-1986 component.

Considerable value lies in the ability to 'fingerprint' the individual organic lead compounds that are present in environmental samples impacted by gasoline. To our knowledge, this is only possible in free phase gasoline samples (since extracting organic lead quantitatively from soils is problematic; see analytical discussion above). TEL was the lone organic lead compound added to automotive gasoline until 1960, when tetramethyl lead (TML) was formulated, and various lead 'packages' were developed and introduced to the gasoline market (Gibbs, 1990). Packages included both physical mixtures and reacted mixtures of TEL and TML. Reacted mixtures are the end products of a catalyzed TEL:TML reaction that results in the formation of five organic lead compounds, namely:

- tetramethyl lead (TML)
- trimethylethyl lead (TMEL)
- diethyldimethyl lead (DEDML)
- methyltriethyl lead (MTEL)
- tetraethyl lead (TEL)

The relative amounts of the five individual organo-lead compounds in any lead package used in gasoline were dependent upon the molar ratios of the two starting materials (TEL and TML). Reacted mixtures (RMs) were marketed typically as RM25, RM50 and RM75, with the numerical designation referring to the molar percentage of TML in the mixture (Table 6.11). Physical mixtures are an unreacted combination of TEL and TML in known percentages (e.g., 20:80, 50:50, and 80:20). Included in the commercially available lead packages for automotive gasolines were lead scavengers, 1,2-dibromoethane and 1,2-dichloroethane, at constant weight percentages of the package (17.9 and 18.8 wt%, respectively). These scavengers have long been used to minimize the precipitation of non-volatile lead oxides in an automobile engine's combustion chambers through formation of volatile lead chloride and bromide salts. Worth

Table 6.11

Approximate weight percentage of total lead for individual lead alkyls in commercial reacted and physical mixes.

	Reacted Mixes			Physical Mixes			TEL only
	RM25	RM50	RM75	PM20	PM50	PM80	
TEL	28.80	4.83	0.09	80.00	50.00	20.00	100
MTEL	49.51	25.59	3.61	0.00	0.00	0.00	0
DEDML	18.60	42.40	20.51	0.00	0.00	0.00	0
TMEL	2.99	23.40	49.60	0.00	0.00	0.00	0
TML	0.10	3.79	26.19	20.00	50.00	80.00	0

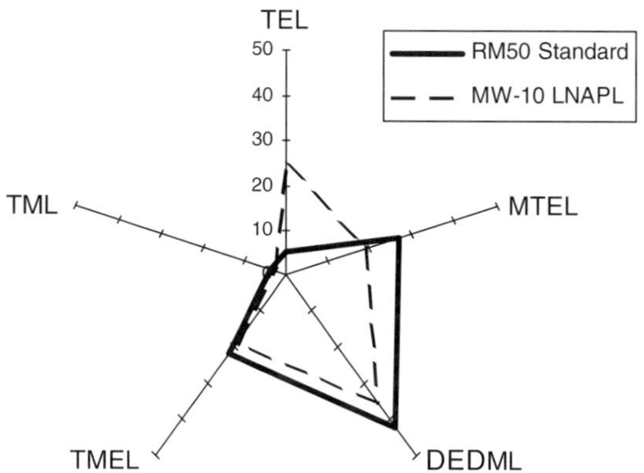

Figure 6.21

Star-diagram showing the alkyl lead distributions in a free product sample and an RM50 standard. Free product appears to be mixture of RM50 and TEL-only gasolines.

noting here is the fact that 1,2-dichloroethane was/is not used in leaded aviation gasolines.

Certain forensic investigations can benefit from determining the lead alkyl distributions since it provides an additional level of 'fingerprinting,' which the concentration of 'total organic lead' cannot. Comparisons among samples can be easily visualized on five-point star diagrams that can be compared to the standard, commercially available RMs (Table 6.11). For example, Figure 6.21 shows the distribution of lead alkyls determined in a free phase gasoline sample from a forensic investigation involving a service station with two past operators. The distribution of lead alkyls present in the free phase product (MW-10) closely matches that of a gasoline containing an RM50 lead package (Table 6.10), with the exception of an elevated level of TEL. This distribution argues that the MW-10 gasoline sample most likely contains a mixture of (at least) two leaded gasolines, one containing RM50 and one containing TEL only. With this knowledge in hand, records research of the types of lead packages historically used by each operator in this market may be able to help determine liability for the contamination.

The amount of a particular lead alkyl package that was used depended on the refiner's needs and economics. To our knowledge, mixing of different lead packages in a single gasoline did not occur. However, very often different lead packages were added to different grades of gasoline produced at a single refinery. Corporate records of particular additive packages, which can become available in the course of a litigation matter's discovery process, can sometimes help unravel issues of leaded gasoline source(s).

Related to the fingerprinting of lead alkyls is the perception that after 1980 RM and PM lead packages were no longer in use (e.g., Hurst, 1996). However in our experience, too many exceptions exist to defensibly utilize 1980 as a 'cut-off' after which lead alkyls other than TEL should not be present in leaded gasolines. It seems more likely that after 1980 the use of lead mixes was reduced, rather than eliminated. Again, careful records research is the only way to be sure what a refiner may have been adding to their gasoline at a given time.

Methylcyclopentadienyl manganese tricarbonyl (MMT) is another anti-knock additive that was first commercially available in 1959, when it was blended with TEL and marketed in 'Motor Mix 33' or 'AK-33X' (57.5% TEL and 6.97% MMT; Gibbs, 1990). Its use was limited until the initial lead phase-down in the early 1970s. At that time, MMT was first independently added to gasoline to offset the octane loss accompanying the removal of alkyl leads and was in widespread use in unleaded and leaded gasolines by 1976. The maximum allowable concentration at that time was 0.125 g Mn per gallon. The Clean Air Act Amendments of 1977, however, banned the use of manganese additives in unleaded gasolines after October 1978, unless the EPA granted a special waiver (Zayed *et al.*, 1999; Morrison, 2000). MMT's use continued in leaded gasoline and its use was even temporarily allowed in unleaded gasolines in the summer of 1979 (during the oil embargo). Its use in leaded gasolines ended in 1996 with the elimination of leaded gasoline. In 1995, EPA granted a waiver for MMT's use in unleaded gasolines, but limited the maximum concentration to 0.031 g Mn per gallon. California regulations have continued to ban the use of manganese additives in unleaded gasoline. In our experience, the presence of MMT in fresh unleaded gasolines is quite rare. Its presence in selected environmental samples (e.g., free phase gasoline), however, may help to determine the age of (at least) a component in a gasoline mixture.

6.4.2.1.2 Methyl Tertiary Butyl Ether

As the EPA mandated the lowering, and eventual removal of lead from automotive gasoline, new classes of octane-boosting additives, with fewer (presumed) negative impacts than organic lead, were being developed and used in increasing frequency to augment gasoline performance. The additives included a variety of alcohols and ethers.

Alcohols, predominantly *tert*-butyl alcohol (TBA) and ethanol, were respectively introduced as a gasoline additive/supplement in the late 1960s and late

1970s, the latter being primarily in response to rising oil prices. (Notably, ethanol was actually first introduced, unsuccessfully, in Nebraska in the 1940s.) Throughout the 1970s a variety of alcohol blends and ether-based additives were developed which further 'extended' the gasoline supply, and boosted octane (Guetens *et al.*, 1982). The most volumetrically important of the ethers, MTBE, was first added to commercially available gasolines in the United States in 1979 (Squillace *et al.*, 1995).

In the early 1980s MTBE was used primarily as an octane booster, and was added at 3–8% vol/vol into a minority of the gasoline pool (usually higher octane, premium blends). In the late 1980s when the need to reduce automobile tailpipe emissions became more obvious, alcohols and ethers proved even more beneficial, and their use expanded. In 1988, Colorado first mandated the use of oxygenated gasoline during winter months. By 1991, Arizona, Nevada, New Mexico and Texas initiated similar requirements and in November 1992 39 CO non-attainment areas required oxygenated gasolines be sold during the winter months (November–February). In some of these areas, MTBE was being blended at up to 15% vol/vol to produce highly oxygenated, cleaner burning gasoline (i.e., 'oxy-fuel').

In 1995, nine areas with severe ozone levels were required to sell reformulated gasoline (RFG), which contained a minimum of 2% oxygen (by weight), year round, which was often met by the addition of MTBE. Many other areas of the country 'opted-in' to the federal RFG program; meanwhile California initiated their own requirements (Phase II, 1.8–2.2 wt% oxygen). At that time it is estimated that 60–70% of all gasoline sold in the US and 95% of gasoline sold in California contained MTBE at some concentration. The recent backlash against the use of MTBE already has led to a reduction in its use and a re-evaluation of its need (EPA Blue Ribbon Panel, 1999).

For the forensic investigator the history of MTBE's (or other oxygenate) use in gasoline would seem to provide a gross method of age dating. However, the environmental behavior of this chemical compared to hydrocarbons (e.g., aqueous solubility) and its convoluted use (e.g., winter only versus year round) often makes its presence/absence or concentration difficult to defensibly interpret. The most useful application of MTBE is in terms of its concentration gradients or spatial distribution in groundwater, which can reveal multiple sources/releases of the MTBE (e.g., Davidson and Creek, 1999).

6.4.2.2 Aviation Fuels

Aviation fuels include two broad categories of fuels, namely, aviation gasoline (avgas) and turbine (jet) fuels, the general fingerprinting characteristics of which are discussed in the following sections.

6.4.2.2.1 Aviation Gasoline

Avgas developments prior to and during World War II continually sought to improve performance of the spark ignition engines of aircraft. The product specifications of avgas have changed little over the past 50 years, largely because of diminishing demand due to the introduction and proliferation of jet engines. Because of this reduced demand, the various pre-existing grades of avgas have been gradually phased out such that modern avgas comes in only three grades, (1) Grade 80, (2) Grade 100, and (3) Grade 100 low-lead (ASTM D910; Bishop and Henry, 1993). Grades 80 and 100 are in only minor use today by civilian and military aircraft, respectively. Grade 100 low-lead (100LL) is by far the most common avgas in use in the US today, and therefore the most likely to be encountered in forensic investigations. In fact, manufacturers of all new single engine aircraft sold in the US today require Grade 100LL avgas, and many older aircraft have been converted to run on 100LL.

Forensic investigations involving avgas often face the difficulty of differentiating among the three grades. Differentiation among the three avgas grades in forensic investigations, particularly on free phase samples that have not been in the environment too long, can occasionally be made simply by color, since each grade in use contains different colored (anthraquinone- or azo-based) dyes (Table 6.12). However, since these dyes will quickly break down in the environment, the forensic investigator must more often rely upon (1) detailed volatile hydrocarbon fingerprinting techniques and (2) lead alkyl measurements (as described earlier in this chapter) in order to distinguish between avgas grades. Fortunately, the stringent specifications for avgas (understandably much stricter than for automotive gasoline) invoke a rather uniform chemical fingerprinting character to avgas of a given grade. Only Grade 80 avgas contains a significant straight-run gasoline component (including naphthenes, paraffins, and some aromatics), which when weathered can sometimes be confused with automotive gasoline. The two higher grades require significantly more octane (than Grade 80 or than automotive gasoline), which is most commonly obtained by blending alkylate (enriched in iso-octane and other iso-paraffins) with small amounts of straight-run gasoline. Reformate can also be a blending component of higher grades of avgas, which can introduce

Table 6.12

General characteristics of modern aviation gasolines used in the US.

Aviation Gasoline Grade	Dye	glpg[a]
Grade 80	Red	0.14
Grade 100	Green	1.12
Grade 100 LL	Blue	0.56

[a]Maximum allowable grams lead per gallon (as TEL).

mono-aromatics (particularly toluene), whose presence tends to improve performance when engines are operated at a rich mixture rating. Non-hydrocarbon blending agents (other than lead additives, dyes, and antioxidants) are prohibited in avgas (Bishop and Henry, 1993). Thus, oxygenates (MTBE, etc.) will not occur in avgas.

As indicated in Table 6.12, alkyl lead additives are still used in avgas in the US. Even the low-lead grade (100LL) can contain up to 0.56 glpg. Unlike automotive gasolines (described above), the form of alkyl lead added to avgas has uniformly been TEL (only). No reacted or physical mixed of TEL and TML were permitted for use in avgas. In addition, chlorine-based lead scavengers (ethylene dichloride) are not used in avgas lead additive packages, only bromine scavengers (ethylene dibromide) are allowed; chlorine in the engine yields more corrosive combustion products detrimental to engine parts.

6.4.2.2.2 *Jet Fuel*

Developers of the first jet engines during World War II sought a fuel source that would not further reduce the valuable gasoline supply. Their attention focused on kerosene. Consequently, the jet fuel still in use today largely resembles kerosene in terms of its gross chemical and physical properties. This is particularly true for the commercial grade of jet fuel (Jet A; Figure 6.2). In fact, Jet A is one of a few petroleum products that can be directly distilled from sweet crude oil feedstock (straight-run jet, Table 6.9). Straight-run jet distilled from sour or aromatic crudes normally requires additional refining (e.g., hydrotreating) before meeting Jet A specifications. In either instance, Jet A is comprised almost entirely of hydrocarbons in the C_8 to C_{18} range. Jet A specifications, however, are broad enough to allow considerable variability in the molecular distributions within this boiling range.

Additives are not required in Jet A; however, antioxidants (optional in most, but required in Jet A refined using hydro-treating), metal deactivators, corrosion inhibitors, anti-icing additives, anti-static additives, and lubricity additives are optional (Henry, 1988). Alkyl lead and dyes are prohibited in Jet A.

Military jet fuels are slightly more widely varying than commercial Jet A. Table 6.13 lists the US military jet fuels, their use and category. It is observed that military fuels fall into three categories, namely (1) wide cut-type fuels, (2) kerosene-type fuels, and (3) selected component fuels. GC/FID chromatograms of examples of each category are shown in Figure 6.22. The wide cut military jet fuels are manufactured by blending a heavy straight-run gasoline (naphtha) or cracked gasoline fraction with straight-run jet, thereby reducing the lower boiling point of these fuels. Kerosene-type military jet fuels, as the name implies, are generally comparable to kerosene and commercial Jet A. These consist of distilled hydrocarbons in the C_8 to C_{18} range. One exception

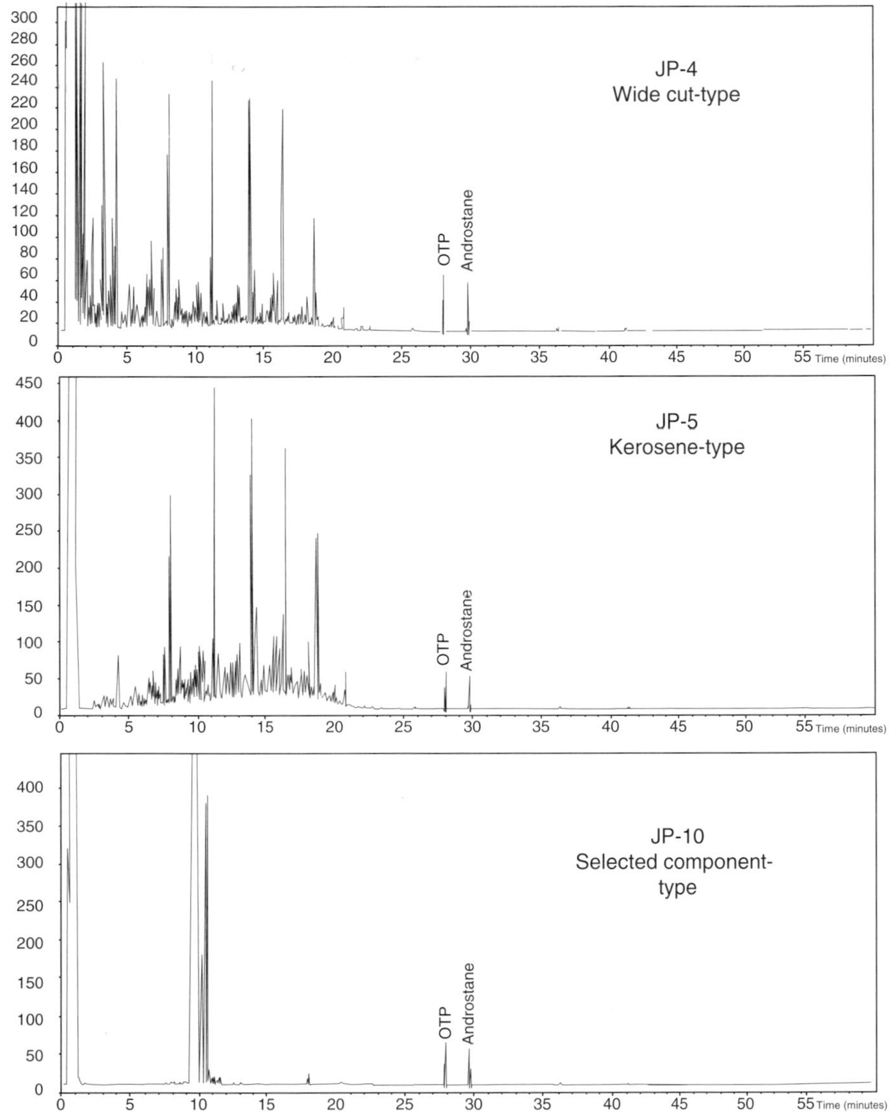

Figure 6.22
GC/FID showing three types of military jet fuels.

is JP-7, a 'low volatility' kerosene-type jet fuel that is depleted in hydrocarbons boiling below C_{12}. Finally, specialty fuels (rocket fuel, etc.) may be rather unique and comprised of only a few selected hydrocarbons (Table 6.13).

6.4.2.3 Land and Marine Diesel Engine Fuels

Diesel engines combust petroleum mixtures by the heat generated during compression of a fuel–air mixture, rather than by a spark (as is the case in gasoline engines). Consequently, diesel engines will tend to (safely) burn most petroleum mixtures that boil between about 150 and 350°C. However, some

Table 6.13
Inventory of US military jet fuels.

Grade	First Issued	Use	Characteristics
JP-1	1944	Obsolete	Kerosene-type
JP-2	1945	Obsolete	Kerosene-type
JP-3	1947	Obsolete	Kerosene-type
JP-4 (Jet B)	1950	Air force standard	Wide cut-type
JP-5	1950	Naval carrier standard	Kerosene-type
JP-6	1956	Obsolete	Wide cut-type
JP-7	1965	Very high performance	Low vol. Kerosene-type
JP-8 (Jet A-1)[a]	1976	Air force standard	Kerosene-type
JP-9	NA	Missile fuel	Methylcyclohexane/JP-10/ norbornadiene blend
JP-10	NA	Propellant	Dicyclopentadiene-dominant

[a]Reduced maximum freezing point (−47°C) compared to Jet A (−40°C).

slow operating diesel engines (e.g., marine diesels and stationary industrial heaters) can combust petroleum mixtures with much higher upper boiling points. ASTM has established specifications for land and marine diesel fuels (ASTM D975) and fuel oils (ASTM D396), and it is from these classifications that the designations of #1, #2, #4, #5, and #6 are obtained (Jewitt *et al.*, 1993). Specifications for a #3 Grade were dropped in 1948 and replaced with those for #2 (Schmidt, 1969), thus essentially eliminating use of #3.

Note that in the US the terms 'diesel fuel' and 'fuel oil' are often used interchangeably whereas in Europe, the term 'fuel oil' is reserved for non-distillate fuels, i.e., residual fuel oils. In practice, it is most common to refer to the distillate fuels (#1, #2, and #4; i.e., fuels that have been vaporized and re-condensed) as *diesel fuels* if they are to be used in diesel engines. These would include diesel fuels #1, #2, and #4 (Table 6.14). If these distillate fuels are not used in diesel engines they are commonly referred to as fuel oils #1, #2, and #4 (see the next section). Similarly, it is most common to refer to the residual fuels (#5 and #6; i.e., fuels that have not been wholly vaporized) as *fuel oils*, regardless of whether they are burned in a diesel engine or not. To complicate matters, fuel oil #6 is referred to as 'bunker C' when it is to be used in marine diesel engines.

Diesel types #1 and #2 are most commonly used in (on road) transport vehicles and (off road) locomotives whereas #4 diesel is most common in stationary diesel motors or furnaces requiring constant loads (Jewitt *et al.*, 1993). While fuel oil #5 and #6 are often used in marine diesel engines, these fuels will be described in the next section dealing with industrial heating and stationary electric power generation. Selected examples of diesel fuel #1, #2, and #4 are shown in Figure 6.23. Each of the diesel fuels shown are dominated by *n*-alkanes that exhibit a normal distribution over the boiling range of each product. The *n*-alkanes reach a maximum near the mid-boiling point, generally between

Table 6.14

Terminology commonly applied to distillate and residual fuels.

For use in	Diesel Engines	Other
Distillate fuels	Diesel Fuel #1 Diesel Fuel #2 Diesel Fuel #4	Fuel Oil #1 (kerosene) Fuel Oil #2 (home heating oil) Fuel Oil #4
Residual fuels	Diesel Fuel #5 Diesel Fuel #6 (bunker C)	Fuel Oil #5 Fuel Oil #6

n-C_{13} and n-C_{16} (~230 to 280°C). This normally distributed *n*-alkane envelope is expected due to the nature of the distillation process.

n-Alkanes are readily ignitable upon compression, and thus, diesel fuels are typically enriched in *n*-alkanes (Figure 6.23). Isoprenoids and aromatic hydrocarbons are slower to ignite upon compression than *n*-alkanes, which introduces an ignition delay within the diesel engine. Ignition delay is an undesirable quality (measured in terms of the cetane index), indicating that excess isoprenoids or alkyl-naphthalenes in diesel fuel are undesirable qualities. Excess *n*-alkanes are also problematic in terms of wax precipitation in cold weather. Thus, all fresh diesel fuels generally exhibit a predominance of *n*-alkanes with lesser amounts of isoprenoids and aromatic hydrocarbons. In cold climates, where wax precipitation is particularly problematic for diesel engines (e.g., Alaska), diesel fuels normally contain smaller relative percentages of *n*-alkanes and more isoprenoids/aromatics (along with special additives to minimize rigidity of wax), as well as, a lesser proportion of n-C_{15}+ compounds.

Diesel fuels #1, #2, and #4 were traditional straight-run distillate products of a crude oil feedstock. However, given the flexibility in chemical composition the diesel engine allows, and the availability of refining intermediate streams, many modern diesel fuels will include a cracked product in the same distillation range (Jewitt *et al.*, 1993). Addition of cracked distillate generally lowers the API gravity of a diesel fuel (relative to straight-run) due to the addition of olefinic and aromatic hydrocarbons (formed by the dehydrogenation that accompanies cracking) and, for diesel fuel #2, can lower the initial boiling point (Schmidt, 1969). The properties of a given diesel fuel, however, are largely a function of the crude oil feedstock. As discussed previously, some primary features expressed within the diesel range (~C_{10}–C_{25}) of crude oil are normally directly transferred to diesel fuels (e.g., Pr/Ph and n-C_{17}/Pr ratios), often permitting the forensic investigator to distinguish among diesel fuels from different sources using primary fingerprinting features within the diesel range (e.g., sesquiterpanes, diterpanes, isoprenoids; Figure 6.10). An example of this approach is demonstrated in Figure 6.24, which shows representative GC/FID chromatograms for weathered diesel fuel/fuel oil #2 from adjacent terminal sites. The degree of weathering and comparable isoprenoid ratios (e.g., Pr/Ph ~ 1.2)

Figure 6.23
GC/FID of selected middle distillate fuel oils.

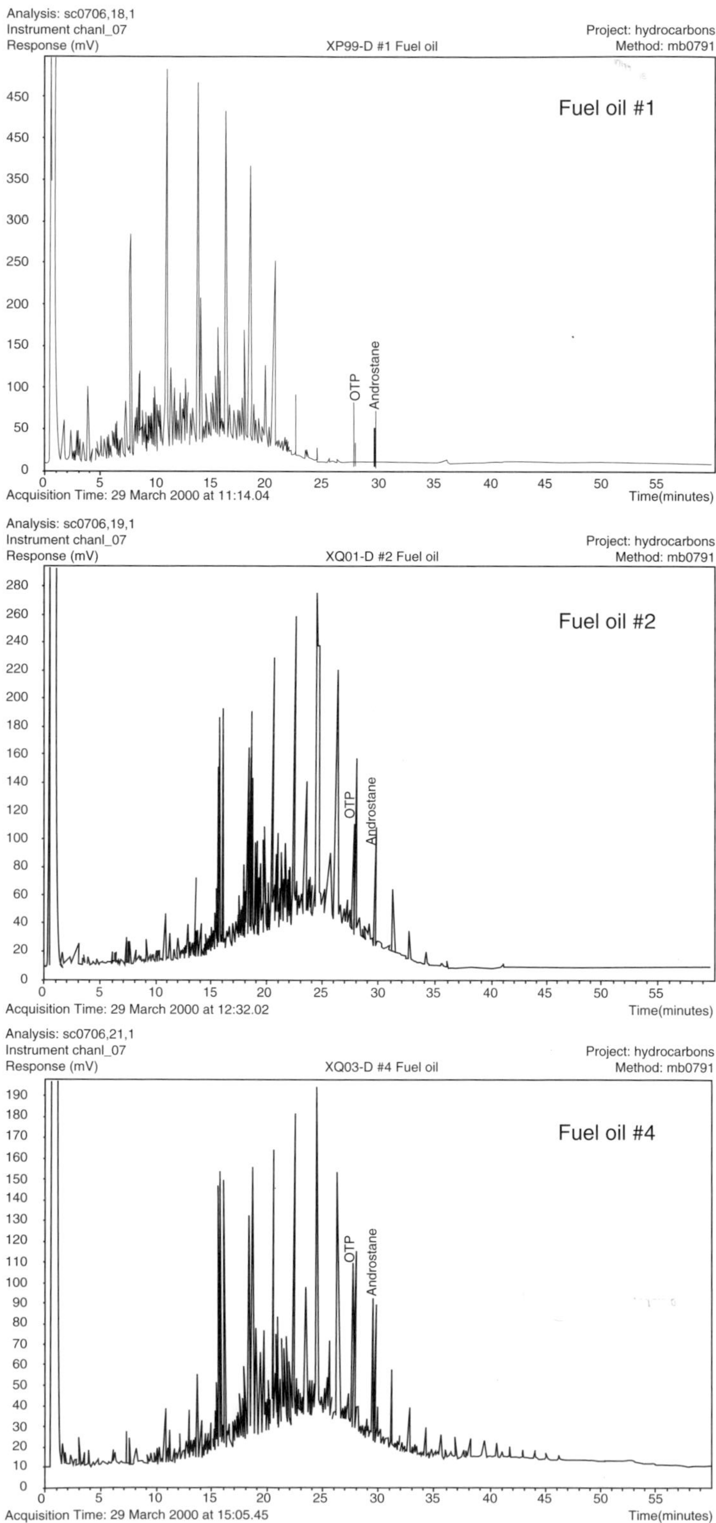

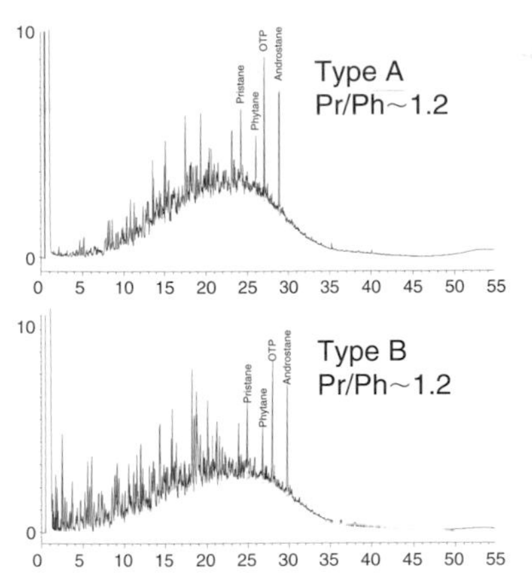

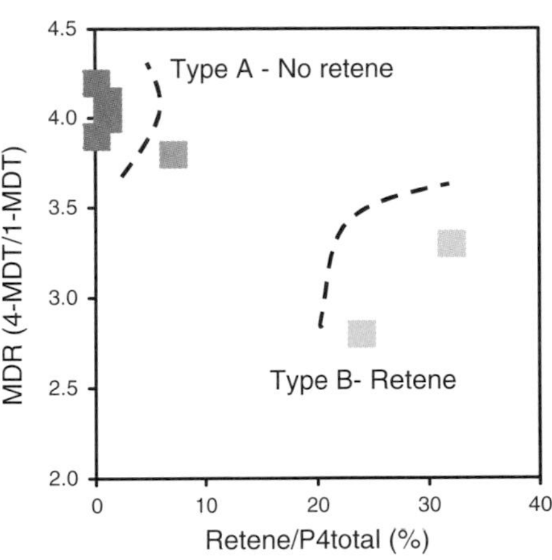

Figure 6.24

GC/FID and cross-plot of weathered diesel fuel/fuel oil #2 from adjacent terminals. Cross-plot shows how PAH analytes allowed for differentiation where GC/FID had not.

suggested a common source. However, GCMS revealed some important diagnostic differences among the weathered products related to the presence/absence of retene (an aromatic diterpenoid biomarker) and the distribution of methyl-dibenzothiophenes, both primary fingerprinting properties inherited from the crude oil feedstocks for each fuel type. A cross-plot of these diagnostic parameters for the weathered diesel fuel/fuel oil #2 samples from the sites reveals the two types of diesel/fuel oil, and a likely mixture at one location.

Fuel oil #4 is not a particularly common fuel. It is sometimes used in larger boilers (e.g., schools and other large buildings). The specifications for fuel oil #4 focus largely on viscosity, which allows for a variety of compositions using different blends. Fuel oil #4 is often blended from whatever refinery intermediates may be available (straight-run distillates, cracked distillates, residual fuels, or even reclaimed oils). Consequently, there is no such thing as a 'typical' fuel oil #4, when it comes to fingerprinting studies. Even though they are used in marine diesel engines, the properties of fuel oil #5 and fuel oil #6 (bunker C) are discussed in the next section.

6.4.2.4 Heating and Power Generating Fuels

The terminology surrounding distillate and residual fuels presented in the previous section indicated that the intended use of the petroleum product dictated its name. In this section, the fingerprinting characteristics of fuels used for heating and power generation are discussed. These include the distillate fuels, (1) kerosene (also known as fuel oil #1, range oil, or stove distillate) and (2) domestic heating oil (fuel oil #2), and the residual fuels, (3) fuel oil #5 and (4) fuel oil #6, also termed bunker C (Table 6.14).

Because of the historic demand for kerosene for heating, cooking, and illumination, it was at one time the most desirable product obtainable from crude oil. This product is now in limited demand (in the US) and constitutes only a minor product at most refineries; other petroleum products within the same boiling range (Jet A) are much more profitable. Like Jet A (Figure 6.2), kerosene is typically a straight-run distillate of a paraffinic feedstock oil. Consequently, it is generally enriched in *n*-alkanes, which reduces the smoking during its combustion, and does not contain excessive aromatics (as might exist in a cracked distillate). If additional refining is necessary, it is usually focused on the removal of odor-causing sulfur compounds.

Fuel oil #2 (home heating oil) is normally obtained as a straight-run distillate from paraffinic crude oils (Martin and Gray, 1993). Domestic heaters are more susceptible to variations in fuel quality (than diesel engines), thus fuel oil #2 can be less variable than diesel fuel #2. This does not mean distinguishing these fuel types is ever easy, since the specifications for the two products are identical. Fuel oil #2 exhibits the same features as diesel fuel #2, many of which are inherited from the crude oil feedstock.

During the winter months in cold climates, it is the practice to reduce the viscosity and pour point of home heating oil through the addition of a kerosene fraction. The improved viscosity allows for ease of handling and improved atomization in domestic oil burners. Figure 6.25 shows the GC/FID fingerprints for two home heating oils, one that has been 'cut' with kerosene. The fuel oil 'cut' with kerosene contains an estimated 30% kerosene; however, blending to reduce viscosity and pour point of fuel oil #2 can require between 10–50% kerosene (Schmidt, 1969). This practice is of considerable importance in forensic investigations involving spills of both fuel oil #2 and kerosene.

The specifications for (1) kerosene and diesel fuel #1 and (2) home heating oil and diesel fuel #2 are identical (ASTM D396); therefore these product pairs share most fingerprinting features. In most situations, the forensic investigator will be unable to distinguish products within these pairs based on the hydrocarbon fingerprints alone.

Fuel oils #5 and #6 (bunker C) are residual fuels (Table 6.14) that are used in marine diesels and industrial power generation. They require pre-heating prior to their use due to their high viscosity, relative to fuel oils #2. As the need for gasoline in distillate fuels increased and refining operations 'squeezed' more of these products from crude oil, the nature of the residual hydrocarbon mixtures has become increasingly 'heavier' in character. As a result, it is necessary to blend a residual fuel hydrocarbon mixture with 20 to 40% of a lower viscosity mixture (from kerosene to crude oil) in order to achieve the desired properties of a marketable fuel oil #5 or #6 (Schmidt, 1969). Blending disparate hydrocarbon mixtures often promotes precipitation of asphaltenes within the residual fuel leading to production of tank bottom sludge, which, of course, must be disposed of.

Figure 6.26 shows the GC/FID chromatograms for six fresh bunker C fuels. The variety in chemical composition among these is remarkable and exemplifies the lack of any such thing as a 'typical' bunker C (fuel oil #6). This type of

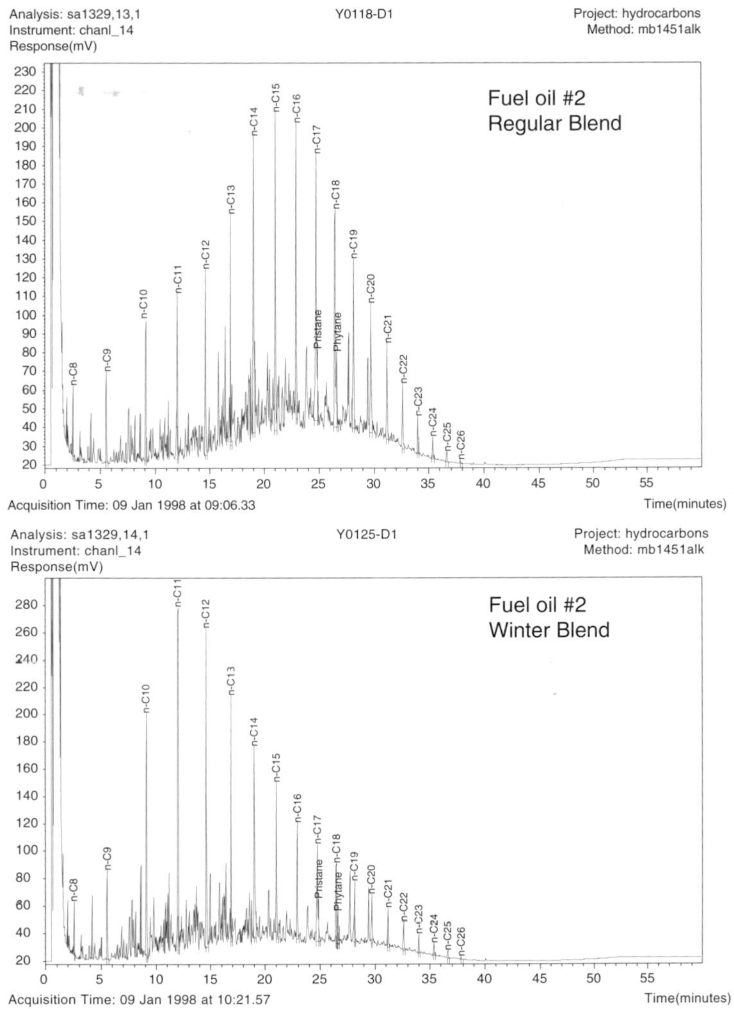

Figure 6.25

GC/FID comparison of regular and winter blends of diesel fuels. Winter blend contains 20–30% kerosene.

variability, however, provides the forensic investigator with opportunity to use a variety of fingerprinting characteristics to distinguish or correlate different bunker C sources. A case study in which biomarkers and PAHs were used to correlate a spilled bunker C to its suspected sources was recently presented by Stout *et al.* (2000a).

6.4.2.5 Lubricating Oils

Petroleum-derived lubricating oils (mineral oils) reduce friction on engine parts and other machinery and/or serve to deliver anti-corrosion agents. They are produced through the atmospheric distillation of crude oil and subsequent vacuum distillation of atmospheric residuals. The heavier fractions produced during vacuum distillation boil between about 300°C and 600°C, which corresponds to a carbon range of approximately C_{17} and C_{45}. These fractions serve as feedstocks to solvent extraction, solvent dewaxing, or hydrofinishing, which

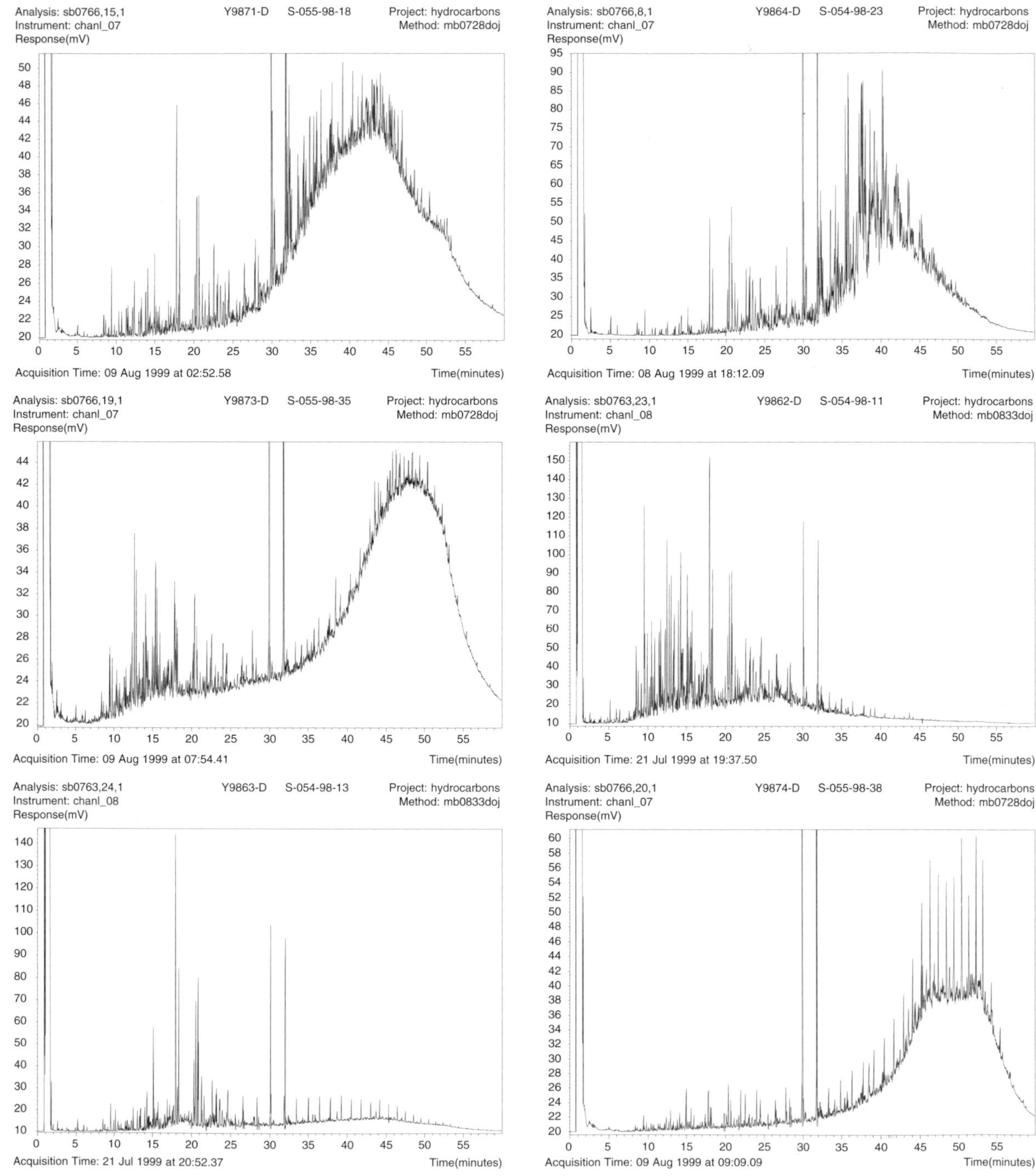

Figure 6.26

GC/FID demonstrating the considerable variability that can exist among bunker C (marine) fuels.

are used to produce a wide variety of lubricating oils, some of which are subsequently blended with specialty chemicals with anti-corrosion, anti-oxidation, anti-rust, or surfactant properties.

Terminology among lubricating oils is largely linked to the application. Common types of lubricating oils include crankcase oil (gasoline and diesel), transmission oil (fluid), hydraulic oil, heat-transfer oils, cutting oils, electrical oils, rolling oils, gear oils, and steam turbine oils (Vazquez-Duhalt, 1989).

Crankcase oil is perhaps the most common lubricating oil encountered by the forensic investigator. In addition to the pervasive leakage of crankcase oil from vehicles (a major contributor to urban runoff), large volumes of waste crankcase oil are produced annually in the US (estimated at 400 million gallons). Recycling programs have reduced the volume of crankcase oil that can find its way into the environment, though improper disposal is still a problem.

Used crankcase oil will have been influenced by the heat of operation as well as, in modern vehicles, the addition of some exhaust components through the positive crankcase ventilation (PCV) value and unburned fuel (both gasoline and diesel fuel #2). The effect of excessive heat can significantly increase the concentration of PAHs in used motor oil. Figure 6.27 compares the PAH concentration and distribution for a new and used motor oil (of the same type; Restek™ Standards). The total concentration of measured PAHs has increased from 6450 mg/kg to 87 400 mg/kg and the distribution has shifted to include numerous 5- and 6-ring PAHs that were absent in the new motor oil. These types of differences can help distinguish new and used motor oil, which could be relevant in certain investigations. In addition, heavy metal (lead, cadmium,

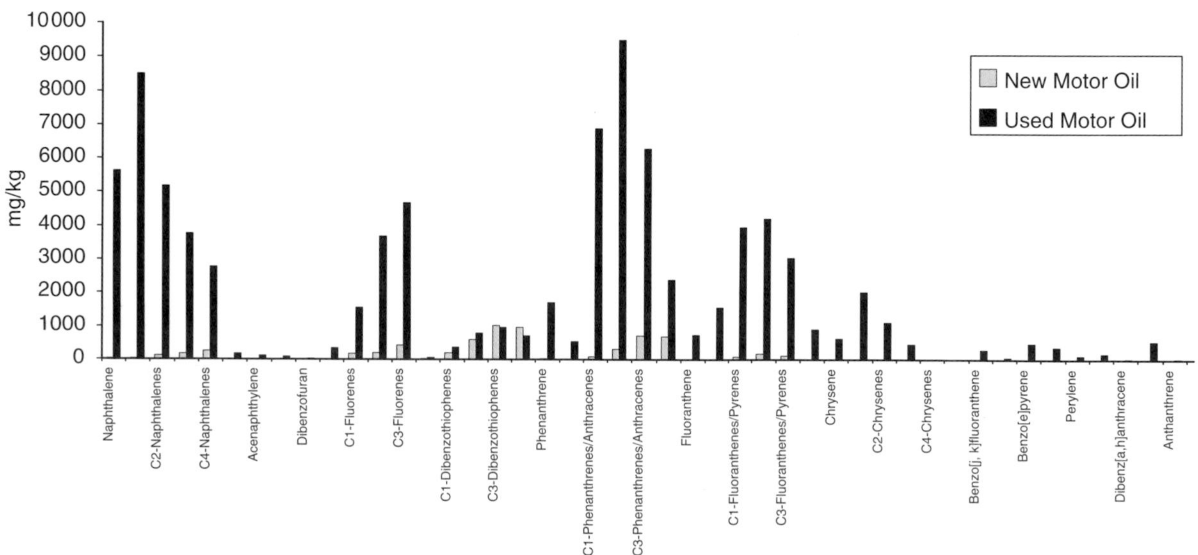

Figure 6.27 Histogram of PAH analytes demonstrating differences between a new and used motor oil.

zinc, and barium), iron steel particles, and copper are reportedly enriched in used motor oil (Vazquez-Duhalt, 1989). It is notable that while the PAH concentration is drastically increased upon use (Figure 6.27), the concentrations and distributions of most biomarkers remains largely unchanged. Thus, biomarker determinations should allow for successful correlation studies between new and used motor oils of the same type, or between used motor oils from different sources.

6.4.3 Manufactured Coke and Gas Production

The secondary effects that the refining of crude oil has on the chemistry of the petroleum products discussed above are certainly an important aspect of most forensic investigations. Equally important are the secondary effects brought about during the production of manufactured coke and gas from (1) coal and (2) crude oil feedstocks. The processes, products, and by-products of these important technologies are discussed in the following sections. The emphasis throughout is placed on the effects that these feedstocks and processes have on the PAH distributions in the products and by-products.

6.4.3.1 Overview of PAH Fingerprinting

A discussion of PAH fingerprints is essential for identifying petroleum (petrogenic material) in the presence of combusted or partially combusted organic matter (pyrogenic material). The differences in PAH distributions between these two anthropogenic PAH sources (petrogenic versus pyrogenic) were first recognized in modern sediments, primarily on the basis of the PAH homologue distributions (Youngblood and Blumer, 1975; Lee *et al.*, 1977; Laflamme and Hites, 1978; Wakeham *et al.*, 1980a). Indeed, the environmental forensic interpretation of petrogenic, pyrogenic and biogenic PAHs is a lengthy topic itself. Therefore, this section focuses on some of the differences between petrogenic and pyrogenic forms of contamination, as a practical means of identifying sources of PAHs in environmental matrices.

One obvious difference between petrogenic and pyrogenic material is the bulk chemical composition. Straight, branched and cycloalkanes comprise the bulk of unweathered petrogenic materials such as crude oil (Figure 6.28). Although pyrogenic materials, like coal tar (Figure 6.28), can contain more than 500 compounds (Novotny *et al.*, 1980; Zander, 1995), aromatic hydrocarbons, particularly parent (non-alkylated) PAH compounds, normally constitute the bulk of coal tar (Lao *et al.*, 1975). Thus, on a gross level PAHs can comprise a much higher mass percentage in most pyrogenic source materials than in most petrogenic source materials.

As noted above, early workers recognized the value of characterizing PAH homologues when attempting to distinguish PAH sources. The relative

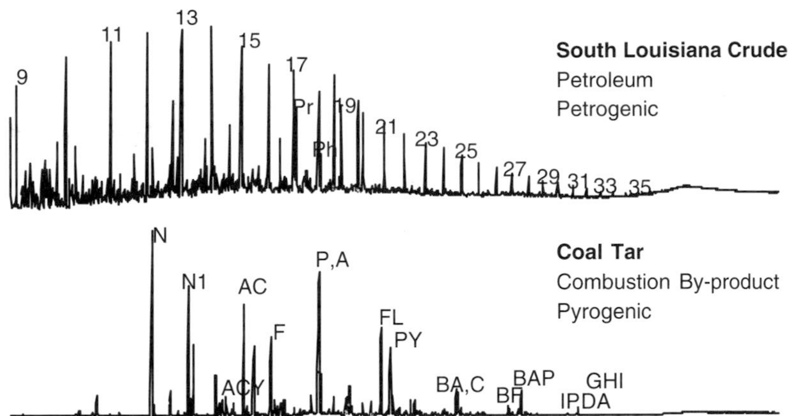

Figure 6.28

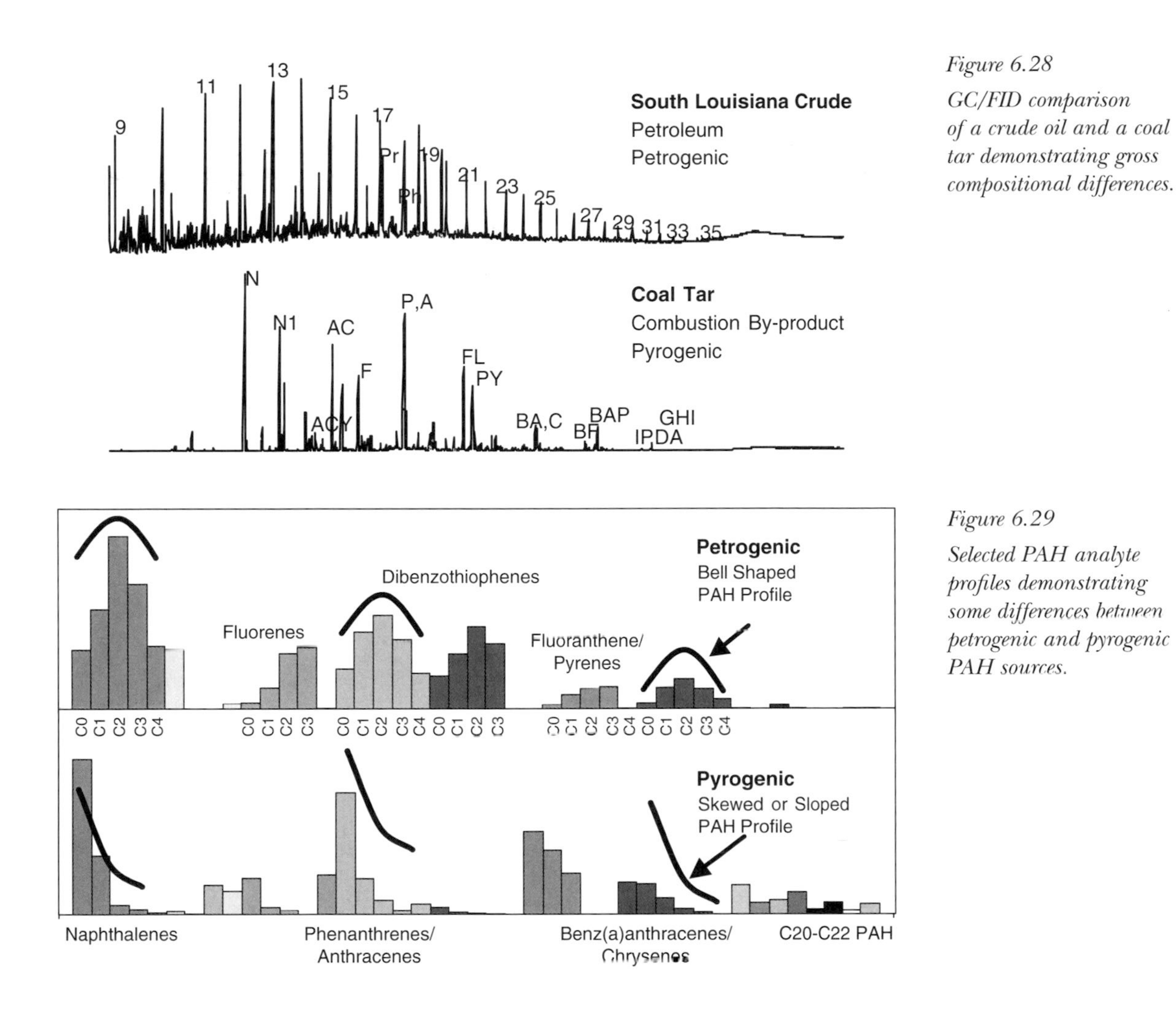

GC/FID comparison of a crude oil and a coal tar demonstrating gross compositional differences.

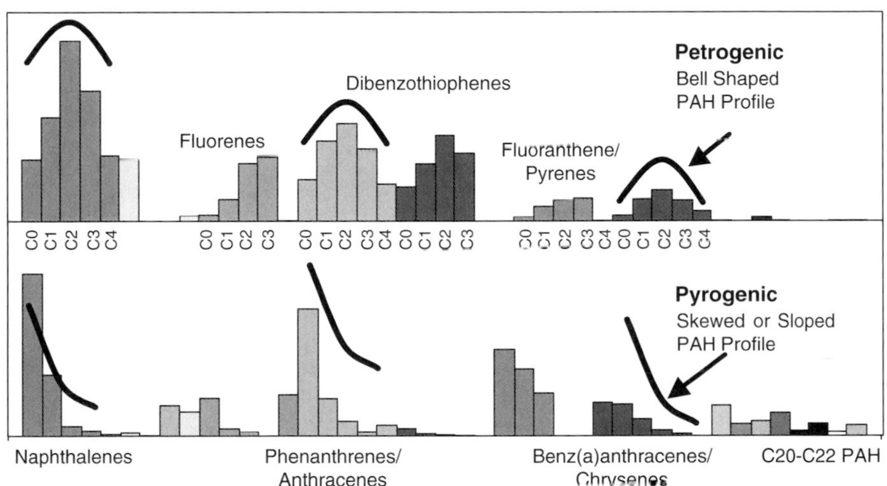

Figure 6.29

Selected PAH analyte profiles demonstrating some differences between petrogenic and pyrogenic PAH sources.

distribution of a significant number of parent and alkylated PAH compounds (determined using GCMS as described above; Table 6.6) still provides a suitable means for differentiating PAHs derived from petrogenic and pyrogenic materials, even when the hydrocarbon mixture in question is substantially weathered (Figure 6.29).

The PAH patterns exhibited within a given homologous PAH group are particularly useful. Generally, petrogenic source materials exhibit PAH homologue patterns where the 2- and 3-carbon alkyl groups (C_2- and C_3-alkylated PAHs) are more abundant than the parent (non-alkylated) PAH. This petrogenic PAH profile has been termed 'bell shaped' (Sauer and Uhler, 1994–1995; Figure 6.29). Pyrogenic source materials generally exhibit PAH homologue patterns in which the parent PAH is most abundant, and there is an inverse relationship between relative abundance and PAH alkylation level (i.e., $C_0 > C_1 > C_2 > C_3 > C_4$). This pyrogenic PAH profile has been generically termed, 'skewed' or 'sloping' (Sauer and Uhler, 1994–1995; Figure 6.29). Weathering of PAH (discussed later in this chapter) can tend to convert a pyrogenic (skewed) profile toward a more petrogenic (bell-shaped) profile. Thus, application of the simple bell-shaped versus skewed homologue patterns is most appropriate within homologue series considered to be unaffected by weathering; e.g., naphthalene patterns should be avoided.

Interpretations surrounding PAH source(s) using the homologue patterns alone are less clear when mixtures are present and PAH profiles are inconclusive. Nonetheless, the high toxicity of these compounds still often mandates remediation, which necessitates the identification and allocation among multiple PAH sources and responsible parties.

In addition, many site investigations require the differentiation of 'background' PAH from the contaminant's signature, or the differentiation between two pyrogenic sources, such as a former manufactured gas plant (MGP) from a tar (by-product) processing facility. Again, problems like these can be very difficult to unravel because the (GC/FID fingerprints and) PAH profiles can be inconclusive. The solution in these instances often requires the synthesis of historical data and chemical data from many different chemical and physical tests performed on many samples representative of all areas on the site.

Previous sections in this chapter evaluated in detail the composition of crude oil and petroleum (petrogenic) products. Therefore, the balance of this section is dedicated to a review of the industrial production of pyrogenic PAHs found in the environment. This review will highlight aspects of PAH formation during the production of metallurgical coke and manufactured gas and the forensic methods useful for deconvoluting the presence of multiple pyrogenic sources.

6.4.3.2 Industrial Production of Pyrogenic PAHs

The historic increase of PAHs in the environment is primarily attributed to the industrial revolution. Hites *et al.* (1980) studied a sediment core collected in New England and documented the production of total PAH (TPAH) as follows: TPAH in coastal sediments was less than 0.3 ppm between 1822 and 1895, rose to 7.3 ppm between 1895 and 1928, declined to 7 ppm between 1935 and 1942, rose again to 14 ppm by 1955 and declined to 10 ppm by 1975. This profile of TPAH concentration in the sediment over time matches closely the onset of the industrialization, the peak industrial outputs of World Wars I and II and the declines in industrial production during the Great Depression and the introduction of clean fuels and environmental regulations beginning in the Cold War era.

Two essential components of the US industrial history were manufactured gas and steel production. Although manufactured gas was produced from about 1816 to 1970 in the US, its widespread use began in the late 1800s (Harkins *et al.*, 1988). As the industrial revolution developed, the pyrolysis of organic matter was the primary means by which both metallurgical coke (a necessary component in the production of steel) and manufactured gas were produced. Like the development of the petroleum industry, the manufactured gas and coke industries evolved regionally (Harkins *et al.*, 1988). In the northeast, residential and industrial consumers demanded manufactured gas for light, heat and power. This demand was most efficiently satisfied by the carburetted water gas (CWG) and coke-oven gas processes. In the coal-rich, mid-continent (Ohio Valley and Illinois) region, the steel industry promoted the production of coke-oven gas over other processes. In the west, oil-gas processes dominated because suitable coal was rare and considerably more expensive than local oil. These gross regional trends can provide a context for evaluating PAH sources in different parts of the US.

Crude oil genesis was described earlier in this chapter. This paragraph describes the genesis of coal. Coal is derived from the remains of land plants, which originally accumulated as peat. Upon burial, peat is converted to coal by prolonged exposure to elevated temperature and pressure in the subsurface, a process known as coalification. The chemical and physical changes during coalification are complex but fairly well understood (Taylor *et al.*, 1998). Simply put, coalification converts resistant plant biopolymers (e.g., lignin) into a highly aromatic, three-dimensional, cross-linked, macromolecular matrix. The number of conjugated aromatic rings per structural unit within the matrix increases with increasing coalification (ultimately resulting in graphite). The average number of aromatic rings per structural unit in most coals is three to five with some individual units containing up to 10 aromatic rings (Davidson, 1982). Upon heating during the gasification or carbonization process, the linkages between some of these units are 'broken,' thereby releasing the conjugated aromatic rings into the volatile liquid fractions (e.g., coal tar). The volatile fractions are comprised, in large part, of 2- to 6-ring PAH compounds.

Coals of different geologic age, or rank, possess different chemical compositions and structures commensurate with the environment in which they formed. The coal rank was an important parameter in the manufacture of coke and gas because the industrial technology was designed for a feedstock with a reliable quantity of volatile matter and BTU/lb, which determined the coking properties and heating capacity respectively (Figure 6.30). Due to the physicochemical requirements of coke, the best 'coking' coals were low sulfur, low ash, medium volatile, bituminous coals (C = 86–89%; Berkowitz, 1979). Other rank coals, when carbonized, tended to yield non-coherent, carbon chars that are unsuitable for coke. On the other hand, technological advances in gas manufacturing permitted the use of a wide range of rank (and quality) among coals

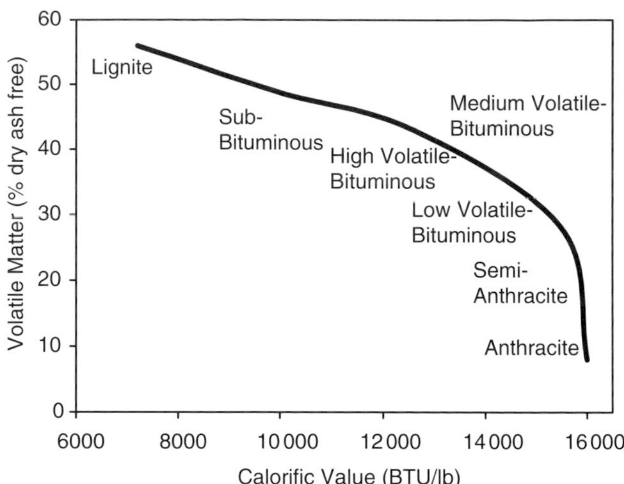

Figure 6.30
Cross-plot showing general coal rank series as characterized by calorific value and volatile matter.

used in gas manufacturing. Lower rank coals (lignite through high volatile bituminous) were preferred but, in fact, peat through anthracite have been used for the manufacture of gas. This enabled the use of less costly and more (locally) abundant coal types and blends, as dictated by the market availability, in gas manufacturing (than in coke manufacturing).

The evolution of manufactured coke and gas technologies influenced the type of facility constructed and the upgrades that occurred over time. For example, beehive coke ovens were often replaced by retorts and by-product coke-ovens depending on the year of construction. Similarly, gas production evolved from (1) coal carbonization, (2) to carburetted water gas (CWG), (3) to oil gas, and finally, (4) to natural (petroleum) gas. The evolution of technology, the local resources, and the commercial demand for manufactured coke and gas influenced greatly the construction and renovation of every facility in the US. In this context, every environmental forensic investigation of these operations is, to varying degrees, site-specific.

The environmental legacies of the coke and gas manufacturing industries include PAHs and other by-products, which are at the heart of many forensic investigations. Several aspects of these industries' technological history relate directly to the identification of the PAH sources and, in turn, the potentially responsible parties (PRPs). One obvious association between contemporary liability and technological history stems from the fact that over time many coke and gas manufacturing companies were absorbed by gas distribution companies, tar processing companies, utilities, steel mills and chemical manufacturers. Consequently, the delineation of the boundaries between the coke plant, gas plant and the by-products refining plant assumes paramount importance. A technical description of the coke and gas manufacturing processes is required to better understand the differences extant among the physical and chemical residuals (see below).

Coal carbonization was initially used for producing charcoal and later metallurgical coke. The latter is of interest herein. Morgan (1926) described the coal carbonization technologies as variations on the destructive distillation of bituminous coal at high temperature. The beehive oven, which gets its name from the shape of the oven, was the oldest form of this technology used on an industrial scale. It proliferated in the mid-to-late 1800s due to the high demand for metallurgical coke and low cost of plant construction. Standard beehive ovens heated several tons (~5 tons) of coal for two to three days, while venting or consuming almost all gases and volatiles (generally known as tar today). Coke was removed by hand or machine and quenched with water.

In the 1850s and 1860s, advancements in refractory materials enabled horizontal retorts to employ a higher carbonization temperature and generate a greater volume of gas (Morgan, 1926). This technology proliferated until World War I due to the decreased coking time and the ability to recover (rather than consume) the coal gas and tar. The horizontal retorts also improved upon the beehive design by facilitating the charging of coal and removal of coke in batches from a closed vessel. Typically, the gas was sold and the tar was combusted by way of heating the retort. Vertical retorts arose after 1910 in the interest of using gravity to more efficiently move the coal and coke through the oven. After its introduction in 1892, the unparalleled utility of the by-product coke oven eventually displaced the use of the beehive and retort designs. Production capacity was increased significantly with standard coal charges of about 25 tons and carbonization times of around 12 to 20 hours. By 1937, only about 5% of the US coke production still came from the less efficient beehive design (Moore, 1940). By-product coke ovens reached generally higher and more homogeneous temperatures (greater than approximately 900°C) than the beehive designs, which improved coke quality and strength. By-product coke ovens relied on machines to charge coal and remove coke while enabling the efficient recovery of the volatile gases, coal tar, ammonia liquor. These by-products were put to various uses. Gas could be used for industrial heating or, after some scrubbing, illumination.

The coal tar was often burned as a fuel or sold to tar refiners, where it was distilled into products for road-making, waterproofing, wood preservation, paints, pipe coatings, or briquetting (described below). Ammonia liquor was further processed for use in the manufacture of fertilizers. Notably, a diagnostic feature of tars produced by the by-product coking process that is useful for identifying PAHs derived from this form of coal carbonization, is the presence of oxygen- and nitrogen-containing hydrocarbons within the by-product tars, hereafter, referred to as tar acids and tar bases.

Gas manufacturing from coal in the US was first reported in 1813 near Providence, Rhode Island (Eavenson, 1942) and in 1816 in Baltimore, Maryland (Morgan, 1926). Innovations in plant design and materials resulted in the clay retort around 1850 and the development of the carburetted water gas design in 1875. Like the industrial revolution around them, gas plant designers constantly experimented with plant configurations, combustion conditions, feedstock types and by-product recovery systems. Between approximately 1885 and 1920, the carburetted water gas process grew from a popular to a dominant position in the gas manufacturing industry. Morgan (1926) described the carburetted water gas process as the alternate flow of air and steam on a bed of coke or bituminous coal for the production of water gas (sometimes called blue gas). This water gas was directed into a carburetter. Here, light hydrocarbons were generated by spraying an enriching oil on hot brick where the oil was 'cracked' and mixed with the water gas. The light hydrocarbons raised the BTU content of the manufactured gas from 300 BTU/ft^3 to about 530 BTU/ft^3 (Edison Electric Institute, 1984). The combined gas was passed through a superheater that facilitated the permanent cracking of oil vapor. The tar generated from the carburetted water gas process was often sold for further processing (see below). However, if the tar was highly emulsified it was often simply discarded (Ehleringer and Rundel, 1989).

With the increased supply of domestic crude oil through the early half of the 1900s, the manufacturing of gas from petroleum (rather than coal) became more widespread in some parts of the US, particularly in the 'coal-poor' Pacific Northwest. Even when later being displaced by natural gas in California, petroleum-derived manufactured gas (i.e., oil gas) remained popular in the Pacific Northwest, whereas its implementation elsewhere was short-lived (1930s and 1950s) and relatively uncommon (Harkins *et al.*, 1988). Consequently, our description of oil gas production will focus on the Pacific Northwest industry. Behar and Pelet (1984) described this process as the atomization of oil with high-pressure steam and subsequent exposure to very hot brick. The resulting blue or water gas mixed with the light hydrocarbons from the cracked oil. The by-products of this process (light oils, tar, and lampblack) were often sold to by-product processing plants. Like the CWG process, the tar was often discarded if it was highly emulsified (Ehleringer and Rundel, 1989).

The enriching oil cracked during the CWG and oil gas processes varied based on the evolving demand for automotive fuel and the resources available during World Wars I and II. Initially, petroleum refining generated naphtha, kerosene (lamp oil), gas oil and paraffin stock (used to make paraffin wax and lube oil). When the demand for naphtha and kerosene switched to automotive fuel (~1910), the cracking of straight-run gas oil became common

(1920s). At this time, the definition of gas oil shifted to performance-based criteria that could be satisfied by blending residual oils and adapting the gas manufacturing process for heavier cracking oils. By 1925, oil sprayed in the carburetter could have been straight-run gas oil, cracked gas oil, heavy asphaltic fuel oil, heavy crude oil, topped crude oil or other residual oil. On the West Coast, crude oil from California was used until about 1919 when residual fuel oil was used almost exclusively. Notably, one distinctive feature of the tars produced by the CWG and oil gas processes was the absence of tar acids and tar bases.

Regardless of the manufacturing process used (CWG versus oil gas), impurities rendered raw manufactured gas as non-marketable. These impurities varied widely in terms of vapor pressure and chemical properties. Consequently, the purification of manufactured gas occurred throughout the whole post-generative process and not in separate, well-defined steps. Oil gas and to a lesser extent, CWG, employed a wash box immediately after the gas generation equipment in order to remove lampblack (elemental carbon fines formed by the 'overcooking' of gas oil). Condensers removed a majority of the tar vapor, water vapor, and naphthalene by reducing the heat content of the gas to atmospheric temperature and isolating the impurities with low vapor pressure. Subsequently, dry scrubbers (tar extractors, shaving scrubbers and electrical precipitators) removed solid and liquid impurities. Wet scrubbers removed the acidic and basic impurities based on chemical principles. Before about 1925, water-washing removed the ammonia, gas oil washing removed naphthalene, straw oil washing removed the light oil and ferrous sulfate reactions removed the cyanogen. After 1925, other types of scrubbers were introduced (Harkins *et al.*, 1988). Coke gas contained all of the impurities described above while CWG did not generally contain ammonia and oil gas did not generally contain ammonia or cyanogen. Hydrogen sulfide was removed by various washing methods (Seaboard, Nickel, Thylox and Ferrox), some of which used arsenic, iron and nickel catalysts (Harkins *et al.*, 1988). The various condensing and scrubbing processes often explain the presence of peculiar and localized assemblages of contaminants and structures found at former coke and gas plants.

Though hydrocarbon (PAH) fingerprinting is essential, the presence of other by-products from the carbonization of coal and gasification of coal or oil can further help identify the source(s) of pyrogenic material as well. In addition to the materials removed from the gas, all plants generated tar. Crude tar primarily consisted of varying amounts of light oil (BTEX, styrene and other volatile organic compounds), PAHs, aliphatic hydrocarbons, heavy molecular weight carbon assemblages (benzene soluble and benzene insoluble) and heteroatomic (sulfur, oxygen, and nitrogen) molecules. Facilities using coal as the feedstock often exhibited particulate coal and coke fines (e.g., coke breeze) throughout the site, and entrained within the tar. Plants using oil as the feedstock often exhibited entrained carbon-rich particles (i.e., lampblack) in the surficial soils and in the tar. Forensic laboratories specializing in tar chemistry employ custom methods for measuring and distinguishing among the various types of tars.

6.4.3.3 Industrial Processing of By-products

Light oil and tar processing plants emerged as industrialists realized the commercial value of the by-products of coke and gas manufacturing. Light oil recovered from the volatile gas and tar was often distilled into benzol, toluol, xylol (light solvent naphtha) and heavy solvent naphtha. Olefins and other impurities were removed by exposure to sulfuric acid. After neutralization with sodium hydroxide, the light tar distillates were either sold for industrial use or refined into benzene, toluene, xylene or solvent naphtha. These refined products constituted the raw materials for gasoline, aviation gas, explosives, isocyanates, phenolic resins, waterproofing, synthetic fibers, floor tiling, paints, polishes and other products (Lowenheim and Moran, 1975).

Tar processing involved the isolation of light oil ($\sim< 220°C$), middle oil ($\sim$220–375°C), and heavy or anthracene oil ($\sim$375°C to 450 or 550°C; Berkowitz, 1979). The tar residue that did not distill into these fractions was classified as pitch. Light oils were discussed above. Middle oils were quite valuable. Crude naphthalene was often crystallized, separated from the oil, washed with water, distilled and sold for the manufacture of phthalic anhydride plasticizers and resins (Lowenheim and Moran, 1975). Phenol-rich tar acids were often extracted from the middle oil using a solution of sodium hydroxide and sold for the manufacture of insecticides, tannery surfactants, resins, disinfectants, antioxidants, phosphate esters and other products. Nitrogen-enriched tar bases were often removed for the manufacture of dyestuff, pharmaceuticals, munitions and other products. Finally, the residual middle oil was often sold as creosote. The heavy oil was processed in a fashion similar to the middle oil. In fact, the middle and heavy oils were often combined or not separated in the first place. When isolated from the middle oil, heavy oil was sold as fuel oil, blended with pitch for specific grades of road tar, or manufactured into paint or insulating products. Pitch was used for the manufacture of waterproofing, roofing, electrodes (aluminum smelting), coal briquettes (as a binder), sealing and lubrication products.

The by-product known as creosote warrants additional attention because wood treating facilities were frequently located near tar generating or processing plants. The pharmaceutical industry originally used the term creosote in reference to the phenolic material derived from wood tar for use as an antiseptic.

Later, chemical engineers isolated tar acids from coal tar for both the pharmaceutical and wood treating industries. As the demand for creosote by the wood treating industry grew with the burgeoning railroad, electricity, telephone and maritime industries, chemical engineers experimented with its formulation in order to maximize the following performance features: permanence, penetration, toxicity, and hydrophobicity. Over time, the chemical composition of creosote changed from a straight-run distillate (approximate boiling point range of ~200°C to ~300°C) to a reformulated product containing non-marketable tar by-products (pressed anthracene cake oil and phenanthrene), enhancement blends (unprocessed coal tar improved permanence), and/or bulking agents (unprocessed water gas or oil tars). In other words, the composition of creosote depended largely on the industrial practices at the site and varied at any given time across the US. Distinguishing creosote from other processed or unprocessed tar products is often confounded by weathering and the commingling of native material. Consequently, the forensic investigator often relies on numerous forensic techniques for the differentiation of creosote in the presence of other pyrogenic material.

The signatures of tar generating and processing facilities can sometimes be differentiated in terms of the presence, absence or alteration of the pyrogenic constituents that can be found on these sites today. Naturally, the two operations were often located near one another because the tar and light oil by-products from the coke/gas plant constituted the feedstock of the processing plant. Therefore, materials found on both locations are likely to be chemically similar and difficult to distinguish. However, the combination of careful historical research with high-quality, chemical characterization of samples providing adequate spatial coverage of the area can sometimes unravel these complex sites. For example, the production of creosote was often associated with the removal of light oil, naphthalene and heavy molecular weight material. Environmental forensic investigations focused on these indicator compounds can collectively assist in the identification of PAHs known to have originated from one of these sources. In addition, many light oil and tar processing plants imported feedstock from other regions when local supplies were diminished. Physical and chemical dissimilarities between the coke/gas manufacturing plant(s) and the tar processing plant are sometimes evident for this reason.

6.4.3.4 Identification of Pyrogenic Sources using Hydrocarbon Fingerprinting

A physical and chemical characterization of environmental media impacted by pyrogenic materials from coke and gas manufacturing industries is often required during environmental forensic investigations for PAHs. A physical characterization is often required of the tar and feedstocks collected from the site. It might include measurements of specific gravity, viscosity, moisture content, ash content, total organic carbon, benzene soluble fractions, molecular weight and organic petrographic analyses (microscopic analysis of particulate structure). This type of investigation should also involve thorough chemical characterization of soil, groundwater, tar and other relevant media. Depending on the objectives and other circumstances, the chemical characterization can include VOCs, PAHs, biomarkers, metals, cyanide, tar acids, tar bases, elemental composition (carbon, hydrogen, sulfur, nitrogen and oxygen), sulfur forms (pyritic, organic and sulfate), aromatic sulfur compounds (thiophenes, benzothiophenes and dibenzothiophenes), simulated distillation curves (aliphatic and aromatic fractions), bulk and compound specific carbon isotopes ($\delta^{13}C$ and ^{14}C activity for PAHs and, possibly, *n*-alkanes). A discussion of the use of the hydrocarbons among these, namely the PAHs and biomarkers, is appropriate in this chapter.

The processes associated with the production of coke and gas from a coal feedstock versus the production of gas from a petroleum feedstock can lead to some striking differences in the nature of the contamination derived from these processes. For example, coal carbonization often involved the gradual and inconsistent heating of a coal charge, particularly in the earlier beehive designs (see above). By contrast, the cracking of gas oil during gas manufacturing was regulated relatively tightly to ensure the consistency and efficiency of the combustion process. Other differences in combustion were plant- and site-specific. These included the variation in rank of the coal feedstock (e.g., anthracite, bituminous coals, or a coal blend), petroleum feedstock type (e.g., crude oil, gas oil, residual oil or other), and volumes of water and air used in gas-quenching/cleaning operations. The means by which the raw gas temperature was reduced could have affected the condensation reactions of PAHs and other molecular assemblages. In short, we believe that the plant type, feedstock material(s), and combustion/quenching kinetics resulted in variable chemical signatures, e.g., particular PAH ratios, which can be useful for recognizing different pyrogenic sources in forensic investigations at these facilities.

Researchers have developed a number of PAH indicator compounds and diagnostic ratios that attempt to 'classify' the different types of tars produced by these industries (Behar and Pelet, 1984; Wakeham *et al.*, 1986; Sutherland *et al.*, 1991; Trust *et al.*, 1995; Whittaker *et al.*, 1995; Mansuy *et al.*, 1997; Jarman *et al.*, 1998; Hilkert *et al.*, 1999; Reddy *et al.*, 2000). Caution is warranted, however, as no single PAH source indicator is universally applicable to all situations. In fact, any forensic investigation requires that the forensic investigator use groups of indicators compounds and ratios systematically, and in concert with other pyrogenic features and historical facts, before reaching a defensible source identification conclusion. In particular, it is prudent to rely upon a broad spectrum of PAHs including alkylated PAHs when identifying a source of release (rather than distributions or ratios derived from selected, parent PAH compounds).

Similarly, we believe that biomarkers can assist the forensic investigator in assessing sources of PAH at former coke and manufactured gas facilities. Biomarkers are generally believed to be thermally destroyed when heated above 800°C (J.M. Moldowan, personal communication, 2000); therefore the absence of biomarkers in tar suggests that the tar was produced by a high temperature(s) process. However, in our experience triterpane and sterane biomarkers are often present in contaminated samples collected from former coke and gas plants. Since biomarkers occur in both petroleum and coal, the presence or absence of biomarkers cannot be used *a priori* to distinguish the type of feedstock material. The presence of biomarkers could indicate a lower carbonization temperature, an incomplete combustion at a higher carbonization temperature, or simply the presence of uncombusted coal or petroleum. In fact, correlating suspected source and fugitive materials can be aided by the co-occurrence of selected biomarkers and/or heavy molecular weight PAH (Emsbo-Mattingly *et al.*, 2000). Furthermore, the specific distributions and concentrations of biomarkers, relative to PAHs, can sometimes aid in distinguishing coal- from petroleum-derived PAH assemblages. Similarly, trace petroleum fingerprints generated from the aliphatic fraction of tar residuals can assist the identification of PAH origin(s) (Emsbo-Mattingly and Mauro, 1999).

Collectively, these physical and chemical tests comprehensively characterize the principal forensic indicator compounds at former coke and gas plants. However, the restrictions of time and cost dictate that these tests be exercised judiciously. Consequently, PAH forensic investigations often rely on total hydrocarbon (fingerprinting) and detailed PAH data from many samples to identify the generic features of the source(s) and boundary areas. Subsequently, a subset of these samples can be subjected to specific tests designed to illustrate the key differences between the pyrogenic materials believed to originate from multiple sources as suggested by the site history and existing data.

Distinguishing between point and non-point sources of PAHs in a study area can be difficult on the basis of chemical fingerprinting alone. Few, if any, methods exist for easily distinguishing all point source PAHs from non-point (e.g., urban runoff, atmospheric fallout, and natural pyrogenic) sources. Consequently, the distinguishing features of the source PAH are usually determined on a site specific basis after numerous samples from the suspected source area(s), suspected boundary and background areas are characterized using hydrocarbon fingerprinting, detailed PAHs, biomarkers, and possibly other source indicators. Historical records and the spatial distribution of the PAH can further reveal the most likely, if any, point sources of pyrogenic PAH within the study area. In addition, the TPAH concentrations in some studies can be important in identifying the most likely source of PAH. That is to say that non-point sources of pyrogenic PAH (or naturally occurring pyrogenic PAH, e.g., fire residues) typically occur in much lower concentrations than those derived from point sources of pyrogenic PAH associated with coke/gas manufacturing or tar processing. In addition, ratios of background signatures (plant waxes, residual/lubricating oil (i.e., unresolved complex mixture), biomarkers and/or isotopic composition) and source signatures (PAH homologue patterns or ratios, tar acids, tar bases and/or combustion by-products) can further help establish PAH sources at these types of sites.

6.5 TERTIARY CONTROLS ON PETROLEUM COMPOSITION

6.5.1 ENVIRONMENTAL EFFECTS ON HYDROCARBON FINGERPRINTING

Understanding the primary and secondary controls and effects on hydrocarbon fingerprinting (above) are important, but no more so than the tertiary controls that will affect the composition of petroleum- and coal-derived contamination after it has been released into the environment. These controls were listed in Table 6.1 (above) and are discussed in this section.

The first tertiary control considered is related to the potential presence of naturally occurring hydrocarbons in selected environmental matrices, mostly soils and sediments. This phenomenon was briefly touched upon earlier in relation to influence of plant waxes in the soils impacted by an oil well blowout (Figure 6.7). In some studies, it may even be necessary to expand the definition of 'naturally occurring' to include a sometimes-pervasive form of 'background' hydrocarbons, e.g., particulates derived from urban air. The second tertiary control to consider is related to the effects of weathering, the process by which changes in chemical composition are brought about through the influences of exposure to air, water, sunlight, soil/sediment, or biota. The final tertiary effect to consider can simply be summarized as 'other fractionation processes.' Sorption of some individual compounds to organic particles in soil and sediment may alter to composition of petroleum in the environment. Another important fractionation process is the remediation activities that have occurred in the past at some sites. In our experience, the most commonly encountered example of this is the remarkable effect that soil vapor extraction (SVE) can have on the fingerprinting of gasoline-impacted soils. These three tertiary effects are discussed in the following sections.

6.5.1.1 Naturally Occurring 'Background' Hydrocarbons

Environmental samples are extraordinarily complex matrices. Forensic investigations often rely, at least in part, upon the hydrocarbon fingerprinting data obtained from these matrices (i.e., soils, sediments, water, air particulates, or biological tissues). Upon solvent extraction in the laboratory (in preparation for fingerprinting analysis), all extractable organic compounds will be collectively removed, including those not derived from anthropogenic contamination. And while some of the compounds extracted can be removed from a soil/sediment extract prior to fingerprinting analysis through various extract 'clean-up' procedures (e.g., EPA Method Series 3600), any naturally occurring *background hydrocarbons* cannot be separated from *true hydrocarbon* contamination. In fact, in some settings even free phase petroleum can contain background hydrocarbons, i.e., compounds that the petroleum itself (acting as a 'solvent') has 'extracted' from the soil through which it has migrated. Therefore, the forensic investigator must be able to recognize, and thereby account for the influence of, any 'background' hydrocarbons on the chemical fingerprint under consideration (Stout *et al.*, 2000b).

Recognizing background hydrocarbons can be facilitated with the analysis of true background samples from the local/site under investigation. Unfortunately, this may not always be possible and furthermore, who can be sure as to what constitutes true 'background' at any given location. Naturally occurring hydrocarbons can take many forms (Table 6.15; Volkman *et al.*, 1992). The influence of these types of materials is typically greatest in investigations involving the petroleum fingerprinting of soils and sediments. For example, most soils and sediments contain some fraction of organic matter derived from modern plant, algal, bacterial, or fire debris. Anyone familiar with the measurement of fraction of organic carbon (f_{oc}) knows that even the 'cleanest' sands from the 'driest' environments contain a small percentage

Modern Sources of Background Hydrocarbons	
Vascular plant debris	Cuticular waxes (*n*-alkanes odd C_{25}–C_{33})
	Triterpenes (e.g., sesquiterpenes, oleanenes, etc.)
	Sterenes
	Selected PAH (e.g., retene, etc.)
Microbial and algal debris	*n*-Alkanes C_{15}–C_{20}
	Triterpenes (e.g., hopenes)
	Sterenes
	Selected PAHs (e.g., perylene)
Wood charcoal/fire debris	Selected PAHs (e.g., fluoranthene, pyrene, etc.)
Ancient Sources of Background Hydrocarbons	
Natural oil seeps	All petroleum hydrocarbons
Detrital coal/shale	Most petroleum hydrocarbons

Table 6.15

Inventory of potential sources for naturally occurring (background) hydrocarbons in environmental samples.

(< 0.25 wt%) of organic matter that, upon solvent extraction, can yield hydrocarbons in measurable concentrations. Just imagine the potentially confounding contribution of naturally occurring hydrocarbons that some sediments, e.g., peat, can make to the 'total petroleum hydrocarbon.'

6.5.1.2 Vascular Plant Debris

The most common confounding influence in forensic studies of soils and sediments is due to the presence of vascular (land) plant debris. Plant debris is pervasive in most soils and coastal sediments and can easily confound conventional chemical fingerprints. The GC/FID of the extract from a Pacific Northwest sediment, dated to the pre-industrialization of the area (*circa* 1823) using ^{210}Pb, contains naturally occurring hydrocarbons (Figure 6.31). The primary features of this type of fingerprint are the odd-carbon-dominated *n*-alkanes (C_{25}–C_{31}), which as described above, are indicative of waxes derived from terrestrial plant debris. There are a few other discrete peaks that correspond to selected non-alkylated PAHs (fluoranthene, pyrene, benzo[a]pyrene), which could be derived from natural pyrogenic (fire) debris in these sediments (Simoneit and Elias, 2000). Additional unidentified peaks in the C_{20}–C_{25} range may be naturally occurring sesquiterpenes, diterpenes, and aromatic diterpenoids (e.g., retene) that are derived from certain plant resins (Laflamme

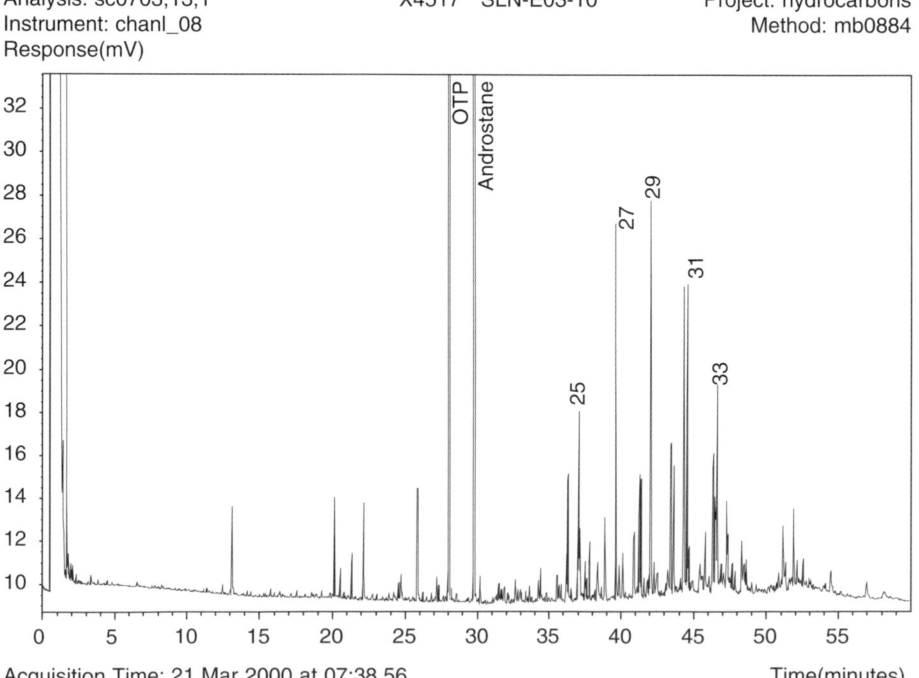

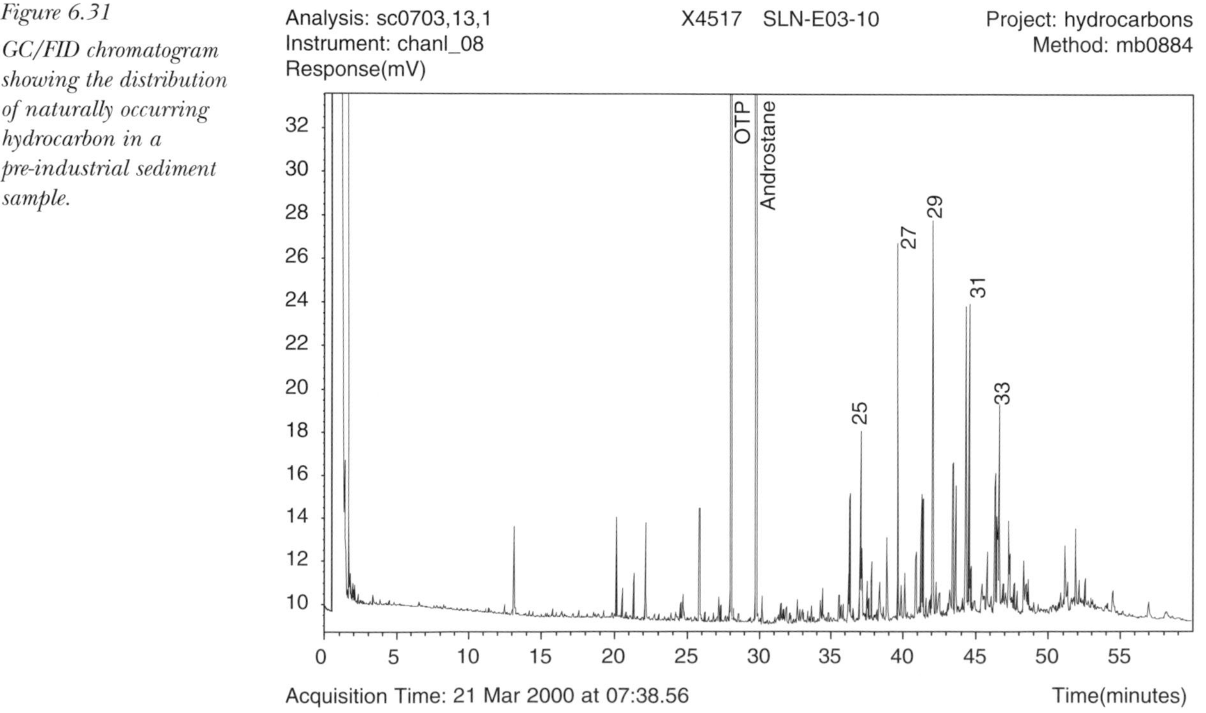

Figure 6.31
GC/FID chromatogram showing the distribution of naturally occurring hydrocarbon in a pre-industrial sediment sample.

and Hites, 1978; Barrick *et al.*, 1980; Wakeham *et al.*, 1980b; LaFleur *et al.*, 1998), as are common in the gymnosperm flora of the area (Simoneit *et al.*, 1986).

In some environments, entire stratigraphic layers of plant debris (i.e., peat) can play havoc with fingerprinting studies (not to mention TPH concentrations). Thus, the forensic investigator must be aware of the nature and potential influence of natural plant debris on hydrocarbon fingerprints.

6.5.1.3 Microbial Biomass

On a mass basis the concentration of bacteria in soils and sediments is extremely small ($\sim 10^8$–10^{10} cells per gram dry weight or ~3% of f_{oc}; Turco and Sadowsky, 1995). Nonetheless, the potential for the soluble microbial biomass in modern soils and sediments to contribute to an environmental sample's chemical signature cannot be ignored since freely extractable microbial lipids (particularly 'biomarkers') can confound some forensic results, especially those dependent upon matching biomarker patterns. For example, Figure 6.32 shows the triterpane biomarker fingerprint (*m/z* 191) of a crude oil-impacted soil. Evidence for the presence of the crude oil lies in the distribution of regular 17α(H),21β(H)-hopanes in a common distribution. However, two unusual (and large) peaks are also recognized that were determined to be immature 17β(H),21β(H)-hopanes. These compounds are known to be early diagenetic products of bacterial cell

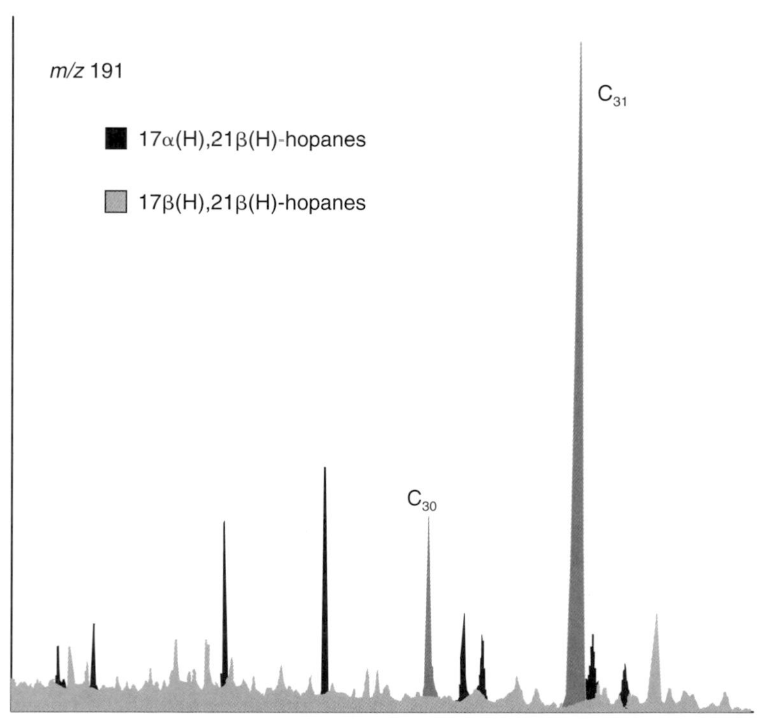

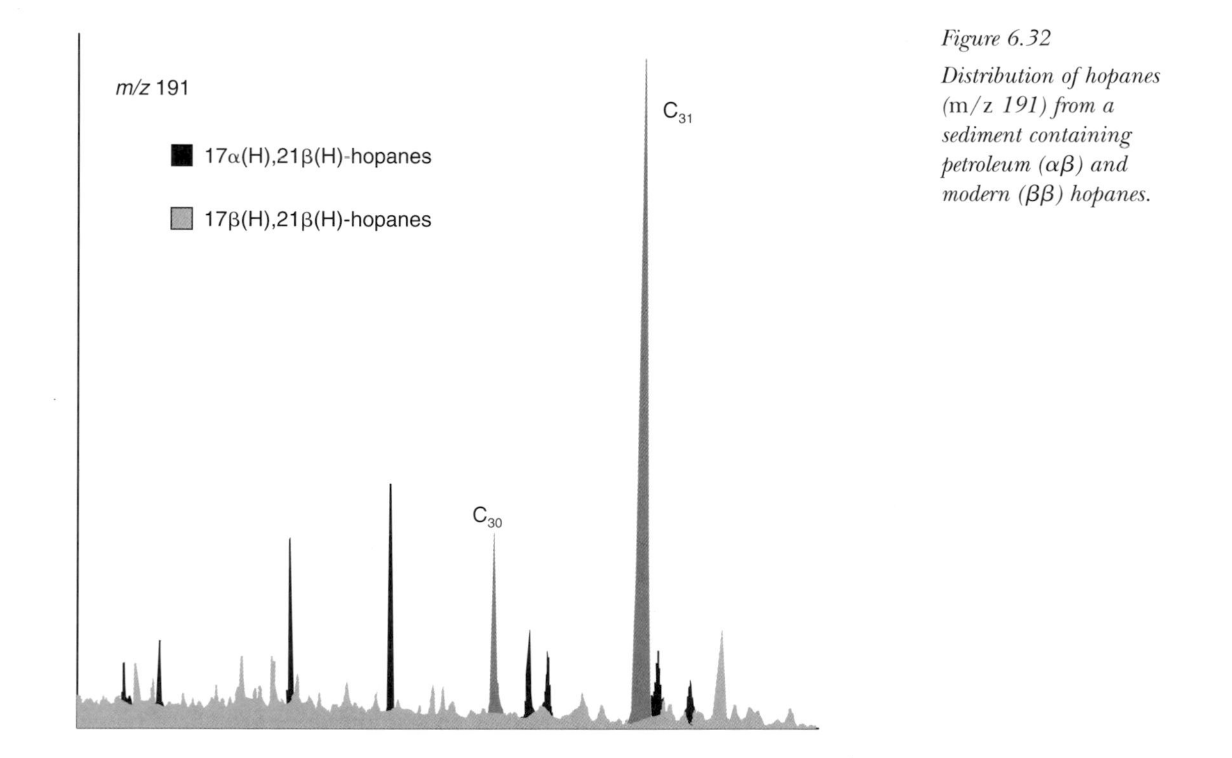

Figure 6.32
Distribution of hopanes (m/z 191) from a sediment containing petroleum (αβ) and modern (ββ) hopanes.

membranes indicating recent bacterial activity in the soil. Thus, this soil's biomarker fingerprint exhibited a 'mixed' signature of the spilled crude oil and the microbial 'background.' Other soils from the site that were impacted by the same crude oil did not contain the 'ββ' hopanes, which could easily have been misinterpreted and led to a 'negative' correlation among crude oil at this site.

The presence of modern microbial biomass can also impart a mixed thermal maturity signature to a petroleum contaminant also present. For example, the presence of excess 22S 17α(H),21β(H)-homohopane in modern sediment could reduce the 22S/(22S + 22R) ratio below equilibrium (<0.55) while the C_{29} sterane 20S/(20S + 20R) ratio remains at equilibrium (~ 0.60). This sort of mixed signal could further confound correlation studies of petroleum contamination relying upon the affected biomarker ratios.

6.5.1.4 Sedimentary Coal and other Organic Particulates

In many coastal environments, coal is eroded from sedimentary rock outcrops and minute coal particles are transported and deposited in coastal sediments by rivers. Alternatively, many industrial properties formerly housed industries that utilized coal. In either case, coal can occur as a 'background' contaminant to petroleum and other hydrocarbon mixtures. In addition, charcoal and other fire residues are not uncommon in both soils and coastal sediments.

The contribution of hydrocarbons that particulate coal may impart is largely a function of the precursor plant material (e.g., Mesozoic gynmosperm- versus Tertiary angiosperm-dominant) and the coal's rank (Chaffee *et al.*, 1986). Lower rank coals (lignite and sub-bituminous) typically yield higher concentrations of non-hydrocarbons during solvent extraction, whereas middle rank (high-to-medium volatile bituminous) yield more hydrocarbons (Radke *et al.*, 1980; White and Lee, 1980). The occurrence of coal particles in a soil or sediment could confound interpretations surrounding TPHs, PAHs, and biomarkers in the sample. Figure 6.33 shows the GC/FID distributions of hydrocarbon extracted from a series of coals of increasing rank (obtained from different US coal fields). The extractable (C_8+) TPH concentration and distributions vary widely among the coals. The distribution of hydrocarbons in the lower rank coals is largely a function of the types of ancient land plant debris that had been incorporated into the coals (e.g., gynmosperm versus angiosperm). The distribution of extractable hydrocarbons in the mid-to-high rank coals is primarily a function of aromatic structure of the coal matrix, and those 'mobile phase' (mostly) aromatic compounds that can be extracted from this matrix. Not recognizing the contribution that dispersed coal particles can make to soils and sediments could easily confound any forensic interpretations. In some instances, organic petrographic inspection of the soil/sediment may be necessary to recognize the presence of coal. In fact, the presence of sedimentary coal in Prince William Sound sediments has been recently argued to have confounded

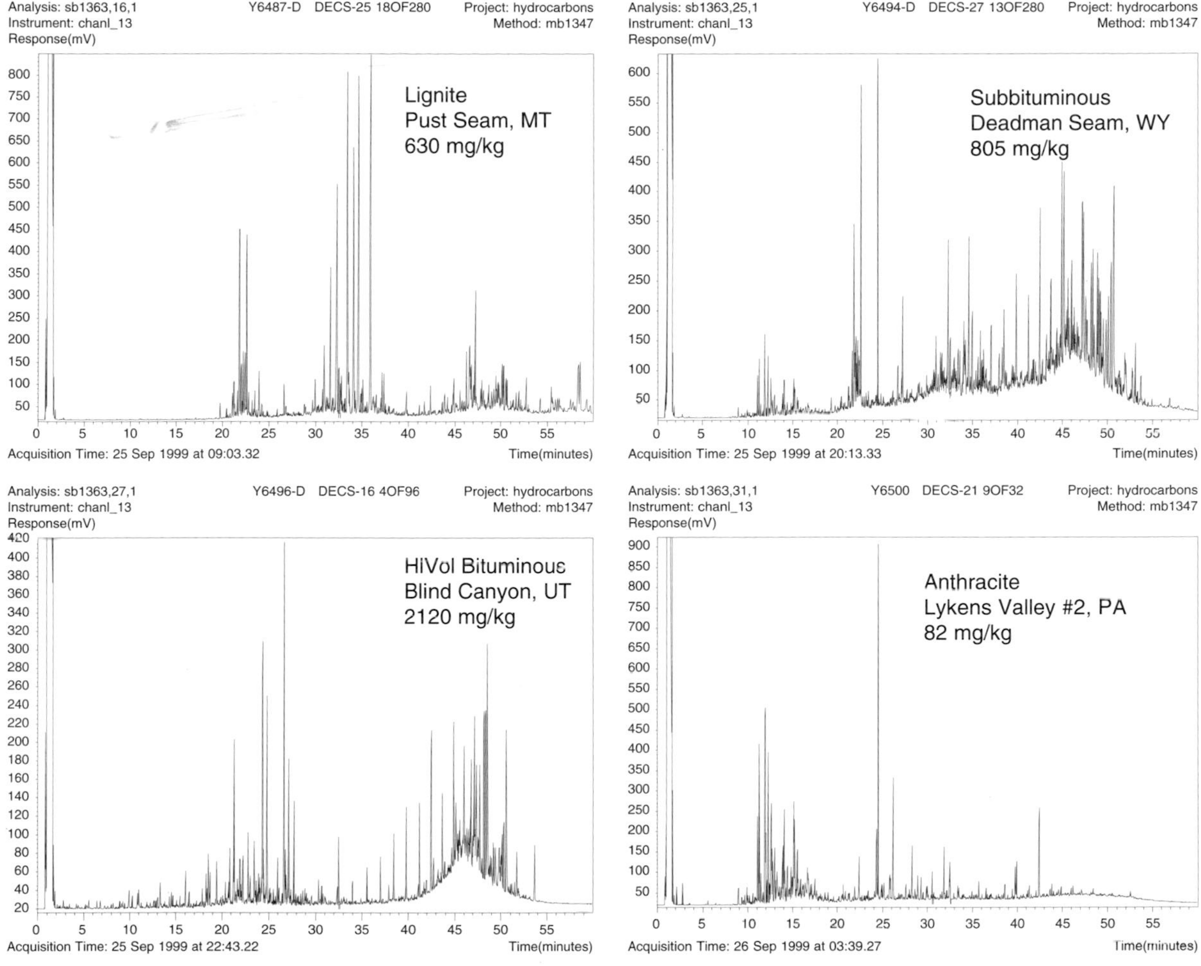

Figure 6.33

Comparison of GC/FID fingerprints of the hydrocarbon extracts from selected coals of different rank.

the allocation of the different PAH sources in Prince William Sound following the *Exxon Valdez* oil spill (Short *et al.*, 1999), although this contention has been refuted (Boehm *et al.*, 2000).

6.5.1.5 Natural Oil Seeps

In some petroleum-related forensic investigations, the occurrence and influence of natural oil seeps must be considered. The reason for this is obvious – *is there any 'real' contamination or only background contamination?* Natural oil seeps are a worldwide phenomenon. In coastal southern California the phenomenon is particularly prolific (e.g., Mikolaj *et al.*, 1972; Straughan, 1979) and must always be considered when investigating the origin of forms of coastal contamination, e.g., tarballs (Wang *et al.*, 1993; see Figure 6.12 above.) The forensic investigator must either eliminate naturally occurring oil as a potential confounding factor (through a thorough comparative study of the local crude oil chemistry versus contaminant chemistry and/or the geology of the area) or establish a defensible means to account for the natural crude oil within the contaminant/spilled oil. Of course, in some instances when the suspected spill oil is from the same geologic formation as the naturally occurring seeped oil (and therefore shares identical or nearly identical primary fingerprinting features), the forensic investigator's use of chemical fingerprinting is vastly reduced. In such instances an investigator is forced to rely upon other forensic tools (e.g., spill trajectory modeling).

6.5.1.6 Urban Runoff

Urban runoff is an important source of contamination to the environment, particularly in coastal waterways (Eganhouse *et al.*, 1982). It consists of a mixture of fugitive debris in water – dust, dirt, particulate matter, soot, solid wastes – that are transported to rivers or coastal waters and sediments via non-point (general runoff) and point (end-of-pipe) sources. In some forensic investigations, the presence of hydrocarbons derived from non-point sources, such as urban runoff, can confound the assessment of a particular point source contribution. In such instances, the hydrocarbon signal associated with urban runoff is essentially a 'background' form of contamination that must be accounted for in the forensic investigation.

The presence and composition of the hydrocarbons within urban runoff have been long recognized (Eganhouse *et al.*, 1982). Lubricating oils (discussed above) contained within urban runoff can be an important source of unresolved residual range hydrocarbons and biomarkers in urban sediments. PAHs are also well-known components of urban runoff. The sources of PAH in urban runoff vary, but the most common sources are (1) urban dust containing combustion-related PAH (principally arising from internal combustion engines, especially diesel-based e.g., Harrison *et al.*, 1996; Marr *et al.*, 1999), (2) street runoff containing traces of used lubricating oils (principally arising from releases from automobiles), and (3) illegal or unintentional discharging of waste oil and petroleum products into storm drain systems.

Numerous studies of PAHs in urban runoff have been conducted around the US over the last two decades (e.g., Eganhouse *et al.*, 1982; Boehm and Farrington, 1984; O'Connor and Beliaeff, 1995; Caricchia *et al.*, 1999; Hostettler *et al.*, 1999). The PAHs in urban runoff and, in turn, in receiving sediments of rivers and harbors are complex mixtures that tend to be dominated by higher molecular weight 4- to 6-ring PAHs (Durell *et al.*, 1991; Peven *et al.*, 1996). This fact is explained by two reasons: (1) the PAHs in storm water runoff often have a very strong pyrogenic PAH signature to begin with (Uhler *et al.*, 1994) and (2) the 2- and 3-ring PAHs are more water soluble and degradable than higher ring PAHs (Neff, 1979; Durell *et al.*, 2000). The influence of this 'background' source of hydrocarbons cannot be ignored in forensic investigations of urban sediments.

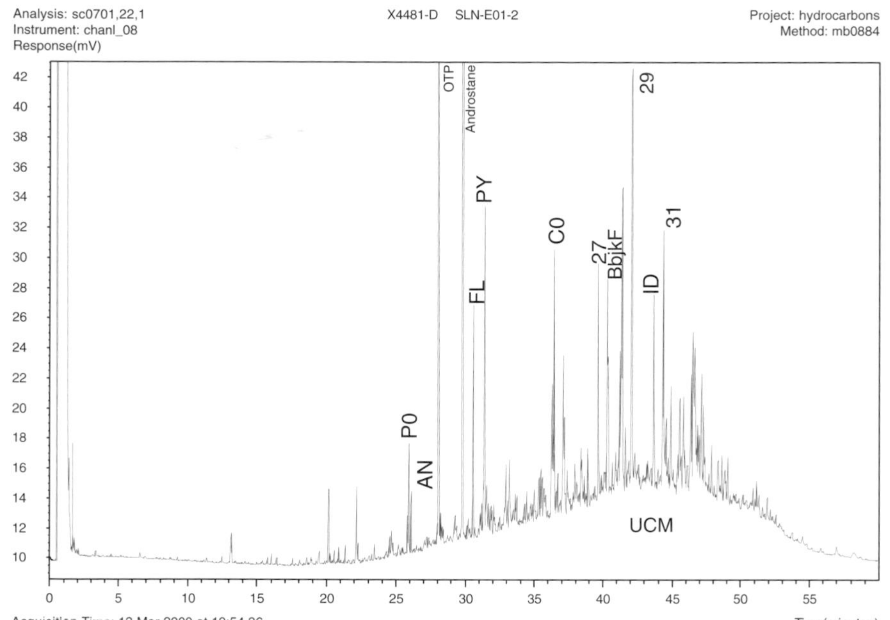

Figure 6.34

GC/FID chromatogram of a contemporary urban sediment.

An example of the 'typical' urban runoff fingerprint obtained from an urban sediment is shown in Figure 6.34. The fingerprint is dominated by discrete resolved peaks and an unresolved complex mixture (UCM) 'hump' in the baseline. The UCM 'hump' ranged from about C_{17} to C_{35}, reaching a peak around C_{31}. Many of the discrete peaks are recognized as various non-alkylated 4- and 5-ring PAHs, among these were a marked abundance of phenanthrene, anthracene, fluoranthene, pyrene, chrysene, benzo[*b,j,k*] fluoranthene, and indeno(1,2,3-*c,d*)pyrene (Figure 6.34). All of these features are consistent with those typical of sediments in urban/industrialized areas (Wade and Quinn, 1979; Barrick *et al.*, 1980; Eganhouse *et al.*, 1982; Hostettler *et al.*, 1999). The prominence of 4- and 5-ring, non-alkylated PAHs is indicative of a combustion (pyrolysis) product (Laflamme and Hites, 1978), as is typical in motor exhaust (Westerholm *et al.*, 1988) or wood smoke (Oahn *et al.*, 1999; Simoneit and Elias, 2000). The prominence of the UCM is typical of biodegraded petroleum, uncombusted petroleum or lubricating (crankcase) oils, all of which may be components of urban runoff (Gogou *et al.*, 2000).

6.5.1.7 Other Sources of 'Background' PAH

The ubiquitous nature of PAH in the environment is attributed to natural and anthropogenic sources (Garrigues *et al.*, 1995). Naturally occurring PAHs originate from the oxidation of biologically produced chemicals (biogenic; e.g., Wakeham *et al.*, 1980a,b; Venkatesan, 1988; Silliman *et al.*, 1998), natural oil seeps (petrogenic), and forest/prairie fires (pyrogenic). In addition, as described above, anthropogenic emissions absorbed to particles and captured in storm water runoff can cause elevated PAH levels in sediments, especially in urban areas. Examples of background levels of TPAH include 330 mg/kg in diesel exhaust (Takada et al., 1991), 100 mg/kg in gasoline exhaust (Takada *et al.*, 1991), 59 mg/kg in road dust (Rogge *et al.*, 1993), 170 mg/kg in surface soil (Bradley *et al.*, 1994) and 17 mg/kg in sediment (Hites *et al.*, 1980). It is understood that these studies only approximate the conditions extant at any particular site. However, they offer benchmarks for identifying situations in which the influence of background PAH might play a significant role.

6.5.2 EFFECTS OF WEATHERING ON HYDROCARBON FINGERPRINTING

Weathering is the term commonly used to describe the influence of physical, chemical, and biological forces on the physical and chemical composition of petroleum-, coal-, and combustion-derived hydrocarbons released into the environment (McCarthy *et al.*, 1998). The three most recognized and well-understood weathering processes that affect the fingerprinting aspects of petroleum and other hydrocarbon mixtures released in the environment are (1) evaporation, (2) water-washing or solubilization, and (3) microbial degradation. Other processes (sorption to soils, photo-oxidation) can have an influence, but in most circumstances, to a lesser degree than the three process discussed below.

6.5.2.1 Evaporation

Petroleum and other hydrocarbon liquids that interface with an atmosphere (whether its ambient air or the air in soil pores) are subject to evaporation. Normally this involves selective losses of lower molecular weight compounds (< ~150 atomic mass units; amu) and lower boiling point (< ~175°C). Thus, the lighter the spilled product is (e.g., gasoline as opposed to crude), the more likely that evaporative loss will constitute a significant portion of the overall weathering (Bruce, 1993). Recent studies have demonstrated and quantified the changes in composition in artificially evaporated fuels (Rodgers *et al.*, 2000). Gasoline, evaporated to 25% of its original mass, lost compounds primarily below ~125 amu while the same degree of evaporation in diesel fuel #2 resulted in the loss of compounds up to and beyond 150 amu. However, as might be expected, this study also demonstrated that more than a compound's mass and boiling point influences the degree to which it may evaporate. In fact, evaporation is largely a function of a compound's vapor pressure and the temperature (Schwarzenbach *et al.*, 1993), properties for which mass and boiling point can act as suitable, but not perfect proxies. Figure 6.35 demonstrates the

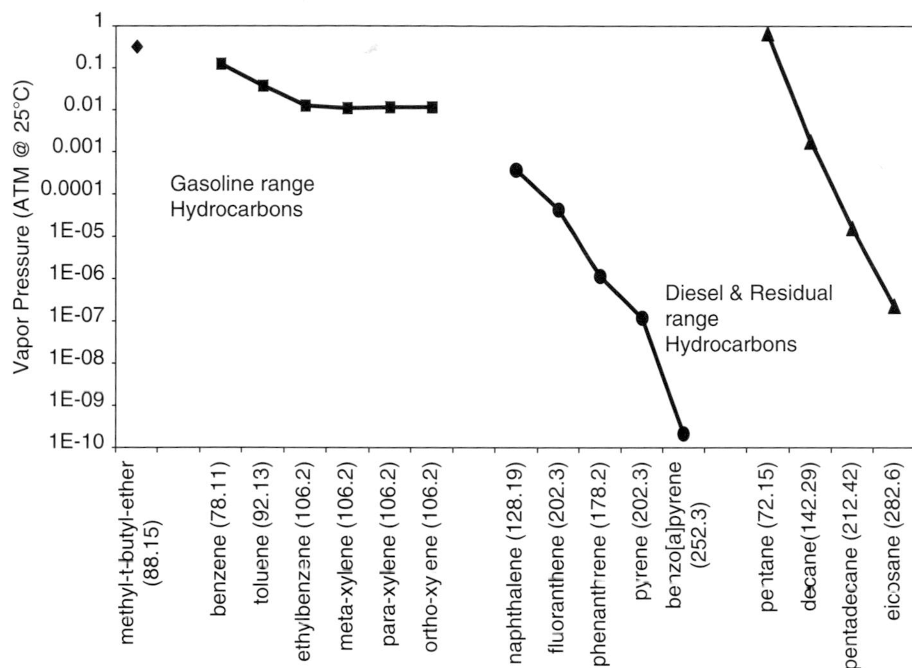

Figure 6.35

Vapor pressure of selected hydrocarbons and MTBE.

variation in vapor pressure (at 25°C) for a variety of compounds of different compound class and mass. Gasoline range hydrocarbons, particularly MTBE, are particularly prone to evaporation due to their higher vapor pressures. Among PAHs and *n*-alkanes there is an obvious inverse relationship between vapor pressure and mass.

In addition to composition, the specific conditions under which a release of petroleum occurs will collectively determine to what degree evaporation will affect the fugitive material. For example, petroleum products present in the subsurface with an underground storage tank (UST) leak/release as the source, rarely show evidence of significant evaporative weathering since they are not directly exposed to the atmosphere. The loss of volatile compounds to vadose zone soil gas does occur following a subsurface release, but rarely to the degree to significantly alter composition and confound forensic interpretations. Even during surface releases, the temperature, wind speed, nature of the impacted surface (water, coarse soils, cement, etc), viscosity of the spilled oil, and the rate of release (e.g., catastrophic flood versus spray) will greatly influence the degree of evaporation that will be experienced by spilled petroleum.

In terms of the effect on hydrocarbon fingerprints, evaporative weathering results in a very pronounced reduction in the relative abundance of the lower molecular weight hydrocarbons. A systematic loss of chromatographic signal from approximately C_{15} down to the lowest hydrocarbons measured is good

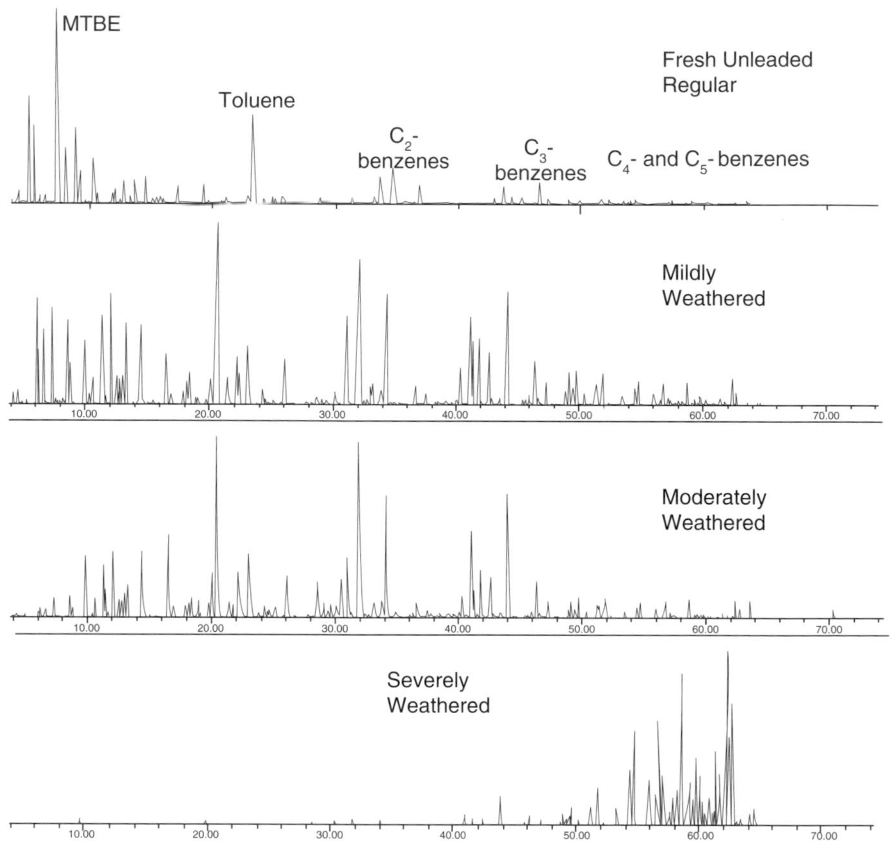

Figure 6.36
GCMS (TIC) patterns showing evaporative weathering of gasoline impacted soils.

evidence for evaporative weathering. Figure 6.36 shows a series of PIANO (GCMS) fingerprints for vadose zone soils that have been impacted by automotive gasoline. The series demonstrates the variable degrees to which evaporation has impacted the distribution of hydrocarbons. Of course, it is difficult to demonstrate evaporative weathering completely independently of other weathering processes. For example, these soils may have been subject to a certain degree of solubilization due to infiltrating water. Nonetheless, most of the changes observed in Figure 6.36 are reasonably attributed to evaporation, which progressively has removed the more volatile compounds leaving a contaminant dominated by C_4- and C_5-alkyl benzenes.

6.5.2.2 Solubilization

As the saying goes, 'oil and water do not mix.' However, there is certainly the potential for the transference of compounds from within an oil phase into water that it may contact. This transference is termed solubilization (or water-washing) and is a function of (1) molar concentration of a given compound in the mixture and (2) the relative solubility of a given compound in water versus

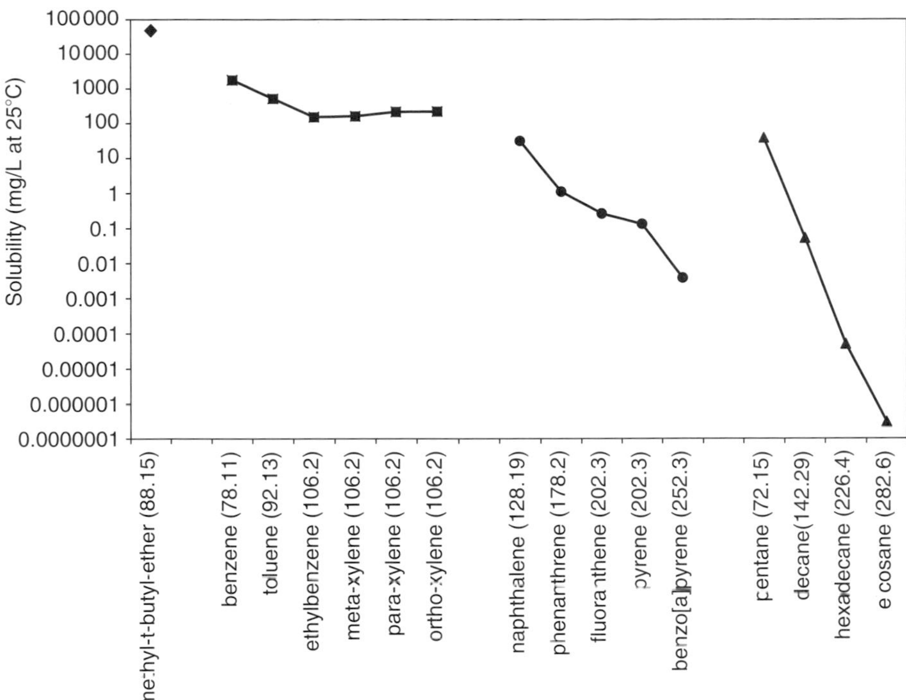

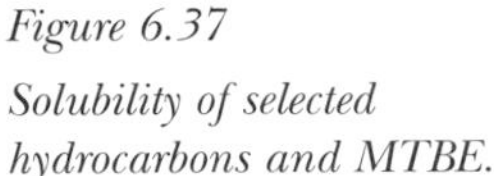

Figure 6.37
Solubility of selected hydrocarbons and MTBE.

its solubility in the hydrocarbon mixture. The degree to which solubilization affects hydrocarbon fingerprint is largely a function of the fuel–water partitioning coefficients (K_{fw}) of the individual compounds in the hydrocarbon mixture. K_{fw} is fuel-specific, meaning that the K_{fw} of toluene in gasoline will be different from the K_{fw} of toluene in Jet A. K_{fw} values are not commonly available but, according to Bruce *et al.* (1991), they can be estimated from the theoretical maximum aqueous solubility (S in mg/L) according to the following equation:

$$\text{Log } K_{fw} = 6.099 - 1.15 \log S \tag{6.1}$$

Given this relationship it can be seen that aqueous solubility (S) can *generally* serve as a proxy for estimating the effects of solubilization for a hydrocarbon mixture in contact with water. As such, hydrocarbons with the highest solubilities are preferentially removed from hydrocarbon mixtures in contact with water. This results in changes in the composition of the mixture that must be recognized in hydrocarbon fingerprinting studies.

Figure 6.37 shows the pure compound solubilities for a selected set of common hydrocarbons. Comparison among these reveals some general 'rules' such as (1) aromatic hydrocarbons are generally more soluble than aliphatic hydrocarbons, (2) increasing degrees of alkylation generally reduce the solubility of similar hydrocarbons, (3) lower molecular weight hydrocarbons within a compound class are generally more soluble than higher molecular weight hydrocarbons in that class, and (4) as demonstrated by MTBE, non-hydrocarbons (polars) are generally more soluble than hydrocarbons.

As might be predicted from these 'rules', light distillate products, especially gasoline, are particularly susceptible to water-washing, and their chromatographic fingerprints can be severely altered by the removal of some of their major chemical constituents that are water-soluble.

Solubilization primarily affects petroleum that is in direct contact with surface water or groundwater. However, given the right surface and soil conditions and rainfall/infiltration rates, water-washing of hydrocarbons within the vadose zone can influence the composition sufficiently to affect the hydrocarbon fingerprinting of the contamination. In many situations, free phase hydrocarbon mixtures reaching the saturated zone (water table) are subject to water-washing that is enhanced by the vertical movements of the water table due to infiltration and recharge fluctuations or tidal 'pumping,' which forms residual oil droplets within the saturated zone, thereby providing a greater opportunity (surface contact area) for solubilization.

Demonstrating the effect of water-washing on petroleum and other hydrocarbon mixtures independently of other forms of weathering is difficult since most weathering processes occur simultaneously. However, Figure 6.38 shows a series of creosote samples recovered from sediment cores that express increasing loss of 2-ring and 3-ring PAHs relative to the 4- to 6-ring PAHs. Although biodegradation can play a role in PAH weathering (Ghoshal *et al.*, 1996), given the recovery of these samples from sub-aqueous sediment cores, at least part of the loss in lower molecular weight PAHs can be attributed to water-washing of the more soluble PAHs in the creosote. Pore water analysis of these sediment cores is currently underway, which would confirm this hypothesis.

6.5.2.3 Biodegradation

There are indigenous microbial populations in most environmental media that are capable of degrading petroleum given favorable conditions such as availability of nutrients (carbon, nitrogen, and phosphorous sources) and an energy source (Turco and Sadowsky, 1995; Kaplan *et al.*, 1996). Energy is obtained from the transfer of electrons from donors to receptors, thus requiring the existence of a suitable donor–acceptor pair. For hydrocarbon biodegradation, molecular oxygen (O_2) is generally the required electron acceptor (Schwarzenbach *et al.*, 1993). Oxygen can be available within the air-filled pores of the vadose zone or dissolved in water. Thus, biodegradation of petroleum and other hydrocarbon mixtures can occur throughout a soil column.

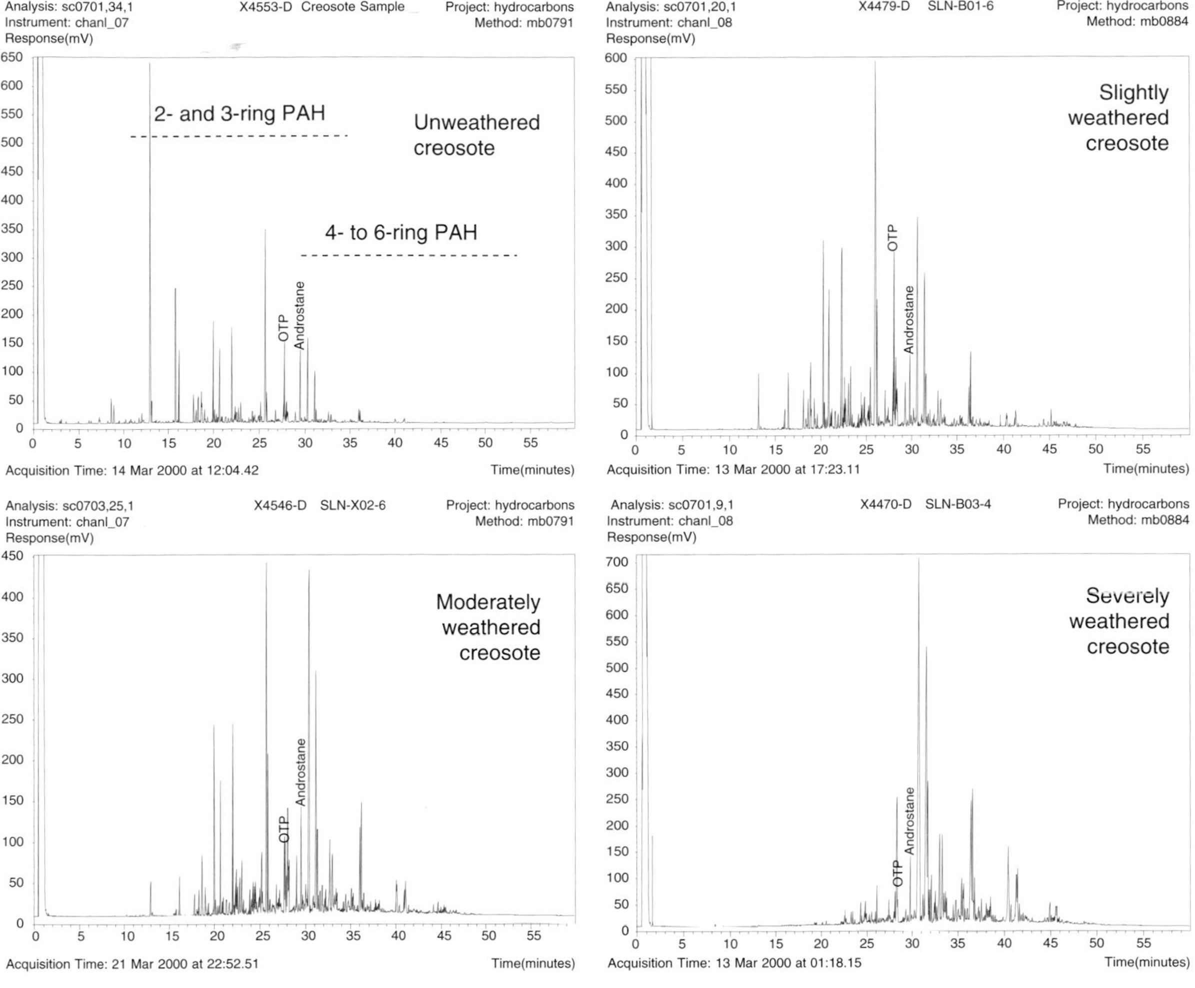

Figure 6.38

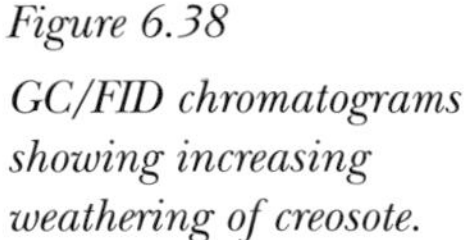

GC/FID chromatograms showing increasing weathering of creosote.

The rate of biodegradation of hydrocarbons is governed by many factors, including:

1. oxygen content,
2. temperature and pH of soil/sediment/water,
3. nature of the indigenous microbial population,
4. concentration of phase separated hydrocarbons, and
5. the chemical structure and properties of the hydrocarbons themselves.

The latter two items are of particular interest to hydrocarbon fingerprinting. First, oxygen is precluded or quickly depleted from soil pores and groundwater wherever separate phase hydrocarbon accumulations exist. Thus, biodegradation in separate phase petroleum pools is significantly retarded due to the excessive concentrations of toxic hydrocarbons in the biologically active aqueous phase. In our experience, biodegradation in separate phase hydrocarbons can be limited to the upgradient edge of the accumulation, i.e., in the vicinity where dissolved oxygen is constantly replenished with fresh groundwater. Certainly, the thickness of the soil containing separate phase hydrocarbons influences the rate of biodegradation also.

Second, although almost all of the major chemical classes of petroleum are biodegradable to some degree, the *n*-alkanes are most susceptible. Consequently, the loss of *n*-alkanes from petroleum and other hydrocarbon mixtures is the most pronounced expression of biodegradation in fingerprinting studies. Branched (isoprenoid) or cyclic (naphthenic and aromatic) hydrocarbons are markedly less biodegradable due to steric hindrance effects that protect carbon-to-carbon bonds from microbial enzymes. As petroleum and other hydrocarbon mixtures become more severely weathered, these more recalcitrant compounds also begin to degrade. Chromatographic evidence of the biodegradation include diminution of isoprenoids and higher molecular weight PAH compounds and a more pronounced unresolved complex mixture ('hump') in the GC chromatogram. These gross features are exemplified in Figure 6.39.

The general sequence of increasing resistance to biodegradation among different compound classes is summarized in Figure 6.40 (Seifert *et al.*, 1984; Peters and Moldowan, 1993; Kaplan *et al.*, 1996). This scale is not intended to represent a series of independent biodegradation steps, e.g., C_1-alkyl PAHs are not completely biodegraded before C_2-alkyl PAH biodegradation commences. Instead, there are notable regions of overlap in the susceptibility of the compound groups listed. For example, PERF (1996) followed the degradation of *n*-alkanes and isoprenoids in a diesel fuel #2 under laboratory degradation conditions (Figure 6.41). The preferential loss of *n*-alkanes relative to isoprenoid hydrocarbons was evident throughout the early phases of biodegradation process. However, as biodegradation proceeded the isoprenoid concentrations were reduced prior to the complete removal of the *n*-alkanes. Thus, the *n*-alkanes need not be absent before the isoprenoids are affected by biodegradation.

Utilizing the relative susceptibility scale shown in Figure 6.40, the degree of weathering for fugitive petroleum products can be semi-quantified and thereby yield a consistent fingerprinting expression to terms such as 'unweathered,' 'slightly weathered,' etc. Petroleum geochemists have developed their own 'scale' of biodegradation of crude oil (on a geologic timescale in the original petroleum reservoir) that relies largely upon the changes in the biomarkers (Peters and Moldowan, 1993). However, we have found that biodegradation on an environmental timescale (rather than a geologic timescale) normally does

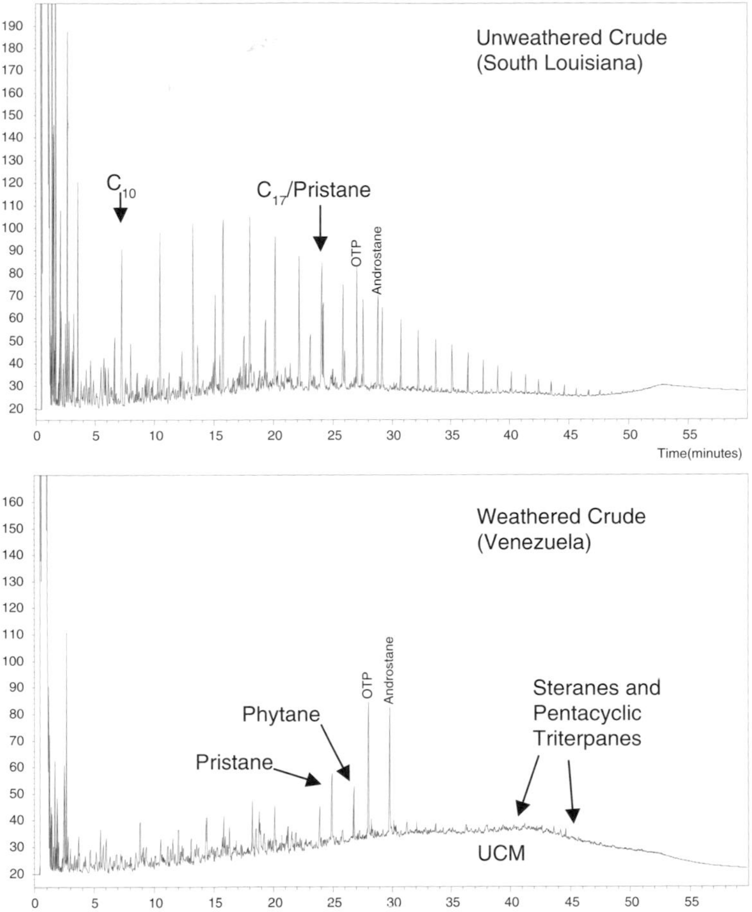

Figure 6.39

GC/FID chromatograms showing comparison of an unweathered and weathered crude oil.

not result in the biodegradation of biomarkers (e.g., Prince *et al.*, 1994). This is certainly a benefit to the forensic investigator since these primary features of petroleum can be utilized in fingerprinting highly weathered petroleum. However, we believe it is necessary to establish a suitable 'scale' for describing the degrees of biodegradation that are encountered in environmental samples. We consistently apply the following terms as indicated in our descriptions of biodegraded petroleum.

Unweathered. A sample that does not appear to have been altered in the environment (i.e., it appears to be typical and comparable to a petroleum or other product standard). Unweathered samples exhibit only primary or secondary features (see sections above).

Slightly weathered. A sample in which only the most susceptible compounds appear to have been reduced relative to less susceptible compounds.

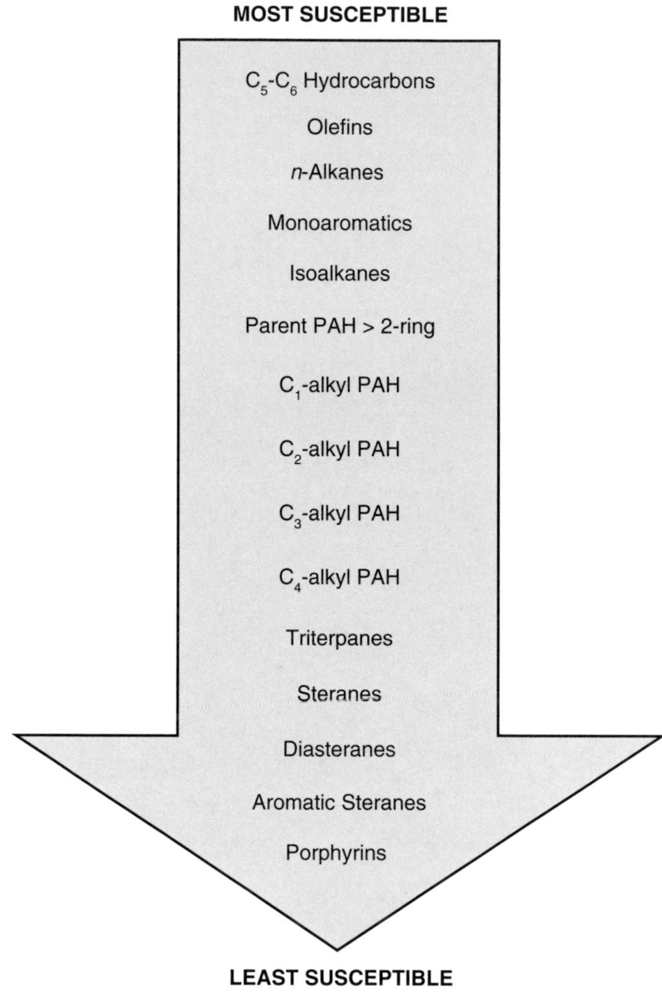

Figure 6.40

Generalized relative susceptibility of hydrocarbon classes to biodegradation.

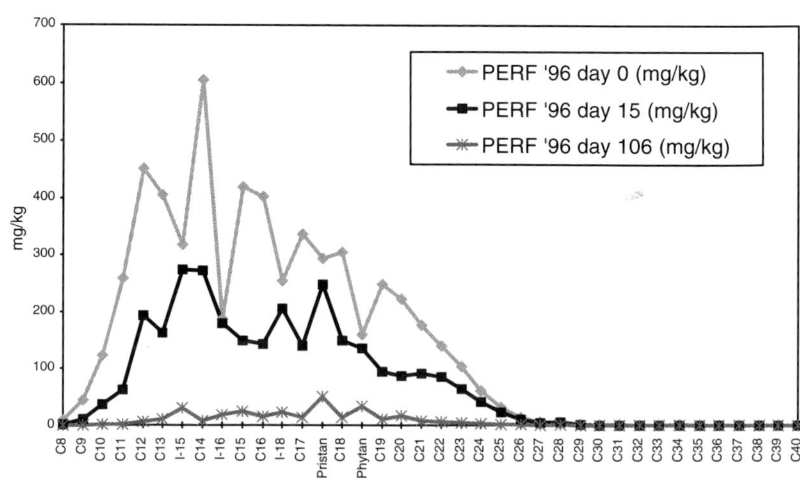

Figure 6.41

Degradation of normal and isoprenoid hydrocarbons determined in a diesel fuel #2 degraded in the laboratory. Data from PERF (1996) study.

For example, in middle distillates and residual fuels the *n*-alkanes appear slightly reduced relative to iso-alkanes and/or the 2-ring PAHs appear enriched. In gasoline, the most volatile compounds are reduced relative to unweathered equivalents.

Moderately weathered. A sample in which a substantial portion of the resolved compounds (e.g., *n*-alkanes and lower boiling aromatic compounds) have been lost or reduced. In the case of distillate products and residual fuels this is often seen by a substantial increase in the chromatographic UCM. The biomarker region becomes obvious through the appearance of resolved peaks.

Severely weathered. A sample in which most of the volatile and semi-volatile compounds are lost, no *n*-alkanes are present, even the isoprenoids appear substantially depleted or absent. PAHs and biomarkers are the only remaining measurable resolved compounds, and there is a significant increase in UCM.

6.5.2.4 Weathering Indices

In addition to the semi-qualitative scale (terminology) described above, there are a variety of comparative indices that are useful to an environmental forensic investigator when trying to determine and compare the weathering state of fugitive hydrocarbon mixtures. These rely upon some of the general properties described above, but can be measured quantitatively using quantitative fingerprinting data. Some examples of commonly utilized weathering indices for gasoline and similar light distillates are presented in Tables 6.16 (gasolines) and 6.17 (non-gasolines). These weathering ratios can express subtle differences in the chemical compositions of otherwise similar fugitive petroleum mixtures versus each other and their unweathered equivalent(s) (or standard

Table 6.16

Selected weathering indices appropriate for gasolines.

Evaporation
n-pentane/*n*-heptane
2-methylpentane/2-methylheptane
wt% < *n*-pentane
Solubilization
Benzene/cyclohexane
Toluene/methylcyclohexane
Total aromatics/total paraffins
Total aromatics/naphthenes
Benzene/toluene
Toluene/xylenes
Biodegradation
3-Methylhexane/*n*-heptane
C_4–C_8 *n*-alkanes + iso/C_4–C_8 olefins
Methylcyclohexane/*n*-heptane
Isoparaffins + naphthenes/total *n*-alkanes

Table 6.17
Selected weathering parameters appropriate for distillate and residual fuels and crude oil.

n-C_{17}/Pristane n-C_{18}/Phytane C_8–C_{15} n-alkanes/C_8–C_{44} n-alkanes
PAH distribution patterns ($C_0 < C_1 < C_2 < C_3 < C_4$) C_3-dibenzothiophenes/C_3-chrysenes[a]
Percentage depletion[b] Percentage depletion of individual target analytes [1 – (C1/C2) * (H2/H1)] * 100 Percentage depletion of total petroleum hydrocarbons (TPH) [1 – (H2/H1)] * 100

[a]Douglas *et al.* (1996)
[b]Douglas *et al.* (1992)
Where: C1 = concentration of analyte in the sample; C2 = concentration of analyte in the source (time zero); H1 = hopane concentration in the sample; H2 = hopane concentration in the source (time zero).
Note: All depletion estimate calculations are done on an oil weight basis; oil weights are obtained during sample preparation.

reference materials). This type of quantitative comparison can be used to establish biodegradation trends within a site, which can sometimes be useful in unraveling sources for the contamination. For example, when comparing a released gasoline sample to a suspected source gasoline (Table 6.16) the degree of evaporation can be monitored by comparing lower molecular weight compounds with corresponding higher molecular weight analogues (i.e., *n*-pentane/*n*-heptane). A similar strategy can be used to develop ratios more prone to change due to solubilization, i.e., compounds with similar structures and boiling points, but dissimilar compound classes (e.g., toluene/methylcyclohexane). Finally, degradation of gasoline can be monitored using ratios of less/more susceptible compounds with similar boiling points and solubilities (e.g., methylcyclohexane/*n*-heptane).

Examples of some commonly utilized weathering indices for middle distillate fuels, residual fuels, crude oils and other petroleum wastes are presented in Table 6.17. The chemical components used in these ratios focus on compounds contained within the boiling range of diesel and residual range hydrocarbons. These product types are often dominated by *n*-alkanes in their unweathered condition, compounds that are prone to biodegradation (Figure 6.37). Thus, simple ratios of n-C_{17}/Pr or n-C_{18}/Ph are useful in monitoring the relative degree of biodegradation in these fuels. Other parameters can focus on the changes in the distribution of alkylated PAHs within a given PAH group. Generally, due to all processes of weathering, the more highly substituted PAHs are less susceptible to weathering (i.e., a parent PAH is more susceptible than the C_4 equivalent isomers). This is an important factor in examining the distribution of PAHs in fingerprinting studies intended to distinguish pyrogenic ('skewed') from petrogenic ('bell-shaped') PAH patterns (Sauer and Uhler,

1994–1995). Similarly, heavier PAHs are less susceptible than lighter PAHs, allowing a ratio like C_3-dibenzothiophenes/C_3-chrysenes to monitor weathering (Douglas *et al.*, 1996).

6.5.2.5 Rate of Weathering

Petroleum and other hydrocarbon mixtures released into the environment will weather at widely varying rates depending upon the site- or incident-specific conditions. Sometimes the conditions are such that weathering of a hydrocarbon mixture is rapid and overwhelmingly affected by a single process. For example, consider a small volume, surface spill of gasoline on a warm day that evaporates in a period of hours. Alternatively, under a different set of circumstances, e.g., a large volume, subsurface release due to catastrophic tank failure, the same gasoline can be only very slowly altered by all forces of weathering, over a period of many years. These two simple (but not uncommon) scenarios demonstrate the variety of weathering effects and rates that must be woven into any interpretation of petroleum fingerprints.

As a result, we caution against the use of weathering alone to establish an age, either *relative* or *absolute*, since there is usually no defensible basis to quantify the rate of weathering (e.g., biodegradation of *n*-alkanes) for a given site. Recall from the discussions above that the rate of weathering depends upon the:

1 spill/release conditions (rate/volume/composition),
2 depth (surface, subsurface, depth to groundwater), and
3 soil conditions (moisture, O_2, microbial populations, stratigraphy).

This summary does not even consider the difficulties in using the degree of weathering to establish age at sites subject to multiple release events and/or mixtures. Thus, while it is reasonable to examine a soil containing a moderately weathered diesel fuel #2 and conclude that it was not spilled yesterday or last week, perhaps not even last year, establishing an age any further is probably unreasonable and indefensible.

While it is sometimes reasonable to compare a single product type (established by criteria independent of weathering, e.g., biomarkers) from a single location (thought to have a limited release history) to attempt to determine a *relative* age, even this approach can be easily confounded by heterogeneities in the site-specific conditions. For example, Stout and Lundegard (1998) examined a soil core taken through an interval of free phase diesel fuel/fuel oil #2 believed to be derived from a single source. The level of weathering, mostly due to biodegradation, increased substantially near and below the oil–water interface, as might be expected given the availability of oxygen at the groundwater–oil interface. Meanwhile the upper part of the core, well above the water table, the petroleum was virtually unweathered. Thus, establishing a relative age for free phase contamination using the degree of weathering alone is difficult at best.

In our opinion, the use of the degree of weathering to establish an *absolute* age for petroleum in the environment is extremely challenging. Nonetheless, this approach has been popularized for dating diesel fuel releases by Christensen and Larsen (1993). Given the implication that this publication has had on environmental forensics, the following discussion is offered.

6.5.2.6 Age Dating Middle Distillate Fuels

A consequence of the disparity in the rate of biodegradation among hydrocarbon classes is the fact that the isoprenoids (pristane and phytane) will degrade more slowly than comparable *n*-alkanes (*n*-C_{17} and *n*-C_{18}; Figure 6.40). This disparity has been long recognized by petroleum geochemists (e.g., Connan, 1984; Kennicutt, 1988; Cassani and Eglinton, 1991). It is this disparity that has been utilized to 'age date' middle distillate fuels by Christensen and Larsen (1993).

Christensen and Larsen (1993) assembled data from 12 sites in northern Europe where a one-time release of diesel fuel/fuel oil #2 reportedly occurred at a known time. In summary, the data from these sites ultimately yielded a linear correlation between the 'time since the release in years' and 'average *n*-C_{17}/Pr ratio in vadose zone soils (> 1 m depth)' (Figure 6.42). The authors reported that the determination of age from *n*-C_{17}/Pr is accurate to ±2.1 years. Kaplan *et al.* (1995) extracted the plotted data points from Christensen

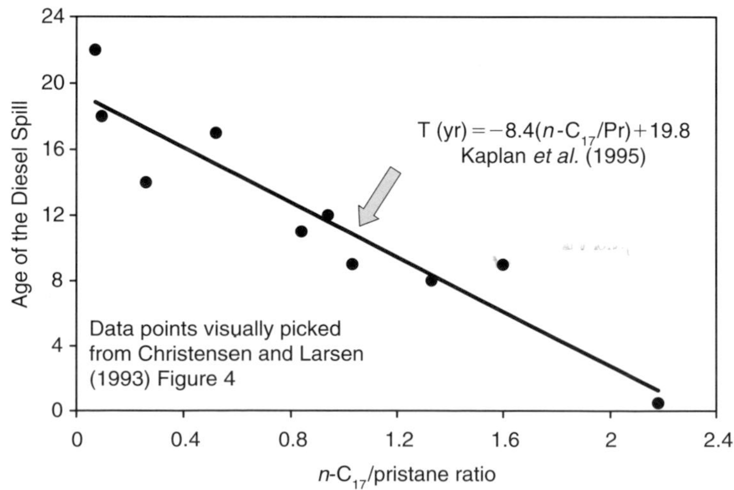

Figure 6.42

Reproduction of the Christensen and Larsen (1993) regression model for age dating diesel fuel oil releases. Formula from Kaplan et al. *(1995).*

and Larsen (1993) and published a linear equation, making the application of this method even simpler. As a result, one need only determine the n-C_{17}/Pr ratio for a soil sample (meeting the criteria stated in the paper), plug it into the formula, to calculate the contaminant's age to within 2.1 years. However, there are numerous caveats that must be considered in this age-dating method that otherwise might lead to oversimplifications of the very complex issue of age dating diesel oil contamination. The caveats that we recognize in the Christensen and Larsen (1993) age-dating method are described below.

Christensen and Larsen (1993) describe the range of conditions (e.g., concentration, oxygen content, moisture content, etc.) that can occur in the subsurface and how these can affect the rates of weathering, particularly degradation. To emphasize this point they describe the lack of weathering that would be expected to occur in an underground storage tank and the excessive weathering that would be expected to occur in agricultural topsoils. They conclude, 'environmental conditions vary considerably in oil-affected soils' and that such variability results in different rates of degradation, even at a single site/plume. Thus, before the method is applied it is necessary to recognize that variable locations within a plume of diesel fuel can give variable n-alkane/isoprenoid ratios (and thus, variable ages). The authors suggest collection of samples in a well-characterized plume, implying that applying the methodology to 'one-time' samples in a poorly characterized diesel plume could easily yield erroneous results. Studies that attempt to apply this method must have some three-dimensional understanding of the plume and its heterogeneity.

The authors' data set was obtained from 12 sites. These 12 sites were selected after a review of more than 200 sites in Denmark and Holland. The reason for their selection was that historical evidence at these 12 sites provided (1) a known date for a sudden diesel release and (2) evidence that a subsequent release(s) had not occurred at these sites after the known release date. This means that the vast majority of the sites reviewed ($> 94\%$) had no evidence for a well-dated, sudden release or no evidence that a subsequent release(s) did not occur. These restrictions serve to emphasize that the age-dating method should be restricted to sites where a release of diesel fuel has occurred suddenly or over a known and limited time interval. If, for example, a slow continuous leak occurs over a period of time or if an existing release is later impacted by a subsequent release, the age-dating method is confounded. The authors acknowledge this restriction but unfortunately, in practice, this restriction is sometimes overlooked. For example, if a 'fresh' diesel fuel mixes with an older, degraded diesel the 'apparent' age of the mixture (as determined by this method) would be intermediate between that of the two separate releases.

Another potential restriction of the age-dating method is that it was based upon sites in northern Europe. Furthermore, the sites selected were paved (minimizing the effects of infiltration), all of the samples were obtained from within the vadose zone (i.e., above the water table), and all samples were obtained from at least 1 meter below ground surface. At the very least, these restrictions deserve consideration before the age-dating method can be applied at any particular site. For example, the soil temperatures at the 12 northern European sites selected were reported to be approximately constant 10°C. The soil temperature (and soil moisture) at a site needs to be considered comparable to those in northern Europe; otherwise the rate of degradation could vary and, in turn, the rate at which the *n*-alkanes are degraded will vary. A change in the rate of degradation at a given site will alter the *slope* of the line relating $n\text{-}C_{17}/Pr$. Thus, application of the Kaplan *et al.* (1995) formula (and its slope; Figure 6.42) could yield erroneous results. The minimum depth of sampling (1 m) was apparently important to the authors since, as discussed above, degradation can be accelerated in near-surface soils (leading to artificially 'older' results). Furthermore, daily and seasonal variations in soil temperature in the upper 1 meter of soils (especially beneath paving) may affect the rate of degradation in this zone. Again, in practice, these restrictions on the age-dating method are sometimes ignored.

One advantage of the method described is that it requires minimal and, what should be, easily obtained chemical data. Many laboratories can provide data suitable to determine the ratio of $n\text{-}C_{17}$/pristane, (and the other aliphatic hydrocarbons discussed in the paper). The authors do emphasize, however, that the age-dating method requires that all relevant compounds be chromatographically resolved. In the case of $n\text{-}C_{17}$ and pristane, many laboratories do not achieve adequate chromatographic separation of these peaks because they employ a GC oven temperature program with too high a heating rate. Without resolution of $n\text{-}C_{17}$ and pristane it can be difficult to determine the amount of each individual compound, therefore, the ratio determined can be imprecise. The chromatographic data used to generate $n\text{-}C_{17}/Pr$ ratios must be of high enough quality to clearly resolve these two compounds. Thus, another important question to ask before applying this methodology is, 'are the data quality (chromatographic resolution) sufficient to determine the $n\text{-}C_{17}/Pr$ ratio precisely?'

Another data consideration is that the authors used 'peak height' data to determine the relevant ratios. They point out that 'peak area' may be more appropriate but, apparently, this type of data was not available to them. Thus, the calibration they develop is based upon the peak heights of $n\text{-}C_{17}$ and pristane. If users of the age-dating method have data generated from the peak areas of $n\text{-}C_{17}$ and pristane (rather than peak height), there should be a concern that the results may vary by some percentage.

The authors surveyed 11 fresh diesel fuels, obtained from different refiners, in order to determine the variability in their chemical composition. The reason this survey was conducted was to determine if the n-C_{17}/Pr (and other) ratios varied among fresh diesel fuels. Any variability that occurred in fresh diesel fuel could have been inherited from the parent crude oil from which the diesel fuel was distilled (i.e., a primary control). If the n-C_{17}/Pr ratio of the fresh diesel fuels surveyed varied significantly, then the method would be confounded. For example, if fresh diesel fuel A had a n-C_{17}/Pr ratio of 1.15 while fresh diesel fuel B had a ratio of 2.2 (e.g., see Figure 6.6 above), then if these two fuels were spilled on the same day (under the same circumstances) diesel fuel A would always appear to be 'older', even though it is not.

Unfortunately, Christensen and Larsen (1993) do not present the raw data for the 11 diesel fuels they surveyed, only the average results (see Table 6.2 in Christensen and Larsen, 1993). Inspection of their Table 6.2 reveals the 11 fresh diesel fuels had an average n-C_{17}/pristane ratio of 1.98 and a standard deviation of 0.83. Assuming the raw data were part of a normal distribution, calculating the 95% confidence interval (95% CI = mean ± (standard deviation/square root of number of samples)) indicates that the 11 fresh diesel fuels had n-C_{17}/Pr ratios measured mostly fell between 1.5 and 2.5. The authors describe these results as evidence of 'low variability' in the fresh diesel fuels from northern Europe. We disagree with this characterization considering the importance that the 'starting ratio' has on application of the curve (Figure 6.42). If we assume the rate of degradation (i.e., the slope of the line) is constant, then the starting ratio of n-C_{17}/Pr is extremely important. They acknowledge this fact but minimize its implication by stating that the diesel fuels in northern Europe originate from only a few crude oil sources (North Sea or Middle East), implying that they should not vary. However, the fresh diesel fuels they surveyed appear to actually have varied quite a bit (n-C_{17}/Pr ratios of 1.89 ± 0.83). Our own experience in characterizing fresh diesel fuels indicates that n-C_{17}/Pr ratios can vary significantly. For example, a n-C_{17}/Pr 'starting ratio' of 1.15 in a fresh middle distillate (see Figure 6.6 above) would, according to the Christensen and Larsen (1993) calibration, indicate that it is about 10 years old. The n-C_{17}/Pr ratio in the fresh fuel oil #2 standard shown in Figure 6.25 is close to 4.0, indicating a 'negative' age according to the method. Obviously, the starting ratio of n-C_{17}/Pr is very important in the application of this method.

The 12 sites selected for study yielded an unknown number of samples, something more than 12, from which the relevant ratios, including n-C_{17}/Pr, were determined. Unfortunately, again these raw data are not reported; only the r^2 values as determined from a correlation between the ratio and the age (in years) of the diesel fuel are reported. An interesting feature of Table 6.3 in Christensen and Larsen (1993) is that the r^2 values for *all* of the data are consistently less than the r^2 values for the *average* data from each site. This suggests that the individual data points from each site varied implying heterogeneity in degradation rates (or analytical precision). (Without the raw data there is no way to evaluate the degree to which the ratios from a single site varied.) The low r^2 value for all of the data (0.66 in the case of n-C_{17}/Pr) serves to emphasize that applying the calibration to data generated from one or a few grab samples from a site could introduce significant error in age dating.

Finally, inspection of the 12 selected sites reveals that only one location had a sudden, known release less than eight years old (the Ejby site). If one inspects the correlation derived, it appears that the result from the Ejby site 'anchors' the calibration line and influences the calculation of the 95% confidence intervals. The quantitative implication of this restriction is difficult to assess since the authors have not reported the raw n-C_{17}/Pr data for the 12 individual sites.

In our opinion, application of the Christensen and Larsen (1993) method for age dating middle distillate fuels requires careful consideration of each of the caveats discussed above. Obviously, more site investigations of this type (data sets) are necessary to evaluate the universal applicability and precision of this method.

6.6 CHEMICAL FINGERPRINTING – ONE COMPONENT OF A FORENSIC STUDY

The information contained in this chapter should help the forensic investigator unravel and defend the issues of source and nature of hydrocarbon contamination in environmental matrices. Chemical fingerprinting is a powerful forensic 'tool,' when it is applied properly. In some instances, chemical fingerprinting can alone meet the objectives of a forensic investigation. However, more often than not, the utility and defensibility of chemical fingerprinting is enhanced when in its incorporated with other forensic 'tools' described in the other chapters in this book.

REFERENCES

Aggarwal, P.K. and Hinchee, R.E. (1991) Monitoring in situ biodegradation of hydrocarbons by using stable carbon isotopes. *Environmental Science and Technology* 25, 1178–1180.

Albaiges, J., Algaba, J., Clavell, E., and Grimalt, J. (1986) Petroleum geochemistry of the Tarragona Basin (Spanish Mediterranean off-shore). *Organic Geochemistry* 10, 441–450.

Atlas, R.M., Boehm, P.D., and Calder, J.A. (1981) Chemical and biological weathering of oil from the Amoco Cadiz spillage within the littoral zone. *Estuarine, Coastal and Shelf Science* 12, 589–608.

Barrick, R.C., Hedges, J.I., and Peterson, M.L. (1980) Hydrocarbon geochemistry of the Puget Sound region – I. Sedimentary acyclic hydrocarbons. *Geochimica Cosmochimica Acta* 44, 1362–1439.

Bastow, T.P., van Aarssen, B.G.K., Alexander, R., and Kagi, R.I. (1999) Biodegradation of aromatic land-plant biomarkers in some Australian crude oils. *Organic Geochemistry* 30, 1229–1239.

Behar, F. and Pelet, R. (1984) Characterization of asphaltenes by pyrolysis and chromatography. *Journal of Analytic and Applied Pyrolysis* 7, 121–135.

Berkowitz, N. (1979) *An Introduction to Coal Technology*. Academic Press, New York.

Bishop, G.J. and Henry, C.P. (1993) Aviation fuels. In *ASTM Manual on Significance of Tests for Petroleum Products* 6th edn. (Dyroff, G.V., ed.). ASTM Publ. No. 28-001093-12, Philadelphia, PA.

Bloom, N.S. and Crecelius, E.A. (1987) Distribution of silver, mercury, lead, copper, and cadmium in central Puget Sound sediments. *Marine Chemistry* 21, 377–390.

Boehm, P.D. and Farrington, J.W. (1984) Aspects of the polycyclic aromatic hydrocarbon geochemistry of Recent sediments in the Georges Bank region. *Environmental Science and Technology* 18(11), 840–845.

Boehm, P.D., Page, D.S., Gifillan, E.S., Bence, A.E., Burns, W.A., and Mankiewicz, P.J. (1998) Study of the fates and effects of the *Exxon Valdez* oil spill on benthic sediments in two bays in Prince William sound, Alaska. 1. Study design, chemistry, and source fingerprinting. *Environmental Science and Technology* 32, 567–576.

Boehm, P.D., Douglas, G.S., Brown, J., Page, D.S., Bence, A.E., Burns, W.A., and Mankiewicz, P.J. (2000) Comment on 'Natural hydrocarbon background in benthic sediments of Prince William Sound, Alaska: oil vs coal'. *Environmental Science and Technology* 34, 2064–2065.

Bradley, L.J.N., Magee, B.H., and Allen, S.L. (1994) Background levels of polycyclic aromatic hydrocarbons (PAH) and selected metals in New England urban soils. *Journal of Soil Contamination* 3, 349–361.

Brooks, P.W. (1986) Unusual biological marker geochemistry of oils and possible source rocks, offshore Beaufort-Mackenzie Delta, Canada. *Organic Geochemistry* 10, 401–406.

Bruce, L.G. (1993) Refined gasoline in the subsurface. *American Association of Petroleum Geologists Bulletin* 77, 212–224.

Bruce, L.G., Hockman, E.L., and Miller, T.G. (1991) Solubility versus equilibrium saturation of gasoline compounds: A method to estimate fuel water partition coefficient using solubility or Koc. In *NWWA/API Conference on Petroleum Hydrocarbons and Organic Chemicals in Groundwater* (Stanley, A., ed.), pp. 571–582.

Caricchia, A.M., Chiavarini, S., and Pezza, M. (1999) Polycyclic aromatic hydrocarbons in the urban atmospheric particulate matter in the city of Naples (Italy). *Atmospheric Environment* 33, 3731–3738.

Cassani, F. and Eglinton, G. (1991) Organic geochemistry of Venezuelan extra-heavy crude oils. 2. Molecular assessment of biodegradation. *Chemical Geology* 91, 315–333.

Chaffee, Alan L., Hoover, David S., Johns, R. B., and Schweighardt, F. K. (1986) Biological markers extractable from coal. *Biological Markers in the Sedimentary Record*, pp. 311–345. Elsevier Press, Amsterdam.

Christensen, L.B. and Larsen, T.H. (1993) Method for determining the age of diesel oil spills in the soil. *Ground Water Monitoring and Remediation*, Fall, 142–149.

Colombo, J.C., Pelletier, E., Brochu, C., and Khalil, M. (1989) Determination of hydrocarbon sources using n-alkane and polyaromatic hydrocarbon distribution indexes: Case Study – Rio de La Plata Estuary, Argentina. *Environmental Science and Technology* 23, 888–894.

Connan, J. (1984) Biodegradation of crude oils in reservoirs. In *Advances in Petroleum Geochemistry*, (Brooks, J. and Welte, D., eds), pp. 299–335. Academic Press, London.

Curiale, J.A. and Odermatt, J.R. (1989) Short-term biomarker variability in the Monterey Formation, Santa Maria Basin. *Organic Geochemistry* 14, 1–13.

Davidson, J. and Creek, D.N. (1999) Using the gasoline additive MTBE in forensic environmental investigations. *International Journal of Environmental Forensics* 1, 57–67.

Davidson, R.M. (1982) *Molecular Structure of Coal*, pp. 84–155. Academic Press, New York.

Didyk, B.M., Simoneit, B.R.T., Brassell, S.C., and Eglinton, G. (1978) Organic geochemical indicators of palaeoenvironmental conditions of sedimentation. *Nature* 272, 216–222.

Dorn, P., Mourao, A.M., and Herbstman, S. (1983) The properties and performance of modern automotive fuels. Society of Automotive Engineers, Paper No. 861178, pp. 51–67.

Douglas, G.S. and Uhler, A.D. (1993) Optimizing EPA methods for petroleum-contaminated site assessments. *Environmental Testing and Analysis*, May/June 93, 46–53.

Douglas, G.S., Prince, R.C., Butler, E., and Steinhauer, W.G. (1993) The use of internal chemical indicators in petroleum and refined products to evaluate the extent of biodegradation. *In Situ and On-Site Bioreclamation Conference Proceedings*, San Diego, CA, April 5–8.

Douglas, G.S., Bence, A.E., Prince, R.C., McMillen, S.J., and Butler, E. (1996) Environmental stability of selected petroleum hydrocarbon source and weathering ratios. *Environmental Science and Technology* 30, 2332–2339.

Durell, G., Ginsburg, L.C., and Shea, D. (1991) CSO effects on contamination of Boston Harbor sediments. Report to Massachusetts Water Resources Authority, Charleston Yard, Boston.

Durell, G., Neff, J., Melbye, A., Johnsen, S., Garpestad, E., and Gruner, H. (2000) Monitoring and assessment of produced water originating contaminants in the Ekofisk Region of the North Sea. SPE 61132 Prepared for Presentation at the SPE International Conference on Health, Safety, and the Environment in Oil and Gas Exploration and Production. June 26–28, 2000. Stavanger, Norway. Society of Petroleum Engineers, Richardson, TX.

Eavenson, H.N. (1942) *The First Century and a Quarter of American Coal Industry*. Waverly Press, Pittsburgh, PA.

Edison Electric Institute (1984) *Handbook on Manufactured Gas Plants*. Utility Solid Wastes Activity Group, Edison Electric Institute, Pittsburgh, PA.

Eganhouse, R.P., Blumfield, D.L., and Kaplan, I.R. (1982) Petroleum hydrocarbons in stormwater runoff and municipal wastes: input to coastal waters and fate in marine sediments. *Thalassia Jugoslavica* 18, 411–431.

Ehleringer, J.R. and Rundel, P.W. (1989) Stable isotopes: history, units, and instrumentation. In *Stable Isotopes in Ecological Research* (Ehleringer, J.R. and Nagy, K.A., eds), pp. 1–15. Springer Verlag, New York.

Ekweozor, C.M. and Udo, O.T. (1988) The oleananes: origin, maturation and limits of occurrence in Southern Nigeria sedimentary basins. *Organic Geochemistry* 13, 131–140.

Emsbo-Mattingly, S.D. and Mauro, D.M. (1999) Occurrence of Refined Petroleum Products at Former Manufactured Gas Plant (MGP) Sites. International Conference on Contaminated Soil and Water. October 18–21, 1999 Amherst, Massachusetts.

Emsbo-Mattingly, S.D., McCarthy, K.J., Stout, S.A., and Uhler, A.D. (2000) Differentiating coal and petroleum derived MGP residues. International Business Corporation 3rd Annual Executive Forum on Environmental Forensics, Washington DC.

Environmental Research and Technology Inc. and Koppers Corp. (1984) *Handbook on Manufactured Gas Plants*. Edison Electric Institute, Pittsburg, PA.

EPA (1983) Methods for chemical analysis of water and wastes. EPA-600-4-79-020. Environmental Monitoring and Support Laboratory, Cincinnati, OH.

EPA (1997) Test methods for evaluating solid waste (SW-846). Update III US Environmental Protection Agency, Office of Solid Waste and Emergency Response, Washington, DC.

EPA (1999) Guidelines establishing test procedures for the analysis of oil and grease and non-polar material under the Clean Water Act and Resource Conservation and Recovery Act; Final Rule. 40 CFR Parts 136 and 260 [FRL-6341-9] RIN 2040-AC63.

EPA Blue Ribbon Panel (1999) *Achieving Clean Air and Clean Water*. EPA420-R-99-021.

Farrimond, P., Taylor, A., and Telnæs, N. (1998) Biomarker maturity parameters: the role of generation and thermal degradation. *Organic Geochemistry* 29, 1181–1197.

Filby, R.H. and Van Berkel, G.J. (1987) Geochemistry of metal complexes in petroleum, source rocks, and coals: an overview. In *Metal Complexes in Fossil Fuels,* (Fibly, R.H. and Branthaver, J.F., eds), American Chemical Society, Washington, DC.

Frank, J.L. (1999) New mandates present fuel challenges for US refiners. *Oil and Gas* Journal Dec. 13, 118–121.

Gaines, R.B., Frysinger, G.S., Hendrick-Smith, M.S., and Stuart, J.D. (1999) Oil spill identification by comprehensive two-dimensional gas chromatography. *Environmental Science and Technology* 33, 2106–2112.

Gallegos, E.J. (1981) Possible significance of the relative concentrations of terpanes, terpenes, steranes, and cadalene in six United States coals. *Journal of Chromatographic Science* 19, 156–160.

Garrigues, P., Budzinski, H., Manitz, M.P., and Wise, S.A. (1995) Pyrolytic and petrogenic inputs in recent sediments: a definitive signature through phenanthrene and chrysene compound distribution. *Polycyclic Aromatic Compounds* 7, 275–284.

Gary, J.H. and Handwerk, G.E. (1984) *Petroleum Refining*, 2nd edn. Marcel Dekker, New York.

Ghoshal, S., Ramaswami, A., and Luthy, R.G. (1996) Biodegradation of naphthalene from coal tar and heptamethylnonane in mixed batch systems. *Environmental Science and Technology* 30, 1282–1291.

Gibbs, L.M. (1990) *Gasoline Additives – When and Why*. SAE Technical Paper Series, International Fuels and Lubricants Mtg. and Exposition, Tulsa, OK.

Gibbs, L.M. (1993) *How Gasoline Has Changed SAE International Fuels and Lubricants Meeting and Exposition*, pp. S01007-S010025. SAE – The Engineering Society for Advancing Mobility Land Sea Air and Space International, Warrendale, PA.

Gibbs, L.M. (1996) *How Gasoline Has Changed II – The Impact of Air Pollution Regulations*, pp. 1–16. SAE International Fall Fuels and Lubricants Meeting and Exposition. SAE – The Engineering Society for Advancing Mobility Land Sea Air and Space International, Warrendale, PA.

Gogou, A., Bouloubassi, I., and Stephanou, E.G. (2000) Marine organic geochemistry of the Eastern Mediterranean: 1. Aliphatic and polyaromatic hydrocarbons in Cretan Sea surficial sediments. *Marine Chemistry* 68, 265–282.

Grantham, P.J. and Douglas, A.G. (1980) The nature and origin of sesquiterpenoids in some tertiary fossil resins. *Geochimica Cosmochimica Acta* 44, 1801–1810.

Grice, K., Alexander, R., and Kagi, R.I. (2000) Diamondoid hydrocarbon ratios as indicators of biodegradation in Australian crude oils. *Organic Geochemistry* 31, 67–73.

Guetens, E.G., Jr, DeJovine, J.M., and Yogis, G.J. (1982) TBA aids methanol/fuel mix. *Hydrocarbon Processing* May, 113–117.

Harkins, S.M., Truesdale, R.S., Hill, R., Hoffman, P., and Winters, S. (1988) US production of manufactured gases: assessment of past disposal practices. EPA/600/2-88/012. Hazardous Waste Engineering Research Laboratory, Office of Research and Development, US Environmental Protection Agency, Cincinnati, OH.

Harrington, R.R., Poulson, S., Drever, J., Colberg, P.J.S., and Kelly, E.F. (1999) Carbon isotope systematics of monoaromatic hydrocarbons: vaporization and adsorption experiments. *Organic Geochemistry* 30, 765–775.

Harrison, R.M., Smith, D.J.T., and Luhana, L. (1996) Source apportionment of atmospheric polycyclic aromatic hydrocarbons collected from an urban location in Birmingham, UK *Environmental Science and Technology* 30, 825–832.

Henry, C.P. (1988) Additives for middle distillate and kerosine fuels. In *Distillate Fuel: Contamination, Storage and Handling. ASTM STP 1005* (Chesneau, H.L. and Dorris, M.M., eds), pp. 105–113. American Society for Testing and Materials, Philadelphia, PA.

Hilkert, A., Douthitt, C.B., Schlüter, H.J., and Brand, W.A. (1999) Isotope ratio monitoring gas chromatography/mass spectrometry of D/H by high temperature conversion isotope ratio mass spectrometry. *Rapid Communications in Mass Spectrometry* 13, 1226–1230.

Hites, R.A., Laflamme, R.E., and Windsor, J.G., Jr. (1980) Polycyclic aromatic hydrocarbons in an anoxic sediment core from the Pettaquamscutt River (Rhode Island, USA). *Geochimica Cosmochimica Acta* 44, 873–878.

Hoering, T.C. (1977) Olefinic hydrocarbons from Bradford, Pennsylvania, crude oil. *Chemical Geology* 20, 1–8.

Hostettler, F.D., Pereira, W.E., Kvenvolden, K.A., van Green, A., Luoma, S.N., Fuller, C.C., and Anima, R. (1999) A record of hydrocarbon input to San Francisco Bay as traced by biomarker profiles in surface sediment and sediment cores. *Marine Chemistry* 64, 115–127.

Huang, W.-Y. and Meinschein, W.G. (1979) Sterols as ecological indicators. *Geochimica Cosmochimica Acta* 43, 739–745.

Hurst, R.W. (1996) Age dating gasoline releases using stable isotopes of lead (ALSA model) In *Geochemical Fingerprinting in Environmental Geology* (Hurst, R.W. and Fisher, J.B., eds). American Association of Petroleum Geologists Short Course No. 9.

Hurst, R.W. (1998) Determining ages and sources of hydrocarbon releases using lead isotopes. *Proceedings of the American Academy of Forensic Science* 76, 304–307.

Hurst, R.W. (2000) Applications of anthropogenic lead archaeostratigraphy (ALAS model) to hydrocarbon remediation. *Environmental Forensics* 1, 11–23.

Jarman, W.M., Hilkert, A., Bacon, C.E., Collister, J.W., Ballschmiter, K., and Risebrough, R.W. (1998) Compound-specific carbon isotopic analysis of Aroclors, clophens, kaneclors, and phenoclors. *Environmental Science and Technology* 32, 833–836.

Jewitt, C.H., Westbrook, S.R., Ripley, D.L., and Thornton, R.H. (1993) Fuels for land and marine diesel engines and for nonaviation gas turbines. In *ASTM Manual on Significance of Tests for Petroleum Products*, 6th edn. (Dyroff, G.V., ed.). ASTM Publ. No. 28-001093-12.

Johnson, M.D. and Morrison, R. (1996) Petroleum fingerprinting: dating a gasoline release. *Environmental Protection* Sept., 37–39.

Kaplan, I.R. and Galperin, Y. (1996) How to recognize a hydrocarbon fuel in the environment and estimate its age of release. *Groundwater and Soil Contamination: Technical Preparation and Litigation Management*. John Wiley & Sons, New York.

Kaplan, I.R., Alimi, M., Galperin, Y., Lee, R.P., and Lu, S. (1995) Pattern of chemical changes in fugitive hydrocarbon fuels in the environment. SPE Paper No. 29754. Presented at the 1995 SPE/EPA Exploration and Production Environmental Conference, March 27–29, 1995 Houston, TX.

Kaplan, I.R., Galperin, Y., Alimi, H., Lee, R.P., and Lu, S. (1996) Patterns of chemical changes during environmental alteration of hydrocarbon fuels. *Ground Water Monitoring and Remediation*. Fall, 113–124.

Kaplan, I.R., Galperin, Y., Lu, S.-T., and Lee, R.-P. (1997) Forensic environmental geochemistry: differentiation of fuel-types, their sources and release time. *Organic Geochemistry* 27, 289–317.

Kelley, C.A., Hammer, B.T., and Coffin, R.B. (1997) Concentrations and stable isotope values of BTEX in gasoline-contaminated groundwater. *Environmental Science and Technology* 31, 2469–2472.

Kennicutt, M.C. II (1988) The effect of biodegradation on crude oil bulk and molecular composition. *Oil and Chemical Pollution* 4, 89–112.

Kinghorn, R.R.F. (1983) *An Introduction to the Physics and Chemistry of Petroleum*. John Wiley & Sons, Chichester.

Kvenvolden, K.A., Hostettler, F.D., Carlson, P.R., Rapp, J.B., Threlkeld, C.N., and Warden, A. (1995) Ubiquitous tar balls with a California-source signature on the shorelines of Prince William Sound, Alaska. *Environmental Science and Technology* 29, 2684–2694.

Laflamme, R.E. and Hites, R.A. (1978) The global distribution of polycyclic aromatic hydrocarbons in recent sediments. *Geochimica Cosmochimica Acta* 42, 289–303.

LaFleur, L., Tatum, V., Borton, D., Gholson, A., and Louch, J. (1998) Current state of knowledge concerning selected sesquiterpenes, diterpene hydrocarbons, and related compounds. National Council of the Paper Industry for Air and Stream Improvement, Inc. (NCASI), Research Triangle Park, NC.

Lao, R.C., Thomas, R.S., and Monkman, J.L. (1975) Computerized gas chromatographic-mass spectrometric analysis of polycyclic aromatic hydrocarbons in environmental samples. *Journal of Chromatography* 112, 681–700.

Larter, S.R., Bowler, B.F.J., Li, M., Chen, M., Brincat, D., Bennett, B., Noke, K., Donohoe, P., Simmons, D., Kohnen, M., Allan, J., Telnæs, N., and Horstad, I. (1996) Molecular indicators of secondary oil migration distances. *Nature* 383, 593–597.

Lee, M.L., Prado, G.P., Howard, J.B., and Hites, R.A. (1977) Source identification of urban airborne PAHs by GC/MS and high resolution MS. *Biomedical Mass Spectrometry* 4, 182–186.

Lee, M.L., Vassilaros, D.L., White, C.M., and Novotny, M. (1979) Retention indices for programmed-temperature capillary column gas chromatography of polycyclic aromatic hydrocarbons. *Analytical Chemistry* 51(6), 768–773.

Lewan, M. (1984) Factors controlling the proportionality of vanadium to nickel in crude oils. *Organic Geochemistry* 48, 2231–2238.

Lowenheim, F. and Moran, M. (1975) *Faith, Keyes, and Clark's Industrial Chemicals*, 4th edn. John Wiley & Sons, New York.

Mackenzie, A.S. (1984) Applications of biological markers in petroleum geochemistry. In *Advances in Petroleum Geochemistry* 1 (Brooks, J. and Welte, D., eds), pp. 115–214. Academic Press, London.

Mackenzie, A.S., Brassell, S.C., Eglinton, G., and Maxwell, J.R. (1982). Chemical fossils: the geological fate of steroids. *Science* 214, 491–504.

Mansuy, L., Philp, R.P., and Allen, J. (1997) Source identification of oil spills based on the isotopic composition of individual components in weathered oil samples. *Environmental Science and Technology* 31, 3417–3425.

Marr, L.C., Kirchstetter, T.W., Harley, R.A., Miguel, A.H., Hering, S.V., and Hammond, S.K. (1999) Characterization of PAH in motor vehicle fuels and exhaust emissions. *Environmental Science and Technology* 33, 3091–3099.

Martin, C.J. and Gray, R.L. (1993) Heating and power generation fuels. In *ASTM Manual on Significance of Tests for Petroleum Products*, 6th edn (Dyroff, G.V., ed.). ASTM Publ. No. 28-001093-12.

McCarthy, K.J. and Uhler, R.M. (1994) Optimizing gas chromatography conditions for improved hydrocarbon analysis. In *Proceedings Ninth Annual Conference on Contaminated Soils*, Amherst, MA.

McCarthy, K.J., Uhler, A.D., and Stout, S.A. (1998) Weathering affects petroleum identification. *Soil and Groundwater Cleanup*, Aug/Sept.

McHugh, L.D. (1997) Remedial Action: Phase III. *Site Auditing: Environmental Assessment of Property*. Specialty Technical Publishers, North Vancouver, BC, Canada.

McRae, C., Love, G.D., and Snape, C.E. (1997) Application of gas chromatography–isotope ratio mass spectrometry to source polycyclic aromatic hydrocarbon emissions. American Chemical Society, Division of Environmental Chemistry – Preprints of Extended Abstracts 37, p. 315.

Mikolaj, P.G., Allen, A.A., and Schlueter, R.S. (1972) Investigation of the nature, extent and fate of natural oil seepage off Southern California. OTC 1549 Prepared for Presentation at the Fourth Annual Offshore Technology Conference. Offshore Technology Conference, Dallas, TX.

Moldowan, J.M., Seifert, W.K., and Gallegos, E.J. (1985) Relationship between petroleum composition and depositional environment of petroleum source rocks. *American Association of Petroleum Geologists Bulletin* 69, 1255–1268.

Moore, E.S. (1940) *Coal – Its Properties, Analysis, Classification, Geology, Extraction, Uses and Distribution*, John Wiley & Sons, New York.

Morgan, J.J. (1926) *Manufactured Gas: A Textbook of American Practice*. J.J. Morgan Publisher (self-published), New York.

Morrison, R. (2000) *Environmental Forensics*. CRC Press, Boca Raton, FL.

Mulroy, P.T. and Ou, L.-T. (1998) Degradation of tetraethyllead during the degradation of leaded gasoline hydrocarbons in soil. *Environmental Toxicology and Chemistry* 17, 777–782.

Murray, A.P., Summons, R.E., Boreham, C., and Dowling, L.M. (1994) Biomarker and *n*-alkane isotope profiles for Tertiary oils: relationship to source rock depositional setting. *Organic Geochemistry* 22, 521–542.

Needleman, H.L. (1998) Review. Clair Patterson and Robert Kehoe: Two views of lead toxicity. *Environmental Research, Section A* 78, 79–85.

Neff, J. (1979) *Polycyclic Aromatic Hydrocarbons in the Aquatic Environment.* Applied Science Publishers, London.

Noble, R.A. (1986) A geochemical study of bicyclic alkanes and diterpenoid hydrocarbons in crude oils, sediments, and coals. PhD Thesis, Department of Organic Chemistry, University of Western Australia.

Noble, R.A., Alexander, R., Kagi, R.I., and Knox, J. (1986) Identification of some diterpenoid hydrocarbons in petroleum. *Organic Geochemistry* 10, 825–829.

Nordtest (1991) Nordtest Method. Oil spill identification. Nordtest, Esbo, Finland.

Novotny, M., Strand, J.W., Smith, S.L., Wiesler, D., and Schwende, F.J. (1980) Compositional studies of coal tar by capillary gas chromatography/mass spectrometry. *Fuel* 60, 213–220.

O'Connor, T.P. and Beliaeff, F.J. (1995) Recent trends in environmental quality: Results from the Mussel Watch Project (and data therein). US Dept. Commerce, National Ocean. Atmosph. Admin., Silver Spring, MD.

O'Malley, V.P., Abrajano, T., and Hellou, J. (1996) Stable carbon isotopic apportionment of individual polycyclic aromatic hydrocarbons in St. John's Harbour, Newfoundland. *Environmental Science and Technology* 20, 634–639.

Oahn, N.T.K., Reutergardh, L.B., and Dung, N.T. (1999) Emission of PAH and particulate matter from domestic combustion of selected fuels. *Environmental Science and Technology* 33, 2703–2709.

Page, D.S., Boehm, P.D., Douglas, G.S., and Bence, A.E. (1995) Identification of hydrocarbons sources in the benthic sediments of Prince William Sound and the Gulf of Alaska following the Exxon Valdez oil spill. In (Wells, P.G., Bulter, J.N., and Hughes, J.S., eds) *Exxon Valdez Oil Spill: Fate and Effects in Alaskan Waters.* STP 1219. American Society for Testing and Materials, Phila., PA, USA, pp. 347–397.

PERF (Petroleum Environmental Research Forum) (1996) Techniques to measure the extent of biodegradation in soils/groundwater. PERF Project 93-04, Final Report.

Peters, K.E. and Moldowan, J.M. (1993) *The Biomarker Guide*. Prentice Hall, Englewood Cliffs, NJ.

Peters, K.E., Scheuerman, G.L., Lee, C.Y., Moldowan, J.M., Reynolds, R.N., and Pena, M.M. (1992) Effects of refinery processes on biological markers. *Energy Fuels* 6, 560–577.

Peters, K.E., Moldowan, J.M., McCaffrey, M.A., and Fago, F.J. (1996) Selective biodegradation of extended hopanes to 25-norhopanes in petroleum reservoirs. Insights from molecular mechanics. *Organic Geochemistry* 24, 765–783.

Peven, C.S., Uhler, A.D., and Querzoli, F.J. (1996) Caged mussels and semipermeable membrane devices as indicators of organic contaminant uptake in Dorchester and Duxbury Bays, Massachusetts. *Environmental Toxicology and Chemistry* 15, 144–149.

Philp, R.P. (1985) Biological markers in fossil fuel production. *Mass Spectrometry Reviews* 4, 1–54.

Philp, R.P. and Gilbert, T.D. (1986) Biomarker distributions in Australian oils predominantly derived from terrigenous source material. *Organic Geochemistry* 10, 73–84.

Pines, H. (1981) *The Chemistry of Catalytic Hydrocarbon Conversions*. Academic Press, New York.

Prince, R.C., Elmendorf, D.L., Lute, J.R., Hsu, C.S., Haith, C.E., Senius, J.D., Dechert, G.J., Douglas, G.S., and Butler, E. (1994) 17α(H),21β(H)-hopane as a conserved internal marker for estimating the biodegradation of crude oil. *Environmental Science and Technology* 38, 142–145.

Radke, M. (1988) Application of aromatic compounds as maturity indicators in source rocks and crude oils. Paper presented at the meeting *Advances in Petroleum Geochemistry*, May 19, 1987. Butterworth, London.

Radke, M., Schaefer, R.G., and Leythaeuser, D. (1980) Composition of soluble organic matter in coals: relation to rank and liptinite fluorescence. *Geochimica Cosmochimica Acta* 44, 1787–1800.

Radke, M., Leythaeuser, D., and Teichmuller, M. (1984) Relationship between rank and composition of aromatic hydrocarbons for coals of different origins. *Organic Geochemistry* 6, 423–430.

Radke, M., Welte, D.H., and Willsch, H. (1986) Maturity parameters based on aromatic hydrocarbons: influence of the organic matter type. *Organic Geochemistry* 10, 51–63.

Radke, M., Garrigues, P., and Willsch, H. (1990) Methylated dicyclic and tricyclic aromatic hydrocarbons in crude oils from the Handil field, Indonesia. *Organic Geochemistry* 15, 17–34.

Radke, M., Welte, D.H., and Willsch, H. (1991) Distribution of alkylated aromatic hydrocarbons and dibenzothiophenes in rocks of the upper Rhine Graben. *Chemical Geology* 93, 325–341.

Radler, M. (1999) 1999 Worldwide Refining Survey. *Oil and Gas Journal*, Dec. 20, 45–89.

Reddy, C.M., Heraty, L., Holt, B., Sturchio, N., Eglinton, T.I., Drenzek, N.J., Xu, L., Lake, J.L., and Maruya, K.A. (2000) Stable chlorine isotopic compositions of Aroclors and Aroclor-contaminated sediments. *Environmental Science and Technology* 34, 2866–2870.

Requejo, A.G., Sassen, R., McDonald, T., Denoux, G., Kennicutt, M.C. II, and Brooks, J.M. (1996) Polynuclear aromatic hydrocarbons (PAH) as indicators of the source and maturity of marine crude oils. *Organic Geochemistry* 24, 1017–1033.

Ries-Kautt, M. and Albrecht, P. (1989) Hopane-derived triterpenoids in soils. *Chemical Geology* 76, 143–151.

Robert, J.C. (1983) *Ethyl: A History of the Corporation and the People who Made it*. University of Virginia Press, Charlottesville, VA.

Rodgers, R., Blumer, E.N., Freitas, M.A., and Marshall, A.G. (2000) Complete compositional monitoring of the weathering of transportation fuels based on elemental compositions from Fourier transform ion cyclotron resonance mass spectrometry. *Environmental Science and Technology* 34, 1671–1678.

Rogge, W.F., Hildemann, L.M., Mazurek, M.A., Cass, G.R., and Simoneit, B.R.T. (1993) Sources of fine organic aerosol. 2. Noncatalyst and catalyst-equipped automobiles and heavy-duty diesel trucks. *Environmental Science and Technology* 27, 636–651.

Sauer, T.C. and Boehm, P.D. (1991) The use of defensible analytical chemical measurements for oil spill natural resources damage assessment. *Proceedings of the International Oil Spill Conference*. American Petroleum Institute, Washington DC.

Sauer, T.C. and Boehm, P. D. (1995) Hydrocarbon chemistry analytical methods for oil spill assessment. Marine Spill Response Corporation Technical Report Series 95-032.

Sauer, T.C. and Uhler, A.D. (1994–1995) Pollutant source identification and allocation: advances in hydrocarbon fingerprinting. *Remediation*, Winter Issue, 25–50.

Schmidt, P. (1969) *Fuel Oil Manual*, 3rd edn. Industrial Press, New York.

Schwarzenbach, R., Gschwend, P.M., and Imboden, D.M. (1993) *Environmental Organic Chemistry*. John Wiley & Sons, New York.

Seifert, W.K. and Moldowan, J.M. (1978a) Applications of steranes, terpanes and monoaromatics to the maturation, migration and source of crude oils. *Geochimica Cosmochimica Acta* 42, 77–95.

Seifert, W.K. and Moldowan, J.M. (1978b) Applications of steranes/terpanes and monoaromatics to the maturation, migration and source of crude oils. *Geochimica Cosmochimica Acta* 42, 77–95.

Seifert, W.K. and Moldowan, J.M. (1986) Use of biological markers in petroleum exploration. In *Biological Markers in the Sedimentary Record* (Johns, R.B., ed.), pp. 261–290. Elsevier Press, Amsterdam.

Seifert, W.K., Moldowan, J.M., and Jones, R.W. (1979) Application of biological marker chemistry to petroleum exploration. Special Paper SP 8. In *Tenth World Petroleum Congress*, pp. 425–440. Bucharest, 1979. Heyden & Son Limited.

Seifert, W.K., Moldowan, J.M., and DeMaison, G.J. (1984) Source correlation of biodegraded oils. *Organic Geochemistry* 6, 633–643.

Sessions, A.L., Burgoyne, T.W., Schimmelmann, A., and Hayes, J.M. (1999) Fractionation of hydrogen isotopes in lipid biosynthesis. *Organic Geochemistry* 30, 1193–1200.

Short, J.W., Kvenvolden, K.A., Carlson, P.R., Hostettler, F.D., Rosenbauer, R.J., and Wright, B.A. (1999) Natural hydrocarbon background in benthic sediments of Prince William Sound, Alaska: oil vs. coal. *Environmental Science and Technology* 33, 34–42.

Silliman, J.E., Meyers, P.A., and Eadie, B.J. (1998) Perylene: an indicator of alteration processes or precursor materials? *Organic Geochemistry* 29, 1737–1744.

Simoneit, B.R.T. (1986) Cyclic terpenoids of the geosphere. In *Biological Markers in the Sedimentary Record* (Johns, R.B., ed.), pp. 43–99. Elsevier Press, Amsterdam.

Simoneit, B.R.T. and Elias, V.O. (2000) Organic tracers from biomass burning in atmospheric particulate matter over the ocean. *Marine Chemistry* 69, 301–312.

Simoneit, B.R.T., Grimalt, J., Wang, T.G., and Cox, R.E. (1986) Cyclic terpenoids of contemporary resinous plant detritus and of fossil woods, ambers and coals. *Organic Geochemistry* 10, 877–889.

Sofer, Z. (1984) Stable carbon isotope compositions of crude oils: application to source depositional environments and petroleum alteration. *American Association of Petroleum Geologists Bulletin* 68, 31–49.

Speight, J.G. (1991) *The Chemistry and Technology of Petroleum*, 2nd edn. Marcel Dekker, New York.

Squillace, P., Pope, D.A., and Price, C.V. (1995) Occurrence of the gasoline additive MTBE in shallow ground water in urban and agricultural areas. FS-114-95. United States Geological Survey, National Water Quality Assessment Program, Rapid City, SD.

Stout, S.A. (1995) Resin-derived hydrocarbons in fresh and fossil dammar resins and Miocene rocks and oils in the Mahakam Delta, Indonesia. In *Amber, Resinite, and Fossil Resins* (Anderson, K.B. and Crelling, J.C., eds), pp. 43–75. American Chemical Society, Washington, DC.

Stout, S.A. and Lundegard, P.D. (1998) Intrinsic biodegradation of diesel fuel in an interval of separate phase hydrocarbons. *Applied Geochemistry* 13, 851–859.

Stout, S.A., Davidson, J., McCarthy, K.J., and Uhler, A.D. (1999a) Gasoline additives – usage of lead and MTBE. *Soil and Groundwater Cleanup*, Feb/Mar, 16–18.

Stout, S.A., McCarthy, K.J., Seavey, J.A., and Uhler, A.D. (1999b) Application of low boiling biomarkers in assessing liability for fugitive middle distillate petroleum products. *Proceedings of the 9th Annual West Coast Conference on Contaminated Soils and Water*, March, 1999 (Abstract).

Stout, S.A., Uhler, A.D., and McCarthy, K.J. (1999c) Biomarkers – Underutilized components in the forensic toolkit. *Soil and Groundwater Cleanup*, June/July, 58–59.

Stout, S.A., Naples, W.P., Uhler, A.D., McCarthy, K.J., Roberts, L.G., and Uhler, R.M. (2000a) Use of quantitative biomarker analysis in the differentiation and characterization of spilled oil. SPE 61460. Prepared for Presentation at the *SPE International Conference on Health, Safety, and the Environment in Oil and Gas Exploration and Production*. June 26–28, 2000, Stavanger, Norway. Society of Petroleum Engineers, Richardson, TX.

Stout, S.A., Uhler, A.D., and McCarthy, K.J. (2000b) Recognizing the confounding influences of 'background' contamination in 'fingerprinting' investigations. *Soil, Sediment and Groundwater* Feb./Mar, 35–38.

Strachan, M.G., Alexander, R., and Kagi, R.I. (1988) Trimethylnaphthalenes in crude oils and sediments: effects of source and maturity. *Geochimica Cosmochimica Acta* 52, 1255–1264.

Straughan, D. (1979) Distribution of tar and relationship to changes in intertidal organisms on sandy beaches in Southern California. In *1979 Oil Spill Conference*, pp. 591–601. American Petroleum Institute, Washington, DC.

Sundararaman, P. (1985) High-performance liquid chromatography of vanadyl porphyrins. *Analytical Chemistry* 57, 2204–2206.

Sutherland, J.B., Selby, A.L., Freeman, J.P., Evans, F.E., and Cerniglia, C.E. (1991) Metabolism of phenanthrene by *Phanerochaete chrysosporium. Applied Environmental Microbiology* 57, 3310–3316.

Takada, H., Tomoko, O., Mamoru, H., and Norio, O. (1991) Distribution and sources of polycyclic aromatic hydrocarbons (PAHs) in street dust from the Tokyo metropolitan area. *The Science of the Total Environment* 17, 45–69.

Taylor, G.H., Teichmuller, M., Davis, A., Diessel, C.F.K., Littke, R., and Robert, P. (1998) *Organic Petrology*. Gebrüder Borntraeger, Berlin, Stuttgart.

ten Haven, H.L., Rohmer, M., Rullkotter, J., and Bisseret, P. (1989) Tetrahymanol, the most likely precursor of gammacerane, occurs ubiquitously in marine sediments. *Geochimica Cosmochimica Acta* 53, 3073–3079.

Tissot, B.P. and Welte, D.H. (1984) *Petroleum Formation and Occurrence*, 2nd edn. Springer-Verlag, Berlin.

Trust, B.A., Mueller, J., Coffin, R.B., and Cifuentes, L.A. (1995) The biodegradation of fluoranthene as monitored using stable carbon isotopes. *Battelle Conference on In Situ and On-Site Bioreclamation Program Publication*, San Diego CA, April 24–27, vol. 3, issue 5, pp 233–239.

Turco, R.F. and Sadowsky, M.J. (1995) *The Microflora of Bioremediation*. Soil Science Society of America Special Publ. No. 43.

Uhler, A.D., West, D.E., Peven, C.S., and Hunt, C.D. (1994) Trace metal and organic contaminants in Deer Island treatment plant effluent; June–November 1993. Poster presentation at Ninth Annual Boston Harbor Symposium, Boston, March 24–25, 1994.

Uhler, A.D., Stout, S.A., and McCarthy, K.J. (1998–1999) Increased success of assessments at petroleum sites in 5 steps. *Soil and Groundwater Cleanup* Dec/Jan, 13–19.

van Aarssen, B.G.K., Bastow, T.P., Alexander, R., and Kagi, R.I. (1999) Distributions of methylated naphthalenes in crude oils: indicators of maturity, biodegradation and mixing. *Organic Geochemistry* 30, 1213–1227.

van Aarssen, B.G.K., Alexander, R., and Kagi, R.I. (2000) Higher plant biomarkers reflect palaeovegetation changes during Jurassic times. *Geochimica Cosmochimica Acta* 64, 1417–1424.

Vazquez-Duhalt, R. (1989) Environmental impact of used motor oil. *Science of the Total Environment* 79, 1–23.

Venkatesan, M.I. (1988). Occurrence and possible sources of perylene in marine sediments – A review. *Marine Chemistry* 25, 1–27.

Volkman, J.K. (1986) A review of sterol markers for marine and terrigenous organic matter. *Organic Geochemistry* 9, 83–99.

Volkman, J.K., Holdsworth, D.G., Neill, G.P., and Bavor, H.J., Jr. (1992) Identification of natural, anthropogenic and petroleum hydrocarbons in aquatic sediments. *Science of the Total Environment* 112, 203–219.

Wade, T.L. and Quinn, J.G. (1979) Geochemical investigation of hydrocarbons in sediments from mid-Narragansett Bay, Rhode Island. *Organic Geochemistry* 1, 157–167.

Wakeham, S.G., Schaffner, C., and Giger, W. (1980a) Polycyclic aromatic hydrocarbons in recent lake sediments – I. Compounds having anthropogenic origins. *Geochimica Cosmochimica Acta* 44, 403–413.

Wakeham, S., Schaffner, C., and Giger, W. (1980b) Polycyclic aromatic hydrocarbons in recent lake sediments – II. Compounds derived from biogenic precursors during early diagenesis. *Geochimica Cosmochimica Acta* 44, 415–429.

Wakeham, S.G., Canuel, E.A., and Doering, P.H. (1986) Geochemistry of volatile organic compounds in seawater: mesocosm experiments with ^{14}C-model compounds. *Geochimica Cosmochimica Acta* 50, 1163–1172.

Wang, Z. and Fingas, M. (1995) Use of methyldibenzothiophenes as markers for differentiation and source identification of crude and weathered oil. *Environmental Science and Technology* 29, 2842–2849.

Wang, Z. and Fingas, M. (1999) Identification of the source(s) of unknown spilled oils. Paper #162. In *1999 International Oil Spill Conference*. American Petroleum Institute, Washington DC.

Wang, Z., Fingas, M., Landriault, M., Sigouin, L., Castle, B., Hostetter, D., Zhang, D., and Spencer, B. (1993) Identification and linkage of tarballs from the coasts of Vancouver Island and Northern California using GC/MS and isotopic techniques. *Journal of High Resolution Chromatography* 21, 383–395.

Wang, Z., Fingas, M., and Li, K. (1994a) Fractionation of a light crude oil and identification and quantitation of aliphatic, aromatic, and biomarker compounds by GC-FID and GC-MS, part I. *Journal of Chromatographic Science* 32, 361–382.

Wang, Z., Fingas, M., and Sergy, G. (1994b) Study of 22-year-old *Arrow* oil samples using biomarker compounds by GC/MS. *Environmental Science and Technology* 28, 1733–1746.

Wang, Z., Fingas, M., and Page, D.S. (1999) Oil spill identification. *Journal of Chromatography* A 842, 369–411.

Waples, D.W. and Machihara, T. (1990) Application of sterane and triterpane biomarkers in petroleum exploration. *Bulletin of Canadian Petroleum Geology* 38, 357–380.

Warton, B., Alexander, R., and Kagi, R.I. (1997) Identification of some single branched alkanes in crude oils. *Organic Geochemistry* 27, 465–476.

Westerholm, R.N., Alsberg, T.E., Frommelin, A.B., Strandell, M.E., Rannug, U., Winquist, L., Grigoriadis, A., and Egebäck, K.-E. (1988) Effect of fuel polycyclic aromatic hydrocarbon content on the emissions of polycyclic aromatic hydrocarbons and other mutagenic substances from a gasoline-fueled automobile. *Environmental Science and Technology* 22, 925–930.

White, C.M. and Lee, M.L. (1980) Identification and geochemical significance of some aromatic components of coal. *Geochimica Cosmochimica Acta* 44, 1825–1832.

Whittaker, M., Pollard, J., and Fallick, T.E. (1995) Characterisation of refractory wastes at heavy oil-contaminated sites: a review of conventional and novel analytical methods. *Environmental Technology* 16, 1009–1033.

Youngblood, W.W. and Blumer, M. (1975) Polycyclic aromatic hydrocarbons in the environment: homologous series in soils and recent marine sediments. *Geochimica Cosmochimica Acta* 39, 1303–1314.

Zander, M. (1995) Aspects of coal tar chemistry/a review. *Polycyclic Aromatic Compounds* 7, 209–221.

Zayed, J., Hong, B., and L'Esperance, G. (1999) Characterization of manganese-containing particles collected from the exhaust emissions of automobiles running with MMT additive. *Environmental Science and Technology* 33, 3341–3346.

CHAPTER 7

CHLORINATED SOLVENTS: CHEMISTRY, HISTORY AND UTILIZATION FOR SOURCE IDENTIFICATION AND AGE DATING

Robert D. Morrison and Brian L. Murphy

7.1 INTRODUCTION

Chlorinated solvents provide significant opportunities for age dating and source identification. The main solvents of interest in environmental forensics are trichloroethylene (TCE), tetrachloroethylene (PCE), 1,1,1-trichloroethane (TCA), methylene chloride, carbon tetrachloride and their respective transformation (biodegradation and hydrolysis) products. This chapter presents information on chlorinated solvent chemistry, historical applications of chlorinated solvents, and commonly used forensic techniques for age dating and source identification.

7.2 CHLORINATED SOLVENT CHEMISTRY

An understanding of basic physicochemical properties of chlorinated solvents is useful when applying and/or evaluating environmental forensic techniques for age dating and source identification. In cases of multiple releases, measurement of physical and chemical properties evaluated via multivariate analysis can provide a basis for source discrimination (Morrison, 2000a). Key properties evaluated and/or measured in many of these techniques include the Henry's law constant, liquid density, solubility, viscosity, vapor pressure and density, and the boiling point of the solvent (Appendix C lists key chemical properties of selected chlorinated solvents). Physical and chemical reactions affecting chlorinated solvent chemistry include hydrolysis, sorption and biodegradation. An understanding of these physicochemical properties is important for evaluating age dating and source identification techniques that rely on these processes.

7.3 HISTORICAL APPLICATIONS OF CHLORINATED SOLVENTS

The historical use of chlorinated solvents often provides pivotal evidence for source identification and age dating. Chemical use patterns associated with specific applications can provide a blueprint for where to sample to confirm a hypothesis regarding the source of a release. The following text provides a historical overview of the products or techniques and uses of TCE, PCE, 1,1,1-TCA, carbon tetrachloride, and methylene chloride.

7.3.1 TRICHLOROETHYLENE

TCE is manufactured via an acetylene process, by the chlorination of ethylene, and by the oxychlorination of ethylene and/or 1,2-dichloroethane (Mertens, 1993). The acetylene process developed in Austria between 1903 and 1905 consists of the chlorination of acetylene to 1,1,2,2-tetrachloroethane that is dehydrohalogenated to TCE in aqueous bases such as calcium hydroxide or by thermal cracking over a catalyst such as barium chloride or activated carbon or silica or aluminum gels at 300–500°C (Doherty, 2000a,b).

TCE is also produced by the high temperature chlorination of ethylene or 1,2-dichloroethane. Catalysts used include potassium chloride, aluminum chloride, Fuller's earth, graphite and activated carbon (Aviado *et al.*, 1976). The acetylene process was not used after about 1978.

The oxychlorination of ethylene or dichloroethane consists of chlorinating these compounds to form a mixture of PCE or TCE in the presence of oxygen and catalysts (mixtures of potassium and cupric chlorides). Less common synthesis techniques include the chlorination of dichloroethylene followed by thermal cracking to produce trichloroethylene and hydrochloric acid. Another process developed by the Taogosei Chemical Company in Japan chlorinates ethylene directly in the absence of oxygen to produce tetrachloroethanes and pentachloroethane. These products are thermally cracked to produce a mixture of TCE, PCE and hydrochloric acid.

TCE is the predominant chlorinated solvent used for vapor degreasing in the United States due to its aggressive solvent action on oils, greases, waxes, tars, gums, resins and certain polymers. The Vietnam War accelerated TCE use for degreasing in the aerospace and automotive industries in the United States. The demand for TCE in the United States peaked in 1968 at about 261 000 metric tons. In 1970, TCE accounted for 82% of all of the chlorinated solvents used in vapor degreasing; by 1976, its share had declined to 42%. TCE was a favorite solvent as it could safely be used with iron, steel, aluminum, magnesium, copper, brass and various plating metals without harm to the parts or to the degreasing equipment (American Society for Metals, 1996). In 1975 the National Cancer Institute reported to the National Institute of Occupational Safety and Health (NIOSH) that TCE was a suspected human carcinogen and that exposure should be minimized. This finding resulted in many industries transitioning to TCA with TCA gradually replacing TCE from about 1963 through 1988. In February 1995, the International Agency of Research on Cancer (IARC) concluded that there was sufficient epidemiological and animal testing data to classify TCE as a probable human carcinogen. A chronology of TCE usage is summarized in Table 7.1 (Aviado *et al.*, 1976; Halogenated Solvents Industry Alliance, 1996; Morrison, 1999, 2000a–d; Doherty, 2000a,b).

7.3.2 TETRACHLOROETHYLENE (PCE OR PERCHLOROETHYLENE)

Tetrachloroethylene is the historical solvent of choice in the dry cleaning industry. PCE was initially produced via the chlorination of acetylene followed by lime dehydrochlorination and chlorination. The acetylene chlorination process became obsolete due to the high price of acetylene in the 1970s (Seiler, 1960). Other processes include the high temperature chlorination of ethylene or 1,2-dichloroethane with TCE as a by-product, the chlorinolysis of light

Table 7.1
Chronology of TCE production and usage.

Year	Historical Information
1864	TCE synthesized by E. Fischer via the reductive dehalogenation of hexachloroethane (Fisher, 1864).
1908	TCE production begins in Austria, the United Kingdom and Yugoslavia (Gerhartz,1986).
1909/10	TCE formulated in Germany; production begins in 1910 (Mellan,1957).
1921	Dow Chemical Company begins synthesizing TCE in the United States (Doherty, 2000b).
1922–1935	Carbide & Carbon Chemicals Corporation produce TCE in the United States.
1925	Roessler & Hasslacher Chemical Company produces TCE in the United States (Hardie, 1964).
1926	TCE manufacturers in the United States include DuPont de Nemours, Carbide & Carbon Chemicals Corporation and Westvaco Chlorine Productions Company.
1928	TCE used to treat trigeminal neuralgia (Oljenick, 1928).
1933	TCE considered for use as a general anesthetic.
1935	TCE produced in Japan. TCE used as a general anesthetic in humans (Aviado *et al.*, 1976).
1935–1937	TCE mentioned as a poisonous chemical.
1940	TCE used for the extraction of oils from oil seed and as a solvent in petroleum refining.
1941–1945	United States government controls the TCE manufacturing during World War II. Manufacturers include Dow, DuPont and Westvaco Chlorine. TCE is used primarily for degreasing machinery parts (Lowenheim and Moran, 1975).
1942	TCE is used as a reagent in synthetic dye production, as an ingredient in rubber cement glue, in insecticides, in paint and varnish removers, in water proofing compositions and as a refrigerating ingredient.
1943	TCE used in the vapor cleaning of motors.
1946/7	In the United States, Dow Chemical Company, DuPont de Nemours, Food Machinery & Chemical Company and Hooker-Detrex manufacture TCE (*Chemical Engineering News*, 1950).
Late 1940s	TCE replaces carbon tetrachloride in metal degreasing and Stoddard Solvent as the solvent of choice for dry cleaning (Ram *et al.*, 1999).
1949	TCE detected in groundwater near Reading, England (Love, 1951). A method capable of detecting TCE in groundwater as low as 1 ppm is reported by Lyne and McLachlan (1949), who states: 'From these two cases it is evident that contamination by compounds of this nature is likely to be very persistent and there is some evidence of toxicity at very low concentrations.'
1949–1955	Niagara Alkali produces TCE (*Chemical Industry*, 1949).
1950	A United States government pamphlet on environmental cancer lists chlorinated aliphatic hydrocarbons as suspected carcinogens (Hueper, 1950).
1950s	Hemorrhagic diseases in the early 1950s are traced to animal feed containing TCE extracted soybean meal. Most manufacturers in the

Table 7.1 Continued

Year	Historical Information
	United States voluntarily withdraw soybean oil meals defatted with TCE in 1952 (Huff, 1971).
1952	TCE used in the prevention of post-harvest decay of fruit and as a degreasing solvent. About 92% of TCE is used in vapor degreasing (*Chemical Week,* 1953).
1953	A method for detecting chlorinated solvents at 10 μg/L is reported (Melpolder *et al.*, 1953).
1954	TCE used in soybean oil extraction. Prior to 1954, amines are the most popular stabilizer used in TCE.
1955	TCE used for vapor degreasing, especially on metallic items with phosphate finishes where strong alkaline solutions cannot be used, and as a freezing point depressant for carbon tetrachloride, usually in fire extinguishers. TCE also used as a solvent for crude rubber, dyes, bitumen, pitch, sulfur, oils, fats, waxes, tar, gums and resins.
1956	Annual sales of TCE in the United States estimated at 250 million pounds (113 000 metric tons) (Kircher, 1957).
1959/60	DuPont's Niagara Falls facility accounts for half the total United States production of 485 million pounds (220 000 metric tons) (*Chemical Engineering News*, 1960).
1961	TCE is listed in a survey as a groundwater contaminant (Middleton and Walton, 1961).
1966	Rule 66 promulgated limiting TCE emissions in Los Angeles County, California.
1967	Detection limits for chlorinated solvents reported at 10 μg/L (Montgomery and Conlon, 1967).
1969	TCE production in the United States peaks at 596 million pounds (270 000 metric tons).
1970	TCE used for 82% of all vapor degreasing in the United States. Diamond Shamrock adopts DuPont's use of trade name Triclene for marketing TCE (*Chemical Engineering News*, 1970).
Early 1970s	TCE detection limits in groundwater is in the low nanograms per liter (Grob and Grob, 1974).
1972	Rhode Island bans the use of TCE (*Chemical Marketing Reporter,* 1975).
1974	The major TCE producers in the United States include Dow Chemical, Ethyl Corporation, Occidental Petroleum Company, PPG Industries and Diamond Shamrock (1969–1977). Dow and PPG produce about 70% of total output of TCE in the United States (Kroschwitz and Howe-Grant, 1991).
1975	The National Cancer Institute reports that massive oral doses of TCE cause liver tumors in mice but not in rats (National Institutes of Health, 1982).
1976	The National Cancer Institute publishes the results of a cancer bioassay that concludes TCE is an animal carcinogen (National Cancer Institute, 1976).
1977	The United States Department of Agriculture (Food and Drug Administration) bans the use of TCE as a general anesthetic, grain fumigant, skin, wound and surgical disinfectant, as a pet food additive, and as an extractant in spice oleoresins isolation, hops and coffee decaffeination. TCE is banned as a food additive and as an ingredient in cosmetic and drug products (Linak *et al.*, 1990).

Table 7.1
Continued

Year	Historical Information
1978	General Foods Corporation ceases using TCE in the decaffeination of its Sanka and Brim brands and switches to methylene chloride (Doherty, 2000b).
1980	Occidental Petroleum Company ceases TCE production in the United States.
1981	Ethyl Corporation, PPG Industries, Diamond Shamrock, and Hooker Chemical Company synthesize TCE in the United States.
1982	Ethyl Corporation ceases TCE production.
1983	TCE primarily used for vapor degreasing in the automotive and metals industries.
1986	Usage of TCE is estimated in the United States at 80% for vapor degreasing, 5% as a chemical intermediate, 5% in miscellaneous uses and 10% exported.
1988	TCE listed on April 1, 1988 as a chemical known to cause cancer by the State of California.
1989	TCE sales are forbidden to all persons under the age of 18 in France.
1991	The use of TCE for metal cleaning and degreasing is estimated at 90% of the total United States production (*Chemical Marketing Reporter*, 1992).
1995	The International Agency for Research on Cancer (IARC) classifies TCE as a probable human carcinogen.
1996	Sweden prohibits the professional use of TCE although the National Chemicals Inspectorate can permit exceptions (European Chlorinated Solvents Association, 2000).
1997	Only Dow Chemical Company and PPG synthesize TCE in the United States (*Chemical Marketing Reporter*, 1997).

hydrocarbon feed stocks, and the oxychlorination of ethylene or other chlorinated hydrocarbons (Doherty, 2000a).

PCE is effective in removing soil containing high melting waxes of low solubility. By the late 1940s or early 1950s, PCE replaced other synthetic solvents, such as carbon tetrachloride, in the dry cleaning industry. The demand for PCE peaked in 1975 when about 348 000 metric tons were produced. In Western Europe, tetrachloroethylene production declined about 4.8% between 1993 and 1997 from 84 000 metric tons in 1993 to 68 000 metric tons in 1997, respectively, due to the replacement of obsolete open dry cleaning machines with closed systems (Halogenated Solvents Industry Alliance, 1998a). In Western Europe PCE is produced by Elf Atochem (France), Dow Europe/Switzerland (produced in Germany), EniChem (Spain), ICI Chemicals and Polymers (United Kingdom) and Solvay/Belgium (France and Italy) (European Chlorinated Solvents Association, 1996). A chronology of historical information for PCE is summarized in Table 7.2.

Table 7.2

Chronology of PCE production and usage.

Year	Historical Information
1821	PCE discovered by Michael Faraday as a thermal decomposition product of hexachloroethane (Partington, 1964).
1840	Regnault creates PCE by passing carbon tetrachloride vapor through a hot tube and by the reduction of hexachloroethane with alcoholic potassium hydrosulfide (Kirk and Othmer, 1949).
1887	Combes synthesizes PCE via the prolonged heating of chloral with anhydrous aluminum chloride.
1894	Meyer isolates PCE as a by-product of the synthesis of carbon tetrachloride from carbon disulfide.
1910s	Production ceased during World War I.
1914	PCE is manufactured in the United States as by-product of carbon tetrachloride synthesis (Doherty, 2000a).
1930s	PCE is introduced as a solvent of choice in the dry cleaning industry (Halogenated Solvents Industry Alliance, 1994a, 1999).
1933	DuPont synthesizes PCE at its Niagara Falls, NY facility (*Chemical Engineering News*, 1970).
1940s	A gradual shift from petroleum derivatives (Stoddard Solvent) to PCE begins in the late 1940s (Halogenated Solvents Industry Alliance, 1999).
1948	PCE replaces carbon tetrachloride as the leading solvent in the dry cleaning industry (*Chemical Week*, 1953).
1952	Approximately 80% of the PCE in the United States is used in dry cleaning and about 15% for metal cleaning, of which about 10% is for vapor degreasing (*Chemical Week*, 1953).
1959	Dow and DuPont supply about half of the PCE in the United States (Doherty, 2000a).
1960	Dow Chemical is the leading PCE manufacturer in the United States at about 120 million pounds (55 000 metric tons) and produces PCE at their Freeport, Texas; Pittsburg, California, and Plaquemine, Louisiana plants (*Chemical Engineering News*, 1962).
1967	The dry cleaning industry is responsible for about 88% of PCE consumption in the United States (*Chemical Engineering News*, 1967).
Late 1970s	The United States Consumer Product Safety Commission designates PCE as a carcinogen because of positive animal study results.
1975–1980	PCE production peaks in the United States.
1980	Spent PCE still bottoms are classified as a hazardous waste under the Resource Conservation and Recovery Act (RCRA) in the United States.
Late 1980s	Dry cleaners consume about 56% of the PCE in the United States (Izzo, 1992).
1988	PCE is listed by the State of California as a chemical known to cause cancer.
1990	PCE production in the United States is about 383 million pounds (173 000 metric tons) of which 55 million pounds (25 000 metric tons) is exported while about 72 million pounds (32 000 metric tons) is imported. PCE is one of 189 substances listed as a hazardous air pollutant under Section 112 of the amended Clean Air Act (Halogenated Solvents Industry Alliance, 1994). PCE use in the United States is estimated to be about 50% in for dry cleaning

Table 7.2
Continued

Year	Historical Information
	and/or textile processing, 25% as a chemical intermediate, 15% in metal cleaning and degreasing while miscellaneous uses accounted for about 10% (Halogenated Solvents Industry Alliance, 1998a).
1992	Approximately 28 000 dry cleaning businesses in the United States use PCE (Wolf, 1992).
1996	The use of PCE as a precursor in chlorofluorocarbon 113, a banned substance under the Montreal Protocol, stops in 1996 (Doherty, 2000a).
1998	The total PCE demand in the United States is about 344 million pounds (156 000 metric tons), of which about 30 million pounds (13 000 metric tons) is imported. PCE usage in the United States includes about 50% as a chemical intermediate, 25% for dry cleaning/textile processing, 10% for automotive aerosols, 10% for metal cleaning/degreasing and 5% for miscellaneous uses (Halogenated Solvents Industry Alliance, 1999).

7.3.3 1,1,1-TRICHLOROETHANE

TCA became the cold cleaning solvent of choice in the mid-1950s when it replaced carbon tetrachloride, although aliphatic naphtha and methylene chloride were also used (Evanoff, 1990; Halogenated Solvents Industry Alliance, 1994a). TCA was initially synthesized via the chlorination of 1,1-dichloroethane and by the addition of hydrochloric acid to 1,1-dichloroethylene in the presence of a ferric chloride catalyst. Another technique involved the noncatalytic chlorination of ethane, which yielded various chlorinated ethanes and ethenes (Gerhartz, 1986). In the United States, TCA is generally synthesized via the chlorination of vinyl chloride derived from 1,2-dichloroethane or through the thermal chlorination of ethane. In Japan, TCA is produced by the chlorination of vinyl chloride.

The demand for TCA increased substantially from 1967 to 1994 in the United States. This rise in TCA demand resulted in part as information became available regarding the potential carcinogenicity of TCE in animals in the late 1960s. The historical demand for TCA in the United States peaked in 1988 when about 298 000 metric tons were consumed. In June 1990, amendments to the original 1987 Montreal Protocol designed to limit chlorofluorocarbon production were signed in London and listed TCA and carbon tetrachloride as ozone-depleting substances with phase-out dates of 2000 and 2005, respectively. As a result, TCA usage declined throughout the 1990s. On a worldwide basis, TCA production declined to 50% of its 1988/1989 levels by 1994/1995.

In Western Europe, most manufacturers ceased TCA production by the end of 1995 except as a chemical intermediary and for some specialized permitted uses. Table 7.3 describes the use and production history of TCA.

Table 7.3

Chronology of TCA production and usage.

Year	Historical Information
1840	TCA synthesized by Regnault via the reaction of chlorine with 1,1-dichloroethane (Kroschwitz and Howe-Grant, 1991).
1914	TCA listed as a rubber solvent manufactured in Europe (Doherty, 2000b).
1924	TCA listed in the Census of Dyes and Synthetic Organic Chemicals.
1931	A German patent is issued for the production of TCA via the hydrochlorination of 1,1,-dichloroethylene in the presence of a ferric chloride catalyst (Farbenindustrie, 1931). Another process published in 1931 was the chlorination of 1,1-dichloroethane (Doherty, 2000b).
1940	A patent is issued to Dow Chemical Company for the production of TCA via a modification to the 1,1-dichloroethylene chlorination process (Nutting and Huscher, 1940).
1946	TCA first used in the United States.
1950s	TCA used for cold cleaning as 1,4-dioxane is found to be an effective corrosion inhibitor (Ram *et al.*, 1999). Dow's capacity is limited to about 20 million pounds (9000 metric tons) per year until about 1961 (Doherty, 2000b).
1954	The first widespread commercialization of 1,1,1-trichloroethane for vapor degreasing is marketed at Dow Chemical Company under the trade name Chloroethene in September 1954 and commercially listed in August 1954 (Archer, 1973).
1957	TCA is widely used in cold-cleaning applications as a replacement for carbon tetrachloride (Barber, 1957; Doherty, 2000b).
1960	The Dow Chemical Company introduces Chlorothene NU for cold-cleaning applications with 1,4-dioxane as a stabilizing agent (Doherty, 2000b).
Early 1960s	TCA replaces carbon tetrachloride as a commonly used solvent.
1961	Dow Chemical is the sole producer of TCA in the United States and introduces Dowclene EC to its product line and doubles its production capacity to about 40 million pounds (18 000 metric tons) per year (Doherty, 2000b).
1962	Vulcan Materials Company patents TCA synthesis via the non-catalytic chlorination of ethane.
1965	Dow introduces Dowclene WR to its product line. TCA production in the United States is about 200 million pounds (90 000 metric tons).
1967	TCA is combined with methylene chloride to create a propellant mixture marketed under the trade name Aerothene (Aviado *et al.*, 1976).
1969	TCA production in the United States is 324 million pounds (147 000 metric tons).
1970	TCA with stabilizers is used for vapor degreasing.
1973	TCA production in the United States is 548 million pounds (248 000 metric tons). Twenty-one deaths result from the abuse of decongestant aerosol sprays containing TCA and fluorocarbon propellants (*Federal Register*, 1973).
1976	TCA production in the United States is 631 million pounds (286 000 metric tons); Ethyl Corporation ceases TCA production.
1979	Vulcan stops TCA production at its Geismar, Louisiana facility.

Table 7.3
Continued

Year	Historical Information
1980	Dow is the leading TCA manufacturer followed by Vulcan and PPG Industries in the United States.
1987	TCA production peaks at 723.8 million pounds in the United States.
1987	Three times more TCA is produced in the United States than TCE (Jackson, 1998).
1990	TCA is phased out due to its ozone depletion properties by 2002 by the amended Montreal Protocol except for essential uses such as metal fatigue and corrosion testing of airplane engines and parts (Randle and Bosco, 1991).
1991	TCA production in the United States is 292.3 million pounds (132 585 metric tons).
1992	The Copenhagen Amendments to the Montreal Protocol requires complete phase out of TCA by January 1, 1996.
1993	Demand for TCA in the United States is 159 000 metric tons; 60% used as a chemical intermediate, 25% in cold cleaning and degreasing, 5% in adhesives and 10% for miscellaneous uses.
1994	TCA is produced in the United States by Dow, PPG and Vulcan Materials Company (Halogenated Solvents Industry Alliance, 1994b).
1995	TCA is used in Europe as a precursor chemical.

7.3.4 CARBON TETRACHLORIDE

Carbon tetrachloride was a popular solvent with widespread use in the early 1900s. Carbon tetrachloride was used for dry cleaning until about 1940, along with TCE (Norge Corporation, no date). Prior to about 1950, carbon tetrachloride synthesis in the United States was via the chlorination of carbon disulfide. After the 1950s, the pyrolytic chlorination of methane and propane was used (Kroschwitz and Howe-Grant, 1991). A variation of this process included the photochemical chlorination of methane (Skeeter and Cooper, 1954). Carbon tetrachloride synthesis outside of the United States was primarily produced by the oxychlorination of hydrocarbons (Doherty, 2000a). Carbon tetrachloride was also a by-product of methylene chloride and/or chloroform production (Kroschwitz and Howe-Grant, 1991). Table 7.4 is a chronology of historical information for carbon tetrachloride.

7.3.5 METHYLENE CHLORIDE (DICHLOROMETHANE)

Methylene chloride is a versatile solvent initially introduced as a replacement for more flammable solvents in the 1920s (European Chlorinated Solvents Association, 1997). Because of methylene chloride's low boiling point (~40°C), it is frequently used to degrease temperature-sensitive materials (ASTM, 1989, 1995; National Institute of Standards and Technology, 2000). Methylene

Table 7.4

Chronology of carbon tetrachloride production and usage.

Year	Historical Information
1839	Carbon tetrachloride synthesized by Regnault using the reaction of chlorine on chloroform in sunlight (Doherty, 2000a).
1890s	Carbon tetrachloride is produced in Germany and in England by United Alkali Corporation.
≈1900	Carbon tetrachloride synthesized in the United States by Dow Chemical (Haynes, 1954).
1902/1905	The first commercial-scale manufacturing in the United States commences when Warner Chemical Company of Carteret, New Jersey produces about 15 000 pounds (6800 kg) (Kirk and Othmer, 1949).
1908	Dow Chemical Company begins commercial production of carbon tetrachloride (Whitehead, 1968).
1921	Dow Chemical, Great Western Electro-Chemical and Warner Klipstein manufacture carbon tetrachloride in the United States. Carbon tetrachloride is used to treat hookworm in animals (Doherty, 2000a).
Early 1930s	Carbon tetrachloride replaces gasoline as a dry cleaning agent (Pankow *et al.*, 1996).
1950s	Carbon tetrachloride synthesized via the pyrolytic chlorination of hydrocarbons such as methane or propane (i.e., chlorinolysis).
1954	A patent assigned to the Diamond Alkali Company discusses the synthesis of carbon tetrachloride by the photochemical chlorination of methane (Skeeter and Cooper, 1954). About one-half of the total demand for carbon tetrachloride is for the production of Freon 11 and other chlorofluorocarbons.
1972	Studies identify carbon tetrachloride as an animal carcinogen (Doherty, 2000a).
1975	Carbon tetrachloride is produced by Allied Chemical, Dow Chemical, DuPont, FMC, Stauffer Chemical and Vulcan Materials (Doherty, 2000a).
1990	The Clean Air Act Amendments designate carbon tetrachloride as a hazardous air pollutant.
1992	Amendments to the Montreal Protocol (London Amendment) include a complete ban on the production and use of carbon tetrachloride by January 1, 2000. Dow Chemical, Vulcan and Occidental produce carbon tetrachloride in the United States (*Chemical Marketing Reporter*, 1992).
1996	The emissive use of carbon tetrachloride is banned, effective in January 1996.

chloride replaced TCA in the mid-1990s in nonflammable adhesive formulations including the fabrication of upholstery foam. In 1986, the United States National Toxicology Program concluded that methylene chloride was a carcinogen in mice and rats (National Toxicity Program, 1986).

The applications of methylene chloride in the United States in 1996 and Western Europe in 1994 are shown in Table 7.5 (European Chlorinated Solvents Association, 1997; Halogenated Solvents Industry Alliance, 1998b).

Table 7.5
Use of methylene chloride in the United States and Western Europe for 1996 and 1994, respectively (%).

Applications	United States (1996)	Western Europe (1994)
Paint stripping	30	19
Adhesives	16	
Aerosols and coatings	11	12
Foam manufacturing	10	
Pharmaceuticals	10	41
Chemical processing	9	9
Metal cleaning	8	
Miscellaneous	6	19

Manufacturers of methylene chloride include: Aragonesas and Erkimia (Spain), Elf Atochem (France), Dow Chemical and Vulcan Materials (United States), Dow Europe/Switzerland (produced in Germany), LIL Europe (Germany), ICI Chemical and Polymers (United Kingdom), and Solvay/Belgium (synthesizes methylene chloride in France and Italy).

7.4 USES OF CHLORINATED SOLVENTS

Activities associated with chlorinated solvents use include vapor degreasing, cold cleaning, dry cleaning, and electronics manufacturing. The presence of chlorinated solvents as an ingredient in septic tank cleaners is included as it has been identified as a source of chlorinated solvent in groundwater in the United States.

7.4.1 VAPOR DEGREASING

Vapor degreasing uses chlorinated or fluorinated solvent vapors to remove soils, greases and waxes from a surface (American Society for Metals, 1996). While numerous equipment designs are available, general components include a heated chamber where the solvent is boiled to produce vapor, a vapor zone, horizontally placed condensers that control the vapor zone height and produce clean solvent for cleaning, and an area above the vapor (freeboard zone). Figure 7.1 illustrates four vapor degreaser design variations. The simplest is a single stage vapor degreaser. In a single stage vapor degreasing unit, sufficient heat is introduced into the sump to boil the solvent (~150–250°F/60–120°C) and generate a vapor. Because the heated vapor is heavier than air, it displaces the air and fills the tank to the cooling coils. The hot vapor condenses when it reaches the cooling zone, thus maintaining a fixed vapor level and creating a thermal balance. As a part is lowered into the solvent vapor, the solvent condenses on the colder part(s) and dissolves the degreased material into the condensate. The condensate/degreased material

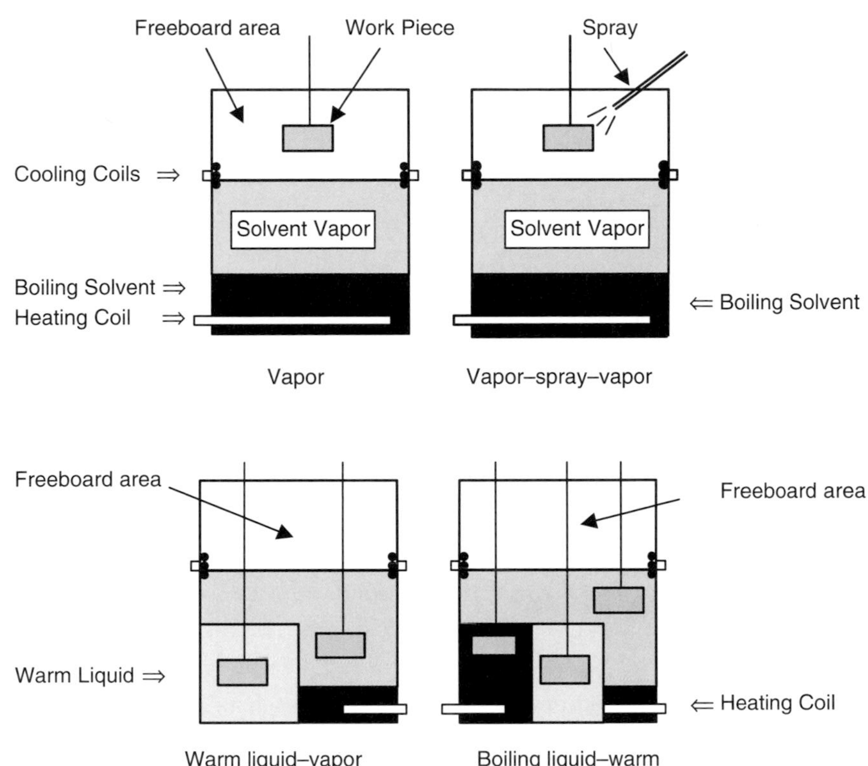

Figure 7.1
Common vapor degreasing designs (after American Society for Metals, 1996).

then drains to a water/solvent separator where the solvent sinks to the bottom and the condensed water containing dissolved solvents is discharged, often into the sanitary sewer. Materials degreased with single stage degreaser include steel spark plugs, aluminum kitchen utensils, steel aircraft and automotive valves, small bore aluminum tubing, gas meters, cold rolled stainless steel and titanium continuous strips (0.01–0.16 inches/0.25–4.06 mm) and steel automatic transmission components (American Society for Metals, 1964).

Some vapor degreaser designs include a spray wand that saturates the part with a solvent prior to lowering it into the solvent vapor. The spray forces the solvent into crevices and blind holes in the part. Parts cleaned with spray wand designs include spark plugs, aluminum kitchen utensils, automatic transmission components, gas meters, automotive and aircraft steel valves, and brass hardware.

For equipment using a warm liquid vapor system, thin sections of material are degreased to attain temperature equalization before placement in a warm liquid bath. The part is held in the vapor zone until condensation ceases and is then lowered into the warm liquid, often with agitation, to facilitate cleaning. From the warm liquid, the piece is transferred to the vapor zone for the final rinse. A warm liquid vapor system is used for degreasing magnesium aircraft castings, steel and brass speedometer shafts and gears, zinc-based automotive die casting, steel and brass screws, steel electron-tube components, aluminum wire (diameters of 0.030–0.125 inches/0.76–3.18 mm), and cast iron and aluminum hand power-tool components.

A boiling liquid–warm liquid–vapor system holds the part to be degreased in the vapor zone until condensation ceases and it is lowered into the boiling solvent. In some cases, the part is first lowered into the boiling solvent. The violent boiling action removes most of the heavy deposits as well as metal chips and insoluble materials. The part is then transferred to the warm solvent and finally to the vapor zone for a final rinse. Once the part is in the vapor zone and temperature equilibrium reached, the part is dry. This system is used to degrease gold and tin plate transistors, steel automotive and aircraft valves, steel and brass screw machine products, zinc-based hand-tool housings, zinc die casts, steel knife blades, steel cable fittings, silicon wafers, steel stampings, and aluminum tubing.

Vapor degreasing equipment may be equipped with vacuum systems, multiple chambers, distillation units for solvent recycling, ultrasonics and mechanized basket trays. Plate 11 is a photograph of a conveyorized degreaser where parts are placed into baskets and cycled through multiple cleaning and degreasing cycles.

Detailed operational information for a vapor degreaser or solvent still is useful in identifying the compatibility of solvents with the equipment. The manufacturer's operating manual contains this information. The term 'kauri butanol value' is frequently encountered in product vapor degreaser product literature and describes the ability of a solvent to dissolve kauri gum. This qualitative measure serves as an estimate of the strength of a solvent. For example, TCE (129), TCA (124) and methylene chloride (136) are considered strong kauri gum solvents while PCE (90) is a moderate solvent and CFC-113 (30) is a weak solvent (Halogenated Solvents Industry Alliance, 2000).

7.4.2 COLD CLEANING

Cold cleaning is identical to vapor degreasing except that the solvent is maintained at room temperature or is heated to a temperature below the solvent's boiling point. Ultrasonic agitation is sometimes used in conjunction with cold cleaning to loosen and remove soils, such as abrasive compounds, from deep recesses or other difficult to reach areas. Cold cleaning is less effective than vapor degreasing because the solvent is not boiled clean (Jackson and Dwarakanath, 1999).

TCA is the dominant historical solvent used for cold cleaning. Other compounds include aliphatic petroleum compounds (e.g., kerosene, naphtha, mineral spirits, Stoddard Solvent), chlorinated hydrocarbons (methylene chloride, PCE, TCE), chlorofluorocarbons (trichlorotrifluoroethane), alcohols (ethanol, isopropanol, 2-ethoxyethanol and methanol), and miscellaneous solvents such as acetone, benzene, and toluene (American Society for Metals, 1996). Stoddard Solvent, mineral spirits and naphtha are used because of their low cost and relatively high flash point. Alcohols used alone or in conjunction with chlorocarbons or chlorofluorocarbons are used for special applications such as degreasing activated soldering fluxes (American Society for Metals, 1996).

7.4.3 DRY CLEANING

O'Hanlon (1977) reports that the earliest known dry cleaner operation began in Paris, France in 1840, although a Norge Corporation Dry Cleaning System Handbook relates a more colorful history (Norge Corporation, no date). According to legend, in 1849 a Paris tailor named Jolly Belin used a common lamp fuel, camphene, to remove soil from garments for Louis Napoleon, who was serving as President of the second French Republic. After successfully washing the garments in camphene, the clothing had much the same appearance as before the wash, resulting in the term *dry cleaning*. By late 1850, camphene was used in dry cleaning shops throughout Europe.

In about 1870, benzene replaced camphene and was in turn replaced with *raw* or (white) gasoline. Given the objectionable odors associated with camphene, benzene and raw gasoline, the National Association of Dryers and Cleaners (now the National Institute of Dry Cleaning) adopted the petroleum product *Stoddard*, named after the man who developed it as a dry cleaning solvent. In the United States, dry cleaners began operating at the turn of the century using gasoline, Stoddard Solvent and naphthas (i.e., 140-F solvent). Gasoline was eventually replaced with Stoddard Solvent, naphtha, carbon tetrachloride, TCE, and PCE. PCE and naphtha are used today for dry cleaning along with petroleum-based solvents with high flash points.

An understanding of the workings of dry cleaning equipment is useful as it is often pivotal in environmental cases dealing with contamination originating from dry cleaning establishments. In the 1930s in the United States, attempts were made by equipment manufacturers to combine the washing, extracting, drying and filtering functions of the small dry cleaner into one unit. Examples include the All-in-One System manufactured in Houston, Texas and a front-loading dry cleaning washer in a cabinet marketed by the Hammond Manufacturing Company of Waco, Texas. One of the first heavy industrial PCE machines was the Prosperity Model 6-A manufactured prior to 1948 in the United States (Prosperity Company Inc., 1948). The basic configuration of dry

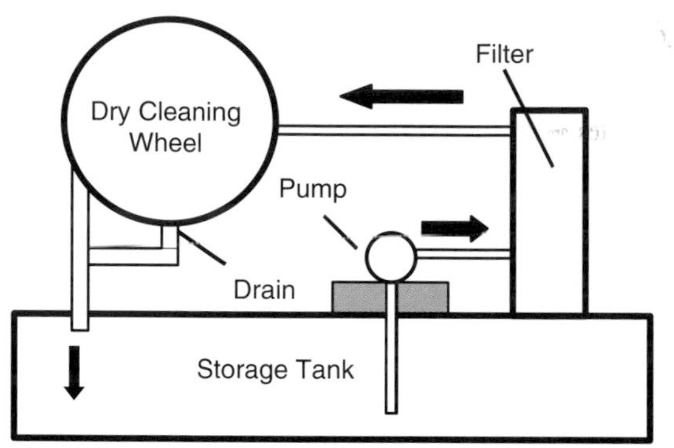

Figure 7.2
Generic dry cleaner design.

cleaning machines used for over the past 100 years is shown in Figure 7.2 and consists of mechanical washing equipment, extracting equipment, drying equipment, a storage tank, filter equipment, and solvent recovery equipment. The soil (e.g., oils, grease, wax, fatty acids that are primarily body oil, tars, vegetable, mineral or animal oil, dust, lint, carbons, skin flakes, ashes, cosmetics, hair, pigments, animal, vegetable and synthetic fibers, and matter generally found in food or drink) is suspended or dissolved in the solvent during the wash cycle and spills over or is pumped from the wash wheel into a storage tank (see Figure 7.2). From the storage tank, the solvent is pumped into the filter where the un-dissolved soil collects on the filtering medium. The strained solvent flows through the filter and into the dry cleaning wheel.

Dry cleaning designs include transfer, dry-to-dry vented, or dry-to-dry closed loop designs. In a transfer machine, the clothing is washed in one unit and physically transferred to a dryer.

Approximately 30% of all dry cleaning machines are transfer units. In a dry-to-dry vented unit, the clothing is washed and dried in the same cylinder. The PCE emitted from the unit is uncontrolled or vented to a control device. Approximately 70% of retail dry cleaners in the United States currently use dry-to-dry units. In a dry-to-dry closed loop unit, the wash and dry cycles occur in the same unit; solvent emissions are controlled with a refrigerated condenser. For forensic investigations, it is useful to obtain specific manufacturer information to determine what solvents are compatible with the equipment. For example, a dry cleaning booklet for the Prosperity Company Model 6-A dry cleaner states: 'Only perchloroethylene can be used in the unit. Never attempt to use any other solvent.' (Prosperity Company, Inc., 1948).

The presence of filters and filter media can be used for age dating and/or source identification. Filters (clarifying systems, bag filters, flat screen filters, tubular screen filters, cartridge filters and regenerative filters) mounted on a still (cooker) are available for straining dry cleaning solvents. Tubular filters consist of tubes, screens or bags pre-coated with a pure grade of inert silica powder such as diatomaceous earth. The filter powder coats the tubes, screens, or bags and forms a pre-coat that retains the soil. When the pre-coat is exhausted, the filters are back washed and the sludge is drained from the bottom of the filter. Regenerative filter designs are pre-coated filters with a large amount of filter powder that are agitated at the end of each wash cycle. The pre-coat mixture is then re-deposited on the filter to provide a porous filter cake on the tubes prior to the next wash cycle. Most regenerative filters contain flexible tubes. The filter cartridge became prominent in the dry cleaning industry in about 1961 (Dow Chemical Company, 1976). Filter cartridges consist of a specially treated, pleated paper perforated outer shell, an interior perforated metal cylinder, a bed of granular carbon, and a perforated center post covered with a mesh wrap. Approximately 90% of dry cleaners use filter cartridges containing activated carbon, although carbonless designs are also available.

The muck cooker distills a portion of the solvent inventory. A muck cooker can recover as much as 95% of a solvent from spent sludge. Distillation units also remove soil from used solvents and sludges. A cooker/still consists of a boiler, cooking or distillation chamber, condensation coils and a separator. Muck cooker and cooker/stills operate by heating the solvent to the boiling point, vaporization, and condensation on water-cooled coils. The liquid is then collected and cycled through a solvent water separator. A muck cooker percolates low-pressure steam through spent filter and still sludge; a high capacity condenser then condenses the steam and the solvent is passed through a solvent water separator. A cooker/still can recover as much as 95% of the solvent from the spent sludge. In the aerospace industry, still bottom sludges typically contain 70–80% solvent and 20–30% oil, grease and solids with traces of water. In the electronics industry, distillation sludges typically contain 85–95% solvents, with oil, flux and traces of water comprising the remaining 5–15% (Evanoff, 1990).

A frequent source of releases to the environment is PCE laden effluent from the water solvent separator. The wastewater (primarily condensate) from the separator is frequently plumbed directly into the sewer or drains into a container that is discharged into the sewer. PCE in wastewater from a water/solvent separator liquid collected from nine dry cleaners in central California averaged 151 parts per million (Izzo, 1992). This study found that leakage through the sewer lines is the primary avenue through which PCE is introduced into the subsurface. Being heavier than water, undissolved PCE settles to the bottom of the sewer line and exfiltrates through joints or cracks in the sewer line. In the city of Santa Barbara, California, for example, the city permitted the disposal of solvent/water separator wastewater into the sewer system through the late 1980s.

The historical disposal practices of distillation sludges and spent solvents provide insight regarding the presence of solvents at locations not associated with dry cleaning operations. The following text from 1964 and 1972 guidelines represents two examples:

> In the absence of any clearly defined ordinances, the sludge is usually poured on dry ground well away from buildings, and the solvents are allowed to evaporate. If the sludge is free flowing, it is placed in shallow open containers and allowed to evaporate before the solids are dumped on the ground
>
> (American Society for Metals, 1964)

> Waste mixtures should not be discharged into drains or sewers where there is a danger that the vapor may be ignited. In cases such as these, the waste should be removed to a safe location (away from inhabited areas, highways, buildings, or combustible structures) and poured onto dry sand, earth, or ashes, then cautiously ignited. Burning of chlorinated hydrocarbon wastes should be done only when permitted by controlling authorities and then under constant supervision.
>
> In other instances, the chlorinated hydrocarbon waste may be placed in an isolated area as before and simply allow the liquid to evaporate
>
> (American Insurance Association, 1972)

Equipment locations where solvent losses occur include the machine door gasket, storage tanks and associated piping, leaky solenoid valves, damper valves, the still or cooker door and pump couplings, especially at the packing glands (Manufacturing Chemists' Association, Inc., 1948; Dow Chemical Company, 1976). Gaskets used in dry cleaning equipment are PCE rated, dating back to World War II when a PCE compatibility rating for plastic gaskets was developed. Vapor clouds can also originate from vapor exhausted from the washer tub. An example of the accumulated volume of PCE lost from even small drips from these equipment locations can be significant (see Table 7.6) (International Fabricare Institute, 1988).

The corrosion of solvent tanks and piping is associated with the presence of water in the system or with the decomposition of PCE. If the water separator is

Table 7.6
Monthly estimated loss from solvent leaks.

	Running Time Each Day		
Drip Sequence	**4 Hours**	**6 Hours**	**8 Hours**
1 second	30 gallons	45 gallons	60 gallons
5 seconds	6 gallons	9 gallons	12 gallons
10 seconds	3 gallons	4.5 gallons	6 gallons
60 seconds	0.5 gallons	0.75 gallons	1 gallon

too warm, the separation of water and the solvent does not occur and water and solvent drains into the storage tank. As the solvent cools in the tank, the water floats and accumulates at the surface, providing an environment conducive for tank corrosion along the interior walls. The presence of a milky white solvent in the tank or floating globules, balls or beads of liquid on the surface of the solvent in the storage tank is an indication of the presence of water in the solvent. A rapid rise in the filter pressure is another indicator as globules of water carried by the circulating solvent seal the gaps between the particles of diatomaceous earth in the filter cake and restrict the passage of circulating liquid solvent from the dirty side of the filter, through the filter cake and to the clean side of the filter. Green water in the water separator indicates corrosion of the copper condensing coils by water in the system. A potential source of exterior corrosion in dry cleaning equipment is the placement of space heaters, which tend to draw solvent vapor into the flames, forming acid vapors. PCE can decompose when heated above 300°F (149°C) and forms an acid that can corrode metal parts in the dry cleaner (Permac Drycleaning, 1978). Decomposition generally occurs in the bottom of the still given the excessive steam temperatures.

7.4.4 ELECTRONICS INDUSTRY

Chlorinated solvents commonly used in the electronics industry include 1,1,2-trichloro-1,2,2-trifluoroethane (Freon 113), TCA, methylene chloride, TCE and PCE. Semiconductor wafer fabrication and printed circuit board production historically used large quantities of chlorinated solvents. Non-chlorinated compounds associated with wafer fabrication, for example, include xylenes, *n*-butyl acetate, ammonium fluoride, hydrofluoric acid, sulfuric acid, copper, zinc, phosphorus, boron, nitric acid, acetic acid, chromic acid and phosphoric acid. The presence of antimony, arsenic, boron, and aluminum in environmental samples may be associated with dopants (used to change the electrical properties of the silicon wafer) and may provide an opportunity for age dating the release of these compounds and/or identifying the source of a contaminant release.

The use of chlorinated solvents in products and processes used in printed circuit board manufacturing provides an example of the use of this information for source identification or age dating. Freon 113 and TCA are used to remove flux from printed circuit boards after the electronic components are soldered to the board. TCA is frequently combined with alcohol because alcohol is an effective flux remover. TCA and methylene chloride are also used as solvents in the photoresist process. In addition, the ingredients in products used in the photoresist process can be used as signature chemicals to age date a contaminant release (Table 7.7) (California Department of Health Services, 1988).

Table 7.7
Compounds associated with the photoresist process.

Commercial Products	Composition and Percentage
Waycoat photoresist	Xylene (85%)
Micro Resist Developer	Stoddard Solvent (95–100%)
Photo Resist Developer I	Aromatic hydrocarbons (10–30%); aliphatic hydrocarbons (>60%)
Photo Resist Developer II	Mixture of petroleum solvents
Negative Photo Resist	2-ethoxyethylacetate (48%); *n*-butyl acetate (5%); xylene (5%)
Positive Photo Resist	2-ethoxyethylacetate (52%); *n*-butyl acetate (6%); xylene (6%)
Ultrasonic Degreaser	Perfluoroisobutylene
Stripper 712D	Dodecyl benzene sulfonic acid; 1,2,4-trichlorobenzene, phenols; pH = 2.4–2.6
Photo Resist Stripper J 100	Methylene chloride (63%), chlorobenzenes; sulfonic acid (23%); phenols and derivatives (14%). May also contain PCE
Micro Strip	Chlorinated solvents; orthocresol, dodecylbenzo sulfonic acid, PCE and dichlorobenzene
Burmar	Phenols, sulfonic acid, aromatic solvents and chlorobenzenes. When the sulfonic acid mixes with water, sulfuric acid is created

In order to effectively utilize the detection of unique ingredients in products for age dating and source identification, an understanding of the chemical use history and changes in chemical usage are required prior to sample collection and analysis. Products containing chlorinated solvents used in the semiconductor industry, for example, changed dramatically in the 1970s. Historical solvent use information for a semiconductor plant in Santa Clara, California, for example, included the use of an aromatic negative resist developer containing xylene, toluene and benzene. The developer was subsequently reformulated to contain only xylenes and ethylbenzene, thereby providing an opportunity to use the presence of benzene for age dating. At the same facility, solvents used prior to 1978 included TCE and methyl ethyl ketone while solvents used after 1978 included cellosolve acetate and hexamethyl-disilazane, which provided another signature chemical for age dating. The detection of these individual compounds in combination with detailed chemical use history underscores the importance of collecting detailed chemical use history prior to sample analysis.

7.4.5 SEPTIC CLEANERS

Chlorinated solvents can be present in products disposed in septic tanks as well as ingredients of septic tank cleaning products. In cases where septic tanks

Septic Tank Cleaners	Ingredients and Concentration by Weight (%)
King of All	Chloromethylbenzenes (100)
Drainz	Methylene chloride (6) Naphthalene (2) Naphthalene (88) Ethylene glycol monophenyleter (4)
Cesstic	Dichloromethane (15) 1,1,1-Trichloroethane (30) Trichloroethene (0.02) Tetrachloroethene (0.1) Carbon tetrachloride (0.01) Alkanes and benzcnes (55)

Table 7.8

Ingredients of three septic tank cleaning products used in the Amherst-North Hampton area of Massachusetts in 1985.

and/or septic tank waste disposal occurred (e.g., sanitary landfills), their potential contribution as a source of chlorinated solvents should be evaluated. In one shelf survey involving 1026 brands of consumer products of 67 product categories, for example, 14% of the brands contained TCA while 34% contained methylene chloride (Westat Inc., 1987). Other products contained TCE and PCE. The detection of PCE, TCE and TCA in groundwater attributable to septic systems includes sites located in Tacoma, Washington (DeWalle *et al.*, 1985); Speonk, New York; Sun Valley, Nevada (Tomson *et al.*, 1984); and Long Island, New York (Pettyjohn and Hounslaw, 1983). Noss *et al.* (1987), estimated that 2% of all septic tank cleaning products purchased in the Amherst-North Hampton area of Massachusetts in 1985 contained chlorinated solvents. Compounds in three septic system cleaners are summarized in Table 7.8 (Noss *et al.*, 1987). The inclusion of TCE in septic tank cleaners appears to have declined by the mid-1980s.

If other indicators of septic system effluent are spatially correlated with solvent contamination, this can assist to confirm the source. In a New Jersey community Maguire and Braids (1990) found that chlorinated solvents were detected coincident with other indicators of septic system effluent. Because of this, and because the presence of chlorinated solvents was not correlated spatially with petroleum compounds, they concluded that a service station, which had been implicated, was not the source.

Metals and other chemical indicators present in septic cleaners may be used for source identification. Background concentrations of these compounds should be included as part of this analysis. The concentration of metals in three septic cleaners is summarized in Table 7.9 (Noss *et al.*, 1987).

Chemicals associated with septic effluent include nitrate, phosphate, ammonium ions, laundry/cleaning agents, caffeine and pharmaceuticals (Seiler *et al.*, 1999). Groundwater samples from three communities in the vicinity of Reno,

Table 7.9

Metals present in three septic tank cleaners.

Description and Major Ingredient	Metal Concentrations in Product (ppm)							
	Cd	Cr	Cu	Pb	Hg	Ni	Ag	Zn
Roto-Sept-X™ ($Al_2(SO_4)_3$)	<2	200	27	22	<0.2	27	3	83
Clobber™ (H_2SO_4)	5	<5	<0.5	<8	<0.1	9	<2	2
Wonder Root-Away™ ($CuSO_4$)	<2	<5	NA[a]	20	<0.2	7	<2	8

[a]Not analyzed.

Nevada, for example, detected the pharmaceuticals chlorpropamide (used in the treatment of diabetes), phensuximide, and carbamazepine (used in the treatment of seizures) as indicators of contamination from septic tanks. Caffeine was also detected in groundwater from a shallow monitoring well in a field irrigated with sewage effluent; the caffeine metabolite 3-methylxanthine and the chlorinated derivates of caffeine (8-chlorocaffeine) were not detected.

The presence of detergent with chlorinated solvents may provide evidence that the chlorinated solvents originated from sewage and well as a potential indicator of the age of the release. Detergents were first used in the United States in 1946 but did not become the dominant cleaning agent until 1953 (LeBlanc, 1984, 1996). Prior to 1965, detergents primarily contained non-biodegradable surfactants, or branched-chain alkylarylsulfonates (Plummer *et al.*, 1993). In 1965, these surfactants were replaced almost exclusively by biodegradable surfactants such as linear alkylarylsulfonates, sodium dodecylsulfates that could be removed by secondary sewage-treatment plants. Shapiro *et al.* (1999) used this historical information in addition to tritium results (used as a marker given the tritium bomb peaks in the 1960s) to confirm the chronology of detergent contamination of a sand and gravel aquifer at Otis Air Force Base in Falmouth, Massachusetts.

7.5 AGE DATING AND SOURCE IDENTIFICATION

Chlorinated solvent data are routinely used for source identification and age dating. Forensic approaches include the commercial availability of the solvent, regulations impacting the use of solvents, equipment design, chemical usage patterns, the presence of unique additives, isotope analysis, and degradation models.

7.5.1 COMMERCIAL AVAILABILITY AND REGULATIONS

The commercial availability of a chlorinated solvent and regulatory impacts on solvent usage can bracket when it was first introduced at a site. Solvent chronologies in Tables 7.1 through 7.4 provide general frameworks when TCE, PCE, TCA and carbon tetrachloride were available. Given that many chemical and commercial synonyms are used in environmental reports, it is important to identify those products containing chlorinated solvent of concern in addition to activities associated with solvent use. Appendix B lists usage and synonyms for selected chlorinated solvents.

An industry-specific approach is also often used in an attempt to delineate when a solvent was used at a facility, although sites can deviate significantly from the industry norm. Key toxicological findings and regulations rarely correlate with solvent transitions and illustrate the difficulty in extrapolating site-specific solvent usage patterns from general industry practices.

International (e.g., Montreal Protocols signed in September 1987 and amended in June 1990), federal, state and/or local regulations or agreements can provide a means to bracket the date of chlorinated solvent usage (Berry and Dennison, 2000). Rule 66 was promulgated by the Los Angeles Air Pollution Control District in July 1966 as a result of concerns regarding the smog-forming potential of TCE (Brunnelle *et al.*, 1966). This rule required the installation of control equipment if TCE emissions from any type of equipment exceeded 40 pounds (18 kg) over 24 hours. This had the consequence that TCE was almost entirely replaced by TCA and PCE as the degreasing solvent of choice in Los Angeles County from 1967 to 1969 (Archer and Stevens, 1977). TCE limits in Los Angeles led, in part, to restrictions included in the Federal Clean Air Act of 1970. As with Rule 3 in San Francisco, Regulation V in Philadelphia and the Clean Air Act Amendments (1990), PCE was determined to be virtually unreactive in the formation of oxidants that contribute to smog and was exempted along with fluorinated hydrocarbons. TCA, however, with a boiling point of 165°F (74°C), was a viable replacement solvent for many degreasers (Table 7.10) since hot water (rather than pressurized steam) could continue to be used to volatilize the solvent.

Table 7.10

Boiling point of selected chlorinated solvents (Montgomery, 1991; Pankow et al., *1996).*

Compound	Boiling Point (°C)
Carbon tetrachloride	76.7
1,1-Dichloroethane	57.3
1,2-Dichloroethane	83.5
1,1,1-Trichloroethane	74.1
Trichloroethene	86.7
Tetrachloroethylene	121.4
1,1,2-Trichloroethane	113.8
Trichlorofloromethane	23.8
1,1,2-Trichlorotrifluoroethane	47.7

Table 7.11

Primary manufacturers of four chlorinated solvents in the United States in the twentieth century.

Manufacturer	Carbon Tetrachloride	TCE	PCE	TCA
Allied Chemical & Dye	1955–1981			
Brown Company	1925–1928			
Carbide & Carbon Chemicals		1922–1935		
Diamond Alkali/Diamond Shamrock	1944–1986		1950–1986	
Diamond Shamrock		1969–1977		
Dow Chemical	1908–2000	1921–2000	1923–2000	1936–1994
DuPont Company	1974–1989		1933–1986	
Eastman Kodak	1925–1927			
Ethyl Corporation	1969–1977	1967–1982	1967–1983	1964–1976
Frontier Chemical/Vulcan Materials	1956–2000		1958–2000	
Great Western Electrochemical	1917–1938			
Hooker Chemical/ Occidental Chemical		1956–1980	1949–1991	
Hooker-Detrex/Detrex Chemical		1947–1972	1947–1971	
LCP	1981–1991			
Mallinckrodt Chemical Works	1956–1960			
Niagara Alkali		1949–1955		
Niagara Smelting/Stauffer Chemical/Akzo	1922–1991			
Occidental Chemical	1987–1994			
Pittsburgh Plate Glass/PPG Industries	1957–1972	1956–2000	1949–2000	
PPG Industries				1962–2000
R&H Chemical/DuPont Company		1925–1972		
Seeley & Company	1941–1943			
Stauffer Chemical			1955–1985	
Taylor Chemical	1933–1944			
Vulcan Materials				1970–2000
Warner Chemical/ Warner-Klipstein/Westvaco Chlorine/Food Machinery and Chemical/FMC Corp	1908–1979			
Westvaco Chlorine		1933–1949	1940–1945	

The date that a manufacturer began synthesizing a chlorinated solvent is frequently used as evidence regarding when it was available at a facility. This approach assumes that potential suppliers and/or products containing chlorinated solvents are known. Table 7.11 identifies manufacturers of four chlorinated solvents in the United States from 1908 to 2000 (after Doherty, 2000a,b).

7.5.2 EQUIPMENT DESIGN AND CHEMICAL USAGE

An understanding of a site's manufacturing processes, equipment and material handling systems is critical in identifying probable sources of a contaminant release. A vapor degreaser may be manufactured to use a solvent within a discrete boiling range thereby narrowing solvents used by a particular degreaser.

Opportunities may also exist to associate a solvent with an activity for bracketing the location and/or timing of a release. For chlorinated solvents, unique applications or impurities in the solvent can assist in this analysis. 1,1,2,2-Tetrachloroethane, for example, is used almost exclusively in military applications, although it also coelutes on a gas chromatogram with PCE and may therefore be misidentified as PCE. Applications of selected chlorinated solvents are summarized in Appendix B (Montgomery, 1991; Mabey, 1995; Pankow *et al.*, 1996; Irwin *et al.*, 1997).

7.5.3 ADDITIVES

The presence of additives in chlorinated solvents presents another opportunity for age dating and source identification. The probability of detecting additives is greatest if a non-aqueous phase liquid (NAPL) sample can be collected for testing. Additives are used for the following purposes (Archer, 1984; Morrison, 2000a):

- As a metal inhibitor that deactivates the metal surface and complexes any metal salts that might form. PCE does not require a metal inhibitor while TCE and TCA contain metal inhibitors and acid acceptors (see below).
- As an acid acceptor that reacts with and chemically neutralizes trace amounts of hydrochloric acid formed during degreasing operations and which may cause corrosion of the degreased part.
- As an antioxidant that reduces the solvent's potential to form oxidation products. TCE requires an antioxidant (Archer, 1996).

The presence of additives in TCA is illustrative of their use for source identification and age dating. Over 100 stabilizer formulations for TCA are reported that have discrete compositions and dates of availability (Doherty, 2000b). Common metal inhibitors blended with TCA include 1,4-dioxane, dioxilane, epichlorohydrin and 1,2-epoxybutane. 1,4-Dioxane is of special interest as it is classified as a toxic pollutant by the United States Environmental Protection Agency and has been detected in groundwater in Japan, Canada and the United States. Structurally similar compounds (1,3-dioxanes) have contaminated drinking water resources in Spain (Beckett and Hua, 2000).

The addition of 1,4-dioxane ($C_4H_8O_2$) to TCA prevents the corrosion of aluminum, zinc and iron surfaces. In 1985, 90% of the dioxane synthesized in the United States was used as a TCA additive. A non-primary alkanol was also added to TCA to allow the storage in steel drums without discoloration. The detection of 1,4-dioxane has been used to distinguish between multiple sources of TCA in groundwater. 1,4-Dioxane can be transported faster than TCA in groundwater along with the TCE additives butyelene oxide and epichlorohydrin. The presence of 1,4-dioxane leading a TCA groundwater plume at the Gloucester chemical-waste landfill near Ottawa, Ontario is such an example. For this site, the retardation rate for dioxane was 1.1 while the retardation rate for TCA was estimated to be greater than 6 (Jackson and Dwarakanath, 1999). When reviewing chemical results for 1,4-dioxane, it should be noted that EPA Method 8240 does not report the presence of 1,4-dioxane while EPA Method 8260 does. It is therefore important to confirm that the presence of 1,4-dioxane is source related and not an artifact of changes in the analytical method (Zemo, 2000).

The fact that manufacturers use different additives for identical solvents is often useful in determining the origin of a solvent. Table 7.12 lists the concentrations of additives in TCA synthesized in the United States, Europe and Japan (Archer, 1984; Jackson and Dwarakanath, 1999).

The initial concentration of most additives in chlorinated solvents is low and varies considerably between manufacturers (ASTM, 1992, 1996). In some instances, specifications regarding the use of a particular solvent grade, such as

Table 7.12
Additives in TCA (% by volume) produced in the United States, Europe and Japan.

Additive	United States	Europe	Japan
1,2-butylene oxide	0.5–0.8	0.6–1.0	0.1–0.6
Nitromethane	0.4–0.7	0.4–1.0	0.1–0.4
1,4-dioxane	2.0–3.5	3.5	3.5
sec-butanol	1.0–2.0		
1,3-dioxolane	1.0		
tert-butanol			3.5–6.5
Methyl butynol			2.0–3.0
Isopropylnitrate			2.0
Acetonitrile			3.0

Department of Defense MIL-T-7003, OT-634C or ASTM D 4080-96, can provide the basis for identifying the initial concentration of an additive. In cases where this information is absent, a general knowledge of the type of stabilizers associated with a solvent can associate a stabilizer with a specific location and/or solvent usage pattern. Additives used in six chlorinated solvents and vinyl chloride are summarized in Table 7.13 (Kircher, 1957; IARC, 1979; Mertens, 1993; Archer, 1996; Jackson and Dwarakanath, 1999; Morrison, 1999).

7.5.4 ISOTOPIC ANALYSIS

Isotope ratios are used in environmental forensics for chlorinated solvents (1) to distinguish between different contaminant sources and/or (2) to demonstrate that biodegradation is occurring (biodegradation can produce a characteristics isotope fractionation) (Hunkeler *et al.*, 1997, 1999; Dayan *et al.*, 1999; Sherwood Lollar *et al.*, 1999; Sturchio *et al.*, 1999). Research indicates that for chlorinated solvents, large and reproducible carbon isotope fractionation associated with biodegradation can be used as an effective indicator to verify intrinsic biodegradation. Isotopic fractionation is a measure of the change in the ratio of heavy to light isotopes because the differences in mass between isotopes result in slight differences in the activation energies during reactions (Ward *et al.*, 2000). An advantage of isotopic analysis versus using chemical concentration data is that changes in concentration from physical processes such as dilution and sorption are frequently difficult to quantify and can complicate chemical data interpretation.

A common area of inquiry in the application of isotope data for source identification is the ability to distinguish between manufacturers of chlorinated solvents, especially for discrimination between sources in a co-mingled groundwater plume. The isotopes most commonly used for this purpose are ^{13}C and ^{37}Cl (Clark and Fritz, 1997). An example of the application of this technique is the identification of multiple sources of TCE. The isotopic range of ACE grade TCE (minimum assay 99.5% TCE), for example, is: $\delta^{13}C = -48.0$ to -27.8‰, $\delta^{37}Cl = -2.54$ to $+4.08$‰, $\delta D = -30$ to $+530$‰ (Poulson and Drever, 1999). The units known as permils (0/00) express the δ values in parts per thousands between the measured and reference isotopic ratios.

Dehydrochlorination reactions are common reactions in the industrial production of many chlorinated hydrocarbons. As a result, many chlorinated hydrocarbons exhibit isotopically enriched δD signatures. A wide range in δD signatures is therefore likely to be associated with different manufacturers (Tanaka and Rye, 1991; Poulson and Drever, 1999). Chemical reactions producing isotopically light hydrochloric acid result in other chemical products becoming isotopically enriched as the yields for these reactions are engineered

Table 7.13

Examples of additives in selected chlorinated solvents.

Chemical	Additive/Impurity
Carbon tetrachloride	Technical grade carbon tetrachloride can contain <1 ppm of carbon disulfide if produced by disulfide chlorination as well as trace amounts of bromine and chloroform (Doherty, 2000a). Corrosion inhibitors include alkyl cyanamides, diphenylamine, ethyl acetate, and ethyl cyanides (McKetta and Cunningham, 1979; Kroschwitz and Howe-Grant, 1991).
Chloroform	Ingredients may include bromochloromethane, carbon tetrachloride, dibromodichloroethane, dibromodichloro-methane, 1,1-dichloroethane, 1,2-dichloroethane, *cis-* and *trans-*1,2-dichloroethene, dichloromethane, diethyl carbonate, ethylbenzene, 2-methoxyethanol, nitromethane, pyridine, 1,1,2,2-tetrachloroethane, trichloroethylene and meta-, ortho- and para-xylenes. In Japan, chloroform has a minimum purity of 99.95% and can contain assorted chlorinated hydrocarbons as impurities.
Dichloromethane	Stabilizers (0.0001–1%) can include phenol, hydroquinone, para-cresol, resorcinol, thymol, 1-naphthol or amines.
PCE	In the presence of water, unstabilized PCE slowly hydrolyzes and forms trichloroacetic and hydrochloric acid that can degrade both metals and rubber materials, such as pump gaskets or seals. Early stabilizers included amines and hydrocarbons (*Chemical Engineering*, 1961). More recent stabilizers include morpholine derivates, and/or mixtures of epoxides and esters (Gerhartz, 1986). Thymol may be present (0.01% wt/wt) as a preservative (Irwin *et al.*, 1997). USP grade PCE contains not less than 99% and no more than 99.5% PCE with the remainder consisting of ethanol.
TCE	Technical grade trichloroethylene TRI-119 and TRI-127 manufactured by PPG Industries, for example, contains the antimicrobial agents thymol (0–20 ppm) and hydrochloro mono-methylether (80–120 ppm). Antioxidants such as amines (0.01–0.001%) or combinations of epoxides, such as epichlorohydrin and esters (0.2–2% total), are often added to TCE. Analgesic grades of TCE can contain thymol that is often dyed with waxoline blue for identification (Huff, 1971). The purity of technical grade TCE is around 99.97% with no free chloride and stabilizers. Compounds that may be present in TCE include acetone, acetylenic compounds, aniline, borate esters, *n*-butane, *p-tert*-butyl catechol, butylguaiacol, butylene oxides, *o*-creosol diisopropylamine, dioxane, epoxy compounds, ethyl acetate, hydroxyanisole derivatives, hydrazine derivatives, hydrazones, isobutyl alcohol (Gerhartz, 1986), isocyanates (aliphatic), lactone, nitro compounds (e.g., *o*-nitrophenol), oxirane, 1-pentanol, phenol, propargyl alcohol, propylene oxide, pyrazoles, pyrazoline derivatives, pyrrole derivatives (*Chemical Engineering*, 1961), sterates, styrene oxide, sulfur dioxide, tetrahydrofuran, tetrahydrothiophene, thiazoles, thymol and triethylamine (Hardie, 1964). Impurities in commercial TCE products include carbon tetrachloride, chloroform, 1,2-dichloroethane, *cis-* and *trans-*1,2-DCE, pentachloroethane, bromodichloroethylene, benzene, TCA and 1,1,2,2-tetrachloroethane (Irwin *et al.*, 1997).

Table 7.13 Continued

Chemical	Additive/Impurity
1,1,1-TCA	Stabilizers (≈3–8%) include acrylonitrile, dialkyl sulfoxides, sulfides and sulfites, methyl ethyl ketone, nitromethane, *n*-methyl pyrrole, 1,4-dioxane, butylene oxide, tetraethyl lead, 1,2-butylene oxide, ethyl acetate, morpholine, tetrahydrofuran, nitriles, 1,3-dioxolane, glycol diesters, monohydric acetylenic alcohols, isopropyl alcohol, toluene and secondary butyl alcohol (McKetta and Cunningham, 1979; Irwin *et al.*, 1997; Doherty, 2000b). 1,4-dioxane may be present in TCA at 0–4% by weight. A Material Safety Data Sheet (MSDS) for Solvent 111® lists the concentration of 1,4-dioxane at a concentration of 25 ppm.
Vinyl chloride	Commercial grades contain 1–2% impurities including acetic aldehyde (≈5 ppm), butane (≈8 ppm), 1,3-butadiene (≈10 ppm), chlorophene (≈10 ppm), diacetylene (≈4 ppm), vinyl acetylene (≈10 ppm), propine (≈3 ppm), methyl chloride (≈100 ppm) (Irwin *et al.*, 1997). Additional impurities can include hydrogen chloride, hydrogen peroxide and acetaldehyde (International Labour Office, 1983). Phenols may be present at concentrations of 25–50 ppm as a stabilizer to prevent polymerization (International Agency for Research on Cancer, 1972).

to be high. Dehydrogenation and dehydrobromination reactions may also produce isotopically enriched compounds, since H_2 and HBr are also isotopically depleted relative to other hydrogen bearing compounds. Dehydrofluorination is unlikely to produce isotopically enriched compounds, as hydrogen fluoride shows relatively small fractionation effects versus other hydrogen bearing compounds. Halogenated compounds are expected to exhibit a wide range of manufacturer dependent isotopic signatures due to various chemical reactions, which may include dehydrochlorination or dehydrogenation reactions and production conditions (e.g., temperature differences, catalysts used, engineering design, etc.) as well as use of different feedstocks. The difference in bond strength results in chlorine isotope fractionation due to temperature and pressure differences during synthesis (Tanaka and Rye, 1991). The ^{37}Cl isotope fraction in organic solvents is bound more tightly to carbon than are ^{35}Cl atoms (Bartholomew *et al.*, 1954).

Van Warmerdam *et al.* (1995) examined the isotopic ratios for $^{13}C/^{12}C$ and $^{37}Cl/^{35}Cl$ for five chlorinated solvents from four manufacturers. Beneteau *et al.* (1996, 1999) continued this work using ^{13}C and ^{37}Cl for comparison between two batches from the Van Warmerdam *et al.* study and five pure-phase chlorinated solvents manufactured by Dow Chemical and PPG. Samples were analyzed using a gas chromatograph with a combustion interface and isotope ratio mass spectrometry. The $\delta^{13}C$ signatures of PCE, TCE and TCA between batches analyzed in 1995 and 1999 from the same manufacturers were found to be variable. For example, the $\delta^{13}C$ of a batch of pure phase TCE produced by PPG in 1995 and 1999 was −27.860.01 and −31.6860.17, respectively. A consistent ^{13}C isotopic difference between PPG and Dow Chemical PCE, however, was observed, suggesting that ^{13}C might still be useful for source identification between manufacturers. $\delta^{37}Cl$ signatures were found to be less sensitive to manufacturing processes than $\delta^{13}C$ signatures. PCE and TCE from PPG in 1995 and 1999, for example, ranged from −2.546 to −2.5260.15‰ and −2.5460.34 and −2.3360.08‰, respectively. The more consistent $\delta^{37}Cl$ signature of PCE and TCE obtained for each batch suggests that $\delta^{37}Cl$ values may provide an opportunity to distinguish between manufacturers. For TCA, $\delta^{37}Cl$ values between batches supplied by PPG in 1995 and 1999 ranged from −2.9060.68 and −0.3660.17‰, respectively, suggesting that the use $\delta^{37}Cl$ values for TCA may not be appropriate. $\delta^{13}C$ results for TCA, however, were similar between 1995 and 1999 PPG batches (−25.860.46 and −25.7860.46‰), suggesting that $\delta^{13}C$ values may be used to distinguish between TCA manufacturers.

Isotopic analysis was used for source identification of a PCE plume located at the Angus site, about 1 kilometer north of the Canadian Forces Base Borden study site, which has been extensively investigated (Kueper *et al.*, 1989; Poulson and Kueper, 1992). The contamination at the Angus site was believed to have originated via leakage from a former above-ground storage tank or by leakage of PCE wastewater through a septic system tile bed. The plume length and width were about 720 and 228 feet (220 and 70 m), respectively; PCE concentrations ranged from the low parts per billion (ppb) to low parts per million (ppm) range. A purge-and-trap method was used for analyzing for ^{13}C and ^{37}C (Vaillancourt, 1998). Beneteau *et al.* (1999) concluded that the $\delta^{13}C$ pattern in the Angus PCE plume at the Borden site suggests one source and/or specific manufacturer associated with the mass of the plume. PCE observed on the fringe of the plume likely originated from minor spills unassociated with the PCE in the central portion of the plume.

Another application of isotope analysis for source identification was at a bulk-chemical transfer site in Toronto, Ontario, Canada (Hunkeler *et al.*, 1999). The previous and current owners stored and transferred similar chemicals (e.g., methanol, methyl ethyl ketone, vinyl and ethyl acetate and butyl acrylate) with the exception of PCE that was stored at the site about 18 years previously. PCE releases were believed to have occurred during the coupling and uncoupling of railway tanker cars and had migrated through the ballast. In addition to the dissolved PCE, groundwater samples contained significant concentrations of methanol, acetate, TCE, *cis*-1,2-dichloroethene (*cis*-1,2-DCE), vinyl chloride and ethene, indicating that dechlorination was occurring, probably stimulated by the methanol or acetate in the groundwater.

Compound specific ratios for the groundwater samples from the Toronto site were determined using a gas chromatograph combustion isotope-ratio mass spectrometry (GCIRMS) system. The mole fraction and carbon isotope ratios of the chlorinated ethenes suggested that the carbon isotope composition of dissolved PCE closest to the suspected release locations (CEL1, CEL2, and CEL3 in Figure 7.3) reflected the original isotopic composition of PCE (Hunkeler *et al.*, 1999). As shown in Figure 7.3, the $\delta^{13}C$ values of all chlorinated ethenes followed a general trend of enriched values with increasing dechlorination, suggesting that the degree of isotope fractionation did not depend on the concentration of the chlorinated ethenes. The comparison of the $\delta^{13}C$ values of dissolved PCE even with dechlorination provided a basis for distinguishing between multiple sources of PCE. The observation that $\delta^{13}C$ values of the ethene reaches the $\delta^{13}C$ value of the initial substrate may also

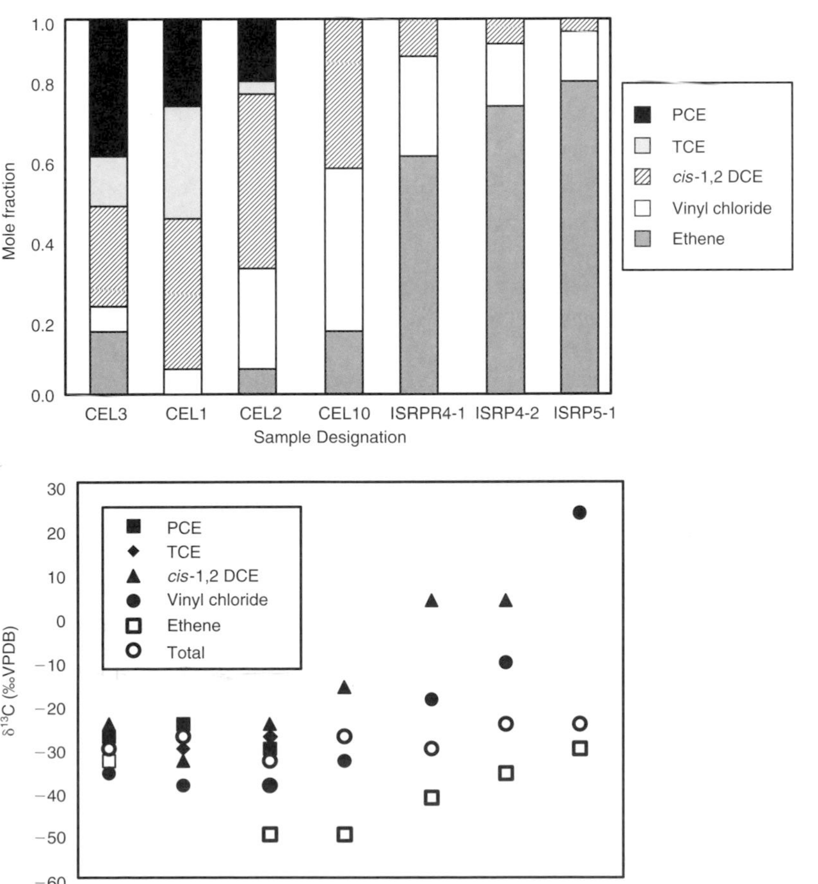

Figure 7.3

Mole fractions and carbon isotope ratios ($\delta^{13}C$) of chlorinated ethenes in groundwater samples taken at a Toronto field site on August 28, 1998. Wells CEL1, 2 and 3 are located closest to a suspected PCE release. VPDB, international reference standard (Pee Dee belemnite) for $\delta^{13}C$ values. (Reprinted with permission from the American Chemical Society. Hunkeler, D., Aravena, R., and B. Butler, 1999. Monitoring microbial dechlorination of tetrachloroethene (PCE) in groundwater using compound-specific stable carbon isotope ratios: Microcosm and field studies. Environmental Science and Technology *33(16), 2733–2738.)*

provide an opportunity for distinguishing ethenes originating from different parent compounds (e.g., PCE versus 1,2-dichloroethane). This opportunity assumes that the parent compounds have different initial $\delta^{13}C$ values and that the dechlorination process has reached completion.

Potential challenges to the use of isotopic interpretations for source identification include whether a qualified laboratory capable of performing precise GCIRMS measurements was used and expert witness qualifications. Data interpretation may also be biased due to the potential for isotopic fractionation in the environment. Slater *et al.* (1998) studied the effects of isotopic fractionation due to volatilization on the isotopic composition of phase separate and dissolved TCE at different concentrations. The stable carbon isotope concentrations indicated that volatilization and dissolution of the TCE did not result in isotopic fractionation but that fractionation occurred due to abiotic dechlorination. Huang *et al.* (1999), however, measured some fractionation (a 3.5‰ depletion in residual ^{13}C after greater than 99% evaporation) that occurred during solvent volatilization. The impact of biodegradation on the isotopic composition of the chlorinated solvent must be assessed when evaluating isotopic data (Abrajano and Sherwood Lollar, 1999). Field data from a PCE spill impacted by biodegradation indicate that a small carbon isotope fractionation appears to occur during the transformation of PCE to TCE and *cis*-1,2-DCE (Beneteau *et al.*, 1999). Aravena *et al.* (1998) also measured significant isotopic fractionation for the dechlorination of DCE to vinyl chloride.

7.5.5 DEGRADATION PRODUCTS AND RATIOS

The detection of chlorinated solvents, degradation models, and chlorinated solvent ratios are used for age dating and source identification. The presence or absence of parent compounds and their breakdown products frequently provide straightforward evidence for age dating and/or source identification. For example, the compound 1,1-dichloroethene (1,1-DCE) is a degradation product of TCA and PCE while chloroethane is a degradation product of TCA or 1,2-dichloroethane. The presence of 1,1-DCE and chloroethane can therefore be argued as evidence of a TCA release (Feenstra *et al.*, 1996). Under strongly reducing conditions, methane is also a degradation product of TCA (Bradley and Chapelle, 1999a,b). The presence of methane may therefore provide an opportunity to trace the migration of TCA although careful consideration is required to filter other possible sources.

Another example of the use of degradation products to indicate the presence of the parent compound is the conversion of TCE by methanotrophs to 2,2,2-trichloroacetaldehyde (e.g., chloral hydrate). Chloral hydrate is a mutagen and acutely toxic. The microbial conversion to chloral hydrate is not really a dehalogenation process but rather a migration of the chlorine to the adjacent carbon (Alexander, 1999). The presence of 2,2,2-trichloroacetaldehyde may therefore implicate the use of TCE at a facility.

Degradation rates of chlorinated solvents have been argued as evidence to age-date a contaminant release (Ram *et al.*, 1999). A common element to many chlorinated solvent models used for age dating is the reliance on a representative half-life value. The validity of each half-life value must therefore be examined, realizing that there is not a single, unique value for each degradation rate but rather a range of reasonable values as shown in Table 7.14 for TCE (Woodbury and Li, 1998).

The heterogeneity of the subsurface physicochemical system at a site and the initial solvent composition introduces significant uncertainty into degradation rates. For example, anaerobic decomposition of solvent can occur in a sewer or drain prior to release into the soil. PCE, for example, can be overheated (~300°F/150°C) resulting in decomposition (Permac Drycleaning, 1978). An example of the influence of the subsurface environment is whether the presence of an iron rich horizon(s) is present as degradation rates can vary over three orders of magnitude between different types of iron (Farrell *et al.*, 2000). If a solvent is preferentially transported via an iron enriched fracture system, reaction rates may be substantially different than for the bulk indigenous soil. In addition, half-life values are usually impacted by the redox potential and microbial activity of the media through which the chlorinated solvent migrates. Another issue is whether the apparent biodegradation rate is an artifact of the distance from the contaminant source, the monitoring well screen length and/or the degree of well desaturation that occurs during purging (Martin-Hayden and Robbins, 1997). The presence of NAPLs can inhibit microbial degradation and/or result in deviations from assumed first order reaction kinetics (Phelps *et al.*, 1988; Barker, 1996; Farrell *et al.*, 2000). These complications introduce additional uncertainty for age dating a chlorinated solvent release (Vogel and McCarty, 1987; Vogel *et al.*, 1987).

Table 7.14
Half-lives of trichloroethylene degradation products.

Transformation	**Range of Half-Life**
TCE to DCE	2.4 days to 0.9 years
cis-1,2-DCE to vinyl chloride	7.2 days to 140 days
trans-1,2-DCE to vinyl chloride	6.2 days to 244 days
1,1-DCE to vinyl chloride	53 days to 132 days
Vinyl chloride to ethene	56 days to 7.92 years
Ethene	~10 years

Another use of chlorinated solvent information is to examine chlorinated solvent ratios versus distance from known or suspected releases (Murphy and Gauthier, 1999). A molar ratio, such as DCE/(TCE + PCE), may be used for this analysis. Ratio analysis generally assumes minimal effects from dispersion, retardation and transport heterogeneities (Ram *et al.*, 1999). When plotted, TCE/PCE and DCE/TCE ratios usually fluctuate with distance due to changes in the subsurface aerobic/anaerobic environments, measurement uncertainty, etc. A DCE/TCE ratio, for example, may increase under anaerobic conditions because TCE degrades faster than DCE but then decrease under aerobic conditions. The presence of additional solvents in combination with abrupt changes in ratios can provide confirmation for identifying an additional release(s), especially if the solvent is mixed (e.g., spent solvents as contrasted with a pure phase solvent). When examining ratio analysis statistically, it should be realized that averaging concentrations over time can mask trends that are otherwise observable. If a ratio analysis has average or normalized groups of data, the data should be dissected on a sampling event basis for forensic analysis.

These general remarks are illustrated in Figure 7.4, which depicts a hypothetical series of PCE releases. In the top diagram groundwater concentrations are shown as a function of downgradient distance. Distances are measured

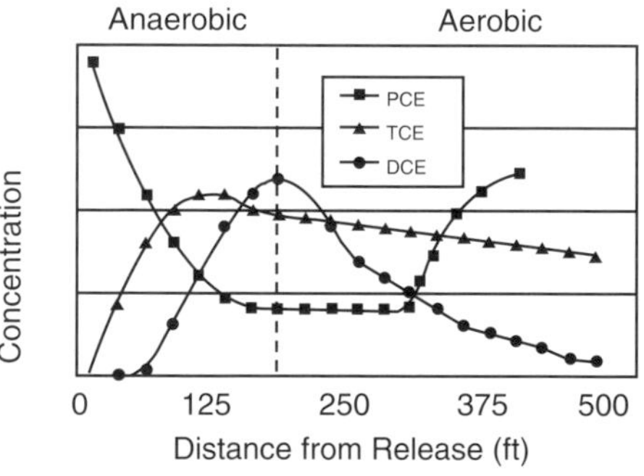

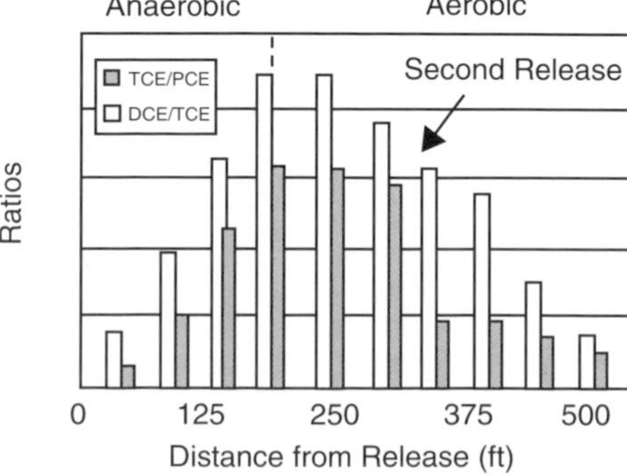

Figure 7.4
Conceptual use of concentration ratio approaches to identify multiple chlorinated solvent releases.

from the location of the first PCE spill. The first PCE spill is assumed to occur in an anaerobic zone. This would likely be the case, for example, if the spill consisted of spent PCE containing dirt, grease and other organics. In the anaerobic reductive dechlorination decay chain depicted in Figure 8.1, PCE produces TCE and then *cis*- and *trans*-1,2-DCE (at about a 30:1 ratio). Although not depicted in Figure 7.4, if conditions are sufficiently reducing DCE will decay to vinyl chloride.

In Figure 7.4 the DCE concentration peaks at about the boundary of the anaerobic/aerobic zone; the TCE concentration peaks at a shorter distance due to the rapid decline in the parent PCE. If the boundary between anaerobic and aerobic zones were changed the relative positions of these peaks would be altered. The slight decline in TCE and the more marked decline in DCE in the aerobic zone are also due to biodegradation. (Any vinyl chloride present would decay even more rapidly than DCE while PCE does not decay aerobically at all.) However, the products of aerobic degradation are short-lived epoxides and other, usually unmeasured products, and there is no additional creation of PCE, TCE, DCE, or vinyl chloride.

The corresponding ratios TCE/PCE and DCE/TCE are plotted in the bottom diagram of Figure 7.4. A second release of PCE is postulated to have occurred at about 325 feet (100 m) from the first release. In this case the release is assumed to be in an aerobic zone. This might be the case, for example if the release was of virgin PCE. As shown in the bar chart the change in ratios is gradual except for the sudden decrease in the TCE/PCE ratio at the location of the second spill.

In reality the ratio data would be more scattered than depicted in Figure 7.4. In a more realistic situation, the issue is whether the change in ratios at the putative location of a second release is statistically significant in terms of the amount of scatter present in the ratios at other distances.

Figure 7.4 depicts only one possible situation amenable to ratio analysis: the releases could be of TCE instead of PCE, the second release could also be in an anaerobic zone, etc.

A trend in ratios can be used to determine the predominant groundwater flow direction for a single release when historical groundwater flow information is absent (Morrison, 2000d). One situation of particular interest in this regard is as when groundwater flows through an anaerobic region. This can produce a sudden change in the DCE/TCE ratio, or the sudden appearance of vinyl chloride, immediately downgradient of the anaerobic zone.

When reviewing chlorinated solvent concentration and/or ratio analysis, the most probable degradation environment through which the solvent is migrating and the most likely degradation pathway, rate and influence of dilution from the source should be defined. Once a pathway is hypothesized, examine this interpretation in the context of other information. For example, TCE degradation varies depending on the amount of hydrogen available, with the most efficient sequence via methanogenesis while a moderately efficient reaction is via sulfate or Fe^{+3} reduction (Lee *et al.*, 2000). The least efficient is via oxidation. The degradation products of TCE, *cis*-1,2-dichloroethene and vinyl chloride are more difficult to degrade anaerobically but are degraded aerobically (Hartmans and deBont, 1992; Chapelle, 1996).

Brugger and Lehmicke (2000) used TCA hydrolysis to date a TCA release as pre or post 1990. The depth to groundwater was about 3–10 feet (1–3 m) and an aquitard was present at a depth of about 13–15 feet (4–4.5 m). Site soils had a high organic matter and concentrations of methane above the saturated zone ranged from 30 to 50% by volume. The dissolved oxygen concentration in the groundwater was less than one part per million. Other potential electron donors in the subsurface included aromatics (BTEX suite), methylene chloride, diesel, and peat layers. Groundwater velocity through the sand was calculated to be about 20–40 feet (6–12 m) per year. Assumptions included the following:

- The half-life for the abiotic decay of TCA to 1,1-DCE is about 4.5 years.
- The ratio of acetic acid to 1,1-DCE is about 4:1 (Cline and Delfino, 1989).
- A DCE/TCA ratio of 0.15 by weight indicates that one half-life (4.5 years) had passed since the TCA entered the groundwater.

Ratios between chlorinated ethenes were also examined. The ratio of *cis*-1,2-DCE/(PCE + TCE) in the groundwater was 0.36 which was argued as evidence of a half-life of 2 years or less and inconsistent with a documented 1986 release (e.g., a half-life of ~7 years would be required). Given the subsurface conditions and current DCE/TCA ratio of 0.033, the authors concluded that the likely residence time was approximately 2–4 years, indicating a post 1990 release of TCA. This conclusion was reached after examining whether the same effects could occur via the dissolution of a dense non-aqueous phase liquid (DNAPL). The argument for a more recent release became compelling because the analytical data were consistent with a series of small recent releases, consistent with the known operational practices at the facility. Other evidence relied upon included the following:

- the size of the reportable spill versus the size of a release needed to generate historical concentrations;
- highly fluctuating concentrations during the previous 9 years;

- TCA concentrations indicative of a DNAPL release that was inconsistent with a release of a 0.2% TCA solution in 1982 and 1983 only;
- maximum soil contaminant concentrations were found near both the surface and at the top of the aquitard, indicating both light non-aqueous phase liquid (LNAPL) and dense non-aqueous phase liquid (DNAPL) releases;
- LNAPL and DNAPL locations and concentrations were inconsistent with continuing dissolution of DNAPL from 1980s spills; and
- the presence of high concentrations of compounds not reported in the 1982 and 1986 releases.

A variation of this ratio method directly uses the ratio of the rate constant for the hydrolysis of 1,1,1-trichloroethane to 1,1-dichloroethene, acetic acid, and hydrochloric acid to estimate when TCA enters groundwater according to the following relationship: TCA $\Rightarrow$ 1,1-DCE (22%) + CH_3CO_2H + HCl. This method assumes that the groundwater temperature (yearly average) and the TCA and 1,1-DCE concentrations in the sample are known (Smith, 1999). Smith *et al.* (1994) suggest that the groundwater temperature can be estimated from the average air temperature at a site when direct measurements are unavailable. The method also assumes that the TCA and 1,1-DCE in the groundwater is known with a high degree of accuracy and precision. Based on Smith *et al.* (1994; Smith and Eng, 1997), Murphy and Gauthier (1999) found the following expression:

$$t = 2.88 \times 10^{-21} \exp[55.604/(1 + 2.175 \times 10^{-3} T)] \ln[1 + 6.25(C_D/C_T)] \quad (7.1)$$

where t is the time in years since TCA entered the groundwater; C_D is 1,1-DCE groundwater concentrations in mg/L; C_T is TCA groundwater concentration in mg/L; and T is groundwater temperature in °F.

The contaminant migration rate is then determined by dividing the horizontal distance to each well by the age of the TCA in the well. Equation 7.1 does not consider the relative affinity of TCA and DCE in the soil. If the organic carbon content in the aquifer is more than a few tenths of a percent, a more accurate estimate is obtained by combining this description of hydrolysis with contaminant transport modeling. A benefit of this approach is that the abiotic decay of TCA occurs independent of pH and redox. Given that TCA can biodegrade and hydrolyze, the presence of the biodegradation product 1,1-DCA or elevated carbon dioxide readings may provide evidence that biodegradation is competing with hydrolysis (Murphy and Gauthier, 1999). Under some conditions, a soil gas survey along the axis of the contaminant plume may provide insight regarding this issue.

Opponents of this technique cite the following factors that may result in inaccurate age estimations:

- Qualitative chemical results (e.g., trace or non-detect) may be difficult to incorporate into the analysis.
- An inadequate number of samples may produce inaccurate results.
- An estimate in warmer climates or for older spills may be inaccurate since C_D/C_T becomes very large.
- If the transformation of TCA to 1,1-dichloroethane by biodegradation is significant, the 1,1-DCE to TCA ratio will be lower than if the ratio were solely the result of the hydrolysis reaction. The age prediction will therefore be older.
- If TCE is also present and TCE biodegradation is significant, then, since some TCE biodegrades to 1,1-DCE, 1,1-DCE concentrations will be higher than what is attributable to TCA hydrolysis alone.

7.6 CONCLUSIONS

The presence of chlorinated solvents in environmental samples offers significant opportunities for age dating and source identification. The dominant solvents available for this purpose include TCE, PCE, TCA, methylene chloride, carbon tetrachloride, and their respective transformation products. The most common age dating and source identification techniques include historical uses, isotopic ratio analysis, degradation relationships, half-life models, and intrusive testing (e.g., soil gas survey, soil sampling, and groundwater testing). When using this information for forensic purposes, sufficient resources to examine the integrity and level of confidence of the analytical data and sampling procedures should be allocated. Ideally, multiple, but independent forensic techniques can be used to develop evidence for age dating or source identification of chlorinated solvents.

ACKNOWLEDGMENTS

The authors wish to thank the following individuals for their manuscript review, assistance in document identification and ideas: Dr. James Picker of Transglobal Environmental Geochemistry, Inc.; Gary Brugger of Exponent; and, Dr. Jay Jones, a private consultant, in Solano Beach, California.

REFERENCES

Abrajano, T. and Sherwood Lollar, B. (1999) Introduction Note: Compound-specific isotope analysis: tracing organic contaminant sources and processes in geochemical systems. *Organic Chemistry* 30(8a), v–vii.

Alexander, M. (1999) *Biodegradation and Bioremediation*, 2nd edn. Academic Press, London.

American Insurance Association (AIA) (1972) *Chemical Hazards Bulletin*. C-86. Chlorinated hydrocarbons. AIA, New York.

American Society for Metals (1964) Vapor degreasing. *Metals Handbook*: *Heat Treating, Cleaning and Finishing*, vol. 2. Metals Park, OH.

American Society for Metals (1996) Guide to vapor degreasing and solvent cold cleaning. *ASM Handbook*, vol. 5: *Surface Engineering*. Metals Park, OH.

American Society for Testing and Materials (ASTM) (1989) *Manual on Vapor Degreasing*, 3rd edn. ASTM, Philadelphia, PA.

American Society for Testing and Materials (ASTM) (1992) *Standard Practice for Solvent Vapor Degreasing Operations*, pp. 131–134. D 3698-92. ASTM, Philadelphia, PA.

American Society for Testing and Materials (ASTM) (1995) Standard specification. Vapor-degreasing grade methylene chloride. ASTM Designation D 4079-95. ASTM, Philadelphia, PA.

American Society for Testing Materials (ASTM) (1996) Standard practice for handling an acid degreaser or still. ASTM Designation D 4579-96. ASTM, Philadelphia, PA.

Aravena, R., Beneteau, K., Frape, S., Butler, B., Abrajano, R., Major, D., and Cox, E. (1998) Application of isotopic finger-printing for biodegradation studies of chlorinated solvents in groundwater. In *Risk, Resource, and Regulatory Issues: Remediation of Chlorinated and Recalcitrant Compounds,* (Wickramanayake, G. and Hinchee, R., eds), pp. 66–71. Battelle Press, Columbus, OH.

Archer, W. (1973) Selection of a proper vapor degreasing solvent. *Cleaning Stainless Steel,* pp. 54–64. STP 538. American Society for Testing Materials, Philadelphia, PA.

Archer, W. (1984) A laboratory evaluation of 1,1,1 trichloroethane-metal-inhibitor systems. *Werkstoffe und Korrosion* 35, 60–69.

Archer, W. (1996) *Industrial Solvents Handbook*. Marcel Dekker, New York.

Archer, W. and Stevens, V. (1977) Chemical profile of chlorinated, aliphatic, aromatic and oxygenated hydrocarbons as solvents. *Industrial Engineering Chemical Production Research and Development* 16, 319–325.

Aviado, D., Zakhari, S., Simaan, J., and Ulsamer, A. (1976) Review of the literature on trichloroethylene. *Methyl Chloroform and Trichloroethylene in the Environment,* Chapter 7. CRC Press, Cleveland, OH.

Barber, J. (1957) Chloroethene in aerosols. *Soap Chemical Specifications* 33, 99.

Barker, J. (1996) Intrinsic plume remediation: Chlorinated solvents and selected pesticides. In *Dissolved Organic Contaminants in Groundwater*. University Consortium Solvents-in-Groundwater Research Program. May 13–16, 1996.

Bartholomew, R., Brown, F., and Lounsbury, M. (1954) Chlorine isotope effect in reactions of tert-butyl chloride. *Canadian Journal of Chemistry* 32, 979–983.

Beckett, M. and Hua, I. (2000) Elucidation of the 1,4 dioxane decomposition pathway at discrete ultrasonic frequencies. *Environmental Science and Technology* 34(18), 3944–3953.

Beneteau, K., Aravena, R., Frape, S., Abragano, T., and Drimmie, R. (1996) Chlorinated solvent fingerprinting using ^{13}C and ^{37}Cl stable isotopes. American Geophysical Union, Section 121, H31B-12. Spring AGU Meeting, Baltimore, MD.

Beneteau, K., Aravena, R., and Frape, S. (1999) Isotopic characterization of chlorinated solvents – laboratory and field results. *Organic Geochemistry* 30, 739–753.

Berry, J. and Dennison, M. (2000) *The Environmental Law and Compliance Handbook*, pp. 718–719. McGraw-Hill Handbooks, New York.

Bradley, P. and Chapelle, F. (1999a) Methane as a product of chloroethene biodegradation under methanogenic conditions. *Environmental Science and Technology* 33(4), 653–656.

Bradley, P. and Chapelle, F. (1999b) Role for acetotrophic methanogens in methanogenic biodegradation of vinyl chloride. *Environmental Science and Technology* 33(19), 3473–3476.

Brugger, G. and Lehmicke, L. (2000) Dating a chlorinated solvent release. 1982 or 1994. *Tenth Annual West Coast Conference on Contaminated Soils and Water*. Session 3: Environmental Forensics. Association for the Environmental Health of Soils. Doubletree Hotel, San Diego, CA. March 20–23, 2000.

Brunnelle, M., Dickinson, J., and Hamming, W. (1966) Effect of organic solvents in photochemical smog formation. Solvent Project, Final Report. Air Pollution Control District, County of Los Angeles. Los Angeles, California.

California Department of Health Services (1988) The reduction of solvent wastes in the electronics industry. Final Report. Waste Reduction Grant Program. Grant Number 86-T0110. California Toxic Substances Control Division. Alternative Technology and Policy Development Section. Sacramento, CA.

Chapelle, F. (1996) Identifying redox conditions that favor the natural attenuation of chlorinated ethenes in contaminated groundwater systems. In *Proceedings of*

Symposium on Natural Attenuation of Chlorinated Organics in Groundwater. United States Environmental Protection Agency. September 11–13, 1996. Dallas TX. EPA/540/R-96/509.

Chemical Engineering (1961) Competition sharpens in chlorinated solvents. 68(22), 62–66.

Chemical Engineering News (1950) The chemical world this week. 28, 3044.

Chemical Engineering News (1960) Trichloro competition gets livelier. 38, 40.

Chemical Engineering News (1962) Coin-op cleaners boost perchloroethylene. 40, 21.

Chemical Engineering News (1967) Wide scope seen for dry cleaning chemicals. 45, 30.

Chemical Engineering News (1970) Solvents: DuPont phasing out. 48, 13.

Chemical Industry (1949) Chemical specialties. 68, 310.

Chemical Marketing Reporter (1975) Chemical profile: Trichloroethylene. 208, 9, 12.

Chemical Marketing Reporter (1992) Chemical profile: Carbon tetrachloride. 241, 7, 50.

Chemical Marketing Reporter (1997) Chemical profile. Trichloroethylene. 252, 23, 37.

Chemical Week (1953) Tri, Per, and Carbon Tet. 72, 56.

Clark, I. and Fritz, P. (1997) *Environmental Isotopes in Hydrogeology*. CRC Press, Boca Raton, FL.

Cline, P. and Delfino, J. (1989) Transformation kinetics of 1,1,1-trichloroethane to the stable product 1,1-dichloroethene. In *Biohazards of Drinking Water Treatment,* (Larson, R.A., ed.), pp. 47–56. Lewis Publishers, Chelsea, MI.

Dayan, H., Abrajano, T., Sturchio, N., and Winsor, L. (1999) Carbon isotope fractionation during reductive dehalogenation of chlorinated ethenes by metallic iron. *Organic Geochemistry* 30(8a), 755–763.

DeWalle, F., Kalman, D., Norman, D., Sung, J., Plews, G., and Lewis, R. (1985) Determination of toxic chemicals in effluent from household septic tanks. Water Engineering Research Laboratory, Office of Research and Development, Cincinnati, OH. USEPA/600/2-85/050.

Doherty, R. (2000a) A history of the production and use of carbon tetrachloride, tetrachloroethylene, trichloroethylene, and 1,1,1, trichloroethane in the United States. Parts I and II. *Journal of Environmental Forensics* 1(2), 69–93.

Doherty, R. (2000b) A historical look at the production and use of TCE and 1,1,1 TCA in the US. IBC's 3rd Annual Executive Forum on Environmental Forensics. Applying Effective Scientific Methods to Decrease Cost and Liability. International Business Communications. June 26–28, 2000. Washington, DC.

Dow Chemical Company (1976) *Top Quality DOWPER Solvent, for Top Quality Cleaning*. Dow Chemical Company, Midland, Michigan.

European Chlorinated Solvents Association (1996) *Chlorinated Solvents in Europe*. Brussels, Belgium.

European Chlorinated Solvents Association (1997) *Solvents Digest* March 1997. Methylene Chloride: an update on human and environmental effects. Brussels, Belgium.

European Chlorinated Solvents Association (2000) *Storage and Handling of Chlorinated Solvents*, 3rd edn. Brussels, Belgium.

Evanoff, S. (1990) Hazardous waste reduction in the aerospace industry. *Chemical Engineering Progress* April, 51–61.

Farbenindustrie Akt. Ges., German Patent 523,426. April 23, 1931.

Farrell, J., Melitas, N., Kason, M., and Li, T. (2000) Electrochemical and column investigation of iron-mediated reductive dechlorination of trichloroethylene and perchloroethylene. *Environmental Science and Technology* 34(12), 2549–2556.

Federal Register (1973) Trichloroethane (1,1,1 trichloroethane, methyl chloroform). 38, 21935. August, 14, 1973.

Feenstra, S. (1996) Characterization of source zones and plumes: selected topics. In *DNAPL Site Characterization and Remediation*. University Consortium Solvents in Groundwater Research Program. November 12–18, San Francisco, CA.

Feenstra, S., Cherry, J., and Parker, B. (1996) Conceptual models for the behavior of dense non-aqueous phase liquids (DNAPLs) in the Subsurface. In *Dense Chlorinated Solvents and other DNAPLs in Groundwater*, (Pankow, J. and Cherry, J., eds), pp. 53–128. Waterloo Press, Guelph, Ontario.

Fisher, E. (1864) Ueber die Einwirkung von Wassestoff auf Einfach-Chlorkohlenstoff. *Jena Z. Med. Naturwiss.* 1, 123.

Gerhartz, W. (1986) *Ullman's Encyclopedia of Industrial Chemistry*, 5th edn. Weinheim, New York.

Grob, K. and Grob, G. (1974) Organic substances in potable water and in its precursor, Part II. Applications in the area of Zurich. *Journal of Chromatography* 90, 303–313.

Halogenated Solvents Industry Alliance, Inc. (1994a) *Perchloroethylene*. White Paper. February 1994. Washington, DC.

Halogenated Solvents Industry Alliance, Inc. (1994b) *White Paper on Methyl Chloroform* (1,1,1-trichloroethane). Washington, DC.

Halogenated Solvents Industry Alliance, Inc. (1996) *Trichloroethylene*. White Paper June 1996. Washington, DC.

Halogenated Solvents Industry Alliance, Inc. (1998a) *Facts about PERC dry cleaning*. Washington, DC.

Halogenated Solvents Industry Alliance, Inc. (1998b) *White Paper on Methylene Chloride*.

Halogenated Solvents Industry Alliance, Inc. (1999) *Perchloroethylene White Paper*. Washington, DC.

Halogenated Solvents Industry Alliance, Inc. (2000) *Typical Properties of Chlorinated Solvents*. Washington, DC.

Hardie, D. (1964) Chlorocarbons and chlorohydrocarbons. Trichloroethylene. In *Kirk-Othmer Encyclopedia of Chemical Technology*, 2nd edn. (Kirk, R. and Othmer, D., eds) pp. 183–195. Wiley-Interscience, New York.

Hartmans, S. and deBont, S. (1992) Aerobic vinyl chloride metabolism in Mycobacterium aurum L1. *Applied Environmental Microbiology* 58(4), 1220–1226.

Haynes, W. (1954) *American Chemical Industry, Background and Beginnings,* vol. 1. Van Nostrand Press, New York.

Huang, L., Sturchio, N., Abrajano, T., and Holt, B. (1999) Carbon and chlorine isotope fractionation of chlorinated aliphatic hydrocarbons by evaporation. *Organic Geochemistry* 30(8a), 777–785.

Hueper, W. (1950) *Environmental Cancer*. United States Government Printing Office. Washington, DC.

Huff, J. (1971) New evidence on the old problems of trichloroethylene. *Industrial Medicine* 40(8), 25–33.

Hunkeler, D., Hoehn, E., Hohener, P., and Zeyer, J. (1997) ^{222}Rn as a partitioning tracer to detect diesel fuel contamination in aquifer: laboratory study and field observations. *Environmental Science and Technology* 31(11), 3180–3187.

Hunkeler, D., Aravena, R., and Butler, B. (1999) Monitoring microbial dechlorination of tetrachloroethene (PCE) in groundwater using compound-specific stable carbon isotope ratios: Microcosm and field studies. *Environmental Science and Technology* 33(16), 2733–2738.

International Agency for Research on Cancer (IARC) (1972) *Monographs on the Evaluation of the Carcinogenic Risk of Chemicals to Man*, vol. 7. IARC World Health Organization, Geneva, Switzerland.

International Agency for Research on Cancer (IARC) (1979) In *IARC Monographs on the Evaluation of the Carcinogenic Risk of Chemicals to Humans. Halogenated Hydrocarbons*, vol. 20. IARC World Health Organization, Geneva.

International Fabricare Institute (1988) *Focus on Dry Cleaning. Solvent Mileage*. Silver Spring, MD. 12(1) 20.

International Labour Office (1983) *Encyclopedia of Occupational Health and Safety*, vols 1 & 2. International Labour Office, Geneva.

Irwin, R., van Mouwerik, D., Stevens, L., Seese, M., and Basham, W. (1997) *Environmental Contaminants Encyclopedia*. National Park Service, Water Resources Division. Fort Collins, CO.

Izzo, V. (1992) *Dry Cleaners – A Major Source of PCE in Groundwater*. State of California Central Valley California Regional Water Quality Control Board, Well Investigation Program. Fresno, CA.

Jackson, R. (1998) The migration, dissolution, and fate of chlorinated solvents in the urbanized alluvial valleys of the southwestern USA. *Hydrogeology Journal* 6, 144–155.

Jackson, R. and Dwarakanath, V. (1999) Chlorinated degreasing solvents: physical-chemical properties affecting aquifer contamination and remediation. *Groundwater Monitoring and Remediation* 19(4), 103–109.

Kircher, C. (1957) *Solvent Degreasing – What Every User Should Know*. Technical Publication 16, pp. 44–49. American Society of Testing Materials (ASTM), Philadelphia, PA.

Kirk, R. and Othmer, D. (1949) *Encyclopedia of Chemical Technology*. Interscience Encyclopedia, New York.

Kroschwitz, J. and Howe-Grant, M. (eds) (1991) *Kirk-Othmer Encyclopedia of Chemical Technology*, 4th edn. John Wiley & Sons, New York.

Kueper, B., Abbot, W., and Farquhar, G. (1989) Experimental observations of multiphase flow in heterogeneous porous media. *Journal of Contaminant Hydrology* 5, 83–95.

LeBlanc, D. (1984) *Sewage Plume in a Sand and Gravel Aquifer*. Cape Cod, Massachusetts. United States Geological Survey Water-Supply Paper 2218.

LeBlanc, D. (1996) Overview of research at the Cape Cod site: Field and laboratory studies of physical, chemical, and microbiological processes affecting transport in a sewage-contaminated aquifer. Untied States Geological Survey Water-Resources Investigations Report 94-4015.

Lee, R., Jones, S., Kuniansky, E., Harvey, G., Sherwood Lollar, B., and Slater, G. (2000) Phreatophyte influence on reductive dechlorination in a shallow aquifer contaminated with trichloroethane (TCE). *International Journal of Phytoremediation* 2(3), 193–211.

Linak, E., Lutz, H., and Nakamura, E. (1990) C_2 Chlorinated Solvents. *Chemical Economics Handbook*. Stanford Research Institute. Menlo Park, CA. 632.30000a-632.3001Z.

Love, S. (1951) Water analysis. *Analytical Chemistry* 23, 253–257.

Lowenheim, F. and Moran, M. (1975) *Faith, Keyes and Clark's Industrial Chemicals*, 4th edn. John Wiley & Sons, New York.

Lyne, F. and McLachlan, T. (1949) Contamination of water by trichloroethylene. *The Analyst* 74, 513.

Mabey, W. (1995) Survey of chemicals encountered in investigation/remediation. Section 6. In *Environmental Chemistry for Investigating and Remediating Soil and Groundwater Contamination*. Department of Engineering Professional Development, College of Engineering, University of Wisconsin at Madison. September 18–20, 1995. Madison, WI.

Maguire, T. and Braids, O. (1990) VOC source delineation through fingerprint analysis of superimposed chemical groundwater contamination. *Groundwater* 28(5), 790.

Manufacturing Chemists' Association, Inc. (1948) *Chemical Safety Data Sheet SD-24 Properties and Essential Information for the Safe Handling and Use of Perchloroethylene*, p. 14. Washington, DC.

Martin-Hayden, J. and Robbins, G. (1997) Plume distortion and apparent attenuation due to concentration averaging in monitoring wells. *Groundwater* 35(2), 339–347.

McCarty, P. (1996) Biotic and abiotic transformations of chlorinated solvents in groundwater. In *Proceedings of the Symposium on Natural Attenuation of Chlorinated Organics in Groundwater*. USEPA/540/R-96/509. Dallas, TX. September 11–13, 1996.

McKetta, J. and Cunningham, W. (1979) *Encyclopedia of Chemical Processing and Design*. Marcel Dekker, New York.

Mellan, I. (1957) *Source Book of Industrial Solvents*, vol. II: *Halogenated Solvents*. Reinhold Publishing Company, New York.

Melpolder, F., Warfield, C., and Headington, C. (1953) Mass spectrometer determination of volatile contaminants in water. *Analytical Chemistry* 25, 1453–1456.

Mertens, J. (1993) Trichloroethylene. In *Kirk-Othmer Encyclopedia of Chemical Technology*, 4th edn, vol. 6, pp. 40–50. Wiley-Interscience, New York.

Middleton, M. and Walton, G. (1961) Organic chemical contamination of groundwater. In *Proceedings of Groundwater Contamination*, pp. 50–56. Cincinnati, OH, April 5–7.

Montgomery, H. and Conlon, M. (1967) The detection of chlorinated solvents in sewage sludge. *Water Pollution Control* 190–192.

Montgomery, J. (1991) *Groundwater Chemicals Field Guide*. Lewis Publishers, Chelsea, MI.

Morrison, R. (1999) *Environmental Forensics. Principles and Applications*. CRC Press, Boca Raton, FL.

Morrison, R. (2000a) *Critical Review of Environmental Forensic Techniques*. United States Environmental Protection Agency In-House Training Course. Sponsored by the University of Wisconsin at Madison Extension Engineering and Professional Development Department. June 19–21, 2000. Lakewood, CA.

Morrison, R. (2000b) Application of forensic techniques for age dating and source identification in environmental litigation. *Environmental Forensics* 1(3), 131–153.

Morrison, R. (2000c) Critical review of environmental forensic techniques. Part I. *Environmental Forensics* 1(4), 157–174.

Morrison, R. (2000d) Critical review of environmental forensic techniques. Part II. *Environmental Forensics* 1(4), 175–196.

Murphy, B. and Gauthier, T. (1999) Forensic analysis of chlorinated solvent contamination data. *Environmental Claims Journal* 11(4), 81–96.

National Cancer Institute (1976) *Carcinogenesis Bioassay of Trichloroethylene*. NCI-CG-TR-2. Washington, DC.

National Institute of Standards and Technology (2000) *Methylene Chloride*. webbook.nist.gov/chemistry/.

National Institutes of Health (1982) *Carcinogenesis Bioassay of Trichloroethylene in F344 Rats and B6C3F1 Mice*. NTD 81-84. National Toxicology Programs. Research Triangle Park, NC. NIH Publication No. 82-1799.

National Toxicity Program (NTP) (1986) *Toxicology and Carcinogenesis Studies of Dichloromethane (Methylene Chloride)*. CAS No. 75-09-2. In F344/N and B6C3F1 Mice (Inhalation Studies). January 1986.

Norge Corporation (no date) *History of Dry Cleaning*. Norge Corporation Dry Cleaning System Handbook. Model KM-G-1. Muskegon, MI.

Noss, R., Drake, R., and Mossman, C. (1987) *Septic Tank Cleaners: Their Effectiveness and Impacts on Groundwater Quality*. The Environmental Institute at the University of Massachusetts at Amherst prepared for the Office of Research and Standards, Massachusetts Department of Environmental Quality Engineering. Publication No. 87-3.

Nutting, H. and Huscher, M. (1940) Preparation of methyl chloroform. United States Patent 2,209,000.

O'Hanlon, A. (1977) Dry cleaning: PERC + water does wonders. *Washington Post*, November 12, 1997.

Oljenick, I. (1928) Trichloroethylene treatment of trigeminal neuralgia. *Journal of the American Medical Association* 91, 1085.

Pankow, J., Feenstra, S., Cherry, J., and Ryan, C. (1996) Dense Chlorinated Solvents and Other DNAPLs in Groundwater: History, Behavior, and Remediation. Chapter 1. In *Dense Chlorinated Solvents and other DNAPLs in Groundwater*, (Pankow, J. and Cherry, J., eds), Waterloo Press, Portland, OR.

Partington, J. (1964) *A History of Chemistry*, vol. 4. Macmillan, London.

Permac Drycleaning (1978) Technical bulletin. How to limit corrosion on drycleaning machines. Number 8/78. Plainview, New York.

Pettyjohn, W. and Hounslaw, A. (1983) Organic compounds and ground-water pollution. *Groundwater Monitoring Review* Fall, 41–47.

Phelps, T., Ringelberg, D., Hedrick, D., Davis, J., Fliermans, C., and White, D. (1988) Microbial biomass and activities associated with subsurface environments contaminated with chlorinated hydrocarbons. *Geomicrobiology Journal* 6, 157–170.

Plummer, L., Michel, R., Thurman, E., and Glynn, P. (1993) Environmental tracers for age-dating young groundwater. In *Regional Ground-Water Quality,* (Alley, W., ed.) Van Nostrand Reinhold, New York.

Poulson, M. and Kueper, B. (1992) A field experiment to study the behavior of tetrachloroethylene in unsaturated porous media. *Environmental Science and Technology* 26, 889–895.

Poulson, S. and Drever, J. (1999) Stable isotope (C, Cl and H) fractionation during vaporization of trichloroethylene. *Environmental Science and Technology* 33(20), 3689–3694.

Prosperity Company, Inc. (1948) *Prosperity Perchloroethylene Model 6-A Dry Cleaning System Handbook*. Syracuse, New York.

Ram, N., Leahy, M., Carey, E., and Cawley, J. (1999) Environmental sleuth at work. *Environmental Science and Technology* 33(21), 464–469.

Randle, R. and Bosco, M. (1991) Air pollution control. *Environmental Law Handbook*, 11th edn, pp. 524–615. Government Institutes, Rockville, MD.

Seiler, W. (1960) Perchloroethylene. *Chemical Engineering News* 38, 124.

Seiler, R., Zaugg, S., Thomas, J., and Howcroft, D. (1999) Caffeine and pharmaceuticals as indicators of waste water contamination in wells. *Groundwater* 37(3), 405–410.

Shapiro, S., LeBlanc, D., Schlosser, P., and Ludin, A. (1999) Characterizing a sewage plume using the ^{3}H-^{3}H dating technique. *Groundwater* 37(6), 861–878.

Sherwood Lollar, B., Slater, G., Ahad, J., Sleep, B., Spivack, J., Brennan, M., and MacKenzie, P. (1999) Contrasting carbon isotope fractionation during biodegradation of trichloroethylene and toluene: Implications for intrinsic bioremediation. *Organic Geochemistry* 30(8a), 813–820.

Skeeter, M. and Cooper, R. (1954) Photochemical process for preparing carbon tetrachloride. United States Patent No. 2,688,592, assigned to Diamond Alkali Company, September 7, 1954.

Slater, G., Dempster, H., Sherwood-Lollar, J., Brennan, M., and MacKensie, P. (1998) Isotopic tracers of degradation of dissolved chlorinated solvents. In (Wickramanayake, G. and Hinchee, R., eds), *Natural Attenuation*: *Chlorinated and Recalcitrant Compounds*, pp. 133–138. Battelle Press, Columbus, OH.

Smith, J. (1999) The determination of the age of 1,1,1-trichloroethane in groundwater. In *Proceedings of Environmental Forensics: Integrating Advanced Scientific Techniques for Unraveling Site Liability*, June 24–25, 1999. International Business Communications, Washington, DC.

Smith, J. and Eng, L. (1997) *Groundwater Sampling: A Chemist's Perspective*. Trillium Incorporated, Coatesville, PA.

Smith, J., Eng, L., and Mill, T. (1994) The determination of age of 1,1,1, trichloroethane in groundwater. Presented at the 46th Annual Meeting of the American Academy of Forensic Sciences, February 14–19, 1994. San Antonio, Texas.

Sturchio, N., Heraty, L., Holt, B., Huant, L., and Abrajano, T. (1999) Stable isotope investigations of the chlorinated aliphatic hydrocarbons. *9th Annual V. M. Goldschmidt Conference*, August 22–27, 1999. Cambridge, MA.

Tanaka, N. and Rye, D. (1991) Chlorine in the stratosphere. *Nature* 353, 707.

Tomson, M., Curran, C., King, J., Wang, J., Dauchy, J., Gordy, V., and Ward, C. (1984) Characterization of soil disposal system leachates. *Report to the US USEPA Office of Research and Development*, USEPA-600/2-84-101. Cincinnati, OH.

Vaillancourt, J. (1998) Chlorine and carbon isotopic trends during the degradation of Trichloroethylene with zero valent metal. MSc Thesis. Department of Earth Sciences, University of Waterloo.

Van Warmerdam, Frape, E., Aravena, S., Drimmie, R., Flatt, R., and Cherry, J. (1995) Stable chlorine and carbon isotope measurements of selected chlorinated organic solvents. *Applied Geochemistry* 10(5), 547–552.

Vancheeswaran, S., Hyman, M., and Semprini, L. (1999) Anaerobic biotransformation of trichlorofluoroethene in groundwater microcosms. *Environmental Science and Technology* 33(12), 2040–2045.

Vogel, T. and McCarty, P. (1987) Abiotic and biotic transformations of 1,1,1-trichloroethane under methanogenic conditions. *Environmental Science and Technology* 21, 1208–1213.

Vogel, T., Criddle, C., and McCarty, P. (1987) Transformation of halogenated aliphatic compounds. *Environmental Science and Technology* 21(8), 722–736.

Ward, J., Ahad, J., LaCrampe-Coulome, G., Slater, G., Edwards, E., and Sherwood Lollar, B. (2000) Hydrogen isotope fractionation during methanogenic degradation of toluene: Potential for direct verification of bioremediation. *Environmental Science and Technology* 34(21), 4577–4581.

Westat, Inc., and Mid West Research Institute (1987) Household solvent products: A 'shelf' survey with laboratory analysis. Prepared for the Exposure Evaluation Division, Office of Toxic Substances, Office of Pesticides and Toxic Substances, United States Environmental Protection Agency. PB88-132899.

Whitehead, D. (1968) *The Dow Story: the History of the Dow Chemical Company*. McGraw-Hill, New York.

Wolf, K. (1992) Case study: pollution prevention in the dry cleaning industry: a small business challenge for the 1990s. *Pollution Prevention Review* Summer, 311–330.

Woodbury, A. and Li, H. (1998) The Arnoldi-finite element method for solving transport of reacting solutes in porous media. In *Nonaqueous-Phase Liquids: Remediation of Chlorinated and Recalcitrant Compounds,* (Wickramanayake, G. and Hinchee, R., eds), pp. 97–106. Monterey CA. May 18–21, 1998 Battelle Press, Columbus, OH.

Zemo, D. (2000) Challenges of developing forensic interpretations using only previous existing site data. *Environmental Litigation: Advanced Forensics and Legal Strategies*. Department of Engineering Professional Development. University of Wisconsin at Madison. April 13–14, 2000. San Francisco, CA.

CHAPTER 8

SUBSURFACE MODELS USED IN ENVIRONMENTAL FORENSICS

Robert D. Morrison

8.1 INTRODUCTION

A model is a representation of reality based on defined mathematical relationships. Models simplify reality for the purpose of predicting outcomes. In environmental litigation, models are used to confirm or challenge when and where a contaminant release occurred. This chapter addresses models used for these purposes and the potential challenges and uncertainties associated with these methods.

The presentation of subsurface models used in environmental litigation is categorized according to whether the contaminant is transported through (1) the ground surface (paved and unpaved), (2) the soil and capillary fringe, and/or (3) the groundwater. This division is useful because each zone uses different mathematics that cumulatively determines the time required for a contaminant to travel from the ground surface to the groundwater.

8.2 CONTAMINANT CHEMISTRY

An understanding of the physicochemical properties of a contaminant is necessary when applying and/or evaluating environmental forensic techniques for age dating and source identification. In cases of multiple releases, a measurement of physical and chemical properties evaluated via multivariate analysis may provide a straightforward basis for source discrimination (Trevizo *et al.*, 2000). Key contaminant properties evaluated and/or measured include the Henry's law constant, liquid density, solubility, viscosity, vapor pressure, vapor density, and boiling point. Physical and chemical reactions affecting contaminant chemistry include hydrolysis, sorption (adsorption and absorption), and biodegradation. An understanding of these physicochemical properties and contaminant reactions is important for evaluating age dating and source identification techniques that rely on these processes.

8.2.1 HENRY'S LAW CONSTANT (K_H)

The *Henry's law constant* (K_H) (also called the *air–water partition coefficient*) is the ratio of a compound's partial pressure in air to the concentration of the compound in water at a given temperature. Values for Henry's law constants are expressed in units of atmospheres for air to moles per cubic meter for water (atm-m^3/mol) or in a dimensionless unit described as $K_{H'} = K_H/(RT)$ where $K_{H'}$ is the dimensionless Henry's law constant, K_H is the Henry's law constant (atm-m^3/mol), R is the ideal gas constant (8.20575×10^{-5} atm-m^3/mol-K) and T is the water temperature (K). As a rule of thumb, compounds with a Henry's law constant greater than 10^{-3} atm-m^3/mol and a molecular weight less than 200 grams per mole are considered volatile (United States Environmental Protection Agency (USEPA), 1996).

A compound with a Henry's law constant less than about 5×10^{-5} atm-m^3/mol is considered soluble and tends to remain in water (Olson and Davis, 1990). Ranges of Henry's law constant values in (atm-m^3/mol) for selected chlorinated solvents are: trichloroethylene ($1.2 \times 10^{-2} - 9.9 \times 10^{-3}$), tetrachloroethylene ($1.3 \times 10^{-2} - 2.9 \times 10^{-3}$), carbon tetrachloride ($2.3 \times 10^{-2} - 3.0 \times 10^{-2}$), methylene chloride ($2.0 \times 10^{-3} - 2.9 \times 10^{-3}$) and 1,1,1-trichloroethane ($1.3 \times 10^{-2} - 1.8 \times 10^{-2}$) (Hine and Mookerjee, 1975; USEPA, 1980; Lyman *et al.*, 1982; Roberts and Dandliker, 1983; Lincoff and Gossett, 1984; Warner *et al.*, 1987; Pankow and Rosen, 1988; Schwille, 1988; Mercer *et al.*, 1990; Montgomery, 1991).

8.2.2 LIQUID DENSITY

The *specific gravity* of a liquid is the ratio of its specific weight to distilled water at 4°C. The fluid density of most organic compounds is greater than $1\,g\,cm^{-3}$. Compounds with densities greater than 1.0, relative to water (e.g., perchloroethylene, trichloroethylene, polychlorinated biphenyls, bromoform) have a greater probability of sinking into the groundwater when introduced as a phase separate liquid. Table 8.1 lists selected contaminants, their chemical formulas and liquid densities (Montgomery, 1991; Ramamoorthy and Ramamoorthy, 1998; Morrison, 1999b).

Compared to other physical properties used in contaminant models, water density changes little. Under a pressure of 0.1 MPa, the density of water is at its maximum at 3.98°C and has a density of $1000\,kg\,m^{-3}$, which is most often assumed in contaminant transport models even though water at 25°C under atmospheric pressure has a density of $997.02\,kg\,m^{-3}$, a relative difference of 2.98% from its maximum value (Grant, 2000). The density of a contaminant as it is transported through the subsurface can change. The density of a solvent mixture, for example, can change due to the selective dissolution of the compounds into soil or rock (Parker *et al.*, 1994). In most cases, the mixture density increases.

8.2.3 VISCOSITY

Viscosity is the property of a substance that offers internal resistance to flow. The viscosity of water increases exponentially with decreasing temperature and is affected by the type and concentration of solutes (Vand, 1948). *Kinematic viscosity* is a term that describes the absolute viscosity of the substance divided by its density. A high-density liquid with a low viscosity has a low kinematic

Table 8.1

Chemical formula and liquid density for selected chlorinated solvents.

Chlorinated Solvent	Formula	Liquid Density (g/cm^3) at 20°C
1,2-Dibromomethane (EDB)[a]	$BrCH_2CH_2Br$	2.18
Tetrachloroethylene (PCE)	CCl_2CCl_2	1.63
1,1,2,2-Tetrachloroethane	$C_2H_2Cl_4$	1.60
Carbon tetrachloride (CT)	CCl_4	1.59
1,1,1,2-Tetrachloroethane	$C_2H_2Cl_4$	1.54
Trichlorofluoromethane (Freon 11)	CCl_3F	1.49
Chloroform[b]	$CHCl_3$	1.49
Trichloroethylene (TCE)	$CHCl{:}CCl_2$	1.46
1,1,2-Trichloroethane	$C_2H_3Cl_3$	1.44
1,1,1-Trichloroethane (TCA)	$C_2H_3Cl_3$	1.35
Dichloromethane (DCM)[c]	CH_2Cl_2	1.34
Dichlorodifluoromethane (Freon 12)	CCl_2F_2	1.33
cis-1,2-Dichloroethylene	$C_2\ H_2Cl_2$	1.28
trans-1,2-Dichloroethylene	$C_2H_2Cl_2$	1.26
1,2-Dichloropropane	$CH_3CHClCH_2Cl$	1.16
1,1-Dichloroethane	$C_2H_4Cl_2$	1.1
Chloroethane	C_2H_5Cl	0.92
Vinyl chloride (VC)	C_2H_3Cl	0.91

[a]Also ethylene bromide and ethylene dibromide.
[b]Also trichloromethane.
[c]Also methylene chloride.

Table 8.2

Kinematic viscosity of selected solvents.

Chemical	Kinematic Viscosity (centistokes)
1,2-Dibromomethane	0.79
1,2-Dichloropropane	0.75
1,1,1-Trichloroethane	0.62
Carbon tetrachloride	0.61
Tetrachloroethylene	0.54
Trichloroethylene	0.39
cis-1,2-Dichloroethylene	0.38
Chloroform	0.38
Methylene chloride	0.32
trans-1,2-Dichloroethylene	0.32

viscosity; such a fluid flows quickly through a porous medium compared with one with a higher kinematic viscosity.

Fluid velocity through porous media is often approximated as being inversely proportional to the kinematic viscosity. A decrease in viscosity therefore increases the velocity of a compound through porous media. The kinematic viscosity of selected chlorinated solvents is summarized in Table 8.2 (Huling and Weaver, 1991; Montgomery, 1991).

Compound	Vapor Pressure (mm at 20°C)	Vapor Density (g/L)
Freon 11	687.0	5.61
Methylene chloride	348.9	3.47
Freon 113	270.0	7.66
1,1,1-Trichloroethane (TCA)	90.0	5.45
Trichloroethene (TCE)	57.8	5.37
Tetrachloroethene (PCE)	14.0	6.86

Table 8.3

Vapor pressure and density of selected chlorinated solvents.

8.2.4 VAPOR PRESSURE AND DENSITY

The *vapor pressure* and *density* of a contaminant, especially chlorinated solvents, are important when modeling solvent transport as a vapor cloud. The vapor pressure and density of selected chlorinated solvents are summarized in Table 8.3 (Montgomery, 1991; Ramamoorthy and Ramamoorthy, 1998).

8.2.5 BOILING POINT AND LATENT HEAT OF VAPORIZATION

The *boiling point* is the temperature at which the vapor pressure of a liquid equals the atmospheric pressure and is usually reported at one atmosphere pressure (760 mm of mercury). The *latent heat of vaporization* is a measure of the energy necessary to maintain a solvent at its boiling point.

The boiling point of a compound, such as a chlorinated solvent, can be used to develop a causal relationship between a specific solvent with equipment and/or activities associated with subsurface contamination. The boiling point of several chlorinated solvents in degrees centigrade are tetrachloroethylene (PCE: 121.4), trichloroethylene (TCE: 87.2), carbon tetrachloride (76.5), trichloroethylene (TCA: 74.1), Freon 113 (47.7), methylene chloride (40.2), and Freon 11 (23.6) (Montgomery, 1991; Ramamoorthy and Ramamoorthy, 1998). Thus equipment used to vaporize PCE, for example, requires pressurized steam while other solvents can be vaporized with hot water. The higher the latent heat of vaporization, the higher the energy needed to keep the solvent at its boiling point.

8.2.6 SOLUBILITY AND NON-AQUEOUS PHASE LIQUIDS (NAPLs)

The *solubility* of a compound is its saturated concentration in water at a known temperature and pressure. In general, the higher the water solubility, the more likely it is for the compound to be mobile while being less accumulative, bio-accumulative, volatile and persistent. The solubility of several pure phase chlorinated solvents in water at 25°C is summarized in Table 8.4 (Huling and Weaver, 1991; Pankow *et al.*, 1996; Ramamoorthy and Ramamoorthy, 1998).

Table 8.4
Solubility of selected chlorinated solvents.

Compound	Literature Solubility (mg/L)
Methylene chloride (MC)	20 000
cis-1,2-Dichloroethylene	3500
1,2-Dibromomethane (EDB)	4200
trans-1,2-Dichloroethylene	6300
Trichloroethylene (TCE)	1100
1,1,1-Trichloroethane (TCA)	1300
Tetrachloroethylene (PCE)	200
Carbon tetrachloride (CTC)	825

The *effective solubility* of a multi-component solvent in water depends on the composition of the mixture. A constituent's solubility within this multi-component mixture may be orders of magnitude lower than the aqueous solubility of the pure chemical in water (Odencrantz *et al.*, 1992).

An issue when developing a conceptual contaminant transport model is whether the contaminant migrated as a non-aqueous phase liquid (NAPL). The contaminant transport model for soil (SESOIL), for example, cannot simulate NAPL transport; another example is FRACTAN, which simulates contaminant transport through fractured porous media (Sudicky and McLaren, 1992). Absent direct observation of a NAPL, several indirect methods are available. One technique is to use the concentration of a dissolved phase single-component NAPL at a threshold concentration as an indicator. In the 1980s, the rule of thumb was that if a dissolved concentration of 10% of the solubility of the compound was detected, the presence of a NAPL was inferred (Feenstra and Cherry, 1998). In the early 1990s, research indicated that concentrations of 1% or more of a compound's solubility constitute a high likelihood of the presence of a NAPL (USEPA, 1992, 1993; Cohen *et al.*, 1993). The current convention is consistent with the 1% rule (Newell and Ross, 1991; Pankow *et al.*, 1996). An examination of 65 plumes with TCE indicated that about 40% are associated with a dense non-aqueous phase liquid (DNAPL) based on a 1% solubility limit while 10% of the plumes are associated with a DNAPL based on the 10% solubility rule-of-thumb. The authors concluded that based on these solubility conventions, DNAPL was suggested in cases where a TCE concentration of 1000 parts per billion (ppb) is detected (McNab *et al.*, 1999).

Other tools available to determine whether a NAPL is indicated include direct measurements or observations and field tests, an example is the addition of a hydrophobic dye powder such as Sudan IV to a soil sample. A method relying upon direct measurements is proposed by Murphy (1999), who assumed that if soil properties and contaminant concentrations are known and equilibrium partitioning is assumed, the presence of a NAPL can be determined.

An illustration of this approach is to use PCE, TCE and TCA and assume a soil with a total porosity of 40%, a water-filled porosity of 15%, a soil organic content ranging from 0.1 to 2%, and a soil dry bulk density of 1.59 g/cm^3. Given this information, PCE, TCE and TCA concentrations indicative of a NAPL are in the range 117–744, 302–1300 and 366–1408 mg/kg, respectively. When the solvent concentration exceeds these values, the contaminant mass in the soil exceeds the capacity of the soil, thereby indicating the presence of a NAPL. If the concentration is significantly less than the estimated concentration range, there is an indication that the contaminant did not enter the soil as a NAPL.

Soil vapor concentrations may provide another means for identifying the presence of a NAPL. A soil gas concentration of 100 parts per million (ppm) by volume is commonly considered to be indicative of the presence of NAPL. Another field technique for NAPL identification is the partitioning tracer method originally developed to measure the residual oil in the subsurface by petroleum engineers (Nelson *et al.*, 1999). The partitioning tracer method compares the transport of one or more partitioning tracers, such as calcium bromide, with a non-reactive tracer. Given that conservative tracers have a partition coefficient of zero relative to a NAPL while partitioning tracers have a non-zero partition coefficient relative to the NAPL, judgments regarding the presence of a NAPL between the injection and recovery wells are possible (Deeds *et al.*, 2000). The retardation and residual NAPL saturation is inferred from breakthrough curves at a monitoring well downgradient of the injected tracer if the NAPL/water partition coefficient for the partitioning tracer is known (Jin *et al.*, 1995, 1997; Annable *et al.*, 1998).

Measurement of naturally occurring radon-222 in groundwater is proposed as a naturally occurring tracer for detecting and quantifying NAPL contamination (Hunkeler *et al.*, 1997; Semprini *et al.*, 2000). In the absence of a NAPL, radon concentrations in groundwater reach a site-specific equilibrium value based on the mineral fraction of the aquifer solids. In the presence of NAPL, the radon concentrations are preferentially reduced due to the preferential partitioning of radon in the organic NAPL phase. At the Borden test site in Ontario, Canada, a mixture of chloroform, TCE and PCE was injected into the shallow sand aquifer. Groundwater samples were collected from monitoring wells upgradient, within and downgradient of the NAPL and analyzed for radon (Feenstra and Cherry, 1998). The radon concentrations in the NAPL zone decreased by a factor of 2 to 3 times compared to values observed in upgradient and downgradient monitoring well samples. Precautions when using the radon method include the significant impact of heterogeneities in aquifer porosity and radon emanations. In the absence of a NAPL, a decrease in porosity results in higher radon concentrations in the pore fluid, thus complicating data interpretation.

Geophysical techniques used to identify NAPLs include electrical impedance tomography (EIT) (Daily *et al.*, 1998), seismic reflection profiling (Waddell *et al.*, 1996), cross radar technology (Wright *et al.*, 1998), negative ion sensors (USEPA, 1998) and complex resistivity techniques. Geophysical techniques when combined with other methods can be applied to provide a higher confidence level regarding the presence of a NAPL. At Hill Air Force Base in Ogden, Utah, for example, electrical impedance tomography was used with fiber optic chemical sensors and neutron logs to verify the presence of NAPLs (USEPA, 1998).

8.3 PROCESSES AFFECTING CONTAMINANT CHEMISTRY

8.3.1 HYDROLYSIS

Hydrolysis is the chemical breakdown of substances by water and depends on the chemistry, solubility, pH, and the oxidation–reduction (redox) potential of compound. Hydrolysis rates in soil may be different than rates reported in water, depending on the effects of pH, redox, sorption and surface catalyzed reactions (Dilling *et al.*, 1975; Mabey and Mill, 1978).

8.3.2 SORPTION

Retardation values for a compound used in contaminant transport models and can exercise a significant impact on the results of the simulations. Retardation is a generic term that includes the sorption of a compound to the material through which it migrates. Sorption encompasses adsorption, absorption, ion exchange, and chemisorption. Absorption (the penetration of substances into the bulk of a solid or liquid) and adsorption (the surface retention of a solid, liquid, or gas molecules by a solid or liquid) are the most important of these sorption processes. The sorption capacity of a compound is described by its *sorption coefficient.* The sorption coefficient is the ratio of an adsorbed chemical per unit weight of organic carbon to its concentration in water. The sorption coefficient is usually estimated via batch isotherm experiments. Predicted correlations are also available that relate the distribution coefficient of a compound to the soil organic content, as expressed by the octanol–water partition coefficient (Fetter, 1994). In general, predicted retardation coefficients are two to five times lower than measured values (Ball and Roberts, 1991).

Retardation values generally increase with increasing fractions of organic carbon and/or clay. A range of retardation values for PCE in a sand and gravel aquifer, for example, is between 1 (no retardation) and 5 (Schwarzenbach *et al.*, 1983; Barber *et al.*, 1988). The retardation value for TCE is reported as less than 10; values between 1 and 2.5 are commonly used in contaminant transport models.

Apparent retardation values can be biased as a function of well design and sampling technique. Apparent retardation rates in one study were found to be inconsistent between monitoring wells depending on the saturated screen length, the degree of screen desaturation during purging and the longitudinal distance from the contaminant source (Robbins, 1989; Martin-Hayden and Robbins, 1997). The selection of a retardation value for an entire well field may therefore be inappropriate due to differences in well construction. A single retardation value may similarly be overly simplifying in cases where concentration averaging is used.

8.3.3 BIODEGRADATION

Biodegradation refers to the transformation of compounds, principally by bacteria, fungi and yeast, into simpler substances. Bacteriological or reductive dechlorination is the principal process resulting in the degradation of chlorinated solvents in the subsurface (Gao *et al.*, 1995). For most biodegradation pathways, the final degradation products are carbon dioxide and water. Figure 8.1 depicts reductive dechlorination degradation pathways for PCE, TCA and carbon tetrachloride (after Vogel *et al.*, 1987a,b; McCarty, 1993, 1994; Pankow *et al.*, 1996; Butler and Hayes, 1999; Jones, 2000; Morrison, 2000c).

As shown in Figure 8.1, the dichloroethene (DCE) isomers for TCE degradation include 1,1-DCE, *cis*-1,2-DCE and *trans*-1,2-DCE. Of these isomers, *cis*-1,2-DCE is produced in the greatest abundance at a rate of about 30 times that of *trans*-1,2-DCE. The decay rate from TCE to *cis*-1,2-DCE is also faster than

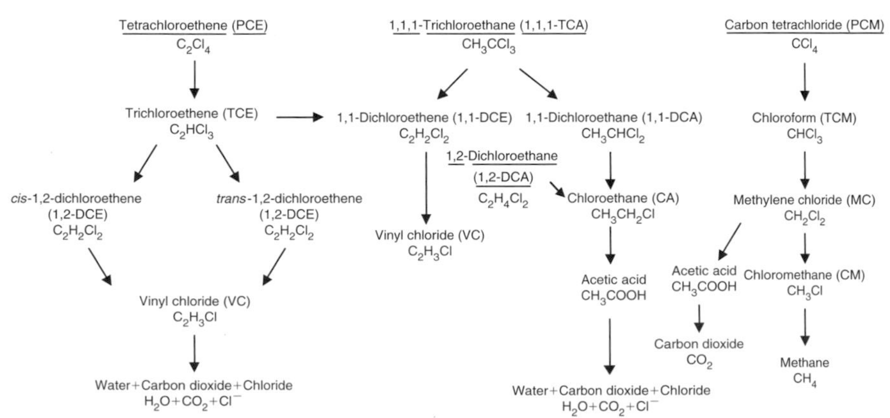

Figure 8.1
Common degradation pathways of PCE, TCA, 1,2-DCA and carbon tetrachloride.

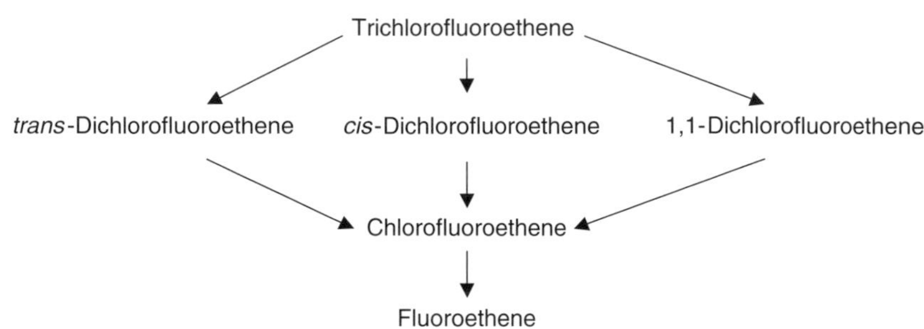

Figure 8.2
Reductive dechlorination pathway of trichlorofluoroethene (TCFE) (after Vancheeswaran et al., *1999).*

from TCE to 1,1-DCE. The decay rate from *cis*-1,2-DCE to vinyl chloride (chloroethene), however, is slower than from *trans*-1,2-DCE and 1,1-DCE to vinyl chloride. Under some conditions, vinyl chloride is degraded to acetate.

Another example of reductive dechlorination is the degradation pathway of trichlorofluoroethene (TCFE) (Figure 8.2). Similarities in the degradation kinetics of TCFE, TCE and PCE, when present at comparable initial concentrations, offer the opportunity of using TCFE and its degradation products as benign reactive tracers to measure *in situ* degradation rates of PCE and TCE. This relationship may provide evidence in support of a chlorinated solvent degradation model used for age dating and/or source identification (see Chapter 7).

Anaerobic degradation occurs via dehalogenation with the release of a chlorine atom. The rate of reductive dehalogenation decreases as more chlorine is removed (Woodbury and Ulrych, 1998). Strong reducing conditions with low dissolved oxygen levels are conducive to dehalogenation.

Aerobic degradation is initiated by enzymes called oxygenases (Tsien *et al.*, 1989). These enzymes initiate degradation after producing a specific organic compound that serves as a carbon and energy source for the bacteria. The rate and ability of microbes to degrade a compound are dependent on the ability of the subsurface environment to support a healthy community of microbes. The necessary nutrients, oxygen content and soil and/or groundwater temperature required to sustain viable microbial communities are highly variable. Biodegradation may not occur if the concentration of the compound is low, although many can be degraded to some extent.

A recent contaminant of concern in the United States is methyl tertiary butyl ether (MTBE). The interpretation and use of MTBE for forensic purposes requires careful analysis especially given the opportunities for cross-contamination. Examples of cross-contamination of MTBE with a non-MTBE fuel can occur during shipping and storage in pipelines, tankers, above-ground storage tanks, and trucks. MTBE has been detected in the presence of jet fuel, diesel fuel, heating oil, aviation gas and waste oil, presumably due to cross-contamination (Hitzig *et al.*, 1998). In Connecticut, 27 of 37 heating oil spill

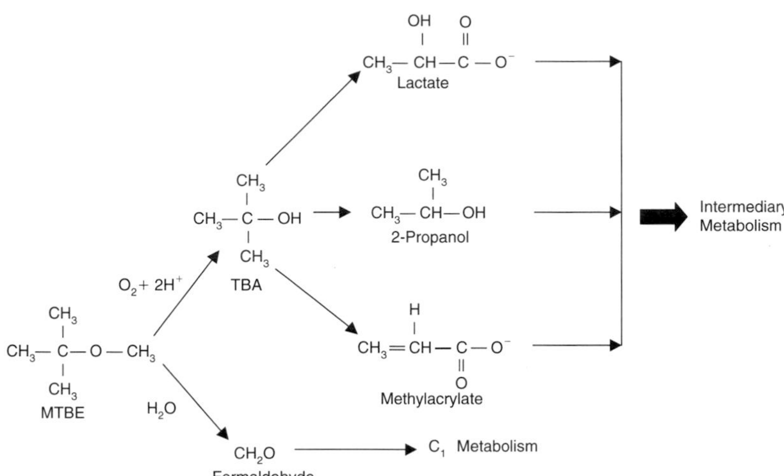

Figure 8.3

Simplified methyl tertiary butyl ether (MTBE) oxidation pathway (after Ramsden, 2000; Drogos, 2000; Jacobs et al.*, 2000).*

sites detected MTBE in the groundwater at concentrations from 1 to 4100 μg/L (Davidson, 2000). An example illustrating the difficulty in MTBE evaluation for age dating and source identification is the results from a Washington Department of Ecology study that detected MTBE in groundwater at about half of 62 leaking underground storage sites, although MTBE is not used as an oxygenate in Washington (Washington State Department of Ecology, 2000).

Biodegradation studies of MTBE in shallow aquifers indicate that MTBE is biodegradable under anaerobic and aerobic environments although it is considered to be inhibited in the presence of BTEX (benzene, toluene, ethylbenzene, and xylene) compounds (Mormile *et al.*, 1994; Landmeyer *et al.*, 1998; Ramsden, 2000). One proposed MTBE degradation pathway is shown in Figure 8.3.

While the author has not encountered a case where MTBE and degradation product ratios are used for age dating and/or source allocation, a greater understanding of the kinetics associated with these pathways may provide a means for developing this argument. A potential difficulty is that while tertiary butyl alcohol (TBA) is a primary metabolite of MTBE via atmospheric and microbial oxidation, it is also a by-product of MTBE synthesis and is often present as an impurity (~0.1–1%) (Drogos, 2000). TBA may also be added to MTBE up to 5% by volume (Brown, 2000). The detection of TBA with MTBE may therefore be independent of its degradation relationship with MTBE (Wilson *et al.*, 2000).

8.4 PAVED SURFACES

Models describing liquid transport through paved surfaces are highly specific to the chemical and physical composition of the paved surface and the physicochemical properties of the contaminant. In order to identify the most likely contaminant transport mechanism (e.g., liquid advection, gas diffusion, liquid diffusion and/or evaporation) to be simulated, a realistic physical and mathematical model is required. The most commonly encountered models through paved surfaces in environmental litigation are liquid and vapor transport models.

A frequent area of inquiry is whether a liquid migrated through a paved or compacted surface (e.g., asphalt, concrete, crushed rock or compacted soil), and if so, the time required (Morrison, 2000a). Absent direct measurements, contaminant transport equations are used to predict a range of values describing when the liquid exited the surface.

8.4.1 LIQUID TRANSPORT THROUGH PAVEMENT

Liquid transport through a paved surface is commonly believed to be a rapid process. This assumption is true if the pavement is cracked allowing unrestricted flow through the pavement or if the spill occurs over an expansion/control or isolation joint filled with permeable wood, oakum or tar. Expansion joints are usually located at the junction of floors and walls, foundation columns, and footings. Isolation joints include pre-molded joint fillers (Kosmatka and Panarese, 1988) and extend the full depth of the slab. Given the sorptivity of expansion and isolation joint materials, testing these materials can establish whether a contaminant was preferentially transported via this pathway.

Absent direct measurements, the presence of cracks and/or the sampling and testing of expansion joint materials, a model is used to estimate the time required for a liquid to permeate a paved surface. Ideally, information available to conceptually formulate a model includes:

- the temporal nature of the release (steady state or transient);
- the saturated and/or unsaturated hydraulic conductivity values of the pavement;
- the physical properties of the contaminant (density, viscosity, vapor pressure);
- the chemical properties of the liquid (pure phase, mixed or dissolved in another fluid);
- air temperature at the time of the release;
- volume released; and
- evaporative flux.

A knowledge of the circumstances of a contaminant release and pavement composition is essential so that reasonable assumptions regarding the release are incorporated into the model. For example, if the model does not account for evaporation and/or assumes that the liquid thickness on the pavement is constant, the transport rate can be overestimated. If cleanup activities are performed coincident with the release (e.g., sawdust, green sand, absorbent socks, Sorball, crushed clay, pumping, etc.) or if the spill occurred in a building with forced air, these competing activities will result in less liquid available for transport through the pavement and hence a slower transport rate. Additional examples of information useful in developing the conceptual model include whether the surface is treated with an epoxy coating or specialty coating to prevent corrosion from acid releases (common in plating shops and dry cleaners), or for a concrete pavement, whether the concrete was mixed with an additive to reduce its permeability to chemicals (e.g., Dow Latex No. 560), pavement thickness, porosity, composition and slope, and the nature of the surface prior to the release (i.e., was it impregnated with oils and dirt, smooth vs. pitted, sloped toward a drain, etc.).

The liquid thickness on the pavement and the duration of time that the liquid is in contact with the pavement are key variables. If liquid containing TCE is released onto a paved surface on a warm sunny day or in a building with forced air, evaporation is rapid. As a consequence little liquid is available to migrate into the pavement. A description for the evaporative loss associated with a release of a compound for a semi-infinite region with a uniform initial concentration and mass transfer at the surface is given by (Choy and Reible, 1999):

$$\partial c_A/\partial t = (D_{A(\text{eff})}/R_f)\partial^2 c_A/\partial z^2 \qquad z \in [0,\infty) \tag{8.1}$$

$$-D_{A(\text{eff})}\partial c_A/\partial z\Big|_{z=0} + (k_a)c_A(z,t)\Big|_{z=0} = 0 \qquad t > 0 \tag{8.2}$$

$$c_A(z,t)\Big|_{z\to\infty} = c_{A0} \qquad t > 0 \tag{8.3}$$

$$c_A(z,t)\Big|_{t=0} = c_{A0} \qquad z \in [0,\infty) \tag{8.4}$$

The concentration profile is given by:

$$\begin{aligned} c_A(z,t) = c_{A0}\Big\{&\text{erf}[R_f z/(4D_{A(\text{eff})}R_f t)^{1/2}] \\ &+ \exp(k_a z/D_{A(\text{eff})} + k_a^2 t/D_{A(\text{eff})}R_f)\,\text{erfc}[R_f z/(4D_{A(\text{eff})}R_f t)^{1/2} \\ &+ k_a(t/D_{A(\text{eff})}R_f)^{1/2}]\Big\} \qquad z \in [0,\infty),\; t > 0 \end{aligned} \tag{8.5}$$

The surface flux out of the system is described by:

$$j_A(t)\Big|_{z=0} = (k_a)(c_{A0})\exp(k_a^2 t/D_{A(eff)}R_f)\operatorname{erfc}[k_a(t/D_{A(eff)}R_f)^{1/2}] \quad t > 0 \tag{8.6}$$

where k_a is the surface mass transfer coefficient; c_A is the concentration of species A in the air phase per unit volume; $D_{A(eff)}$ is the effective diffusion coefficient of species A; R_f is the retardation factor; j_A is the mass flux of the component; z is depth of position; and t is time.

Conversely, if TCE accumulates in a blind concrete sump/neutralization pit or clarifier, the TCE may reside for a sufficient period to allow for the TCE to migrate through the concrete.

Numerous equations and models can be modified to calculate the rate of transport of a liquid through pavement (Ghadiri and Rose, 1992). For saturated flow, a simplified one-dimensional expression for the vertical transport of a liquid is Darcy's law where outflow (q) in time (t) from a paved column with an area (α) and thickness (l) subjected to a head difference (Δh) and the saturated hydraulic conductivity ($K_{saturated}$) is equal to:

$$q/t = [K_{saturated}\,\alpha/l](\Delta h) \tag{8.7}$$

This expression defines the downward velocity (v) of the liquid as equal to the downward flux (q/t) divided by the area α and by the porosity of the pavement. The downward flux is the saturated hydraulic conductivity multiplied by the vertical gradient. Dividing the velocity into the pavement thickness gives the transport time. Saturated hydraulic conductivity and porosity values for paved materials are measured directly or estimated from published values.

Pavement transport models using Darcy's law assume that the pavement is saturated with liquid prior to the release. If the pavement is unsaturated, liquid transport may be dominated by unsaturated flow resulting in contaminant velocities several times slower than for saturated flow. In such cases, the use of unsaturated zone equations such as the Richards equation, may be more appropriate (Richards, 1931; Hillel, 1980). In order to develop the relationship between moisture content, capillary potential and corresponding unsaturated hydraulic conductivity values, a soil moisture characteristic curve is required which is the functional relationship between moisture content and capillary potential. The wetting fluid used to create a soil moisture characteristic curve for the concrete is ideally similar to the released liquid so that the extent of repulsion and/or retardation due to the hydrophobic or hydrophilic properties of the liquid is measured.

8.4.2 VAPOR TRANSPORT THROUGH PAVEMENT

The transport of a vapor cloud through a paved surface is a common area of inquiry, especially with vapor degreasers. Vapor degreasers and dry cleaning equipment are often installed in a concrete catch basin or trough. Catch basins can exacerbate the potential for vapor transport because they accumulate the vapor and minimize vapor dilution with the ambient air.

If both vapor and liquid releases are probable, it may be difficult to determine the more realistic transport scenario. Existing test results may provide this information. One example indicative of vapor transport is the detection of a compound in soil vapor that is absent in the soil. An example is the detection of MTBE in soil vapor that is absent in soil at the same location but is present in the underlying groundwater.

Vapor diffusion through pavement can be more rapid than liquid transport. Ideally, the following information is available for developing a vapor transport model through a paved surface:

- vapor density of the primary compound and/or liquid mixture;
- whether the vapor source is constant or transient above the pavement;
- the Henry's law constant for the contaminant;
- a measured gas permeability of the pavement (American Petroleum Institute Method RP40);
- pavement thickness, porosity and moisture content;
- vapor concentration above the pavement, and
- the vapor concentration within and below the pavement prior to the spill.

Vapor density is a key variable and is approximately equal to the molecular weight (MW) of the fluid divided by the molecular weight of air (≈29). The vapor density of common chlorinated solvents relative to air is TCA (4.6), TCE (4.5), PCE (5.7) and vinyl chloride (2.1) (Montgomery, 1991; Pankow *et al.*, 1996).

Numerous equations are available to estimate the velocity of a vapor through pavement (Crank, 1985; McCoy and Rolston, 1992) including the following expression for the unsteady, diffusive radial flow of vapor from a source (Cohen *et al.*, 1993):

$$\partial^2 C_a/\partial r^2 + [1/r(\partial C_a/\partial r)] = (R_a/D^*)(\partial C_a/\partial t) \tag{8.8}$$

where the air filled porosity (n_a) of the pavement is assumed to be constant, R_a is the vapor retardation coefficient, C_a is the computed concentration of the vapor in air, r is the source radius, and D^* is the effective diffusion coefficient (for TCE = $3.2 \times 10^{-6}\,m^2$ per second and $0.072\,cm^2$ per second for PCE) (Millington and Quirk, 1959; Lyman *et al.*, 1982) that is equal to:

$$D^* = D\tau_a \tag{8.9}$$

where $\tau_a = n_a^{3.333}/n_t^2$ and n_t^2 is the total pavement porosity, which is the sum of the air filled porosity and the volumetric water content. The pavement vapor retardation factor (R_a) in Equation 8.8 is equal to:

$$R_a = 1 + n_w/(n_a K_H) + \rho_b K_d/(n_a K_H) \tag{8.10}$$

where n_w is the bulk water content, n_a is the air-filled pavement porosity, ρ_b is the pavement bulk density, K_d is the distribution coefficient and K_H is the dimensionless Henry's law constant.

Another conceptual option is to consider vapor diffusion through a medium (concrete) bounded by two parallel plates. This approach assumes a pavement thickness (l) and a diffusion coefficient (D) whose surfaces $x = 0$ and $x = l$ are maintained at a constant concentration specified as C_1 and C_2, respectively.

After a time, steady state conditions are reached and concentrations are constant at all locations in the pavement. The diffusion equation in one dimension reduces to (Crank, 1985):

$$d^2C/dx^2 = 0 \tag{8.11}$$

which assumes that the diffusion coefficient (D) is constant. On integrating with respect to x, the following expression arises:

$$dC/dx = \text{constant} \tag{8.12}$$

and by introducing the conditions at $x = 0$, $x = l$, and integrating, then:

$$[C - C_1/C_2 - C_1] = x/l \tag{8.13}$$

Equation 8.13 describes a linear concentration change from C_2 to C_1 through the pavement. The rate of diffusion (F) through the pavement is assumed to be the same and is described by:

$$F = -D\,dC/dx = -D(C_2 - C_1)/l \tag{8.14}$$

If the pavement thickness (l) and the surface concentrations C_1 and C_2 are known, D is deduced from an observed value of F with Equation 8.14. A value for the diffusion coefficient can also be measured experimentally.

If the surface $x = 0$ is maintained at a constant concentration C_1 and at $x = l$, evaporation into the atmosphere is assumed for which the equilibrium concentration immediately within the paved surface is C_2, then with a mass transfer coefficient h:

$$\mathrm{d}C/\mathrm{d}x + h(C - C_2) = 0, \quad x = l \tag{8.15}$$

and

$$(C - C_1)/(C_2 - C_1) = (hx)/(1 + hl) \tag{8.16}$$

and

$$F = Dh(C_1 - C_2)/(1 + hl) \tag{8.17}$$

If the surface conditions of the pavement are:

$$\mathrm{d}C/\mathrm{d}x + h_1(C_1 - C) = 0, \ \ x = 0 \quad \text{and} \quad \mathrm{d}C/\mathrm{d}x + h_2(C - C_2) = 0, \ \ x = l \tag{8.18}$$

then

$$C = [h_1C_1\{1 + h_2(l - x)\} + h_2C_2(1 + h_1x)]/[h_1 + h_2 + h_1h_2l] \tag{8.19}$$

and

$$F = Dh_1h_2(C_1 - C_2)/[h_1 + h_2 + h_1h_2l] \tag{8.20}$$

8.4.3 CHALLENGES TO LIQUID AND VAPOR TRANSPORT THROUGH PAVEMENT

While the most effective challenges to liquid or vapor transport models through pavement are site specific, generic categories include (Morrison, 2000b):

- variability of the model parameters (e.g., are they measured or reasonably assumed?);
- the accuracy of the known circumstances of the spill event(s);
- the environmental conditions at the time of the release(s) (e.g., thickness of the liquid, composition of the liquid, duration of the release, ambient air temperature, etc.);
- consistency of the modeled results with measured contaminant concentrations under the paved surface (if available);
- the impact of potential short circuiting pathways such as expansion joints and/or cracks on the conceptual model; and

- if a physical sample of the paved surface is obtained and the physical properties tested, the consistency of the input parameters used in the model with these direct measurements.

8.5 CONTAMINANT TRANSPORT IN SOIL

A frequent inquiry in environmental litigation is when did a contaminant enter the groundwater or migrate onto another property and/or whether the observed contaminant plume is consistent with an alleged source or release date. Absent direct measurements or testimony regarding the timing of a release, transport equations are relied upon to develop reasonable estimates. Equations for contaminant migration through the underlying soil are available that range in complexity from one-dimensional infiltration to three-dimensional multiphase flow equations (Ghadiri and Rose, 1992; Hughes *et al.*, 1992; Sanders, 1997; Selim and Ma, 1998). A determination is required prior to model selection to identify whether the fluid was transported as a vapor or as a liquid (e.g., saturated or unsaturated flow) through the soil column.

8.5.1 VAPOR TRANSPORT

Vapor is transported in soil via advective or gaseous diffusion. Advective transport is especially important near the ground surface due to atmospheric pressure variations or near buildings that create pressure gradients due to differential heating. Gaseous diffusion describes transport via molecular processes and is the primary transport mechanism for contaminant migration through the soil. In 1855, Adolf Fick described gaseous diffusion as $F = -D(\mathrm{d}C/\mathrm{d}x)$ where F is the mass flux, D is the diffusion coefficient and $\mathrm{d}C/\mathrm{d}x$ is the concentration gradient. This expression is known as Fick's first law (Fick, 1855).

For contaminant vapor transport through soil, the effective diffusion coefficient is the vapor diffusion coefficient corrected for soil porosity. For many vapors, the diffusion coefficient is approximately $0.1\,\mathrm{cm}^2$ per second. A conservative approximation is that porosity reduces vapor diffusivity by a factor of 10. Thus for vapors, the effective diffusion coefficient D_e can be approximated as equal to about $0.01\,\mathrm{cm}^2$ per second. A first order approximation illustrating the distances that a contaminant can travel via vapor diffusion through soil in one year (assuming no adsorption) is:

$$D = [(2)(0.01\,\mathrm{cm}^2\mathrm{s}^{-1})(31\,536\,000\,\mathrm{s})]^{1/2} \approx 800\,\mathrm{cm} \approx 25\,\mathrm{feet} \qquad (8.21)$$

More complex expressions for vapor transport are available in the literature (Carslaw and Jaeger, 1959; Crank, 1985; Ozisk, 1993).

8.5.2 LIQUID TRANSPORT

Numerous mathematical descriptions and approaches are available to estimate the arrival time of a contaminant to the groundwater. Contaminant transport is generally more rapid in coarse as contrasted with fine grained soils although parameters such as the degree of soil compaction/cementation, unsaturated versus saturated flow, and soil organic matter content can present exceptions to this generalization.

A variety of approaches is available for estimating whether a contaminant release on the ground surface entered the groundwater and if so, when. An example of a qualitative approach for estimating if a liquid from an underground storage tank has reached the groundwater is available if the soil characteristics, liquid properties and release circumstance are known. For petroleum hydrocarbons migrating through the unsaturated zone, the depth of penetration is estimated by (Dragun, 1988):

$$D = \mathrm{K}V_{\mathrm{HC}}/A \tag{8.22}$$

where D is the maximum depth of penetration; K is a constant dependent upon the soil's retention capacity for oil and oil viscosity; V_{HC} is volume of infiltrating petroleum hydrocarbon, and A is the area of the spill. Typical values for K are shown in Table 8.5 (USEPA, 1987).

Models used to model the transport of a contaminant through the soil column all rely upon similar mathematics. A common one-dimensional equation describing liquid transport via advection and diffusion in soil is (Jury *et al.*, 1986; Jury and Roth, 1990; Jury and Fluhler, 1992):

$$R_{\mathrm{l}} \partial C_{\mathrm{l}}/\partial t = D_{\mathrm{e}} \partial^2 C_{\mathrm{l}}/\partial z^2 - V \partial C_{\mathrm{l}}/\partial z - \lambda_{\mu} R_{\mathrm{l}} C_{\mathrm{l}} \tag{8.23}$$

where C_{l} is the pore water concentration in the vadose zone; λ_{μ} is the decay constant; R_{l} is the liquid retardation coefficient; D_{e} is the effective diffusion coefficient; and, V is the infiltration rate.

The retardation coefficient (R_{l}) is estimated by:

$$R_{\mathrm{l}} = 1 + \frac{\rho_{\mathrm{b}} K}{\theta} + \left(\frac{\phi - \theta}{\theta}\right) K_{\mathrm{H}} \tag{8.24}$$

where ρ_{b} is the soil bulk density (American Society of Testing Materials Method D3550), K is the distribution coefficient for the contaminant, ϕ is soil porosity, θ is the soil moisture content (American Society of Testing Materials Method D2216), and $\mathrm{K_H}$ is the Henry's constant for the contaminant.

Table 8.5

Typical values for K for various soil textures for three petroleum hydrocarbons (after Dragun, 1988).

Soil Texture	**K**		
	Gasoline	**Kerosene**	**Light fuel oil**
Coarse gravel	400	200	100
Gravel to coarse sand	250	125	62
Coarse to medium sand	130	66	33
Medium to fine sand	80	40	20
Fine sand to silt	50	25	12

For most soils, adsorption occurs on the organic carbon particles; the organic carbon partition coefficient (K_{ow}) for organic chemicals is either measured or estimated. The partition coefficient is estimated by (Lyman *et al.*, 1982):

$$K_d = 0.6 f_{oc} K_{ow} \tag{8.25}$$

where f_{oc} is the fraction of organic carbon in the soil; and K_{ow} is the octanol-partition coefficient of the contaminant of interest.

The degradation rate constant is estimated by:

$$\lambda_\mu = \ln(2)/T_{1/2} \tag{8.26}$$

where $T_{1/2}$ is the degradation half-life of the contaminant of interest. The effective diffusion coefficient is:

$$D_e = \tau_L D_{LM} + K_H \tau_G D_{GM} \tag{8.27}$$

where D_{LM} is the molecular diffusion coefficient in water; τ_L is soil tortuosity to water diffusion; D_{GM} is molecular diffusion coefficient in air; and τ_G is soil tortuosity to air diffusion.

The tortuosity of a compound in water and air is described by (Millington and Quirk, 1959):

$$\tau_L = \upsilon^{10/3}/\phi^2 \quad \text{and} \quad \tau_G = (\phi - \upsilon)^{10/3}/\phi^2 \tag{8.28}$$

where υ is the air-filled porosity. The selection of a contaminant transport model has a profound effect on the estimated time required for a contaminant to reach groundwater, especially since hundreds of models are available (Ghadiri and Rose, 1992; Selim and Ma, 1998). To illustrate the impact of model selection on the simulated results, four contaminant transport models (PESTAN, WAVE, SESOIL, and VLEACH) are considered to illustrate these differences. Similar model comparisons are described in the literature (Oster, 1982; Melancon *et al.*, 1986).

PESTAN (Pesticide Analytical Model) is a one-dimensional model that simulates aqueous phase transport and transformations under steady state flow conditions in a homogeneous soil (Enfield *et al.*, 1982). Solutions are provided for (1) linear sorption and first order decay without dispersion, (2) dispersion and linear sorption without decay, and (3) adsorption using the Freundlich isotherm with first order decay but without dispersion. Solutions 1 and 2 are solved analytically while solution 3 is solved numerically. PESTAN is based on the dispersive-convective differential mass transport equation described by (Bonazountas *et al.*, 1997):

$$\mathrm{d}(\theta c)/\mathrm{d}t = \mathrm{d}/\mathrm{d}z(\theta K_\mathrm{d}\mathrm{d}c/\mathrm{d}z) - \mathrm{d}(vc)/\mathrm{d}z - \rho\,\mathrm{d}s/\mathrm{d}t \pm \sum P \qquad (8.29)$$

where θ is the soil moisture content, c is the dissolved contaminant concentration in the soil, K_d is the apparent diffusion coefficient of the contaminant in soil air, v is the Darcy velocity of the soil moisture, ρ is soil density, s is the adsorbed concentration of the compound on soil particles, $\sum P$ is the sum of sources or sinks of the contaminant within the soil, and z is soil depth. PESTAN has been tested in the laboratory and field (Jones and Black, 1984). The United States Environmental Protection Agency Office of Pesticide Programs, for example, used it to model aldicarb transport in a Florida soil. Donigian and Rao (1986a,b) compared these predictions with aldicarb data reported by Hornsby *et al.* (1983) and found a reasonable agreement between the predicted and measured data.

Model assumptions include: steady state flow conditions, all infiltrated water goes to recharge, the soil is a single, homogeneous layer, contaminant degradation occurs at the ground surface, hydrolysis occurs within the soil column, the rate of liquid-phase degradation does not change with time or depth, and once water enters the soil as a slug, sorption and dispersion influence pesticide transport (Ravi and Johnson, 2000). Input data includes water solubility, an infiltration rate, soil bulk density, a sorption constant, degradation rates, saturated hydraulic conductivity, and a dispersion coefficient.

WAVE (Water and Agrochemicals in the Soil, Crop and Vadose Zone) was developed for the Land and Water Management of the Catholic University of Leuven, Belgium. WAVE is a one-dimensional deterministic model using the finite difference technique to solve differential equations (Vanclooster *et al.*, 1999). WAVE is based on the following expression of the Richards equation:

$$\partial\theta/\partial t = W(h)\partial h/\partial t = \partial[K(h)(\partial h/\partial x + 1)]/\partial x - A(h) \qquad (8.30)$$

where W is the soil water capacity ($\partial\theta/\partial h$), K is the hydraulic conductivity, h is the soil water hydraulic head and A is the soil water abstraction rate. The soil moisture retention characteristic curve is described using the empirical function of Van Genuchten and Alves (1982):

$$\theta = \theta_\mathrm{r} + (\theta_\mathrm{s} - \theta_\mathrm{r})/(1 + |\alpha h|^\mathrm{n})^\mathrm{m} \qquad (8.31)$$

where θ_s is the saturated water content, θ_r is the residual water content, α, n and m are empirical shape factors that account for the hysteresis associated with the wetting and drying curve. The unsaturated hydraulic conductivity factor, K, used in Equation 8.32 is based on the predictive model of Muralem (1976) as described by:

$$K = K_s \phi_s^{\lambda} [1 - (1 - \phi_s^{1/m})^m]^2 \qquad (8.32)$$

where ϕ_s is the relative saturation, K_s is the saturated hydraulic conductivity and λ is a shape parameter. WAVE assumes the existence of immobile or stagnant soil water regions situated at the terminus of soil pores and that contaminant adsorption occurs reversibly and linearly. Contaminant transport is determined by chemical diffusion, convection, and hydrodynamic dispersion. Volatilization is not considered.

In the United States, two commercially available models commonly used in environmental litigation are SESOIL (Seasonal Soil Compartment Model) (Odencrantz *et al.*, 1992; Bonazountas *et al.*, 1997) and VLEACH (One-dimensional Finite-difference Vadose Zone Leaching Model) (Rosenbloom *et al.*, 1993). Given that these models are frequently encountered, each is briefly discussed.

SESOIL was developed as a risk screening-level model that uses soil, chemical and meteorological input values. SESOIL is a one-dimensional vertical transport code for the unsaturated zone and simultaneously models water transport and contaminant sediment transport. SESOIL can simulate contaminant fate and migration on a monthly basis and can estimate the average concentration of the contaminant in the underlying groundwater. The soil column can be divided in up to four layers with each layer possessing different soil properties. Each soil layer can in turn be subdivided into a maximum of 10 sublayers. Table 8.6 lists some of the climatic, soil and chemical input data used in SESOIL (General Sciences Corporation, 1998).

SESOIL uses the following expression to calculate the depth reached by a chemical with a linear equilibrium partitioning between its vapor, liquid, and adsorbed phases (Jury *et al.*, 1984):

$$D = (J_w t_c)/[(\theta + \rho_b K_d) + (f_a H)/R(T + 273)] \qquad (8.33)$$

where D is depth; J_w is water velocity; t_c is convective time; θ is soil water content; ρ_b is soil bulk density; K_d is contaminant partitioning coefficient; f_a is air filled porosity; H is Henry's law constant; R is universal gas constant; and T is soil temperature (°C).

SESOIL is sensitive to the saturated hydraulic conductivity value(s) used and the organic matter content. The selection of the organic matter content and a

Table 8.6

Key input parameters for SESOIL.

Data Category	Parameter Information
Climatic data	Air temperature; cloud cover fraction; relative humidity; short wave albedo; rainfall depth; mean storm duration; number of storms per month; length of rainy season within a month.
Soil	Bulk density; intrinsic permeability; soil disconnectedness index; effective porosity; organic carbon content; cation exchange coefficient; Freundlich equation exponent.
Chemical	Solubility in water, diffusion coefficient in air, Henry's law constant, molecular weight, valence, hydrolysis rate (if appropriate), biodegradation rate in liquid and soil; adsorption coefficient on organic carbon; adsorption coefficient to soil.

representative hydraulic conductivity value can result in considerable differences (years) in the estimated travel time of a contaminant to groundwater.

SESOIL has been used as a screening tool in exposure assessments and in developing cleanup levels in California. Trojanczyk *et al.* (1997), for example, reported its use to estimate when benzene from a woodpole plant entered the groundwater. The author has used the model to provide estimates of the arrival time of a contaminant to groundwater in litigation involving insurance claims.

VLEACH is a one-dimensional finite difference model that describes the movement of an organic contaminant as a dissolved solute, as a gas and as an adsorbed compound (Rong, 1999). The mathematical expression for transport in each phase and model assumptions are described. Liquid phase advection is expressed as:

$$\partial C_l/\partial t = -q/\theta(\partial C_l/\partial z) \tag{8.34}$$

where C_l is the liquid phase concentration, θ is the water filled porosity of soil by volume, $-q$ is the infiltration rate, z is the vertical dimension, and t is time. Gaseous diffusion is defined as:

$$\partial C_g/\partial t = -D_e(\partial^2 C_g/\partial z^2) \tag{8.35}$$

where C_g is the gaseous phase concentration and D_e is the effective diffusion coefficient. The governing equation used for adsorption is:

$$C_s = (K_{oc} f_{oc})C_l \tag{8.36}$$

where C_s is the adsorbed phase concentration, K_{oc} is the organic carbon partition coefficient and f_{oc} is the soil organic carbon content. The governing equilibrium equation used for each vertical simulation cell is:

$$C_T = (\theta C_l) + (\phi - \theta) C_g + (\rho_b C_s) \quad (8.37)$$

where C_T is the total concentration for all three phases at equilibrium, ϕ is the soil porosity and ρ_b is soil bulk density. Recent additions to the model provide the ability to incorporate Monte Carlo simulations so that a range of input values can be used (Rong and Wang, 2000). VLEACH assumptions include: local or instantaneous equilibrium between each phase, linear isotherms describe the partitioning of the contaminant between each phase, the moisture content in the homogenous soil is constant, liquid phase dispersion is neglected causing higher dissolved concentrations and lower travel time estimates than what would occur in reality, no *in situ* degradation or production occurs, volatilization from the soil boundaries is unimpeded or completely restricted, and non-aqueous phase liquids and/or flow conditions derived from fluid variable density are not simulated.

Jolicoeur (2000) used PESTAN, WAVE and SESOIL to estimate the transport of four pesticides (fenamiphos, chlorpyrifos, dimethoate, and oxamyl) through soils in the Lake Naivasha region northwest of Nairobi, Kenya. The chemical properties of these pesticides are summarized in Table 8.7 (after Jolicoeur, 2000).

Soil textures ranged from sandy loams to clay. An annual recharge of about 63 inches (1.6 m) was selected along with a yearly annual evapotranspiration rate of 40 inches (1.016 m). The depth of penetration of these four pesticides after one year was considered assuming that the pesticide was applied on the first day of the year.

Table 8.7

Chemical properties of fenamiphos, chlorpyrifos, dimethoate, and oxamyl.

Parameter and Units	Fenamiphos	Chlorpyrifos	Dimethoate	Oxamyl
Solubility (mg/L)	700	2	25 000	280 000
Molecular weight (g/mol)	303.4	422.9	229.3	219.4
Air diffusion coefficient (cm^2 s)	0.036	0.031	0.0476	0.0497
Henry's constant (atm-m^3/mol)	5.87×10^{-10}	4.45×10^{-6}	1.15×10^{-9}	0.24×10^{-9}
K_{oc}	169.82	6026	19.95	25.12
Liquid phase biodegradation-day	0.005775	0.0231	0.03465	0.05
Molecular diffusion coefficient (cm^2 per day)	0.012	0.012	0.012	0.012
Freundlich exponent (–)	1	1	1	1
Vapor enthalpy (kcal/mol)	20	20	20	20

K_{oc}, soil/sediment partition or sorption coefficient (organic carbon based).

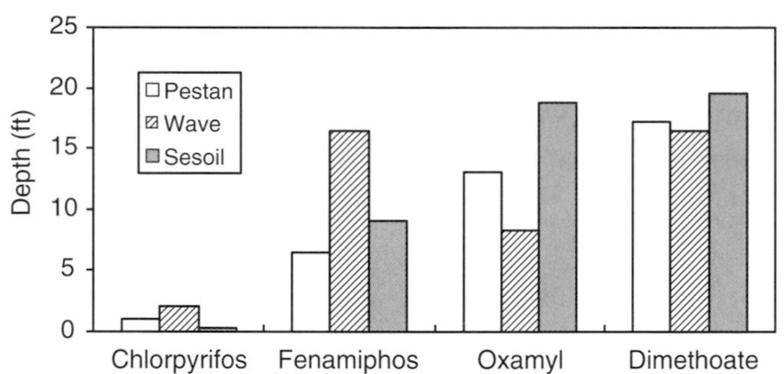

Figure 8.4
Depth of penetration of four pesticides after one year using PESTAN, WAVE and SESOIL.

As shown in Figure 8.4, model selection plays a significant role in estimating the travel time to the groundwater. The estimated transport time of the pesticide fenamiphos (also known as Nemacur) ranged from about 6 feet (1.8 m) using PESTAN to 17 feet (5.18 m) with WAVE. Similarly, if the release is a mixture of contaminants, the selection of a particular compound as a surrogate for the mixture can impact the estimated date of entry into the groundwater.

8.5.3 CHALLENGES TO CONTAMINANT TRANSPORT MODELS IN SOIL

Challenges to contaminant transport models in soil generally include an examination of the presence and significance of preferential pathways and the validity of the model assumptions. Categories of preferential pathways impacting model results include natural and artificial features, colloidal transport and cosolvent transport.

8.5.4 NATURAL AND ARTIFICIAL PREFERENTIAL PATHWAYS

Natural and artificial (dry wells, cisterns, utility line backfill, etc.) preferential pathways provide a means for contaminants to be rapidly introduced at depth (Barcelona and Morrison, 1988). Natural preferential pathways include worm channels, decayed root channels (Plate 12), soil fractures, slickenslides, swelling and shrinking clays, highly permeable soil layers, and insect burrows. The term *preferential flow* encompasses a range of processes with similar consequences for contaminant transport. The term implies that infiltrating liquid does not have sufficient time to equilibrate with the slowly moving water residing in the soil (Jarvis, 1998). Preferential flow includes finger flow (also viscous flow) (Glass and Nicholl, 1996), funnel flow (Philip, 1975; Kung, 1990a,b), and macropore flow (White, 1985; Morrison and Lowry, 1990).

Finger flow (also dissolution fingering) is initiated by small- and large-scale heterogeneities in soil such as a textural interface between a coarse-textured sand and an underlying silt (Fishman, 1998; Miller *et al.*, 1998). The term *finger flow* refers to the splitting of an otherwise uniform flow pattern into fingers.

These fingers are associated with soil air compression encountered where a finer soil overlies a coarse and dry sand layer. The contact interface between the contaminant and the water in the capillary fringe results in an instability. The spacing and frequency of these fingers are difficult to predict although they are at the centimeter scale and are sensitive to the initial water content (Ritsema and Dekker, 1995; Imhoff *et al.*, 1996).

Numerical simulations of fingering suggest that transverse dispersion has a significant impact on the formation and anatomy of fingers. Aspects of fingering phenomena that introduce uncertainty when modeling contaminants such as non-aqueous phase liquids (NAPLs) include the effect of dispersion, the impact of heterogeneity on porous media properties and residual NAPL saturation, the validity of fingering when a NAPL solution is flushed with chemical agents such as surfactants and alcohols, and incorporation of the impacts of fingering on NAPL phase mass transfer models when the model is discretized at scales larger than the centimeter scale.

Funnel flow occurs in soils with lenses and admixtures of particle sizes. For a saturated soil, the coarsest sand fraction is the preferred flow region; for unsaturated flow, finer textured materials are more conductive. Examination of textural descriptions on boring logs and contaminant concentration depth profiles can provide insight to determine if contaminant transport via funnel flow is a viable transport mechanism.

A macropore is a continuous soil pore that is significantly larger than the inter-granular or inter-aggregate soil pores (micropores). While a macropore may constitute only 0.001 to 0.05% of the total soil volume, it may conduct a majority of an infiltrating liquid.

8.5.5 COLLOIDAL TRANSPORT

Colloidal transport describes the sorption of a hydrophobic compound to a colloid particle in water. A colloid is defined as a particle ranging from 0.003 to 10 μm in diameter. Colloids exist as suspended organic and inorganic matter in soil or aquifers. In sandy aquifers, the predominant colloids that are mobile range in size from about 0.1 to 10 μm.

The contaminant mass associated with colloidal transport may be significant. Numerous studies have demonstrated that contaminants are transported at depth via colloidal transport (Sheppard *et al.*, 1979; Robertson *et al.*, 1984; Gschwend and Reynolds, 1987; Puls *et al.*, 1990; Pohlmann *et al.*, 1994). Buddemeier and Rego (1986), for example, found that virtually all of the activity of manganese, cobalt, antimony, caesium, cerium, and europium was associated with colloidal particles in groundwater samples collected from underground nuclear-test cavities at the Nevada Test Site. An example of the transport of polycyclic aromatic hydrocarbons via colloidal transport was examined in two creosote-impacted aquifers on Zealand Island in Denmark (Villholth, 1999). The mobile colloids were composed predominately of clay, iron oxides, iron sulfides, and quartz particles. The researchers concluded that the sorption is hydrophobic and associated with the organic content of the colloids. Creosote associated contaminants (i.e., polycyclic aromatic hydrocarbons, phenols and nitrogen, sulfur or oxygen containing heterocycles (Mueller *et al.*, 1989)) were found to be associated primarily with colloids larger than 100 nm. These findings indicate that colloid-facilitated transport of polycyclic aromatic hydrocarbons may be significant. While Sheppard *et al.* (1980) concluded over 20 years ago that the transport of radionuclides by colloidal clay particles should be considered in contaminant transport models, this transport mechanism is rarely considered and/or simulated in soil or groundwater models.

8.5.6 COSOLVENT TRANSPORT

Hydrophobic compounds are generally considered to be immobile in the soil profile due primarily to their low water solubility and their tendency to be adsorbed by clay, organic matter and mineral surfaces (Odermatt *et al.*, 1993). Soil contaminant transport models tend to predict low velocities for these compounds. Cosolvation of these compounds in a fluid that can introduce these contaminants at depth is rarely included in contaminant transport models. Contaminants such as polychlorinated biphenyls (PCBs) or DDT can be remobilized by the preferential dissolution of the PCBs into a solvent released into the same soil column (Morrison and Newell, 1999). A variation to this scenario is the preferential dissolution of an immobile chemical into a solvent prior to its release (e.g., PCBs dissolved in a dielectric fluid).

A similar transport mechanism occurs when a contaminant sorbed by a soil is washed with a liquid that remobilizes the compound. An example is the presence of copper bound in soil under a leaking neutralization pit. Low pH wastewaters leaking through an expansion joint and contacting the precipitated copper will remobilize and transport the copper with the low pH wastewaters to depth until the acidic wastewater and copper solution is buffered and the copper re-precipitates at a lower depth.

8.5.7 MODEL ASSUMPTIONS

Potential challenges to contaminant transport models in soil include examination of the model assumptions, including the circumstances of the release and whether these are realistically simulated in the model. Most soil transport models, for example, do not use statistical methods to determine the precision of the release parameters (Opdyke and Loehr, 1999). Release parameters are commonly reported deterministically (i.e., as a point estimate) rather than probabilistically (i.e., a distribution of values and their associated probabilities). The reasonableness of the release parameter assumptions or measurements should therefore be considered to determine whether the release assumptions are reasonable.

A simplifying assumption often encountered in environmental forensics and which works successfully at the laboratory scale is that the soil is homogeneous (Miller *et al.*, 1965). Large-scale field studies of solute transport through the unsaturated zone, for example, have revealed enormous variability in pore-water velocities and hydrodynamic dispersion coefficients (Quisenberry and Phillips, 1976; Schulin *et al.*, 1987; Andreini and Steenhuis, 1990). This field variability coupled with the uncertainties associated with estimating other processes relevant to contaminant transport can introduce serious difficulties in the simulation of heterogeneous field-scale transport. A number of approaches, including scaling theories, Monte Carlo simulations, stochasticontinuum models (Russo and Dagan, 1991), and stochastic convective models (Jury and Roth, 1990) have been proposed to deal with these field-scale issues.

One means to establish a confidence level for the various uncertainties in the input parameters is to perform a probabilistic analysis (Rong 1995; Duke *et al.*, 1998). Two common probabilistic methods are the likelihood method that is effective in estimating the confidence of parameters in nonlinear models (Donaldson and Schnabel, 1987; Bates and Watts, 1988) and Monte Carlo simulations (Ang and Tang, 1990; Berthouex and Brown, 1994). Monte Carlo simulations provide a numerical method for generating simulated data using assumed probability distribution. This analysis identifies those parameters that influence the calculated groundwater contaminant concentration and provides an uncertainty analysis, including the most sensitive model variables (Hetrick and Pandey, 1999).

8.6 GROUNDWATER MODELS

There are currently in excess of 400 groundwater flow (advective) and contaminant transport models (Zeng and Bennett, 1995). Contaminant transport models are available in the public and private domain and are available from government sources such as the United States EPA Environmental Center for Subsurface Modeling Support and the Robert S. Kerr Environmental Center at epa.gov/ada/models/html and at the United States Geological Survey at water.usgs.gov/software/ground_water.html. Private domain models can be obtained from trade associations, university research associations, and commercial vendors.

Calculations of the rate of contaminant transport in groundwater are used as evidence in insurance litigation for source identification, age dating and cost allocation purposes. In the United States, it is estimated that computer-based predictions of contaminant transport influence legal and policy decisions involving the allocation of at least 1 billion dollars each year (Eggleston and Rojstaczer, 2000). These models have been primarily limited to the realm of environmental consultants, although some peer-reviewed literature is available on the subject (Gorelick *et al.*, 1983; Uffink, 1989; Bagtzoglou *et al.*, 1992; Wagner, 1992; Skaggs and Kabala, 1994, 1995; Woodbury *et al.*, 1996; Snodgrass *et al.*, 1997; Neupauer *et al.*, 1999).

The origin of inverse modeling for contaminant transport in groundwater is predated by research in the heat transfer literature (Huang and Ozisk, 1992; Bayo *et al.*, 1992; Silva-Neto and Ozisk, 1993). Governing equations in the heat transfer literature, for example, are similar to the advection–dispersion equations used in groundwater models, except that the advection term is seldom used in inverse heat conduction problems (Lattes and Lions, 1969; Alifanov and Artyukhin, 1976; Tikhonov and Arsenin, 1977). Issues regarding parameter confidence are also addressed in many heat transfer equations (Beck and Arnold, 1977; Alifanov and Nenarokomov, 1989; Marquardt and Auracher, 1990; Carasso, 1992).

8.6.1 INVERSE MODELS

The term *backward extrapolation model* was reported in 1991 by Allen Kezsbom and Alan Goldman in an article describing the *Sterling v. Velsicol Chemical Corporation* case (855 F.2d 1188 6th Circuit, 1988) (Kezsbom and Goldman, 1991; Kornfeld, 1992). Subsequent authors describe this approach with the term *reverse or inverse modeling* (Erickson and Morrison, 1994; Woodbury and Ulrych, 1996; Morrison, 1999b; Parker, 2000), and *hindcasting* and *backward random walk* (Chandrasekhar, 1943). In its simplest application, inverse modeling relies upon measured properties or contaminant concentrations to extrapolate to some point in the past, the age and location of a contaminant release, most frequently by using geostatistical and optimization approaches (Gorelick and Remson, 1982; Khachikian and Harmon, 2000). Inverse models are used in environmental litigation to reconstruct complex, multi-parameter release histories with known source locations from spatially distributed plume concentration data (Skaggs and Kabala, 1994, 1995, 1998; Woodbury and Ulrych, 1996; Parker, 2000).

An early application of inverse modeling for contaminant transport in groundwater used a least squares regression technique to identify the locations and discharge rates of contaminants that were simultaneously migrating from five disposal sites over a five-year period (Gorelick *et al.*, 1983). Other applications used backward probability models to establish the most likely location of a contaminant release. Bagtzoglou *et al.* (1992) described a reverse time random particle method to estimate the probability that a particular site was the source of a contaminant plume. These probability approaches usually assume a single, point source release (Birchwood, 1999, 2000). In another application, the vertical concentrations of PCE and TCE in an aquitard were evaluated to reconstruct a release history; only diffusion and sorption were assumed to affect mass transport in the aquitard (Liu, 1995; Ball *et al.*, 1997; Liu and Ball, 1999). Dimov *et al.* (1996) used the adjoint formulation of the one-dimensional advection–dispersion equation to identify a contaminant source and estimate the contaminant loading rates at different nodes. Another approach modeled contaminants from a constant source concentration using two-dimensional diffusion to estimate source characteristics and transport parameters from head and concentration data (Sonnenborg *et al.*, 1996). Sidauruk *et al.* (1997) developed a set of analytical procedures for contaminant source identification that relied upon contaminant concentrations measured within a two-dimensional contaminant plume. Contaminant flow was assumed to be homogeneous and groundwater velocity was uniform for instantaneous and continuous releases. The authors noted the limitations of certain parameters that can only be uniquely determined by sampling at a minimum of two different times.

Skaggs and Kabala (1994) used Tikhonov regularization (Tikhonov and Arsenin, 1977) to transform an algebraically underdetermined inverse problem into a minimization problem with a unique solution as well as using the method of quasi-reversibility to solve the problem (Skaggs and Kabala, 1995). The location of the source was assumed to be known. Wagner (1992) proposed a deterministic approach for the combined estimation of model parameters and source characteristics. Bagtzoglou *et al.* (1992) used backward location probabilities to identify contaminant sources. Probability maps were then developed using a random walk method by reversing the flow field and assuming that dispersion was constant.

The foundational advection/dispersion equation used in traditional groundwater models is also used for backward extrapolation modeling (Wilson *et al.*, 1981; Wilson and Linderfelt, 1991; Liu, 1995; Wilson and Liu, 1995). Governing assumptions are that the porous medium is homogeneous and isotropic, saturated with a fluid and that Darcy's law is valid (Bear *et al.*, 1992). If these assumptions are violated, the applicability or precision of the model is compromised to some degree.

One expression used to run a one-dimensional advection–dispersion model backward in time is described by (Wilson and Liu, 1995):

$$\partial g/\partial t = D(\partial^2 g/\partial x^2) - V(\partial g/\partial x) \quad (8.38)$$

where g is a probability density or cumulative distribution function of interest, D is the dispersion coefficient, t is time, and x is the upgradient distance to the source. The authors hypothesized that by selecting appropriate initial and boundary conditions, Equation 8.38 can be solved for the desired probability function to create travel time and location probability maps when the location of the release is unknown. Neupauer and Wilson (1999) describe the solutions for source and time probabilities for a pumping well as:

$$\partial f_x/\partial \tau = D(\partial^2 f_x/\partial x^2) + v(\partial f_x/\partial x) \qquad (8.39)$$

where

$$f_x \rightarrow 0 \quad \text{as } x \rightarrow -\infty,$$

and

$$v f_x + D(\partial f_x/\partial x) = 0 \quad \text{at } x = 0 \quad \text{and} \quad f_x(x,0) = \delta(x - x_d)$$

where τ is backward time ($\tau = t_d - t$, where t_d is the detection time), x_d is the detection location at a pumping well ($x_d = 0$), D is the dispersion coefficient, v is the groundwater velocity, f_x is the location probability and δ is the Dirac delta function. The solution for Equation 8.39 provides the backward probability for a detection at $x_d = 0$, as (Neupauer and Wilson, 1999):

$$f_x(x;\tau) = 1/(\pi D\tau)^{1/2} \exp\{-(x + v\tau)^2/4D\tau\}$$
$$-(v/2D)\exp\{-(vx)/D\}\,\text{erfc}[(-x + v\tau)/(4D\tau)^{1/2}] \qquad (8.40)$$

For a contaminant parcel removed at the pumping well ($x_d = 0$), Equation 8.40 describes the probability density function of the location of the contaminant parcel at time τ before it was observed at the pumping well. The backward model for the travel time probability map is then (Wilson and Liu, 1995):

$$\partial f_\tau/\partial \tau = D(\partial^2 f_\tau/\partial x^2) + v(\partial f_\tau/\partial x) \qquad (8.41)$$

where

$$f_\tau \rightarrow 0 \quad \text{as } x \rightarrow -\infty,$$

and

$$v f_\tau + D(\partial f_\tau/\partial x) = v\delta(\tau) \quad \text{at } x = 0 \quad \text{and} \quad f_\tau(x,0) = 0$$

where D and v are constant and the pumping well is assumed to be at $x_d = 0$. The solution to Equation 8.41 is the backward-in-time time probability from location x to a detection at $x_d = 0$ (Wilson and Liu, 1995):

$$f_\tau(x;\tau) = 1/(\pi D\tau)^{1/2} \exp\{-(x + v\tau)^2/4D\tau\}$$
$$-(v^2/2D)\exp\{-(vx)/D\}\,\text{erfc}[(-x + v\tau)/(4D\tau)^{1/2}] \qquad (8.42)$$

For a contaminant detected at the pumping well, Equation 8.41 describes the probability density function of the travel time from an upgradient source location (x) to the pumping well at $x_d = 0$.

8.6.2 LNAPL MODELS

In cases where light non-aqueous phase liquids (LNAPLs) are of interest, numerical models are available to predict LNAPL plume migration over time (Kaluarachchi *et al.*, 1990; Parker *et al.*, 1994; Parker, 2000) and to age date the release using a direct estimate and nonlinear parameter estimation approach (Mull, 1970; Dracos, 1987). The direct estimation approach involves the estimation of a release date from a single measurement of travel distance by describing the phase separate model in the form of $\tau = t_x - \Delta t_x$ where t_x is the travel time from the source to observation location x and Δt_x is the time between the reference date and the observation date. The nonlinear parameter estimation method involves the simultaneous estimation of when a contaminant release occurred. Additional unknown model parameters can also be estimated from multiple measurements of travel distances at various time steps. Mean values of the estimated variables and their confidence limits are determined using nonlinear regression methods.

If an inverse model is used to simulate the migration of a phase-separate fluid, different governing equations and parameters are needed (Butcher and Gauthier, 1994; Parker and Lenhard, 1990). The solution of the inverse problem for identifying NAPL location involves estimating dissolution parameters in addition to conventional flow and transport modeling parameters (Khachikian and Harmon, 2000). Descriptions of this approach are thoroughly discussed in the literature (Kool *et al.*, 1987).

A key variable for phase-separate models is whether the three-phase (air, water, and NAPL) expression describing the pressure distributions between these fluids is representative of the site. LNAPL models normally assume that the phase-separate liquid is continuous and/or connected. If this assumption is incorrect, the LNAPL volume in the subsurface can be overestimated. If monitoring wells are separated by hundreds of feet and the LNAPL thickness is thin (less than a foot), the possibility that the LNAPL is not continuous is highly probable. Other challenges include the accuracy of LNAPL or DNAPL (dense non-aqueous phase liquid) fluid viscosity values, the selected interfacial tension values for the different liquids, the values used to represent the soil texture, and the accuracy of any phase-separate measurements. Fluid viscosity and density are rarely measured and are more commonly obtained from the literature. Published numbers can, therefore, deviate significantly from the field values due to weathering and/or co-mingling of the LNAPL with other compounds.

8.6.3 CHALLENGES TO INVERSE MODELS

The successful review of an inverse model includes an analysis of model parameters and conceptual framework. Examples of model parameters incorporated into many inverse models that represent potential sources of uncertainty include (Erickson and Morrison, 1995; Morrison, 1998):

- The reasonableness of the selected porosity and hydraulic conductivity value(s). Soil porosity and hydraulic conductivity within an aquifer and with distance can change dramatically. A representative porosity value, on a field scale, is usually a fitted parameter within a published range of values for the predominant soil type encountered. As a result, differences in the modeling results can be adjusted by slight variations in the selected porosity value. Typical porosity values range from 0.25 to 0.50 for unconsolidated soils.
- The consistency of the groundwater flow direction and velocity over time. The hydraulic gradient value selected is an important variable to scrutinize as its value can vary in time and distance. If regional or vicinity wide data are relied upon to define the hydraulic gradient versus site measurements, considerable differences can occur. Sources of localized variations in the gradient include pumping wells, rivers, streams and groundwater recharge or spreading basins. If pumping wells in the immediate vicinity of the site are present, collecting the extraction rates, representative transmissivity values for the aquifer and calculating the radius of pumping influence is useful to demonstrate the potential impact of pumping wells on the local gradient. Historical variations in the hydraulic gradient can introduce significant uncertainty regarding the historical direction of groundwater flow and hence the time required for the contaminant plume to attain its measured leading edge geometry. Typical hydraulic gradient values range from 0.1 to 0.001.
- The validity of the selected hydraulic gradient(s) values over time and distance from the release (e.g., in *Anderson v. Cryovac* the court found that uncertainties associated with the hydraulic gradient concealed the true direction of the groundwater flow).
- The number of data points and time interval during which the data were collected (Mahar and Datta, 1997). In *Renaud v. Martin Marietta Corporation* 1990, the expert hydrogeologist for the plaintiff constructed a chemical fate and transport model using only one data point to describe a continuous release of contaminants over an 11-year period.
- The nature of the release (steady verses non-steady).
- The loading rate (in the *Velsicol* case, plaintiff's groundwater model assumed a loading rate when production records were available to establish a loading rate, which were used by the defendants).
- The value(s) selected for aquifer(s) thickness (model specific).
- The horizontal and transverse dispersivity values (Mercer, 1991).

- Contaminant retardation and/or degradation rates (Davidson, 2000).
- Identification of the leading edge of the contaminant plume (model specific).
- The effect of recharge/discharge rates (if applicable) of water into the system and its impact on plume geometry and contaminant velocity. In the *Velsicol* case, for example, the court wrote, 'the district court rejected the defendant's water model as inaccurately under representing the extent of chemical contamination in the groundwater supply. In refuting the defendant's model, the court reasoned that Velsicol had failed to factor in the massive dumping of liquid waste, the ponding of water in the trenches, and the draw down on the aquifer caused by new homes.'

Some inverse models require an accurate value for the contaminant plume length, the location of the release, or when a contaminant first entered a monitoring well downgradient from the source. If the source location and/or the plume length are assumed, significant variations in the estimated age of the release can occur.

Another issue is whether the compound or class of compounds used to determine the plume length biases the measurement. The selection of benzene, for example, may reflect a stabilized or decreasing length while the use of MTBE for the same plume may indicate an expanding plume, depending on the source strength and effects of biodegradation. A similar dilemma exists for transformation products. Among likely parent–daughter compound combinations for chlorinated solvents, distances between the respective maximum concentration locations were found to be on the order of 10–25% of the maximum plume length at most sites (McNab *et al.*, 1999). For multiple releases, compound selection, release location and potential cosolvation impacts require scrutiny (Morrison and Newell, 1999). Examples of possible challenges or areas of inquiry when evaluating inverse models results include (Morrison, 1999a,b,c, 2000b):

- Adjust the model-input parameter, within a reasonable range, so that results are produced consistent with measured values but inconsistent with the alleged released date. For example, the distribution of simulated head measurements depends to some degree on the recharge rate to saturated hydraulic conductivity value ratio.
- Determine the confidence level associated with simulated or assumed model parameters (e.g., groundwater velocity, analytical bias, hydraulic conductivity, temporal and spatial variations of the data, time interval between sampling dates, etc.).
- Collect water level measurements coincident with an activity, such as a pumping well, that stresses the aquifer. Use the model to simulate this activity and compare the actual measurements with those predicted by the model. As stated by an expert in *United States* v. *Hooker Chemical & Plastics Corporation* (1985), 'The accuracy of these

equations [referring to those used in a mathematical model] is tested by comparing them with what actually occurs within the groundwater system.'

- Identify whether the water mass balance is conserved. A well-calibrated model exhibits a small difference in the water balance, which is often included as output to the model.
- Evaluate model calibration via the preparation of scattergrams comparing values for measured heads with the simulated heads for each well. If the calibration is perfect, the data sets should fall along a 45-degree line. Data points farther from this line reflect a greater deviation from actual flow conditions. Plotting the residuals (the difference between the simulated and observed heads) can also provide insight regarding model calibration.
- Compare the input parameters and compare them with those used for other site models (e.g., models used for designing a remediation system, health risk models used for site closure, etc.). It is not uncommon that physical and chemical parameters used in a model in support of a risk based corrective action (RBCA) study are different than those used for inverse models for the same site.

Inverse models rarely incorporate the time required for the contaminant to flow through a paved surface. If the depth to groundwater is shallow or the contaminant is introduced quickly into the groundwater via a preferential pathway (i.e., a dry well, cistern, well casing, etc.) age dating the release with an inverse model may be a reasonable approach for estimating the age and/or location of a well defined, discrete release. If the release occurred at the ground surface on impermeable soils, and/or the depth to groundwater is significant (>50 ft), the cumulative time required for contaminant migration through the soil column can be significant. Ignoring this portion of the conceptual model with an inverse model can significantly underestimate age of the release. In the *Velsicol* case, the plaintiffs' experts assumed that carbon tetrachloride and chloroform entered the aquifer almost immediately after commencement of disposal in October 1964. This was inconsistent with a United States Geological Survey study in 1967.

Boundary conditions are required for inverse models. For a program such as MODFLOW, general head boundaries are used to define the lateral boundary conditions that determine the flux of water recharge or discharge along these boundaries. The boundary conditions are a function of the hydraulic conductivity, groundwater flow gradient, and the absolute difference in water level elevations between the block elements located on the lateral boundaries and those outside of the model grid. Examples of boundary conditions include no-flow, specified flux and fixed head boundaries. While model boundary conditions are fixed and cannot be changed during a single simulation, they can be adjusted between simulations. It is conceptually undesirable to alter the boundary flux conditions to assist in calibration of each stress period, versus accounting for these differences by adjustments in dynamic features such as pumping wells or recharge of surface water bodies located within the grid. The impact of a model boundary can be examined if all model input files and software are available to reproduce the modeling result using different boundary conditions.

Grid selection is important. For numerical models, finite difference and finite element grids are used. Block centered and mesh centered grids are used for finite element grids. Finite element grids are generally more versatile than finite difference grids. For a finite difference model, the grid density should be examined to ascertain whether the data support finer mesh nodes or whether higher grid densities are selected in areas of interest but which contain insufficient data to warrant a higher grid density.

In finite difference modeling, numerical dispersion is inherent due to errors associated with the computational algorithms, especially in areas of varying grid size. The three-dimensional block size selected must be examined to determine the relative horizontal and vertical element aspect ratio. Aspect ratios less than 4 are generally acceptable; the horizontal and vertical aspect ratios that are greater than 4 become more susceptible to numerical dispersion.

For multiple layered models, it is important to determine whether the vertical gradients between the layers are measured or estimated. To confirm measured vertical gradients, the head difference between a shallow and deep well should be divided by the vertical distance between the bottom of both well screens. A negative value indicates a downward flow component. If the vertical gradients are estimated, an attempt should be made to determine the level of uncertainty associated with these values and their overall impact on solute transport between layers.

There is usually some arbitrariness in defining source loading rates as well as when loading began (e.g., at the commencement of facility operations, 10 years after commencement, etc.) in a groundwater modeling grid. The analyst should remove as much arbitrariness as possible by examining the site operating history and the characteristics of transport through the vadose zone. Issues regarding the validity of a particular loading rate and its location include:

- whether the soil and groundwater chemistry justifies the selected location and input rate;
- whether the mass loading rate is continuous or is transient in response to groundwater fluctuations or remediation activities; and
- whether the start date for the mass loading is consistent with the operational history of the contributing surface sources.

Most contaminant transport models perform a mass balance; this output should be obtained and reviewed. A significant error in the mass balance calculation indicates that the solution is imprecise.

The selection of an appropriate contaminant transport model should be carefully examined. If the conceptual model does not represent the relevant flow and contaminant transport phenomena, the subsequent modeling effort is wasted. Model complexity is determined by the quality of the data available for its design and verification. An example of an inappropriate model selection is a simple two-dimensional model used for a complex three-dimensional system. Conversely, a complex three-dimensional model may be selected that is overpowered relative to the available data. Model selection considerations include the following issues (Cleary, 1995):

- It is advisable to select a model with the prestige of a state or federal governmental agency development or official use, a substantial published history in peer-reviewed journals and/or one that has already been tested in court. If the model code is obscure, it should be thoroughly scrutinized.
- The model should include a user's manual listing its governing assumptions, advantages, and capabilities.
- The model should be validated against analytical solutions for comparison.
- The analytical solution should have the same number of space dimensions as the numerical model.
- The model should be benchmarked against a numerical code.
- A model that has an available source code should be selected.
- A three-dimensional model may better represent reality than a two-dimensional approximation.

The model simulation ultimately selected to create a trial still or animation should be evaluated in the context of all the simulations. It is not unusual that hundreds of simulations are performed until a simulation is obtained for use as evidence for a particular allegation. Discarded simulations should be compared with the selected simulation so that the legitimacy of the selected simulation, as the best representation, can be evaluated.

Of the hydrogeologic parameters, the saturated hydraulic conductivity value generally introduces the most significant variability in the computer simulations (Rong *et al.*, 1998). It is generally recognized that the most representative measurements for determining the saturated hydraulic conductivity of a high permeability formation is via a pump test. Hydraulic conductivity values that rely upon slug tests, sieve analysis, and laboratory measurements of soil cores are considered less reliable. Measurements from a slug test are generally reliable to about one or more orders of magnitude with its

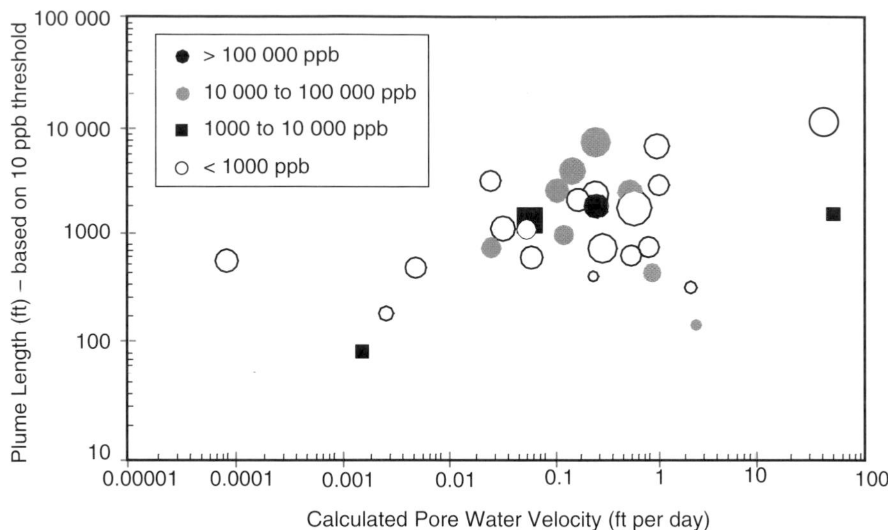

Figure 8.5
Distribution of TCE relative to plume length, concentration and pore water velocity (after McNab et al., *1999).*

rate prior to the point in time that it is measured. This assumption may be invalid.

8.7 CONCLUSIONS

Contaminant transport models for age dating and source identification must be carefully selected and evaluated to determine the probability that the resulting information will be useful to all parties involved in the engagement. In the context of environmental litigation, the simulation results should be coupled with other corroborative evidence. Model results should be able to withstand intense scientific scrutiny, relative to the purpose for which the model was used.

REFERENCES

Alifanov, O. and Artyukhin, E. (1976) Regularized numerical solution of nonlinear inverse heat-conduction problem. *Journal of Engineering Physics* 29, 934–938.

Alifanov, O. and Nenarokomov, A. (1989) Effect of different factors on the accuracy of the solution of a parameterized inverse problem of heat conduction. *Journal of Engineering Physics* 56, 308–312.

Andreini, M. and Steenhuis, T. (1990) Preferential paths of flow under conventional and conservation tillage. *Geoderma* 46, 85–102.

Ang, A. and Tang, W. (1990) *Probability Concepts in Engineering Planning and Design*, vol. II: *Decision, Risk and Reliability*. Published by A. Ang, University of California at Irvine and W. Tang, University of Illinois at Urbana-Champaign.

Annable, M., Rao, P., Hatfield, K., Graham, W., Wood, A., and Enfield, C. (1998) Partitioning tracers for measuring residual NAPL: Field-scale test rests. *Journal of Environmental Engineering* 124, 498–503.

Baca, E. (1999) On the misuse of the simplest transport model. *Groundwater* 37(4), 483.

Bagtzoglou, A., Doughery, D., and Thompson, A. (1992) Application of particle methods to reliable identification of groundwater pollution sources. *Water Resources Management* 6, 15–23.

Ball, W. and Roberts, P. (1991) Long term sorption of halogenated organic chemicals by aquifer material. 1 Equilibrium. *Environmental Science and Technology* 24, 1223–1236.

Ball, W., Liu, C., Xia, G., and Young, D. (1997) A diffusion-based interpretation of tetrachloroethene and trichloroethane concentration profiles in a groundwater aquitard. *Water Resources Research* 33, 2741–2757.

Barber, L., Thurman, M., Schroder, M., and LeBlanc, D. (1988) Long-term fate of organic micropollutants in sewage contaminated groundwater. *Environmental Science and Technology* 22, 205–211.

Barcelona, M. and Morrison, R. (1988) Sample collection, handling and storage: water, soils and aquifer solids. In *Proceedings of a National Workshop on Methods for Groundwater Quality Studies*, November 1–3, 1988. Arlington, VA, pp. 49–62. Agricultural Research Division, University of Nebraska at Lincoln, Lincoln, NE.

Bates, D. and Watts, D. (1988) *Nonlinear Regression Analysis and its Applications*. John Wiley & Sons, New York.

Bayo, E., Moulin, H., Crisalle, O., and Gimenez, G. (1992) Well-conditioned numerical approach for the solution of the inverse heat conduction problem. *Numerical Heat Transfer, Part B* 21, 79–98.

Bear, J. (1979) *Hydraulics of Groundwater*. McGraw-Hill, New York. 569.

Bear, J., Beljin, M., and Ross, R. (1992) Fundamentals of ground-water modeling. EPA Groundwater Issue. United States Protection Agency, Office of Research and Development, Office of Solid Waste and Emergency Response. EPA/540/S-92/005, April 1991.

Beck, J. and Arnold, K. (1977) *Parameter Estimation in Engineering and Science*. John Wiley & Sons. New York.

Berthouex, P. and Brown, L. (1994) *Statistics for Environmental Engineers*. Lewis Publishers, Boca Raton, FL.

Birchwood, R. (1999) Identifying the location and release characteristics of a groundwater pollution source using spectral analysis. In *Proceedings of the 19th Annual American Geophysical Union Hydrology Days Conference*, Colorado State University, Fort Collins, Colorado, August 16–18, 1999, pp. 37–50.

Birchwood, R. (2000) A spectral analysis of sampling design and ill-posedness in the recovery of the release history of a groundwater contaminant plume. *Water Resources Research* (in review/press).

Bonazountas, M., Hetrick, D., Kostecki, P., and Calabrese, E. (1997) *SESOIL in Environmental Fate and Risk Modeling*. Amherst Scientific Publishing. Amherst, MA.

Brown, A. (2000) Treatment of drinking water impacted with MTBE. In *Mealey's MTBE Conference*, May 11–12, 2000, Marina del Rey, CA, pp. 475–518. Mealey Publications, King of Prussia, PA.

Buddemeier, R. and Rego, J. (1986) Colloidal radionuclides in groundwater. *Annual Report*. Lawrence Livermore National Laboratory, Livermore, CA. Report No. UCAR 10062/85–1.

Butcher, J. and Gauthier, T. (1994) Estimation of residual dense NAPL mass by inverse modeling. *Groundwater* 32(1), 71–78.

Butler, E. and Hayes, K. (1999) Kinetics of the transformation of trichloroethylene and tetrachloroethylene by iron sulfide. *Environmental Science and Technology* 33(12), 2021–2027.

Carasso, A. (1992) Space marching difference schemes in the nonlinear inverse heat conduction problem. *Inverse Problems* 8, 25–43.

Carslaw, H. and Jaeger, J. (1959) *Conduction of Heat in Solids*. Clarendon Press, Oxford.

Chandrasekhar, S. (1943) Stochastic problems in physics and astronomy. *Review of Modern Physics* 15(1), 1–43.

Choy, B. and Reible, D. (1999) *Diffusion Models of Environmental Transport*. Lewis Publishers, Boca Raton, FL.

Cleary, R. (1995) Introduction to applied mathematical modeling in groundwater pollution and hydrology with IBM-PC applications. *NGWA IBM-PC Applications in Groundwater Pollution and Hydrology*, Chapter 4. August 13–18, 1995. San Francisco, CA.

Cohen, R., Mercer, J., and Matthews, J. (1993) *DNAPL Site Evaluation*, EPA/600/R-93/022. US Environmental Protection Agency, Ada, OK.

Crank, J. (1985) *The Mathematics of Diffusion,* 2nd edn. Clarendon Press, Oxford.

Daily, W., Ramirez, A., and Johnson, R. (1998) Electrical impedance tomography of a perchloroethylene release. *Journal of Environmental and Engineering Geophysics* 2(3), 189–201.

Davidson, J. (2000) Fate, transport and remediation of MTBE. In *Mealey's MTBE Conference*. May 11–12, 2000, Marina del Rey, CA, pp. 445–474. Mealey Publications, King of Prussia, PA.

Deeds, N., McKinney, D., and Pope, G. (2000) Laboratory characterization of non-aqueous phase liquid/tracer interaction in support of a vadose zone partitioning interwell tracer text. *Journal of Contaminant Hydrology* 41, 193–204.

Dilling, W., Tefertiller, N., and Kallos, G. (1975) Evaporation rates and reactivities of methylene chloride, chloroform, 1,1,1, trichloroethane, trichloroethylene, tetrachloroethylene, and other chlorinated compounds in dilute aqueous solutions. *Environmental Science and Technology* 9(6), 833–838.

Dimov, I., Jaekel, U., and Vereecken, H. (1996) A numerical approach for determination of sources in transport equations. *Computers in Mathematical Applications* 32, 31–42.

Domenico, P. (1987) An analytical model for multidimensional transport of a decaying contaminant species. *Journal of Hydrology* 91, 49–58.

Domenico, P. and Schwartz, F. (1990) *Physical and Chemical Hydrogeology*. John Wiley & Sons, New York.

Donaldson, J. and Schnabel, R. (1987) Computation experience with confidence regions and confidence intervals for nonlinear least squares. *Technometrics* 29(1), 67–82.

Donigian, A. and Rao, P. (1986a) Overview of terrestrial processes and modeling. In *Vadose Zone Modeling of Organic Pollutants*, Chapter 1 (Hern, S. and Melancon, S., eds). Lewis Publishing, Chelsea, MI.

Donigian, A. and Rao, P. (1986b) Example model testing studies. In *Vadose Zone Modeling of Organic Pollutants*, Chapter 5 (Hern, S. and Melancon, S., eds). Lewis Publishing, Chelsea, MI.

Dracos, T. (1987) Immiscible transport of hydrocarbons infiltrating in unconfined aquifers. In *Proceedings of Symposium on Oil in Fresh Water*, Alberta, Canada (van der Molen, J. and Hrudey, S., eds), pp. 159–272. Pergamon Press, Oxford.

Dragun, J. (1988) *The Soil Chemistry of Hazardous Materials*. Hazardous Materials Control Research Institute, Silver Springs, MD.

Drogos, D. (2000) MTBE v. other oxygenates. In *Mealey's MTBE Conference*, May 11–12, 2000, Marina del Rey, CA, pp. 299–334. Mealey Publications, King of Prussia, PA.

Duke, L., Rong, D., and Harmon, T. (1998) Parameter-induced uncertainty in modeling vadose zone transport of VOCs. *Journal of Environmental Engineering* 124(5), 441–448.

Eggleston, J. and Rojstaczer, S. (2000) Can we predict subsurface mass transport. *Environmental Science and Technology* 34(18), 4010–4017.

Enfield, C., Carsel, R., Cohen, S., Phan, T., and Walters, D. (1982) Approximating pollutant transport to groundwater. *Groundwater* 20, 711–722.

Erickson, R. and Morrison, R. (1995) *Environmental Reports and Remediation Plans: Forensic and Legal Review*. John Wiley & Sons, Environmental Law Library, New York.

Feenstra, S. and Cherry, J. (1998) Aqueous concentration ratios to estimate mass of multi-component NAPL residual. In *Nonaqueous Phase Liquids: Remediation of Recalcitrant Compounds* (Wickramanayake, G.B. and Hinchee, R., eds), pp. 55–60. Battelle Press, Columbus, OH.

Fetter, C. (1994) *Applied Hydrogeology*, 3rd edn. Prentice Hall, Upper Saddle River, NJ.

Fick, A. (1855) Ueber diffusion. *Annalen der Physik (Leipzig)* 170, 59–86.

Fishman, M. (1998) DNAPL infiltration and distribution in subsurface: 2D experiment and modeling approach. In *Nonaqueous-Phase Liquids: Remediation of Chlorinated and Recalcitrant Compounds.* (Wickramanayake, G. and Hinchee, R., eds), pp. 37–42. Battelle Press, Columbus, OH.

Gao, J., Skeen, R., and Hooker, B. (1995) Effect of temperature on perchloroethylene dechlorination by a methanogenic consortium. In *Bioremediation of Chlorinated Solvents* (Hinchee, R., Leeson, A., and Semprini, L., eds), pp. 53–59. Battelle Press, Columbus, OH.

Gelhar, L. and Axness, C. (1981) *Stochastic Analysis of Macro-dispersion in Three Dimensionally Heterogeneous Aquifers*. Report No. H-8, Hydraulic Research Program. New Mexico Institute of Mining and Technology, Soccorro, New Mexico.

Gelhar, L. and Axness, C. (1983) Three-dimensional stochastic analysis of macrodispersion in aquifers. *Water Resources Research* 19(1), 161–180.

Gelhar, L., Welty, C., and Rehfeldt, K. (1992) A critical review of data on field-scale dispersion in aquifers. *Water Resources Research* 28(17), 1955–1974.

General Sciences Corporation (1998) *SESOIL Reference Guide and User's Guide*. Version 3.0. General Sciences Corporation, 4600 Powder Mill Road, Beltsville, MD.

Ghadiri, H. and Rose, C. (1992) *Modeling Chemical Transport in Soils Natural and Applied Contaminants*. Lewis Publishers, Chelsea, MI.

Glass, R. and Nicholl, M. (1996) Physics of gravity fingering of immiscible fluids within porous media: An overview of current understanding and selected complicating factors. *Geoderma* 70,133–163.

Gorelick, S. and Remson, I. (1982) Optimal dynamic management of groundwater pollutant sources. *Water Resources Research* 18(1), 71–76.

Gorelick, S., Evans, B., and Remson, I. (1983) Identifying sources of groundwater pollution: An optimization approach. *Water Resources Research* 19, 779–790.

Grant, S. (2000) Physical and chemical factors affecting contaminant hydrology in cold environments. United States Army Corps of Engineers, Engineer Research and Development Center. Cold Regions Research and Engineering Laboratory. Hanover, New Hampshire. ERDC/CREEL TR-00-21.

Gschwend, P. and Reynolds, M. (1987) Monodisperse ferrous phosphate colloids in an anoxic groundwater plume. *Journal of Contaminant Hydrology* 1, 309.

Hetrick, D. and Pandey, A. (1999) A methodology for establishing cleanup objectives in the unsaturated soil zone using sensitivity and uncertainty analysis for chemical fate and transport. *Journal of Soil Contamination* 5(5), 559–576.

Hillel, D. (1980) *Introduction to Soil Physics*. Academic Press, Orlando, FL.

Hine, J. and Mookerjee, P. (1975) The intrinsic hydrophilic character of organic compounds. Correlations in terms of structural contributions. *Journal of Organic Chemistry* 40(3), 292–298.

Hitzig, R., Kostecki, P., and Leonard, D. (1998) Study reports LUST programs are feeling effects of MTBE releases. *Soil and Groundwater Cleanup* August/September, 15–19.

Hornsby, A., Rao, P., Wheeler, W., Nkedi-Kizza, P., and Jones, R. (1983) Fate of Aldicarb in Florida citrus soils. 1. Field and Laboratory Studies. *Proceedings of the NWWA/USEPA Conference on Characterization and Monitoring of the Vadose (Unsaturated) Zone*. Las Vegas, NV. National Water Well Association, Columbus, OH.

Huang, C. and Ozisk, M. (1992) Inverse problem of determining unknown wall heat flux in laminar flow through a parallel plate duct. *Numerical Heat Transfer. Part A* 21, 55–70.

Hughes, B., Gillham, R., and Mendoza, C. (1992) Transport of trichloroethylene vapors in the unsaturated zone: a field experiment. *Proceedings of the Conference on Subsurface Contamination by Immiscible Liquids*, International Association of Hydrogeologists, Calgary, Alberta, pp. 81–88. Balkema, Rotterdam.

Huling, S. and Weaver, J. (1991) Dense nonaqueous phase liquids. United States Environmental Protection Agency, Office of Solid Waste and Emergency Response. USEPA/540/4-91-002.

Hunkeler, D., Hoehn, E., Hohener, P., and Zeyer, J. (1997) ^{222}Rn as a partitioning tracer to detect diesel fuel contamination in aquifer: laboratory study and field observations. *Environmental Science and Technology* 31(11), 3180–3187.

Imhoff, P., Thyrum, G., and Miller, C. (1996) Dissolution fingering during the solubilization of nonaqueous phase liquids in saturated porous media. 2. Experimental observations. *Water Resources Research* 32(7), 1929–1942.

Jacobs, J., Guertin, J., and Herron, C. (2000) *MTBE: Effects on Soil and Groundwater Resources*. CRC Press, Boca Raton, FL.

Jarvis, N. (1998) Modeling the impact of preferential flow on nonpoint source pollution. In *Physical Nonequilibrium in Soils: Modeling and Application* (Selim, H. and Ma, L. eds), pp. 195–221. Ann Arbor Press, Chelsea, MI.

Jin, M., Delshad, M., Dawarakanath, V., McKinney, D., Pope, G., Sepehrnooci, Tilford, C., and Jackson, R. (1995) Partitioning tracer test for detection, estimation and remediation performance assessment of subsurface nonaqueous phase liquids. *Water Resources Research* 31(5), 1201–1211.

Jin, M., Butler, M., Jackson, R., Mariner, P., Pickens, J., Pope, G., Brown, C., and McKinney, D. (1997) Sensitivity model design and protocol for partitioning tracers in alluvial aquifers. *Groundwater* 35(6), 964–972.

Jolicoeur, J.L.C. (2000) Groundwater contamination potential of agriculture around Lake Naivasha. Master's Thesis. Comparison of five unsaturated soil zones models. International Institute for Aerospace Survey and Earth Sciences. Enschede, The Netherlands.

Jones, J. (2000) Physical and chemical properties of common contaminants. *Forensics in Environmental Science and Technical Applications.* EPA In-House Environmental Forensics Course. June 19–21, 2000. Lakewood, CO.

Jones, R. and Black, R. (1984) Monitoring aldicarb residues in Florida soil and water. *Environmental Toxicology and Chemistry* 3, 9–20.

Jury, W. and Fluhler, H. (1992) Transport of chemicals through soil: mechanisms, models and field applications. *Advances in Agronomy,* pp. 141–201. Academic Press, London.

Jury, W. and Roth, K. (1990) *Transfer Functions and Solute Movement through Soil: Theory and Applications*. Birkhauser Verlag, Basel.

Jury, W., Spencer, W., and Farmer, W. (1984) Behavior assessment model for trace organics in soil: III Application of screening mode. *Journal of Environmental Quality* 13(4), 573–579.

Jury, W., Sposito, G., and White, R. (1986) A transfer function model of solute transport through soil. 1. Fundamental Concepts. *Water Resources Research* 22, 243–247.

Kaluarachchi, J., Parker, J., and Lenhard, R. (1990) A numerical model for areal migration of water and light hydrocarbon in unconfined aquifers. *Advances in Water Resources* 13, 29–40.

Kezsbom, A. and Goldman, A. (1991) The boundaries of groundwater modeling under the law: Standards for excluding speculative expert testimony. *Environmental Claims Journal* 4(1), 5–30.

Khachikian, C. and Harmon, T. (2000) Nonaqueous phase liquid dissolution in porous media: Current state of knowledge and research needs. *Transport in Porous Media* 38, 3–28.

Kool, J., Parker, J., and van Genuchten, M. (1987) Parameter estimation for unsaturated flow and transport models: A review. *Journal of Hydrology* 91, 255–293.

Kornfeld, I. (1992) Comment to the boundaries of groundwater modeling under the law: Standards for excluding speculative expert testimony. *Tort and Insurance Law Journal* 28(1), 59–68.

Kosmatka, S. and Panarese, W. (1988) *Design and Control of Concrete Mixtures*, 13th edn. Portland Cement Association, Skokie, IL.

Kung, J. (1990a) Preferential flow in a sandy vadose zone. 1. Field observation. *Geoderma* 45, 51–58.

Kung, J. (1990b) Preferential flow in a sandy vadose zone. 2. Mechanism and implications. *Geoderma* 46, 59–71.

Landmeyer, J., Chapelle, F., Bradley, P., Pankow, J., Church, C., and Tratnyek, P. (1998) Fate of MTBE relative to benzene in a gasoline-contaminated aquifer (1993–98). *Groundwater Monitoring and Remediation* 18(4), 93–102.

Lattes, R. and Lions, J. (1969) *The Method of Quasireversibility, Applications to Partial Differential Equations*. Elsevier Press, New York.

Lincoff, A. and Gossett, J. (1984) The determination of Henry's law constant for volatile organics by equilibrium partitioning in closed systems. In *Gas Transfer at Water Surfaces* (Brutsaert, W. and Jirka, G., eds) pp. 17–25. Reidel Publishing, Germany.

Liu, C. and Ball, W. (1999) Application of inverse methods to contaminant source identification from aquitard diffusion profiles at Dover AFB, Delaware. *Water Resources Research* 35, 1975–1985.

Liu, J. (1995) Travel time and location probabilities for groundwater contaminant sources. Master's Thesis. Department of Earth and Environmental Science, New Mexico Institute of Mining and Technology. Socorro, New Mexico.

Lyman, W., Reehl, W., and Rosenblatt, D. (1982) *Handbook of Chemical Estimation Methods*. McGraw-Hill, New York.

Mabey, W. and Mill, T. (1978) Critical review of hydrolysis of organic compounds in water under environmental conditions. *Journal of Physical and Chemical Reference Data* 7, 383–415.

Mahar, P. and Datta, B. (1997) Optimal monitoring network and ground-water pollution source identification. *Journal of Water Resource Planning and Management* 123, 199–207.

Marquardt, W. and Auracher, H. (1990) An observer-based solution of inverse heat conduction problems. *International Journal of Heat Mass Transfer* 33, 1545–1562.

Martin-Hayden, J. and Robbins, G. (1997) Plume distortion and apparent attenuation due to concentration averaging in monitoring wells. *Groundwater* 35(2), 339–347.

McCarty, P. (1993) In situ bioremediation of chlorinated solvents. *Biotechnology* 4, 323–330.

McCarty, P. (1994) An overview of anaerobic transformation of chlorinated solvents. *Symposium on Natural Attenuation of Groundwater*, pp. 104–108. USEPA/600/R-94/162. Office of Research and Development, Denver CO. August 30–September 1, 1994.

McCoy, B. and Rolston, D. (1992) Convective transport of gases in moist porous media: Effect of absorption, adsorption, and diffusion in soil aggregates. *Environmental Science and Technology* 26(12), 2468–2476.

McNab, W. (2001) The chlorinated volatile organic compound historical case analysis database: A tool for understanding plume behavior using data mining approaches. Submitted Abstracts. *11th Annual West Coast Conference on Contaminated Soils, Sediments and Water.* Association for the Environmental Health of Soils. March 19–21, 2000. San Diego, CA.

McNab, W., Rice, D., Bear, J., Ragaini, R., Tuckfield, C., and Oldenburg, C. (1999) Historical case analysis of chlorinated volatile organic compound plumes. Lawrence Livermore National Laboratory, University of California. Environmental Protection Department. UCRL-AR-133361. 31. [http://www-erd.llnl.gov/library/AR-133361.pdf]

Melancon, S., Pollard, J., and Hern, S. (1986) Evaluation of SESOIL, PRZM, and PESTAN in a laboratory leaching experiment. *Environmental Toxicology and Chemistry* 5(10), 865–878.

Mercer, J. (1991) Common mistakes in model applications. In *Proceedings of the American Society of Civil Engineers (ASCE) Symposium on Groundwater* Nashville, TN. July 29–August 1, 1991. American Society of Civil Engineers, Washington, DC.

Mercer, J., Skipp, D., and Griffin, D. (1990) Basics of pump-and-treat groundwater remediation technology. USEPA Report: USEPA/600/8-90/003. USEPA Center for Environmental Research Information, Cincinnati, OH.

Miller, C., Gleyzer, S., and Imhoff, P. (1998) Numerical modeling of NAPL dissolution fingering in porous media. In *Physical Nonequilibrium in Soils Modeling and Application* (Selim, H. and Ma, L. eds), pp. 389–415. Ann Arbor Press, Chelsea, MI.

Miller, R., Biggar, J., and Nielsen, D. (1965) Chloride displacement in Panoche clay loam in relation to water movement and distribution. *Water Resources Research* 1, 63–73.

Millington, R. and Quirk, J. (1959) Permeability of porous media. *Nature (London)* 183, 387–388.

Montgomery, J. (1991) *Groundwater Chemicals Field Guide*. Lewis Publishers, Chelsea, MI.

Mormile, M., Liu S., and Suflita, J. (1994) Anaerobic biodegradation of gasoline oxygenates: Extrapolation of information to multiple sites and redox conditions. *Environmental Science and Technology* 28(3), 1727–1732.

Morrison, R. (1998) Estimating the timing of a contaminant release via transport modeling. *Environmental Claims Journal* 10(2), 75–90.

Morrison, R. (1999a) Reverse and confirmation models: applications and challenges. *Environmental Claims Journal* 12(1), 103–117.

Morrison, R. (1999b) *Environmental Forensics. Principles and Applications*. CRC Press, Boca Raton, FL.

Morrison, R. (1999c) Overview of environmental forensic techniques. In-House Environmental Protection Agency Training. Forensics in Environmental Science and Technical Applications. November 8–10, 1999. Chicago, IL.

toxic torts involving large numbers of plaintiffs are often based on the putative harm done by an air emission source. In principle, exposures can occur via inhalation of ambient air, contact with soils affected by the deposition of contaminants, or indirectly, e.g., via ingestion of vegetables grown in contaminated soils. Inhalation exposures occur mainly during the time that the source is operating; however, exposures to soils, including inhalation exposures to fugitive dusts, can potentially occur long after a source has ceased operation.

There are two main forensic lines of inquiry: (1) interpreting soil contamination patterns in terms of historical sources, the subject of this chapter, and (2) interpreting air monitoring data in terms of currently operating sources, the subject of Chapter 11.

The source in question may not be responsible for the entirety of the observed soil contamination. Natural or regional background, as well as emissions from other nearby sources, may contribute to observed soil concentrations. When contaminants in soil are alleged to originate from a specific air emission source, it is often possible to determine the relative contribution of this source by comparing observed soil concentrations to concentrations estimated using dispersion modeling.

For a source that is no longer operating or even no longer in existence, soil contamination by long-lived or 'recalcitrant' compounds may be the only physical evidence of the source's impact while it was operating. This chapter describes how soil concentrations and spatial distribution patterns are used to determine contributions from air emission sources. To the extent that soil sampling and data analyses provide evidence regarding the emission characteristics of a past source, these analyses can be used to estimate inhalation exposures that occurred when the source was operating.

In Section 9.2 wet and dry deposition processes are described. In Section 9.3, information required to estimate soil concentrations by air dispersion modeling is discussed. Often, historical source emission parameters needed for dispersion modeling are unknown or unavailable and must be estimated. Techniques for estimating source emission parameters are included in Section 9.3.1 and obtaining and interpreting meteorological data is discussed in Section 9.3.3. The widely used Gaussian plume equation which predicts downwind concentrations based on source and meteorological variables is described in Section 9.3.5. Both absolute soil concentrations and soil concentration patterns can be used to test whether or not a particular source is responsible. The two techniques are discussed in Sections 9.3.6 and 9.3.7. In Section 9.4 redistribution processes that can alter soil contamination patterns subsequent to deposition are described. Section 9.5 contains a discussion of using ratios of soil concentrations for different species to identify sources.

enhancement may occur under drain spouts. This results in higher concentrations in samples collected near building foundations than in samples collected in open areas. Mielke (1994) found drip line enhancement of lead in soils around unpainted buildings in inner city locations compared to open yard samples. The source of lead was vehicles burning leaded gasoline. Of course, some care must be used in interpreting this effect for contaminants such as lead because drip line enhancement can also be caused by deterioration of exterior paint. Not surprisingly, Mielke found that painted buildings had a greater lead drip line enhancement than masonry buildings.

In general, with the exception of paint related materials, a drip line enhancement of contaminants is a signature of an air source. Because of soil removal or addition, the effect may or may not be present at a specific building. It may be necessary to sample a population of buildings in order to see a statistically significant effect. Although we are unaware of any studies that have investigated the following possibility, it seems reasonable to expect that the amount of the enhancement for a population of buildings will vary between the side of the building facing the source and other sides of the building.

9.4.3 SEDIMENT ENRICHMENT

Precipitation may redistribute materials deposited on impervious surfaces such as roads and parking lots. Finer particles, which typically contain higher concentrations of contaminants, as discussed below, are easily suspended in heavy rains, transported along drainage pathways, and collected in sedimentation areas of rivers and streams. For this reason, fine sediments collected from rivers, streams, and drainage ditches may contain higher concentrations of contaminants originating from combustion-related sources compared to open area soils. Sediment samples (and drip line samples) should be considered as populations distinct from open area soil samples.

As previously mentioned, particles emitted from controlled combustion sources are generally smaller than a few micrometers in diameter and behave essentially as a gas. Smaller particles typically have higher concentrations of condensed contaminants, because of their larger surface to volume ratio, and are more susceptible to redistribution processes than larger soil particles.

Higher concentrations associated with smaller particles are often reported in terms of an enrichment ratio or enrichment factor. Schroeder *et al.* (1987) define an elemental enrichment factor (EF) based on a comparison of concentration ratios

$$\mathrm{EF} = \frac{C_{\mathrm{i}} / C_{\mathrm{n}}(\text{ambient})}{C_{\mathrm{i}} / C_{\mathrm{n}}(\text{background})} \tag{9.10}$$

where C_i is the concentration of the element of interest and C_n is the concentration of a ubiquitous background element such as iron or silicon. The ratio of C_i/C_n in ambient air particulate is compared to the ratio of C_i/C_n in background soils. The quotient is defined as the enrichment factor.

Similarly, Ghadri and Rose (1991) define an enrichment ratio for sorbed chemicals (nutrients and pesticides) as the 'ratio of a chemical concentration in eroded sediment to that of the original soil from which the sediment originates.' The authors hypothesize that finer particles, richer in sorbed contaminants, are also created by the action of raindrops stripping the concentrated outer layer of larger particles. In soil aggregate peeling experiments, the authors report enrichment ratios for pesticides in the removed outer layers of natural soils of less than a factor of two. Sheppard and Evenden (1992) review enrichment ratios for metals and conclude that most natural redistribution processes result in enrichment ratios less than 10.

9.4.4 RESUSPENSION

Contaminants attached to particles may be resuspended by winds following deposition. When this occurs repeatedly it is referred to as 'saltation.' Wind tunnel studies have shown that emissions rise by the cubic power of wind speed above some threshold velocity (Cowherd *et al.*, 1985). Thus the redistribution pattern over sufficient time will be determined by the high wind speed portion of the wind rose illustrated in Figure 9.1.

9.5 CONCENTRATION RATIOS

In some cases difficulties in determining sources due to redistribution processes can be obviated by examining ratios of different species. If a source is known to have emitted several different contaminants then, in principle, these can be used to find the source 'signature' in soils. For example, Kimbrough and Suffet (1995) used characteristic ratios of lead : antimony : arsenic in air, soil, and waste samples collected near a secondary lead smelter to identify the source of these elements in off-site soils. Characteristic ratios of these elements were maintained even as absolute concentrations decreased with distance downwind of the plant. A similar analysis has been reported near a smelter in Denver (Christen, 2000). In this case cadmium concentrations in soil decreased with distance as expected for smelter emissions but arsenic concentrations did not. Instead arsenic concentrations varied with land use. These observations were partly responsible for the conclusion that arsenic concentrations were largely due to herbicide use.

Care is required in examining ratios of chemicals which degrade in the environment at different rates. For example, during gas phase transport, dioxins and furans are degraded by sunlight with dechlorination of more highly chlorinated isomers occurring more rapidly than that of less chlorinated isomers. This can result in similar dioxin and furan homologue ratios in soils even when emission profiles from different sources were originally distinct.

Care must also be taken in basing a ratio analysis on contaminants that occur naturally at high concentrations or on materials which are ubiquitous, such as gasoline products, phthalate plasticizers, or wood treating products. Dioxins and furans are also widespread contaminants in the environment as discussed in the next section.

9.6 CASE STUDY

This case study is a toxic tort in which the plaintiffs claimed dioxin and furan exposure. Defendants were the operators of and suppliers to a furniture manufacturing operation not far from Dallas, Texas. Waste solvents, paints, and waxes were burned at the site both in a gas fired boiler and in an open pit used for fire training exercises. Wastes were burned in the boiler on a weekly basis while the burn pit was only used occasionally. Most of the plaintiffs were in an urbanized area to the west of the facility. The highest measured concentrations were in the sediments at one location in a creek to the southwest.

All dioxin measurements were converted to toxic equivalent quantities (TEQs), consistent with EPA's 1994 dioxin reassessment (EPA, 1994a), and all quantitative results reported below are in TEQ. At the creek 'hot spot,' several samples were above 1000 parts per trillion (ppt). At other locations concentrations ranged from 2 to 50 ppt. However, only 2 out of 34 values were above 10 ppt. In EPA's 1994 dioxin reassessment the North American background concentration was estimated as 8 ppt. In an EPA reassessment still undergoing review the urban background is estimated as slightly greater than this value (www.epa.gov/ncea/pdfs/dioxin/dioxreass.htm). Since the site is on the outskirts of a small city the question arises as to whether most of the values are simply representative of background.

The creek hot spot was not in the prevailing wind direction, which is from the south. In fact, according to Figure 9.1, winds blow from the northeast toward the hot spot only about 4% of the time compared with about 20% of the time from the south. The isopleths based on dispersion modeling shown in Figure 9.3 also indicate that the highest values would be found to the north of the facility. The hot spot location was wooded but permitted vehicle access. There was evidence it had been used for dumping. The location was also across the street from a hospital with a recently removed medical waste incinerator.

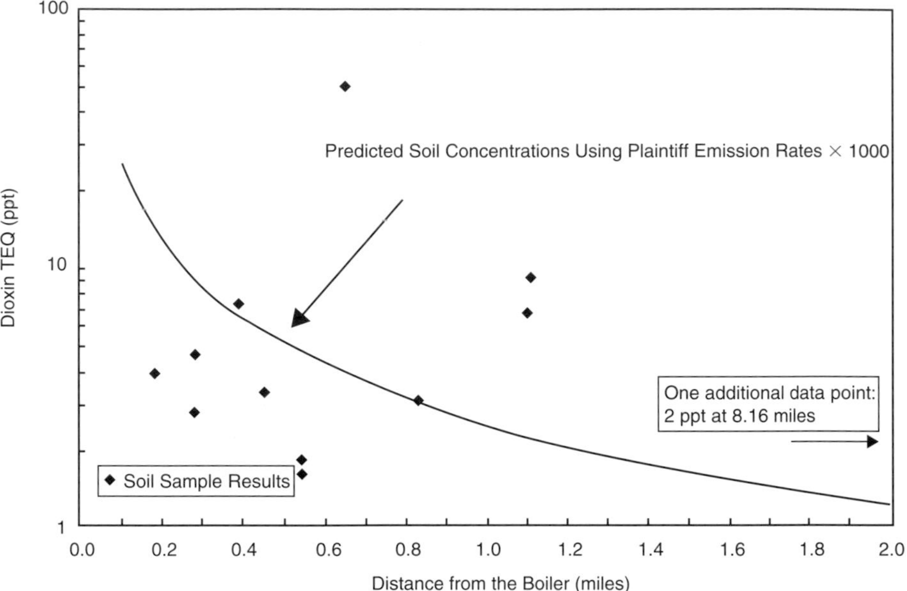

Figure 9.4

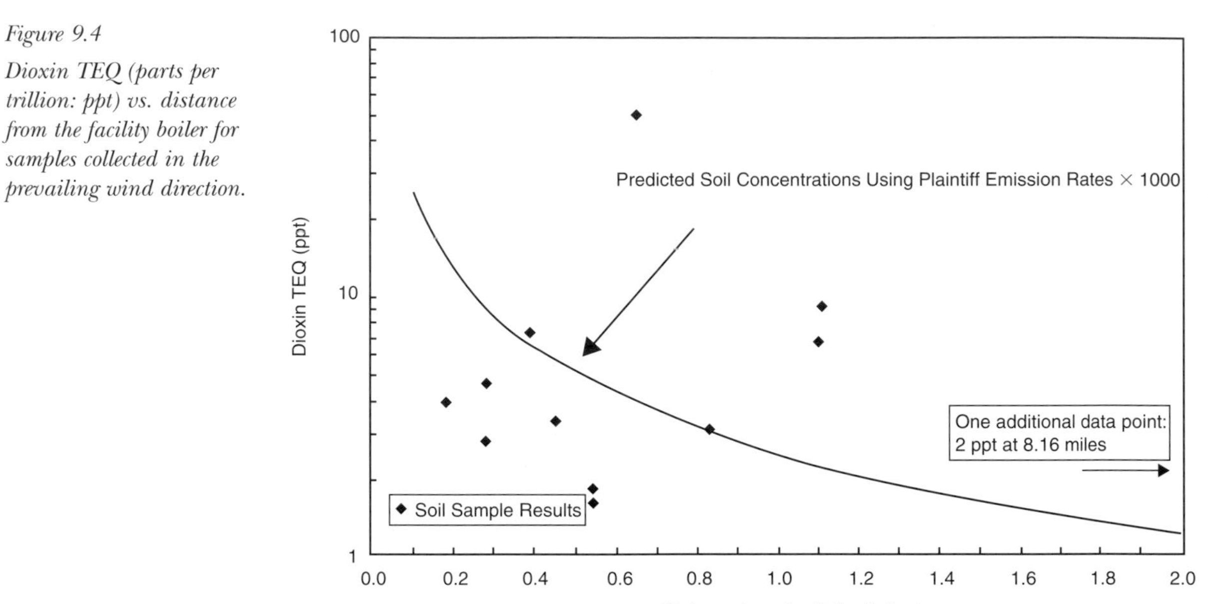

Dioxin TEQ (parts per trillion: ppt) vs. distance from the facility boiler for samples collected in the prevailing wind direction.

Since the furniture manufacturing facility boiler was small with a short stack the peak air concentration for emissions from both the boiler and the burn pit were expected to occur quite close to the facility. A transect of soil measurements was taken going north of the plant in the prevailing wind direction. Results on a semi-log plot are shown in Figure 9.4. Overall there is no indication that concentrations are falling off with distance from the putative source.

The highest concentration shown of nearly 50 ppt occurred at the base on the lee side of a hill where, unlike the other samples, there was a clay patch. The other measurement above 10 ppt, namely 24 ppt, occurred at a drainage ditch for runoff from the facility parking lot. Thus there is some evidence that at least the highest soil concentrations were determined largely by very local conditions.

Note that in Figure 9.4 the curve of predicted soil concentrations has been multiplied by 1000. We used the plaintiff expert's estimated emissions for the dispersion modeling but found that these had to be multiplied by about a factor of 1000 in order to approximate the observed soil concentrations. For the deposition modeling it was assumed that all contaminants reaching the ground would be deposited in the top 2 inches (5 cm) from which the samples were drawn. The plaintiff expert's estimate of emissions was based on calculating the stack gas volume due to burning waste based on stoichiometry and then assuming that dioxins and furans would be present in this stack gas volume at the highest level found in any incinerator in an EPA survey (USEPA, 1994b). The argument was that since emissions from the small industrial boiler were uncontrolled, dioxin levels should be at least this high. However, the reality is counterintuitive, in that the use of control equipment for particles or sulfur dioxide, as was the case for the incinerators surveyed by EPA, necessitates lowering gas temperatures into the dioxin formation range. In facilities without control equipment, stack gas temperature may be above the range of dioxin formation until the stack gases are emitted. In the case of the manufacturing facility boiler the stack gas temperature was at least 900°F (482°C) above the optimum dioxin formation range (250–450°C or 482–842°F). The rapid temperature drop and dilution of stack gases that occurs outside the stack would be effective in limiting dioxin formation.

The plaintiff's expert also noted that the homologue profile or 'fingerprint' shown in Figure 9.5, was the same for all the environmental samples. According to this expert that demonstrated that there was only a single source. However, when the congener profiles for bottom ash samples from the boiler were added, as shown in the figure, these had a distinctly different profile. In fact, the environmental samples probably resemble each other because of the preferential degradation of the lower chlorinated homologues. The bottom ash samples differ, not so much because they come from a different source, although that is true, but because they have not been degraded, in particular photodegraded, in the same way.

As a result of these arguments we were able to conclude that, except for the creek hot spot, dioxins in soils represented background such as would be found in other similar communities. While there were several possible explanations for the creek hot spot, the manufacturing facility operations were not one of the credible ones. The case settled on what we understand to be favorable terms for the defendants who presented these arguments.

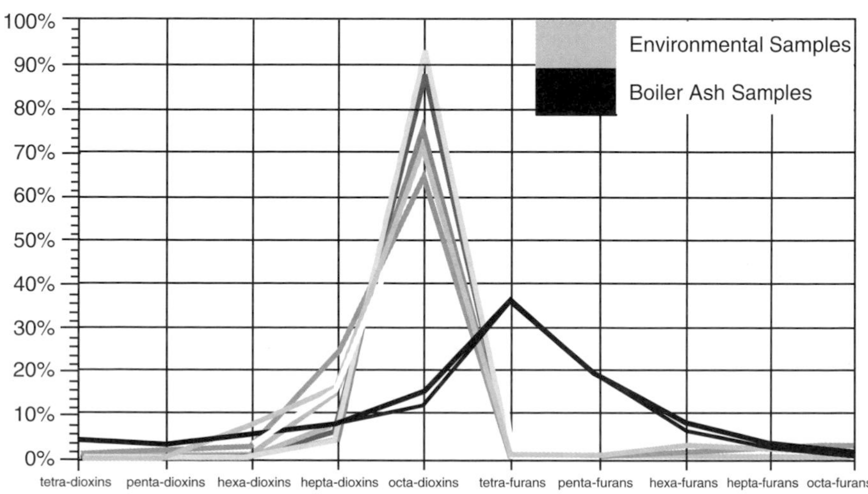

Figure 9.5

Dioxin/Furan homologue fingerprint. (There is no particular rationale for octa-dioxins to be connected by a line to tetra-furans but this is how plaintiff's expert chose to display the data.)

9.7 SUMMARY

For source identification there should be a rough correspondence between measured surface soil concentrations and the concentrations that result from dispersion and deposition modeling. Soil concentrations should be highest downwind in the prevailing wind direction and should be roughly proportional to wind direction frequency in other directions. In the absence of downwash and terrain a single maximum should occur downwind at a distance determined by the typical plume height, which in turn is determined by source operating parameters. Background effects will be minimized if predictions are compared to soil concentrations obtained near this maximum. Comparison will also be facilitated if soil sampling is done to sufficient depth to capture most of the soil loading, particularly in permeable soils.

Redistribution processes also affect soil concentration patterns. Soil concentrations due to airborne sources are enhanced along building perimeters. A concentration enhancement can also occur in sediments relative to soils. In some cases, ratios of different contaminant concentrations can be used to identify sources, even with redistribution, when background and degradation effects are properly accounted for.

ACKNOWLEDGMENT

Certain of the above material appeared in different form in the *Environmental Claims Journal*, Vol. 11, No.2, Winter 1999.

REFERENCES

Bair, F.E. (1992) *Weather of US Cities*. Gale Research, Detroit, MI.

Christen, K. (2000) Arsenic in old herbicides comes back to haunt Denver. *Environmental Science and Technology* 34, 376A–377A.

Cowherd, C., Muleski, G., Engelhart, P., and Gillette, D. (1985) *Rapid Assessment of Exposure to Particulate Emissions from Surface Contamination*. Office of Health and Environmental Assessment, US Environmental Protection Agency, Washington, DC, EPA/600/8-85/002,NTIS PB85-192219.

Ghadri, H. and Rose, C.W. (1991) Sorbed chemical transport in overland flow: 1. Nutrient pesticide enrichment mechanism. *Journal of Environmental Quality* 20, 628–633.

Hanna, S.R., Briggs, G.A., and Hosker, R.P. (1982) *Handbook on Atmospheric Diffusion*. Technical Information Center, US Department of Energy.

Kimbrough, D.E. and Suffet, I.H. (1995) Off-site forensic determination of airborne elemental emissions by multi-media analysis: A case study at two secondary lead smelters. *Environmental Science and Technology* 29, 2217–2221.

Mielke, H.W. (1994) Lead in New Orleans soils: new images of an urban environment. *Environmental Geochemistry and Health* 16(3/4), 123–128.

Schroeder, W.H., Dobson, M., Kane, D.M., and Johnson, N.D. (1987) Toxic trace elements associated with airborne particulate matter: A review. *Journal of the Air Pollution Control Association* 37(11), 1267–1285.

Sheppard, S.C. and Evenden, W.G. (1992) Concentration enrichment of sparingly soluble contaminants (U, Th, and Pb) by erosion and by soil adhesion to plants and skin. *Environmental Geochemistry and Health* 14, 121–131.

Turner, D.B. (1994) *Workbook of Atmospheric Dispersion Estimates – An Introduction to Dispersion Modeling*. CRC Press, Chapel Hill, NC.

US Environmental Protection Agency (1994a) Estimating Exposure to Dioxin-Like Compounds (draft). Office of Research and Development, Washington, DC. EPA/600/6-88/005Ca.

US Environmental Protection Agency (1994b) Combustion Emissions Technical Resource Document (CETRED), Draft. Office of Solid Waste and Emergency Response, Washington, DC. EPA 530-R-94-014.

US Environmental Protection Agency (1995) *Air Quality Criteria for Particulate Matter,* vol. I, EPA/600/AP-95/001a. Office of Research and Development, Washington, DC.

US Environmental Protection Agency (1999) Appendix W to Part 51 – Guideline on Air Quality Models, 40 CFR Ch. I (7-1-99 Edition).

probabilities.' Thus, for example, at $p = 0.1$ and $v = 5$, 90% of the area under the curve lies within ±2.015 standard deviations, 5% lies greater than 2.015 standard deviations, and 5% lies less than −2.015 standard deviations away. More extensive tables of '*t* values' are found in most statistics texts (for example, see Zar, 1984; Kachigan, 1991; or Gilbert, 1987).

Although most statistical tests are based on the assumption that the underlying distribution is normal, most environmental data appear to have frequency distributions that are log-normal. Two advantages of the log-normal distribution in describing environmental data are that it always gives positive values (there are no negative concentrations) and it can account for a small fraction of higher values (hot spot contamination) in the right side or 'tail' of the curve. Ott (1990) has shown that a series of successive random dilutions results in a distribution of values that is log-normal and thus provides a physical explanation for why environmental data often appear to be distributed this way.

By definition, data are said to be log-normally distributed if the log-transformed values are normally distributed. This means log-normal data can be analyzed using statistical techniques requiring normally distributed data if the data are first natural log-transformed. This is an important point because most parametric techniques require an assumption of normality or data that can be transformed to fit a normal distribution. The probability distribution function (PDF) for the log-normal distribution is given in Equation 10.10 where μ_y and σ_y are the mean and standard deviation of the transformed variable $y = \ln(x)$.

$$f(x) = \frac{1}{x\sigma_y\sqrt{2\pi}} e^{-\frac{(\ln x - \mu_y)^2}{2\sigma_y^2}} \qquad (10.10)$$

The transformation step can introduce some complexities, as discussed by Gilbert (1987). For example, estimated quantities in the transformed scale (e.g. μ_y and σ_y) can lead to biased estimates when they are transformed back into the original scale (Wallace, 1997). Also, data presented in the transformed scale can be difficult to interpret. A number of authors have addressed these issues specific to the log-normal distribution (Land, 1975; Parkin *et al.*, 1988; Singh *et al.*, 1997).

10.2.3 CONFIDENCE LIMITS AND HYPOTHESIS TESTS

It is evident from the preceding discussion that with a given distribution assumption, it is possible to determine the probability of randomly selecting an object with a value less than or greater than a selected value (or between two selected values) from that distribution. For example, given a normal distribution with a mean of 10 and standard deviation of 2, we can determine the probability of randomly selecting an object with a value of 7 or less. Using Equation 10.9 we can calculate a z-score of $(7 - 10)/2 = -1.50$ and look up the probability corresponding to a z-score of -1.5. From a standard table of z-scores we find the probability $= 0.0668$. Thus, there is a 6.68% probability of selecting an object with a value of 7 or less from a population of values fitting a normal distribution with a mean of 10 and standard deviation of 2.

Because of the central limit theorem, we can make similar statements about the probability of estimating a mean value from a sample of size n drawn from a normal distribution with a mean of 10 and standard deviation of 2. For example, what is the probability of calculating a sample mean greater than 11 in a sample of 8 objects drawn from that distribution? Using Equation 10.6, we can construct an expression similar to the expression in Equation 10.9.

$$z = \frac{(\bar{x} - \mu)}{\sigma_{\bar{x}}} = \frac{(\bar{x} - \mu)}{\frac{\sigma}{\sqrt{n}}} \tag{10.11}$$

Thus, using Equation 10.11 we can calculate a z-score of 1.41 and look up a corresponding probability of 0.0793. Thus, there is a 7.93% probability of calculating a sample mean of 11 or greater in a sample of 8 objects randomly selected from a population of values fitting a normal distribution with a mean of 10 and standard deviation of 2. The significance of the central limit theorem is that the underlying population does not have to be normal, particularly if n is large.

10.2.3.1 Confidence Limits

This basic procedure can also be used to calculate symmetrical confidence limits about the true mean of the underlying population. For example, from Table 10.2 we know that 95% of the area under a standard normal curve is bound by $z = \pm 1.96$. If we plug this value into Equation 10.11 and solve for $\bar{x}$ we get

$$\bar{x} = \mu \pm 1.96 \frac{\sigma}{\sqrt{n}} \tag{10.12}$$

Thus there is a 95% probability that a sample mean of 8 objects randomly selected from the underlying population will lie between 8.62 and 11.38 (i.e., $10 - 1.38$ and $10 + 1.38$). These upper and lower limits are called confidence limits, and in this case would be referred to as lower and upper 95% confidence limits about the mean. This is commonly written as:

$$P\left\{\mu - 1.96 \frac{\sigma}{\sqrt{n}} \leqslant \bar{x} \leqslant \mu + 1.96 \frac{\sigma}{\sqrt{n}}\right\} = 0.95 \tag{10.13}$$

But we never really know the true mean and true standard deviation (μ and σ) of a given population. In practice, a sample mean is calculated from a given data set and then confidence limits are determined. Then we can make the statement that there is a given probability that the true mean lies somewhere within these limits. Since we do not know the true population standard deviation, σ, we need to use the sample standard deviation, s, to calculate confidence limits. We also need to use the t distribution instead of the normal distribution to account for the uncertainty in our estimation of the standard deviation.

$$P\left\{\bar{x} - t\frac{s}{\sqrt{n}} \leqslant \mu \leqslant \bar{x} + t\frac{s}{\sqrt{n}}\right\} = p \tag{10.14}$$

If we continue with our example, the appropriate t statistic corresponding to a 95% probability level with $v = n - 1 = 7$ degrees of freedom is 2.365 (compared to a corresponding z-score of 1.96 if our sample contained all the measurements). If we assume the sample standard deviation is 2.5, then we can say with 95% probability that the true mean lies somewhere between 8.91 and 13.09 (11 − 2.09 and 11 + 2.09). Note that we can narrow our confidence limits about the mean simply by increasing the number of measurements in our sample.

10.2.3.2 Hypothesis Testing

In hypothesis testing, the convention is to assume that there is no difference between two values (this is referred to as the 'null hypothesis' denoted H_o). This assumption holds unless we can determine that the probability of seeing such a large difference is so small that it is more likely that our initial assumption is wrong than it is we have encountered this rare occurrence. The convention is to conclude that our initial assumption is wrong when the probability associated with the null hypothesis is 5% or less.

When we establish our null hypothesis, it is also important to establish an alternate hypothesis (denoted H_a). For a null hypothesis of no difference between two sample means, i.e., H_o: $\mu_a = \mu_b$ there are three different alternate hypotheses: H_a: $\mu_a \neq \mu_b$; H_a: $\mu_a > \mu_b$; or H_a: $\mu_a < \mu_b$. The *a priori* decision of which alternate hypothesis is selected determines whether we are performing a one-tailed or two-tailed test. If our decision level is set at 5%, and our alternate hypothesis is H_a: $\mu_a \neq \mu_b$, then we do not care if $\mu_a > \mu_b$ or $\mu_a < \mu_b$. In this case, our 5% decision level is interpreted as a 2.5% chance of rejecting our null hypothesis when in fact $\mu_a > \mu_b$ and a 2.5% chance of rejecting the null hypothesis when in fact $\mu_a < \mu_b$. This is referred to as a two-tailed test. If, however, we really would like to know if the average concentration on site is greater than the average background level, i.e., $\mu_a > \mu_b$, and we are willing to accept a 5% chance of wrongly rejecting our null hypothesis, then we would be performing a one-tailed test.

In making this decision to accept or reject the null hypothesis, two types of error are possible, referred to as a Type I error and a Type II error. A Type I error occurs when the null hypothesis is rejected (we cannot conclude that the values are drawn from the same population) when in fact the values are drawn from the same population. A Type II error occurs when we accept the null hypothesis of no difference between means when in fact there is a difference.

The probability of committing a Type I error is equal to α (alpha), the significance level. Alpha is a value that is chosen by the investigator and usually set equal to 0.05. With alpha set equal to 0.05, we are willing to accept a 5% chance of rejecting our null hypothesis of no difference between means when in fact the null hypothesis is true. The probability of committing a Type II error is equal to β (beta), the power. The power of a test is more difficult to determine. It is a function of alpha, the standard error of the difference between the two means, and the size of the effect that we are trying to detect.

The power of the test is often neglected, which can lead to a false confidence in decision-making. For example, if $\alpha = 0.05$, and $\beta = 0.5$, there is a 5% chance of rejecting the null hypothesis when in fact it is true; but, there is a 50% chance of accepting the null hypothesis when it is false. For this reason, some statisticians believe that the only significant conclusion is when the null hypothesis is rejected (for example see Oakes, 1986; Reckhow *et al.*, 1990). If the null hypothesis cannot be rejected, the reason may be that the test had insufficient power to detect a difference (e.g., the sample size was too small or the data are highly variable). Thus, a poorly designed test based on too few measurements can be biased towards accepting the null hypothesis.

10.3 APPLICATIONS IN ENVIRONMENTAL FORENSICS

10.3.1 COMPARING SAMPLE MEANS

Evidence that a facility may have contributed to the level of contamination at a site can be demonstrated by comparing data collected at the site with an appropriate set of background measurements. In comparing two data sets, it is important to keep in mind that it is not possible to prove that two data sets are different; rather, we are only able to assess the likelihood that they are drawn from the same population.

10.3.1.1 Student's *t*-Test

Perhaps the most common test for comparing samples is Student's *t*-test, also known as the *t*-test. The *t*-test is a parametric technique, which as previously described, means that some underlying distribution is assumed for the test. For the *t*-test (and most parametric techniques) it is assumed that each sample is drawn from a normal distribution. It is also assumed that the measurements in the sample were randomly selected from the distribution and that the variances of the two data sets are equal. (The *t*-test can also be used to compare a data set with a single value, for example to compare the mean of a data set to some cleanup level.)

In conducting a *t*-test for comparing means, the underlying hypothesis is that there is no difference between sample means. In other words, we establish the null hypothesis that the difference in the means is zero, i.e., H_o: $\mu_1 = \mu_2$ or, H_o: $\mu_1 - \mu_2 = 0$. In most cases, the alternate hypothesis is that the means are not the same, i.e. H_a: $\mu_1 \neq \mu_2$. The test statistic is given as:

$$t = \frac{\bar{x}_1 - \bar{x}_2}{\frac{s_1}{\sqrt{n_1}} + \frac{s_2}{\sqrt{n_2}}} \tag{10.15}$$

If the variances are equal, as they are assumed to be in the *t*-test, then we can calculate a pooled variance, s_p^2, and pooled standard deviation, s_p where s_p = square root of s_p^2.

$$s_p^2 = \frac{(n_1 - 1)s_1^2 + (n_2 - 1)s_2^2}{n_1 + n_2 - 2} \tag{10.16}$$

The variance ratio test or *F* test can be used to evaluate whether or not the variances are in fact equal, i.e., whether or not it is likely that the two sample variances are drawn from the same population. We test the hypothesis H_o: $\sigma_1^2 = \sigma_2^2$ and assign some probability alpha, of committing a Type I error. The *F* statistic is calculated by simply dividing the larger variance by the smaller variance. This ratio is then compared to the tabular value of $F_{\alpha,v1,v2}$ for a given alpha and the degrees of freedom associated with each data set (tables of *F* values are found in most statistics texts). If we accept the null hypothesis and conclude no difference in variances, the *t* statistic can be calculated using the pooled standard deviation as indicated below.

$$t = \frac{\bar{x}_1 - \bar{x}_2}{s_p \sqrt{\frac{1}{n_1} + \frac{1}{n_2}}} \tag{10.17}$$

An example data set to demonstrate application of the *t*-test is provided in Table 10.4. Summary statistics are listed in Table 10.5. The data consist of 25 measurements of lead in soil collected on-site (on-site data) and 25 measurements collected off-site, representing background measurements (background data).

Table 10.4

Example data set used to demonstrate the t*-test.*

On-site Concentrations (ppm)	Background Concentrations (ppm)
353	86
300	111
288	306
459	111
345	346
324	142
473	73
339	217
236	149
310	417
359	111
420	584
357	68
581	263
762	171
315	176
406	266
279	146
504	155
233	81
299	336
371	106
452	80
384	101
480	178

Table 10.5

Summary statistics for the data presented in Table 10.4.

	On-site Data	Background Data
Number of samples	25	25
Minimum value	233.4	67.8
Maximum value	761.8	583.6
Sample average	385	191
Sample variance	13 515	15 925
Sample standard deviation	116	126
Standard error of the mean	23.3	25.2

A frequency distribution of the data is presented in Figure 10.2. The values on the *x*-axis represent the upper end of the bin range. Thus, the first bin range covers the values from 0 to 75, and the second runs from 76 to 150, etc.

Test assumptions about normality and equal variance should be checked before performing the *t*-test. As indicated in Figure 10.2, both data sets appear to have a distribution that is skewed to the right, resembling a log-normal distribution. The coefficient of variation test, described by the US Environmental Protection Agency (USEPA, 1989), provides a simple and quick check to detect gross non-normality in the data set. The coefficient of variation is equal to the

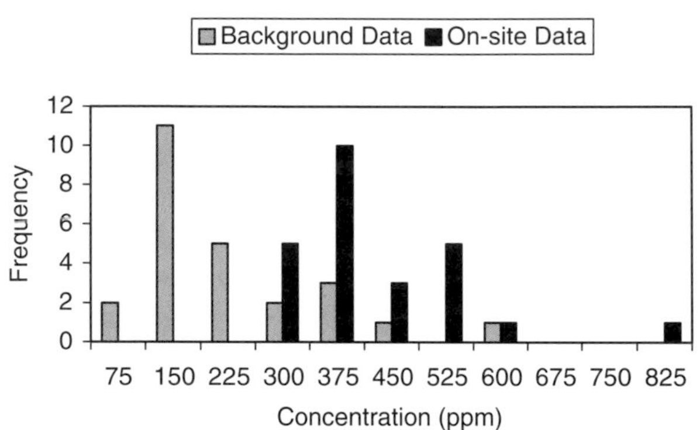

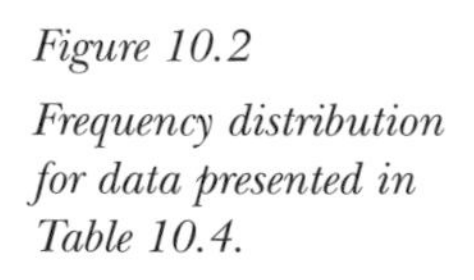

Figure 10.2
Frequency distribution for data presented in Table 10.4.

standard deviation divided by the mean. A value greater than 1.0 is evidence that the data are not normally distributed. In this example, both coefficients of variation are less than 1.0 so there is no evidence of gross non-normality.

A more powerful test that can detect whether or not a data set fits a specified distribution is the χ^2 (chi-squared) goodness of fit test[3]. The χ^2 test compares observed frequencies in a sample data set with expected frequencies if the data set fit some known distribution e.g., normal, log-normal, or some other specified distribution. The χ^2 statistic is calculated as

$$\chi^2 = \sum_{i=1}^{k} \frac{\left(N_i - E_i\right)^2}{E_i} \tag{10.18}$$

where k is the number of bin ranges, N is the observed number of measurements falling within a given bin range, and E is the expected number of measurements falling within the range. If there were no difference between the observed and expected frequencies, the value for χ^2 would be zero. As the calculated value for χ^2 increases, we become less certain that the distributions are similar. If the χ^2 statistic exceeds the tabular value (found in most statistics texts) for a given alpha level with $k - 3$ degrees of freedom, we must reject the null hypothesis of no difference between distributions.

The number of bin ranges should be selected so that E is at least 4. Thus, with $n = 25$, k should be less than $25/4 = 6.25$, so we set $k = 6$. The easiest way to perform this test by hand is to select bin ranges so that $E_1 = E_2 = E_3$ etc. Thus, for $n = 25$ and $k = 6$, E is equal to $25/6 = 4.17$. USEPA (1989) contains a table of appropriate bin ranges for a normal distribution that can be used for any data set if the data are first converted to *z*-scores. Bin ranges for $k = 6$ are presented in Table 10.6 along with the observed frequencies for the on-site and background data sets.

Using Equation 10.18, the calculated χ^2 statistics for the on-site and background data sets are 6.43 and 12.7, respectively. The tabular value for χ^2 with

[3] Note that there are other techniques for testing distribution assumptions such as the Kolmogorov Smirnov test and the Wilk Shapiro test (see Zar, 1984; Gilbert, 1987)

Table 10.6

Observed and expected frequencies calculated in the χ^2 test for normality for the data set in Table 10.4.

Bin Range for $k = 6$	Expected Frequency (E)	Observed Frequency (N) On-site	Observed Frequency (N) Background
< -0.97	4.17	2	1
-0.97–-0.43	4.17	7	9
-0.43–0	4.17	7	7
0–0.43	4.17	2	1
0.43–0.97	4.17	4	3
>0.97	4.17	3	4

Table 10.7

Summary statistics for the natural log transformation of the data presented in Table 10.4.

	On-site Data	Background Data
Number of samples	25	25
Minimum value	5.45	4.21
Maximum value	6.64	6.37
Sample average	5.92	5.08
Sample variance	0.0751	0.345
Sample standard deviation	0.274	0.587
Standard error of the mean	0.0548	0.117

Table 10.8

Observed and expected frequencies calculated in the χ^2 test for normality for the natural log transformation of the data set in Table 10.4.

Bin Range for $k = 6$	Expected Frequency (E)	Observed Frequency (N) On-site	Observed Frequency (N) Background
< -0.97	4.17	5	3
-0.97–-0.43	4.17	5	6
-0.43–0	4.17	4	5
0–0.43	4.17	3	3
0.43–0.97	4.17	3	5
>0.97	4.17	5	3

3 degrees of freedom at the 95% probability level ($\alpha = 0.05$) is 7.815. Since the χ^2 statistic for the background data exceeds the tabular value, we cannot conclude that the data were drawn from a normal distribution.

When the data are found to be significantly different from normal, it may be possible to 'normalize' the data by taking the natural log of each data point. This is a common transformation since most environmental data sets are adequately described by a log-normal distribution. Summary statistics for the log-transformed data are presented in Table 10.7. Observed and expected frequencies for the χ^2 test are found in Table 10.8.

Note that the bin ranges and expected frequencies in Table 10.8 remain the same if the test is performed on data that have been converted to z-scores. The calculated χ^2 statistics for the log-transformed on-site and background data sets

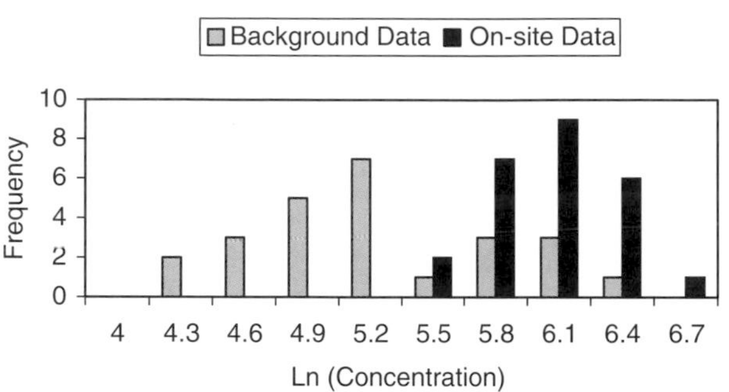

Figure 10.3
Frequency distribution for data presented in Table 10.4 after log transformation.

are 1.16 and 2.11, respectively. The tabular value for χ^2 with 3 degrees of freedom at the 95% probability level is also the same, 7.815. Since the χ^2 statistic for both the on-site and background data are less than the tabular value, we can conclude that the log-transformed data sets are not significantly different from normal. The frequency distributions of the natural log-transformed data are presented in Figure 10.3. Plotting the data is always recommended to get a 'feel' for the data set. The on-site data clearly resemble a normal distribution; however, the background data still exhibit some deviation from normal, but nothing significant according to the χ^2 test.

The ratio of the variances in the log-transformed data set is $0.345/0.0751 = 4.59$. This ratio can be compared to the tabular F value with $n - 1 = 24$ degrees of freedom for the numerator and denominator $F_{0.05,\,24,\,24} = 2.27$. Since the variance ratio exceeds the tabular F value, we cannot conclude that the variances are equal at the 95% probability level. (This is clearly evident in Figure 10.3.) The scenario where both data sets are not significantly different from normal but the variances are unequal is referred to as the 'Behrens–Fisher problem' (Zar, 1984). One solution to this problem is known as Cochran's approximation to the Behrens–Fisher (Zar, 1984; Ross, 1997). The t statistic is calculated according to Equation 10.19 as illustrated below.

$$t = \frac{5.92 - 5.08}{\dfrac{0.274}{\sqrt{25}} + \dfrac{0.587}{\sqrt{25}}} = \frac{0.84}{0.172} = 4.88 \tag{10.19}$$

However, the tabular value is Student's t with degrees of freedom calculated according to Equation 10.20.

$$v = \frac{\left(\dfrac{s_1^2}{n_1} + \dfrac{s_2^2}{n_2}\right)^2}{\dfrac{\left(\dfrac{s_1^2}{n_1}\right)^2}{n_1 - 1} + \dfrac{\left(\dfrac{s_2^2}{n_2}\right)^2}{n_2 - 1}} \tag{10.20}$$

In this case, $v = 17.7$ (or 17, when rounded to the next smallest integer). The tabular value for Student's t with alpha = 0.05 and 17 degrees of freedom is 2.11. Since the calculated value of 4.88 exceeds the tabular value of 2.11, we must reject the null hypothesis and conclude that there is less than a 5% probability that the observed difference between the two means could occur by chance. This is strong evidence that lead concentrations on-site are higher than background levels and suggests that the facility has indeed affected the soils on-site. In addition, the t-test has allowed us to arrive at this conclusion with a quantitative estimate of the confidence in our conclusion.

10.3.1.2 Wilcoxon's Rank Sum Test

Data sets that cannot be transformed to approximate a normal distribution can be compared using a non-parametric technique such as Wilcoxon's rank sum test (also known as the Mann–Whitney U test). With a non-parametric test, there is no requirement that the data fit some underlying distribution. Wilcoxon's rank sum test is designed to test the null hypothesis that the two data sets are drawn from the same distribution (whatever it is), or similar distributions with the same central tendency.

The test is conducted by combining the data points from the two samples, while maintaining their identity, and ranking the data from lowest to highest (or highest to lowest). Ties (data points with the same value) are assigned the average of the ranks they would have been assigned had they not been tied. Gilbert (1987) notes that an advantage of the Wilcoxon rank sum test is that it can handle a moderate number of non-detect values by treating them as ties.

The sum of the ranks from each sample (R_1 and R_2) are tallied and the smaller sum is selected for comparison (n_R is the number of data points associated with the sample with the smaller sum). For samples with $n \leq 20$, R may be compared directly to a table of R values (for example see Langley, 1970). For samples with more than 20 data points, the significance of R can be determined by computing Z_{rs}.

$$Z_{rs} = \frac{R - \dfrac{n_R(n_1 + n_2 + 1)}{2}}{\sqrt{\dfrac{n_1 \cdot n_2(n_1 + n_2 + 1)}{12}}} \tag{10.21}$$

Because R is approximately normal for large n, all we are really doing is calculating a standard normal deviate analogous to Equation 10.9 where:

$$Z_{rs} = \frac{R - E(R)}{\sqrt{\mathrm{Var}(R)}} \tag{10.22}$$

the expected value of R, $E(R)$ is given as

$$E(R)=\frac{n_R(n_1+n_2+1)}{2} \tag{10.23}$$

and the variance Var(R) is estimated using the equation below (Reckhow *et al.*, 1990).

$$\mathrm{Var}(R)=\frac{n_1\cdot n_2(n_1+n_2+1)}{12} \tag{10.24}$$

Gilbert (1987) also gives an equation for calculating Z_{rs} when ties are present.

$$Z_{rs}=\frac{R-\frac{n_R(n_1+n_2+1)}{2}}{\left\{\frac{n_1 n_2}{12}\left[n_1+n_2+1-\frac{\sum_{j=1}^{g}t_j\left(t_j-1\right)^2}{(n_1+n_2)(n_1+n_2-1)}\right]\right\}} \tag{10.25}$$

where g is the number of tied groups, and t_j is the number of tied data in the jth group.

In our example data set presented in Table 10.4, the smaller sum of ranks is equal to 396 and there is one group of ties with $t = 3$. The resulting value of Z_{rs} calculated using Equation 10.25 is 4.68 – corresponding to a <0.001 probability that the two data sets are drawn from the same underlying distribution. The results of this test confirm our earlier conclusion based on Student's t test.

Higgs *et al.* (1999) used the rank sum test to compare patterns of soil lead contamination observed on elementary school properties. The authors found that soils on inner-city school properties had significantly higher lead concentrations than soils on outer-city school properties, consistent with earlier findings that lead tends to cluster within the interior of the largest cities.

10.3.1.3 Paired *t*-Test

If paired data sets are available, for example, upwind/downwind measurements, then the paired t-test can be used. In the paired t-test, the null hypothesis is that the population mean of the differences between each data pair (μ_d) is equal to zero i.e., H_o: $\mu_d = 0$, where d_i is the difference between the ith data pair and $\bar{d}$ is the sample mean of the differences between each data pair. The paired t-test assumes that the values of d_i are normally distributed. The test statistic is calculated as

$$t=\frac{\bar{d}}{\frac{s_d}{\sqrt{n}}} \tag{10.26}$$

where:

$$s_d = \sqrt{\frac{\Sigma\left(d_i - \bar{d}\right)^2}{n-1}} \tag{10.27}$$

The calculated t statistic is compared with the tabular t value with $n - 1$ degrees of freedom. If the calculated value exceeds the tabular value then the null hypothesis is rejected.

Daily upwind–downwind total suspended particulate (TSP) measurements collected at the site of a former gold mine in California during the months of July, August and September are presented in Table 10.9. The downwind–upwind difference between each data pair reflects a contribution from the site. A scatter plot of downwind–upwind measurements for the three months (Figure 10.4) indicates that downwind TSP concentrations are generally greater than upwind measurements.

Table 10.9
Upwind/downwind TSP concentrations ($\mu g/m^3$) for July, August, and September at a former gold mining site in California.

Day	July		August		September	
	Upwind	Downwind	Upwind	Downwind	Upwind	Downwind
1	56	116	59	112		
2	64	78			84	68
3	73	86			75	56
4			53	85	78	69
5			66	165	82	53
6			68	167		
7	36	53	58	145		
8	45	98	56	95	166	56
9	72	115			105	73
10	62	70			84	63
11					90	39
12					180	63
13						
14	51	147	89	148		
15	63	171	75	161	87	38
16	56	157			105	46
17	66	150			35	38
18	47	125	48	78	67	36
19			55	53		
20			53	137		
21	125	113	55	103		
22	51	88	54	76	123	36
23	50	50			93	108
24	32	118			45	15
25	41	169	61	75	69	33
26			65	102	83	36
27			61	55		
28	53	166	62	82		
29	59	75	57	68	41	34
30	86	76				

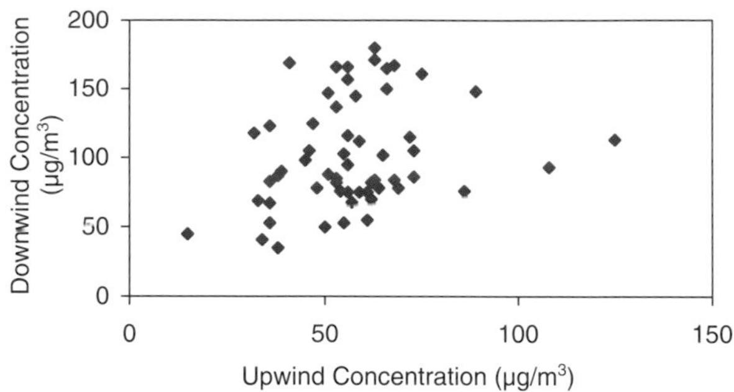

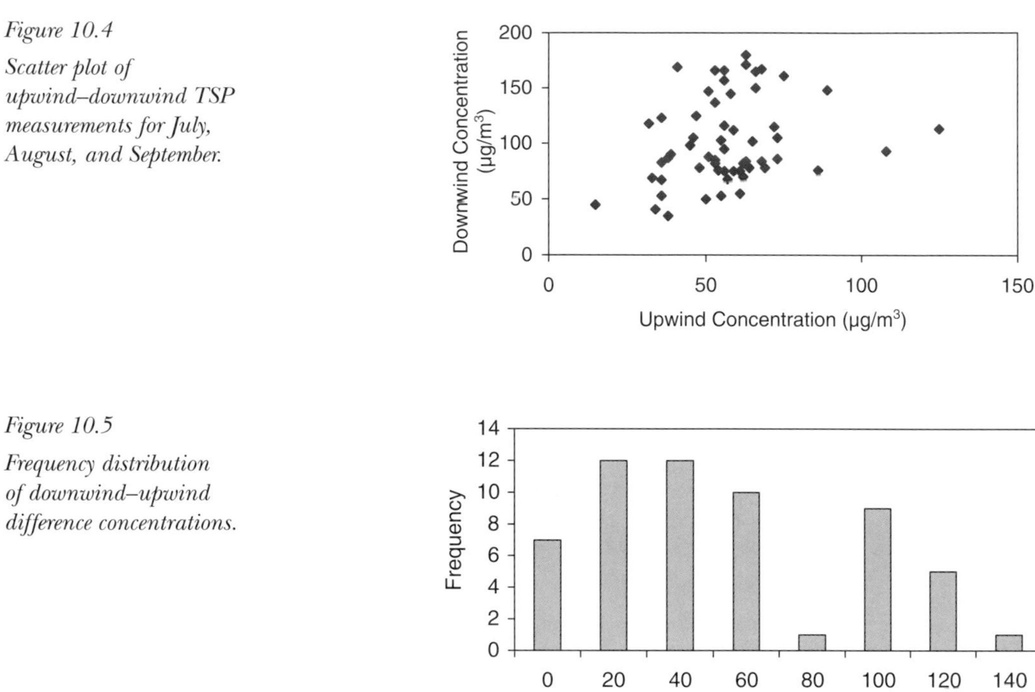

Figure 10.4
Scatter plot of upwind–downwind TSP measurements for July, August, and September.

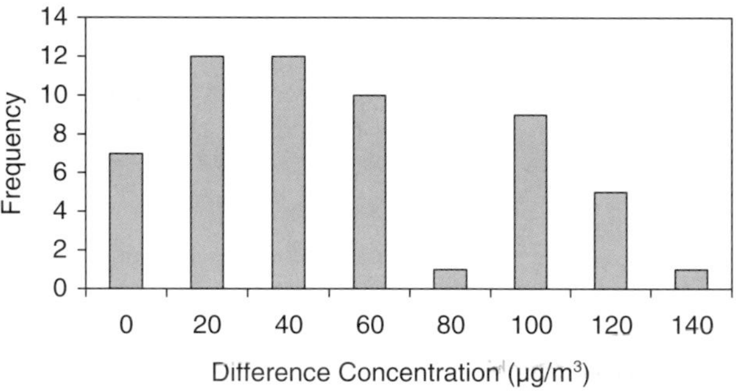

Figure 10.5
Frequency distribution of downwind–upwind difference concentrations.

A frequency distribution plot of the downwind–upwind difference concentrations is shown in Figure 10.5. Note that the first bin includes data in the range of −20 to 0 and represents data pairs where upwind concentrations exceed downwind values.

Once again, we can test the assumption of normality with a χ^2 test. Using Equation 10.18, and dividing the 57 difference measurements into 8 equal bins, the calculated χ^2 statistic is 6.30 compared to a tabular value of 11.07 for $\alpha = 0.05$ and $v = 5$. Thus, we can accept our null hypothesis of no difference between the observed distribution of difference values and a normal distribution. We can now proceed with the paired t-test. The average downwind–upwind difference between the 57 data pairs is 45.2 and the standard deviation is 38.2. Thus using Equation 10.26, we calculate $t = 8.94$.

In comparison, the tabular value for $t_{0.05,\,56} = 2.003$ so we reject our null hypothesis that the mean difference of 45.2 is not significantly different than zero. We can take this analysis one step further and calculate confidence limits about the mean difference using Equation 10.14.

$$P\left\{45.2 - 2.003\frac{38.2}{\sqrt{57}} \leq \mu_d \leq 45.2 + 2.003\frac{38.2}{\sqrt{57}}\right\} = 0.95 \qquad (10.28)$$

Thus there is a 95% probability that the true mean difference for the months of July, August, and September lies somewhere between 35.1 and 55.3 $\mu g/m^3$. Note that only three months out of the year have been selected for this example. This analysis could also be performed for all the available data. If the investigation was being conducted in support of a toxic tort case, then the time period of exposure could be analyzed. It is also possible that exposure could be greater in certain months. This hypothesis could be tested using an analysis of variance technique.

10.3.1.4 Analysis of Variance (ANOVA)

If we wish to compare more than two samples, for example the months of July, August, and September, we could perform multiple *t*-tests. However, as the number of samples to be tested increases, the number of possible comparisons becomes large. For example, comparing five samples would require $5(5-1)/2 = 10$ *t*-tests; and 10 samples would require 45 *t*-tests. Moreover, as we perform more tests, we increase our chances of committing a Type I error, even though each test might be conducted at the 0.05 level. This is because, at the 0.05 level, we would expect to observe a significant difference between samples in 1 out of 20 tests strictly by chance.

Analysis of variance (ANOVA) is a technique that can be used to compare the means of two or more samples. (If there are only two samples, then ANOVA is identical to the two-sample *t*-test.) It is a parametric technique so all of the same assumptions apply. It is assumed that the data are drawn at random from a normal population, and that sample variances are equal.

The one-way analysis of variance technique involves, just as the name implies, an analysis of the variability or variance in a set of measurements. The test is usually conducted on experimental data designed to test the effect of various treatments on the outcome of a *single* variable (hence the one-way). In this case, the null hypothesis is that the means for each treatment group are the same, i.e. H_o: $\mu_1 = \mu_2 = \mu_3$ etc. In our example, we could test the hypothesis that there was no significant difference in operations at the facility between the months of July, August, and September that would affect the contribution of TSPs to ambient air. To test the hypothesis, we determine if the variability between months is any greater than the variability within each month of measurements. The difference measurements, grouped by month, are summarized in Table 10.10.

Two estimates of the population variance are calculated. One estimate is called the 'within groups variance,' also called the error variance. It is simply a pooling of the variances within each group, analogous to the pooled variance described in Equation 10.16. In this case, the error variance is equal to:

$$\text{Error variance} = \frac{19(2013) + 17(1155) + 18(1238)}{19 + 17 + 18} = 1484 \qquad (10.29)$$

Table 10.10

Summary statistics grouped by month for downwind–upwind difference measurements.

	July	August	September
Minimum	−12	−6	−15
Maximum	128	99	117
Mean	51.65	45.11	38.52
Variance	2013	1155	1238
Number of samples	20	18	19

The other estimate is called the 'between groups variance' and is estimated by comparing the mean of each group, $\bar{x}_i$, with the grand mean, $\bar{x}$, for all data points:

$$\text{Between groups variance} = \frac{\sum_{i=1}^{k} n_i (\bar{x}_i - \bar{x})^2}{k-1} \qquad (10.30)$$

In our example the between groups variance is equal to

$$\frac{20(51.65-45.21)^2 + 18(45.11-45.21)^2 + 19(38.52-45.21)^2}{2} = 840 \qquad (10.31)$$

If the null hypothesis is true (i.e., H_o: $\mu_1 = \mu_2 = \mu_3$), then the between groups variance should be essentially equal to the within groups variance and the ratio of these two quantities (between groups/within groups) should be close to one. If the group means are not equal, then the between groups variance will be larger than the within groups variance and the ratio will exceed one. The one-tailed variance ratio test or F test with $(k-1)$ and $(N-k)$ degrees of freedom is used to compare variances. A one-tailed test is performed because our alternate hypothesis is that between groups variance is greater than the within groups variance, not just that it is different.

In our example, the calculated F statistic is equal to $840/1484 = 0.57$. Thus the variance between groups is smaller than the variance within groups. The tabular value for F with $\alpha = 0.05$, 2 degrees of freedom in the numerator and 54 degrees of freedom in the denominator is 3.17. Thus we accept the null hypothesis of no difference between means.

In performing an ANOVA, the convention is to construct an ANOVA table where the variance is separated into two parts, the sum of squares (SS) and the degrees of freedom (DF). The between groups SS is equal to the numerator in Equation 10.30 and the degrees of freedom is equal to the denominator. The within groups SS is defined in Equation 10.32.

$$\text{Within groups SS} = \sum_{i=1}^{k}\left[\sum_{j=1}^{n}\left(x_{ij} - \bar{x}_i\right)^2\right] \qquad (10.32)$$

Source of Variation	SS	DF	MS	F
Between groups	1 680	2	840	0.57
Within groups (error)	80 136	54	1484	
Total	81 816	56		

Table 10.11

Analysis of variance table for the null hypothesis of no difference among average downwind–upwind difference measurements for the months of July, August, and September.

and the degrees of freedom are equal to $N - k$. An ANOVA table for the TSP example is illustrated in Table 10.11.

Finally, we should check our assumptions of equal variance using the F test. The ratio of the larger to smaller variance is 2013/1155 = 1.74. Compared to the tabular value of $F = 2.63$ for $\alpha = 0.05$ and with 19 and 17 degrees of freedom, we can accept our null hypothesis of no difference. Since the variances are constant, we can use the within groups variance = error mean square as a pooled variance for estimating confidence limits about each of the group means. Thus, for example, the 95% confidence limits about the mean downwind–upwind difference observed in July are

$$P\left\{51.6 - 2.093\frac{\sqrt{1484}}{\sqrt{20}} \leq \mu_d \leq 51.6 + 2.093\frac{\sqrt{1484}}{\sqrt{20}}\right\} = 0.95 \qquad (10.33)$$

where the value 2.093 is the Student's two-tailed t statistic for $\alpha = 0.05$ and 19 degrees of freedom. Thus we can say that there is a 95% probability that the true mean difference in July lies between 33.6 and 69.6 $\mu g/m^3$.

If the calculated F statistic exceeded the tabular value, we would reject our null hypothesis and conclude that there is a significant difference between groups; but, we would not know which groups are different. There are procedures like the Tukey multiple comparison test that can be used to determine which groups are different. The Tukey test is discussed by Zar (1984) and Berthouex and Brown (1994).

Boehm *et al.* (1998) used analysis of variance to compare polycyclic aromatic hydrocarbons (PAHs) in sediments from different locations. The authors analyzed the distribution of PAHs in sediments collected from Prince William Sound, Alaska, to distinguish petrogenic, pyrogenic, and biogenic sources. PAH levels in sediments collected from the Bay of Isles, a bay affected by the *Exxon Valdez* oil spill, were then compared to PAH levels in sediment samples collected from a bay receiving minimal impact from the spill using analysis of variance techniques. The authors were able to show that there is no evidence of large-scale offshore transport of spill oil to the subtidal sediments.

The ANOVA is robust to small deviations from equal variance and normality assumptions (Zar, 1984). However, if the deviations are significant, there are non-parametric techniques available such as the Kruskal–Wallis test.

10.3.1.5 Kruskal–Wallis Test

The Kruskal–Wallis test is similar to Wilcoxon's rank sum test in that we are comparing the sum of ranks applied to the data. The test statistic is calculated as

$$H = \frac{12}{N(N+1)} \sum_{i=1}^{k} \frac{R_i^2}{n_i} - 3(N+1) \tag{10.34}$$

where R_i is the sum of ranks for the ith group. For the TSP example, the sum of ranks for July, August, and September are 616, 533, and 504, respectively, and the calculated H value is 0.68. There is a correction for ties. The correction factor, C, is given in Equation 10.35.

$$C = 1 - \frac{\sum_{j=1}^{g} t_j^3 - t_j}{N^3 - N} \tag{10.35}$$

where g is the number of tied groups, and t_j is the number of tied data in the jth group. The value of H corrected for ties, H_c, is equal to H/C. For large data sets (large N), the correction factor is minimal (in our example 0.9996 with $g = 10$ and $t_j = 2$). For larger samples, the calculated H statistic is compared to the tabular value for χ^2 with $k - 1$ degrees of freedom. At $\alpha = 0.05$ and $v = 2$, $\chi^2 = 5.991$. Thus, as expected, we arrive at the same conclusion of no difference between average downwind–upwind difference measurements for the months of July, August, and September.

The most common application of these parametric and non-parametric techniques is for the comparison of concentrations on-site with background levels. Spatial and temporal variations in background can complicate the analysis but these issues can be addressed with proper sampling design and modifications to these basic procedures. Numerous techniques have been published (e.g. Liggett, 1984; Gilbert and Simpson, 1990).

In addition to comparing on-site levels with background, these statistical techniques can also be used to evaluate cost allocation schemes if it is claimed that two distinct plumes are present (by comparing distributions of constituents in each plume, for example), or to test natural resource damage claims if it is claimed that there are a statistically significant fewer number of wildlife species in a given area.

10.3.2 LINEAR CORRELATION AND LINEAR REGRESSION ANALYSIS

Thus far we have discussed techniques for comparing a single variable – e.g. lead in soil, TSP in ambient air, etc. – between or among samples. However, most analytical techniques are designed to measure multiple constituents at once, such as volatile organics in air or metals in soil. Simple techniques for evaluating the relationships among multiple variables in a sample are discussed in this section. The two basic techniques are correlation and regression. Other more sophisticated techniques like principal components analysis and receptor modeling are related to these basic techniques and discussed elsewhere in the book.

There are slight but important differences between linear regression and linear correlation. Simple linear regression seeks to quantify the functional dependence of one variable (the dependent variable, usually designated 'y') on another (the independent variable, usually designated 'x'). The relationship is assumed to be linear; and for each value of x there is assumed to be a normal distribution of y values. A good example of regression analysis is the construction of an analytical calibration curve where the functional relationship between sample concentration and instrument response is quantified. In this case, the instrument response (the dependent variable) depends upon the concentration of analyte (independent variable) sampled by the instrument.

In contrast, linear correlation analysis is used to assess the linear association between two independent variables. In correlation analysis it is assumed that *both* variables are drawn from underlying normal distributions. The above distinction does not preclude one from performing a regression analysis on two independent variables – in fact, this is often done. However, one must be careful about interpretation of the results.

10.3.2.1 Linear Regression Analysis

The general equation for a straight line is given in Equation 10.36:

$$y_i = \alpha + \beta x_i \tag{10.36}$$

where α and β are both population parameters and α is the 'y intercept' and β is the slope of the line. In any given data set, there are a number of straight lines that can be drawn through the data. Rarely do all the values fit on a single straight line. In regression analysis, the 'best fit line' through the data is determined where 'best fit line' is defined as the line which minimizes the square of the deviations between the y_i values predicted by the equation for the line (denoted by $\hat{y}_i$) and the observed y_i values, i.e.,

$$\sum_{i=1}^{n} (\hat{y}_i - y_i)^2 = \text{Minimum} \tag{10.37}$$

Thus, we are trying to minimize the vertical deviations between the observed data points and the fitted line. If we could sample and determine all values of x_i and y_i we could determine both α and β. Since this is rarely possible, we estimate the slope and intercept (denoted b and a, respectively) using the following equations:

$$b = \frac{\sum_{i=1}^{n}(x_i - \bar{x})(y_i - \bar{y})}{\sum_{i=1}^{n}(x_i - \bar{x})^2} \tag{10.38}$$

and

$$a = \bar{y} + b\bar{x} \tag{10.39}$$

The uncertainty in our estimates of the slope and intercept can be addressed using an analysis of variance approach. We are actually testing the null hypothesis that the true slope is zero, H_o: $\beta = 0$. A regression analysis table is constructed where the variances are broken down into a sum of squares (SS) term and a degrees of freedom (DF) term. Construction of a regression analysis table is illustrated in Table 10.12.

The analysis is similar to that in ANOVA except that instead of dealing with discrete groups, the variable x is continuous. If the slope is zero, this is the same as saying that there is no difference between group means for an infinite number of groups. Once again, we have two variance estimates, the regression mean square (MS) and the residual MS. The one-tailed F test with (1) and $(n - 1)$ degrees of freedom is used to compare the two estimates. If there is any slope, positive or negative, the contributions from both ends of the fitted line will increase the regression SS term and the ratio will exceed one.

The coefficient of determination, r^2, is a measure of the percentage of the total variation in y that can be explained by the regression line. The coefficient of determination is equal to the regression SS divided by the total SS as indicated in Equation 10.40.

$$r^2 = \frac{\text{Regression SS}}{\text{Total SS}} \tag{10.40}$$

Table 10.12

Construction of a regression analysis table.

Source of Variation	SS	DF	MS
Total	$\sum_{i=1}^{n}(y_i - \bar{y})^2$	$n - 1$	
Regression	$\sum_{i=1}^{n}(\hat{y}_i - \bar{y})^2$	1	$\frac{\text{Regression SS}}{\text{Regression DF}}$
Residual	$\sum_{i=1}^{n}(\hat{y}_i - y_i)^2$	$n - 2$	$\frac{\text{Residual SS}}{\text{Residual DF}}$

The residual MS is the variance of y after taking into account the dependence of y on x. It is denoted $s^2_{y \cdot x}$. The square root of this term is known as the 'standard error of the regression' or 'standard error of the estimate.' It is a measure of how well the fitted equation predicts the dependence of y on x.

The standard error of the slope, s_b, can be calculated from the standard error of the estimate according to Equation 10.41.

$$s_b = \sqrt{\frac{s^2_{y \cdot x}}{\sum_{i=1}^{n}(x_i - \bar{x})^2}} \tag{10.41}$$

From the standard error of the slope, we can estimate confidence intervals about the true slope.

$$P\left\{b - t_{\alpha,n-2}\, s_b \leq \beta \leq b + t_{\alpha,n-2}\, s_b\right\} = p \tag{10.42}$$

This is an important result because in many environmental forensic investigations we are interested in discovering trends in the data. Trends in concentration that increase or decrease over time or distance can usually be evaluated using simple linear regression techniques. For example, contaminant concentrations that decrease over time at a rate that is proportional to the concentration are said to follow first order kinetics. Most natural attenuation processes in the environment such as biodegradation, photolysis, and volatilization are modeled assuming first order or pseudo first order kinetics. In one example, Kaplan *et al.* (1997) describe how the ratio of (benzene + toluene)/(ethylbenzene + xylenes) can be used to monitor changes in a dissolved gasoline plume over time. The ratio decreases over time due to greater dispersion and more rapid degradation of B + T compared to E + X. At one site, the decrease over time was modeled as a pseudo first order process with a half-life estimated to be 2.3 years.

The basic equation is given below.

$$C = C_0\, \mathrm{e}^{-\mathrm{k}t} \tag{10.43}$$

where C is contaminant concentration at time t, C_0 is contaminant concentration at $t = 0$, and k is the first order rate constant (t^{-1}).

Equation 10.43 can be rearranged into a form resembling Equation 10.36 by taking the natural log of both sides of the equation. The resulting expression is presented below.

$$\ln C = \ln C_0 - \mathrm{k}t \tag{10.44}$$

where the slope is equal to the negative of the first order rate constant and the y intercept is equal to the natural log of the concentration at time $t = 0$.

The half-life (the time required for the concentration to be decreased by one half) is equal to the value 0.693 divided by the first order rate constant.

$$t_{1/2} = \frac{\ln(2)}{k} = \frac{0.693}{k} \tag{10.45}$$

A common practice in forensic investigations is to estimate the date of release of a contaminant to the environment based on half-life values reported in the literature. If sufficient data exist, site data may be evaluated using linear regression techniques to establish a site-specific rate constant. In addition, we can place confidence limits about our estimate to evaluate the uncertainty in release dates. If the confidence limits about the slope include zero, then the trend is not significant.

10.3.2.2 Pearson's Product Moment Correlation Coefficient

Pearson's product moment correlation coefficient, r, also referred to as simply the correlation coefficient, is a dimensionless value that can range from -1 for a perfect negative linear correlation to $+1$ for a perfect positive linear correlation. A value of zero indicates no linear relationship between variables. It is important to remember that the correlation coefficient is a measure of *linear* association between two variables. Data arranged in a perfect circle are clearly associated with each other, but these data would yield a correlation coefficient equal to zero. The correlation coefficient is a parametric statistic and it is assumed that both variables are drawn from an underlying normal distribution.

The correlation coefficient is calculated as

$$r = \frac{\sum_{i=1}^{n}(x_i - \bar{x})(y_i - \bar{y})}{\sqrt{\sum_{i=1}^{n}(x_i - \bar{x})^2 \sum_{i=1}^{n}(y_i - \bar{y})^2}} \tag{10.46}$$

The correlation coefficient can also be written as:

$$r = \frac{\sum_{i=1}^{n}(x_i - \bar{x})(y_i - \bar{y})}{(n-1)s_x s_y} = \frac{\sum_{i-1}^{n} z_x z_y}{n-1} \tag{10.47}$$

In this form, it is apparent that the correlation coefficient is like an average of the products of paired z-scores (Kachigan, 1991). It is also easy to see that when positive z-scores are paired with positive z-scores and negative z-scores are paired with negative z-scores, the correlation coefficient will be positive indicating a positive correlation. And, when positive z-scores are paired with negative z-scores, the correlation coefficient will be negative indicating a negative correlation.

Note that the correlation coefficient is also related to the slope of the best fit line through the data.

$$r\frac{s_y}{s_x} = b \tag{10.48}$$

The square of the correlation coefficient is called the coefficient of determination and reveals the proportion of the variance in '*y*' that can be predicted from '*x*' or vice versa, the proportion of the variance in '*x*' that can be predicted from '*y*.'

Peers (1996) notes that the correlation coefficient can be affected by the presence of outliers, and suggests that it is most appropriate when n is greater than 30. In calculating r, we have estimated the population correlation, ρ, which we could calculate using Equation 10.46 if we knew all values of x and y. We can measure the confidence in our estimate by either testing the null hypothesis that the two variables are independent, i.e., H_o: $\rho = 0$, or calculating confidence limits about r. The test statistic for the null hypothesis is calculated according to Equation 10.49 and compared to the t value with $n - 2$ degrees of freedom.

$$t = \frac{r\sqrt{n-2}}{\sqrt{\left(1-r^2\right)}} \tag{10.49}$$

Confidence limits on r can be estimated by first transforming the correlation coefficient to Fisher's z value (which is different than the normal distribution z value) where:

$$\text{Fisher's } z = 0.5\ln\frac{(1+r)}{(1-r)} \tag{10.50}$$

and the standard error (SE) is given as:

$$\text{SE} = \frac{1}{\sqrt{n-3}} \tag{10.51}$$

The 95% confidence limits are Fisher's $z \pm 1.96$ (SE); where the 1.96 is derived from the standard normal curve (i.e., 95% of the standard normal curve lies within ±1.96 standard deviations). Once the confidence limits (CL) about Fisher's z value are calculated, they must be converted back to correlation coefficients using the formula:

$$r = \frac{e^{2\text{CL}} - 1}{e^{2\text{CL}} + 1} \tag{10.52}$$

The correlation coefficient is a useful tool for exploratory data analysis. For example, Kimbrough and Suffet (1995) calculated correlation coefficients for metals (lead, antimony, arsenic, cadmium, chromium, copper, and nickel) in air, soil, and dust near two secondary lead smelters in California in order to determine those metals best suited for establishing characteristic ratios associated with impacts from the operation. Lead, antimony, and arsenic were selected and ratios of these three species were calculated in on- and off-site samples to determine the source of off-site concentrations. A similar analysis was performed by Amter and Eckel (1996).

When multiple species are compared, it is useful to prepare a correlation matrix. It is prepared by calculating correlation coefficients for each possible pair of variables and displaying them in a matrix as shown in Table 10.13.

The correlation matrix displayed in Table 10.13 was prepared from residential indoor air samples collected at a site in Louisiana where a prior spill had left a significant quantity of jet fuel floating on the water table. Concentrations of benzene, toluene, ethylbenzene, and xylenes (BTEX chemicals) detected in indoor air in some homes caused regulators to suspect the floating product on the water table as a source.

The highest correlation coefficients were reported for the association of ethylbenzene with the xylenes, and for the association of *o*-xylene with *m,p*-xylene. Benzene concentrations were most closely correlated with toluene. Soil gas data were also collected at the site. Significant correlation coefficients were also recorded for BTEX chemicals in soil gas as indicated in Table 10.14.

Soil gas beneath the site was evaluated as a potential source of BTEX chemicals in indoor air by comparing benzene/toluene (B/T) and ethylbenzene/total

Table 10.13

Correlation matrix for BTEX concentrations detected in indoor air using natural log-transformed data.

	Benzene	Toluene	Ethylbenzene	*m/p*-Xylene	*o*-Xylene
Benzene	1				
Toluene	0.82	1			
Ethylbenzene	0.61	0.5	1		
m/p-Xylene	0.56	0.44	0.99	1	
o-Xylene	0.6	0.48	0.96	0.97	1

Table 10.14

Correlation matrix for BTEX concentrations detected in soil gas using natural log-transformed data.

	Benzene	Toluene	Ethylbenzene	Xylenes
Benzene	1			
Toluene	0.66	1		
Ethylbenzene	0.52	0.94	1	
Xylenes	0.47	0.86	0.88	1

xylenes (E/X) ratios. Summary statistics are reported in Table 10.15. Benzene/toluene and ethylbenzene/xylene concentrations are plotted in Figures 10.6 and 10.7. Note that the soil gas data are reported in units of parts per million (ppm) while the indoor air data are reported in units of parts per billion (ppb).

The strong positive linear correlation between ethylbenzene and total xylenes in indoor air is evident in Figure 10.7 as the data clearly can be fitted to a straight line.

B/T ratios in indoor air ranged from 0.16 to 0.98 with an average of 0.59 whereas soil gas B/T ratios ranged from 0.44 to 17 with an average value of

Table 10.15

Comparison of benzene/toluene and ethylbenzene/total xylenes ratios in indoor air and soil gas.

Statistic	**Benzene/Toluene**		**Ethylbenzene/Xylenes**	
	Indoor Air	**Soil Gas**	**Indoor Air**	**Soil Gas**
Average	0.59	3.38	0.19	0.70
Standard Deviation	0.23	3.46	0.03	0.57
Minimum	0.16	0.44	0.14	0.20
Maximum	0.98	17.0	0.26	4.0
No. of samples	17	60	26	48

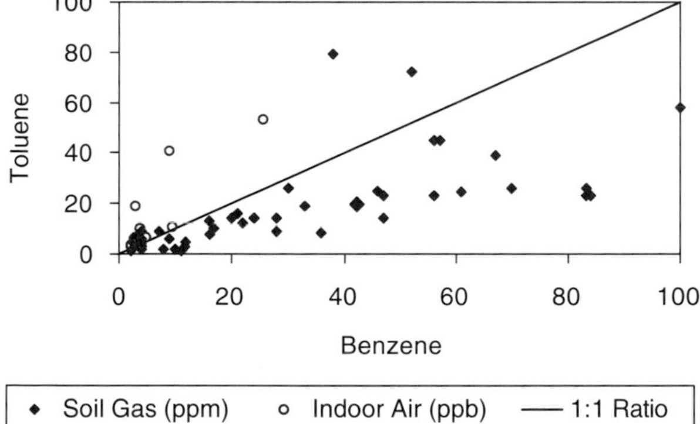

Figure 10.6

Scatter plot of benzene and toluene concentrations in indoor air and soil gas.

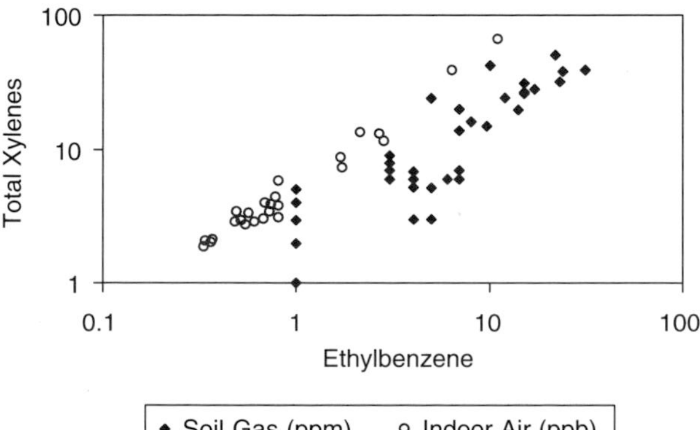

Figure 10.7

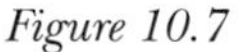

Log–log scatter plot of ethylbenzene and total xylene concentrations in indoor air and soil gas.

3.38. A Student's *t*-test performed on log-transformed data indicated a significant difference between data sets at the 99% confidence level. A similar conclusion may be drawn by comparing E/T ratios. E/T ratios in indoor air ranged from 0.14 to 0.26 with an average of 0.19 compared to soil gas E/T ratios with a range of 0.2 to 4.0 with an average of 0.7.

Once significant ratios have been identified, double ratio plots can be an effective means of displaying the results. A double ratio plot of B/T vs. E/X is shown in Figure 10.8. The indoor air data are neatly clustered in the lower left corner of the plot, and there appears to be no relationship between BTEX concentrations detected in indoor air and soil gas. In contrast, characteristic ratios reported in the literature for gasoline related sources are remarkably similar (Table 10.16).

10.3.2.3 Spearman Rank Correlation Coefficient

If the data are obviously not normal, then a non-parametric technique may be used. The Spearman rank correlation coefficient, r_s, is a correlation technique

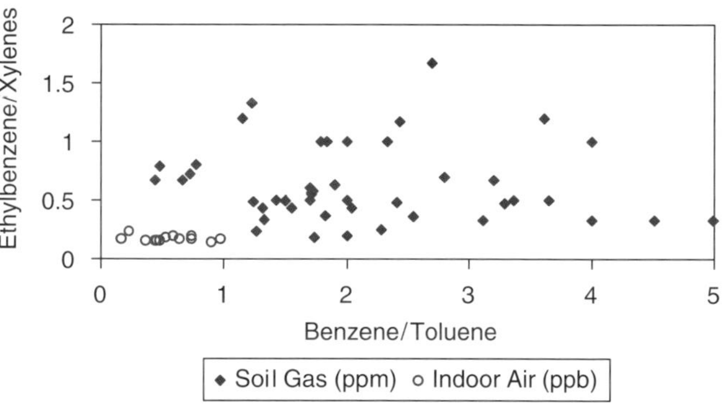

Figure 10.8
Double ratio plot of ethylbenzene/total xylenes vs. benzene/toluene for indoor air and soil gas.

Table 10.16
Benzene/toluene and ethylbenzene/total xylenes ratios reported in gasoline related samples, indoor air and soil gas.

Sample Type	Benzene/Toluene Ratio	Ethylbenzene/Xylene Ratio
Indoor air	0.59	0.19
Ambient air sampled near busy roadway[a]	0.41	0.21
Headspace sampled above gasoline[a]	0.50	0.20
Gasoline engine exhaust (catalyst)[b]	0.42	0.27
Gasoline engine exhaust (no catalyst)[b]	0.62	0.21
Whole liquid gasoline[b]	0.19	0.16
Soil gas	3.38	0.70

[a]Conner *et al.* (1995).
[b]Harley *et al.* (1992).

that operates on the ranks of the data. Each variable is ranked separately and the difference between ranks for each data pair is recorded. The Spearman rank correlation coefficient is calculated according to the following equation:

$$r_s = 1 - \frac{6\sum_{i=1}^{n} d_i^2}{n^3 - n} \qquad (10.53)$$

where d_i is the difference between ranks for each data pair = x_i, y_i. If ties are involved, the equation is more complicated but only makes an appreciable difference if there are a large number of ties.

$$r_s = \frac{\frac{(n^3 - 3)}{6} - \sum_{i=1}^{n} d_i^2 - \sum T_x - \sum T_y}{\sqrt{\left[\frac{(n^3 - 3)}{6} - 2\sum T_x\right]\left[\frac{(n^3 - 3)}{6} - 2\sum T_y\right]}} \qquad (10.54)$$

where g is the number of tied groups, t_j is the number of tied data in the jth group, and

$$\sum T_x = \frac{\sum_{j=1}^{g}\left(t_j^3 - t_j\right)}{12} \quad \text{(for } x \text{ values)} \qquad (10.55)$$

and

$$\sum T_y = \frac{\sum_{j=1}^{g}\left(t_j^3 - t_j\right)}{12} \quad \text{(for } y \text{ values)} \qquad (10.56)$$

Hypothesis testing and setting confidence limits are performed as described for the Pearson product moment correlation coefficient.

In addition to exploratory data analysis, the Spearman rank correlation coefficient is a useful statistic for detecting trends in concentration with time and distance (El-Shaarawi *et al.*, 1983). There are no underlying distribution assumptions and the distance and time measurements need not be uniformly spaced.

As an example, benzene concentration data presented by Zemo (2000) are plotted in Figure 10.9. The data in Figure 10.9 suggest that benzene levels are decreasing. The Spearman rank correlation coefficient can be used to test if the decrease is significant. Using Equation 10.53, the calculated rank correlation coefficient is − 0.586 and, using Equation 10.49, the calculated t statistic is 3.688. The corresponding t value with α = 0.05 and 26 degrees of freedom is

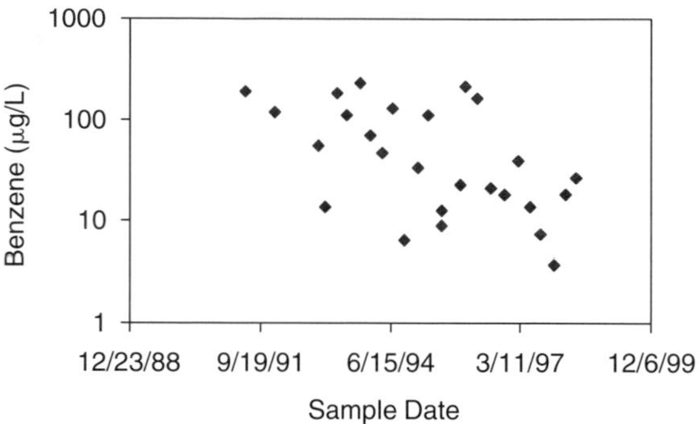

Figure 10.9
Benzene concentration (log scale) vs. time (Zemo, 2000).

2.056. Since the calculated t statistic exceeds the tabular value, we can reject our null hypothesis of $\rho = 0$ and conclude at the 95% probability level that there is a negative linear relationship between benzene concentrations and time.

The same type of analysis could also be used to evaluate trends with distance from a source.

10.3.2.4 Multiple Linear Regression and Correlation

The logical extensions to simple linear regression and correlation models are multiple linear regression and correlation techniques. In most cases, we have simultaneous measurements of more than two variables. We can extend the concepts of linear regression and quantify the dependence of a single variable on multiple independent variables. Equation 10.57 describes the general equation.

$$\hat{y}_i = \alpha + \beta_1 x_{1,i} + \beta_2 x_{2,i} + \cdots + \beta_m x_{m,i} \tag{10.57}$$

It is assumed that y is linearly dependent on x_1 and also linearly dependent on x_2. The β terms are called partial regression coefficients and provide a measure of the relationship between y and a single x variable after removing the effects of all other variables. For example, β_1 is a measure of the relationship between y and x_1 after accounting for effects from all other x variables. The α term is equal to the y intercept when all the β terms are set equal to zero.

Using the basic approach illustrated in Table 10.12, it is possible to evaluate the fit of the regression curve, calculate a coefficient of multiple determination analogous to Equation 10.40, and calculate a multiple correlation coefficient reflecting the overall relationship between all m variables. The equations for performing a multiple regression analysis and calculating the α and β terms are beyond the scope of this chapter (for a discussion of techniques, see Zar, 1984).

One example where linear regression techniques were used to evaluate spatial patterns of soil contamination is presented by Small *et al.* (1995).

The authors modeled residential soil lead concentrations near a former automotive battery recycling facility in Pennsylvania to distinguish contributions from the site from other sources such as lead-based paint. A weighted least squares multiple linear regression technique was used. Contributions from the site were represented by relative deposition rates estimated with an atmospheric dispersion model. The authors used indicator variables to represent potential contributions from lead-based paint and other sources. For example, whether or not the sample was collected within five feet of a home built prior to 1955 was selected as an indicator of potential contributions from lead-based paint. Effects from drainage features that might result in concentration enhancements were also included in the model. With this model the authors were able to predict the spatial distribution of lead in soils and help identify additional properties likely to contain lead in soils above the specified cleanup level.

10.4 CONCLUSIONS

Relatively simple statistical analysis techniques can be used in environmental forensic investigations to compare data sets, characterize associations between variables, evaluate trends, and make predictions. Moreover, it is often possible to assign a degree of confidence to the results. This advantage is particularly useful in litigation scenarios where experts are often asked to assign a probability to the correctness of their opinions.

Of course, the results are not always unambiguous. After all it was Benjamin Disraeli that said 'There are three kinds of lies. Lies, damn lies, and statistics.' The outcome of a test depends upon which data are included. For example, the decrease in benzene concentrations in Figure 10.9 is not significant if we only look at data between 1991 and 1996. Thus, the results of any statistical analysis should be interpreted in relation to which data are included (and which data are *not* included), how the test was performed, what assumptions were made, and the validity of those assumptions.

REFERENCES

Amter, S. and Eckel, W.P. (1996) Comment on 'Off-site forensic determination of airborne elemental emissions by multi-media analysis: a case study at two secondary lead smelters.' *Environmental Science and Technology* 30(7), 2417–2418.

Berthouex, P. and Brown, L. (1994) *Statistics for Environmental Engineers.* CRC Press, Boca Raton, FL.

Boehm, P.D., Page, D.S., Gilfillan, E.S., Bence, A.E., Burns, W.A., and Mankiewicz, P.J. (1998) Study of the fates and effects of the *Exxon* Valdez oil spill on benthic sediments in two bays in Prince William Sound, Alaska. 1. Study design, chemistry, and source fingerprinting. *Environmental Science and Technology* 32, 567–576.

Charrow, R. and Bernstein, D. (1994) *Scientific Evidence in the Courtroom: Admissibility and Statistical Significance after Daubert.* Washington Legal Foundation, Washington, DC.

Conner, T.L., Lonneman, W.A., and Seila, R.L. (1995) Transportation-related volatile hydrocarbon source profiles measured in Atlanta. *Journal of the Air and Waste Management Association* 45, 383–394.

El-Shaarawi, A.H., Esterby, S.R., and Kuntz, K.W. (1983) A statistical evaluation of trends in the water quality of the Niagara River. *Journal of Great Lakes Research* 9(2), 234–240.

Gilbert, R.O. (1987) *Statistical Methods for Environmental Pollution Monitoring.* Van Nostrand Reinhold, New York.

Gilbert, R.O. and Simpson, J.C. (1990) Statistical sampling and analysis issues and needs for testing attainment of background-based cleanup standards. Presented at the Workshop on Superfund Hazardous Waste, Arlington, VA. February 21–22.

Harley, R.A., Hannigan, M.P., and Cass, G.R. (1992) Respeciation of organic emissions and the detection of excess unburned gasoline in the atmosphere. *Environmental Science and Technology* 26, 2395–2408.

Higgs, F.J., Mielke, H.W., and Brisco, M. (1999) Soil lead at elementary public schools: comparison between school properties and residential neighborhoods of New Orleans. *Environmental Geochemistry and Health* 21, 27–36.

Kachigan, S.K. (1991) *Multivariate Statistical Analysis: A Conceptual Introduction.* Radius Press, New York.

Kaplan, I.R., Galperin, Y., Lu, S., and Lee, R. (1997) Forensic environmental geochemistry: differentiation of fuel types, their source and release time. *Organic Geochemistry* 27(5/6), 289–317.

Kimbrough, D.E. and Suffet, I.H. (1995) Off-site forensic determination of airborne elemental emissions by multi-media analysis: a case study at two secondary lead smelters. *Environmental Science and Technology* 29, 2217–2221.

Land, C.E. (1975) Tables of confidence limits for linear functions of the normal mean and variance. *Selected Tables in Mathematical Statistics* III, 385–419.

Langley, R. (1970) *Practical Statistics.* Dover Publications, New York.

Liggett, W. (1984) Detecting elevated contamination by comparisons with background. *Environmental Sampling for Hazardous Wastes*, pp. 119–128. American Chemical Society, Washington.

Oakes, M. (1986) *Statistical Inference*. Epidemiology Resources, Chestnut Hill.

Ott, W.R. (1990) A physical explanation of the lognormality of pollutant concentrations. *Journal of the Air and Waste Management Association* 40, 1378–1383.

Parkin, T.B., Meisinger, J.B., Chester, S.T., Starr, J.L., and Robinson, J.A. (1988) Evaluation of statistical estimation methods for lognormally distributed variables. *Soil Science Society of America Journal* 52, 323–329.

Peers, I. (1996) *Statistical Analysis for Education and Psychology Researchers*. The Falmer Press, London.

Reckhow, K.H., Clements, J.T., and Dodd, R.C. (1990) Statistical evaluation of mechanistic water-quality models. *Journal of Environmental Engineering* 116(2), 250–268.

Ross, D.L. (1997) Multivariate statistical analysis of environmental monitoring data. *Ground Water* 35(6), 1050–1057.

Singh, A.K., Singh, A., and Engelhardt, M. (1997) The lognormal distribution in environmental applications. US EPA Office of Research and Development, EPA/600/R-97/006. December.

Small, M.J., Nunn, A.B., Forslund, B.L., and Daily, D.A. (1995) Source attribution of elevated residential soil lead near a battery recycling facility. *Environmental Science and Technology* 29, 883–895.

US EPA (1989) Statistical Analysis of Ground-Water Monitoring Data at RCRA Facilities – Interim Final Guidance. Office of Solid Waste, Washington, DC. February.

Wallace, J. (1997) Calculating removal efficiencies: confidence limits and hypothesis tests. *Journal of Air and Waste Management Association* 47, 976–982.

Zar, J.H. (1984) *Biostatistical Analysis*, 2nd edn. Prentice Hall, Englewood Cliffs, NJ.

Zemo, D. (2000) Environmental Litigation: Advanced Forensics and Legal Strategies. University of Wisconsin Short Course. April 14.

FURTHER READING

Beers, Y. (1953) *Introduction to the Theory of Error*. Addison-Wesley Publishing Company Inc., Cambridge, MA.

Box, G.E., Hunter, W.G., and Hunter, J.S. (1978) *Statistics for Experimenters: An Introduction to Design, Data Analysis, and Model Building*. John Wiley & Sons, New York.

Dunn, O.J. and Clark, V.A. (1974) *Applied Statistics: Analysis of Variance and Regression*. John Wiley & Sons, New York.

Mandel, J. (1964) *The Statistical Analysis of Experimental Data*. Dover Publications, New York.

Ott, W.R. (1995) *Environmental Statistics and Data Analysis*. CRC Press, Boca Raton, FL.

Thorndike, R.M. (1978) *Correlational Procedures for Research*. Gardner Press, New York.

CHAPTER 11

PARTICULATE PATTERN RECOGNITION

John G. Watson and Judith C. Chow

11.1 INTRODUCTION

Small particles suspended in the atmosphere cause adverse health effects, reduce visibility, and are a nuisance when they deposit onto surfaces. Most suspended particles are too small to be seen with the unaided eye, and even when they can be viewed by magnification, their origins are not often evident. Many sources contribute to a specific event, and the proportions of these contributions change from event to event. Some particles retain the form in which they were emitted, but others are created from emitted gases through chemical reactions, or they undergo chemical transformations that change their chemical and physical characteristics. 'Receptor models' use the variability of chemical composition, particle size, and concentration in space and time to identify source types and to quantify source contributions that affect particle mass concentrations, light extinction, or deposition. This chapter describes several air quality receptor models that have been applied or are under development, identifies measurements that can be taken to apply them, and shows examples of how receptor models were used to make pollution control decisions. Of necessity, these descriptions are brief, and the reader is guided to more complete works on each topic through selected references.

11.2 RECEPTOR MODELS

Receptor models (Cooper and Watson, 1980; Watson, 1984; Hopke, 1985, 1991) provide a theoretical and mathematical framework for quantifying source contributions. Receptor models interpret measurements of physical and chemical properties taken at different times and places to infer the possible or probable sources of excessive concentrations and to quantify the contributions from those sources. Receptor models contrast with source models (e.g., Seigneur, 1998) that combine source emission rates with meteorological transport and chemical changes to estimate concentrations at a receptor. Receptor and source models are imperfect representations of reality, and input data are seldom complete. Using both types of model helps to identify and quantify their inaccuracies and to focus further investigation on the areas of greatest discrepancy. Some receptor models are statistical, establishing empirical relationships between variables that are not necessarily explainable by known physical processes. The most believable and useful receptor models have a sound physical basis, although they may apply some of the same multivariate mathematical solutions that are used in statistical models. Examples from each type are described in the following subsections.

11.2.1 CONCEPTUAL MODEL

As with all forensic investigations, source apportionment by receptor or source models begins with plausible theories about the causes of an excessive concentration or effect. A conceptual model (e.g., Watson *et al.*, 1998a; Pun and Seigneur, 1999) provides reasonable, though not necessarily accurate, explanations of: (1) potential sources, (2) size, chemical, and temporal characteristics of particle and precursor gas emissions from these sources, (3) particle and precursor gas emission rates, (4) meteorological conditions that affect emissions, transport, and transformation, and (5) frequency and intensity of the effect. The conceptual model is derived from previous experience (e.g., tests on similar sources, particle movement under similar meteorological conditions), the nature of the problem (neighborhood complaints, consistently poor visibility over a local area or large region, exceeding an air quality standard), and available measurements (from air quality and meteorological sensors). Sometimes the conceptual model yields an obvious culprit and appropriate remedial actions. If not, the conceptual model guides the location of monitoring sites, sampling periods, sampling frequencies, sample durations, selection of samples for laboratory analysis, and the particle properties that are quantified in those samples to test and discard theories until only one remains.

11.2.2 CHEMICAL MASS BALANCE (CMB)

The CMB model (Watson *et al.*, 1991, 1998b) expresses ambient chemical concentrations as the sum of products of species abundances and source contributions. These equations are solved for the source contributions when ambient concentrations and source profiles are supplied as model input. Source profiles consist of the mass fractions of selected particle properties in source emissions; examples are given below. Several different solution methods have been applied, but the effective variance least squares estimation method (Watson *et al.*, 1984) is most commonly used because it incorporates precision estimates for all of the input data into the solution and propagates these errors to the model outputs. The CMB is the only air quality model that provides quantitative uncertainty values for source contribution estimates. It provides the basic structure for other pattern recognition receptor models that can be derived from physical principles. The CMB has been applied to evaluate particle control effectiveness (e.g., Chow *et al.*, 1990; Engelbrecht *et al.*, 2000), determine causes of excessive human exposure (Vega *et al.*, 1997; Schauer and Cass, 2000), quantify contributions to light extinction (e.g., Watson *et al.*, 1988, 1996), and to estimate sources of particle deposition (Feeley and Liljestrand, 1983). Although most often applied to ambient air samples for particulate matter and organic gases, CMB modeling software (Watson *et al.*, 1990, 1997) has also been used for forensic applications in quantifying contributions to oil spills (Pena-Mendez *et al.*, 2001), toxic components in lakebed sediments (Kelley and Nater, 2000), and water quality (Olmez, 1989).

11.2.3 ENRICHMENT FACTORS (EF)

The EF model (Sturges, 1989; Rashid and Griffiths, 1993) compares ratios of chemical concentrations in the atmosphere to the same ratios in a reference material, such as average crustal composition or sea salt for a region (Mason, 1966; Lawson and Winchester, 1979). These ratios are often normalized to the silicon (Si) concentrations in air and in soil, as Si is the most abundant element in suspended dust. Reference ratios vary substantially between bulk soil and suspendable particles, and among different regions (Rahn, 1976). Differences are explained in terms of other sources, but not with quantitative attribution as in the CMB. Heavy metal enrichments are usually attributed to industrial emitters. Sulfur enrichment is attributed to secondary sulfate particles formed from gaseous sulfur dioxide emissions. Potassium, and sometimes chloride, enrichment is attributed to burning and cooking. This method has also been used to examine different geological strata to determine how specific elements, such as lead, may have deposited over long time periods (Weiss *et al.*, 1999). Vanadium and nickel enrichments indicate contributions from residual oil combustion or refinery catalyst crackers, while enriched selenium can often be attributed to coal-fired power stations. Iron, manganese, copper, zinc, and lead enrichments usually indicate steel mill, smelting, or plating contributions. Calcium is often enriched near cement manufacture or where cement products have been used in construction. Sodium and chloride are enriched near the coast, dry lake beds, and after de-icing materials are applied to streets.

11.2.4 MULTIPLE LINEAR REGRESSION ON MARKER SPECIES (MLR)

The MLR model (Kleinman *et al.*, 1980; White, 1986; Lowenthal and Rahn, 1989) expresses particle mass, a component of mass such as sulfate, light extinction, or other observables as a linear sum of unknown regression coefficients times source marker concentrations measured at a receptor. The markers, which may be chemical elements or compounds, must originate only in the source type being apportioned. The regression coefficients represent the inverse of the source profile chemical abundance of the marker species in the source emissions. The product of the regression coefficient and the marker concentration for a specific sample is the tracer solution to the CMB equations that yields the source contribution. MLR has been used to estimate source contributions to excessive particle concentrations (Morandi *et al.*, 1991), to relate light extinction to power station and smelter emissions (Malm *et al.*, 1990), and as the major epidemiological tool in relating adverse health effects to particle concentrations (Lipfert and Wyzga, 1997; Vedal, 1997; Spix *et al.*, 1999).

11.2.5 TEMPORAL AND SPATIAL CORRELATION EIGENVECTORS

Principal component analysis (PCA), factor analysis (FA), and empirical orthogonal functions (EOF) are in this category, and there is a large variability in how these methods are applied and interpreted (e.g., Hopke, 1981, 1988; Hopke *et al.*, 1983; Henry, 1986, 1987, 1991; Paatero, 1997). The patterns from these methods are empirically derived, as a basic-principles derivation has not yet been formulated. PCA, FA, and EOF are applied to correlations between variables over time or space. Temporal correlations are calculated from a long time series (at least fifty, but preferably several hundred samples) of chemical concentrations at one or more locations. Eigenvectors of this correlation matrix (a smaller matrix of coordinates that can reproduce the ambient concentrations when added with appropriate coefficients) are determined and a subset is rotated to maximize and minimize correlations of each vector with each measured species. These rotated vectors (also called 'factors,' 'principal components,' or 'orthogonal functions') are interpreted as source profiles by comparison of factor loadings with source measurements. Several different normalization and rotation schemes have been used.

For a large spatial network (25 to 50 locations), spatial correlations are calculated from chemical measurements taken on simultaneous samples at the monitoring sites (Henry, 1997a; White, 1999). Eigenvectors of this correlation matrix represent a spatial distribution of source influences over the area, providing that the samplers have been located to represent the gradients in source contributions. As with temporal correlation models, several normalization and rotation schemes have been applied.

Eigenvector receptor models have been used to evaluate contributions to excessive particle concentrations that might affect health (Thurston and Spengler, 1985) and to determine sulfate contributions to light extinction over large regions (Henry, 1997b).

11.2.6 NEURAL NETWORKS

Known inputs and outputs are presented to a neural network that is intended to simulate the human pattern-recognition process (Wienke *et al.*, 1994; Reich *et al.*, 1999; Perez *et al.*, 2000). Training sets that have known source–receptor relationships are used to establish the linkages in the neural net that are then used to estimate source contributions for data sets with unknown relationships. The network assigns weights to the inputs that reproduce the outputs. Neural networks can provide functional relationships that are solutions to the MLR and CMB equations (Song and Hopke, 1996). Neural networks are empirical models and their results are only as good as the known relationships with the training sets and the extent to which the unknown data sets share the same patterns with the training sets.

11.2.7 TIME SERIES

Spectral analysis (Perrier *et al.*, 1995; Hies *et al.*, 2000), intervention analysis (Jorquera *et al.*, 2000), lagged regression analysis (Hsu, 1997), and trend analysis (Somerville and Evans, 1995) models separate temporal patterns for a single variable and establish temporal relationships between different variables. These models have been used to identify sources, to forecast future pollutant concentrations, and to infer relationships between causes and effects. It is especially important to include meteorological indicators in time series models (e.g., Weatherhead *et al.*, 1998; Shively and Sager, 1999) and to use data sets with comparable measurement methods and sampling frequencies (Henry *et al.*, 2000).

11.3 PHYSICAL, CHEMICAL, AND TEMPORAL PROPERTIES

The receptor models described above require multivariate measurements that are as specific as possible as to particle sizes, chemical and physical characteristics, temporal variation, and locations of sources. Chow (1995), McMurry (2000), Willeke and Baron (1993), Cohen and Hering (1995), and Watson *et al.* (1998c) provide details on methods that measure these properties with varying levels of time resolution, accuracy, precision, validity, and expense. Aerosol measurement technology is tending toward *in situ* monitors capable of sequential samples of 1 hour or less duration. However, specific chemical characterization still requires collection of particles on a substrate over periods of 6 to 24 hours to obtain enough material for subsequent laboratory analysis.

11.3.1 PARTICLE SIZE

Figure 11.1, an elaboration on concepts first discovered by Whitby *et al.* (1974), shows how particle mass (PM) concentrations are typically distributed according to size when suspended in the atmosphere. Particles present in different portions of this spectrum indicate their origins as well as their effects. The integral mass quantities $PM_{2.5}$ and PM_{10} are currently regulated by National Ambient Air Quality Standards (NAAQS, US Environmental Protection Agency (EPA), 1997; U.S. Supreme Court, 2001) intended to protect public health and are monitored in urban areas throughout the United States. The $PM_{2.5}$ integral contains the particles that most efficiently scatter and absorb light (Watson and

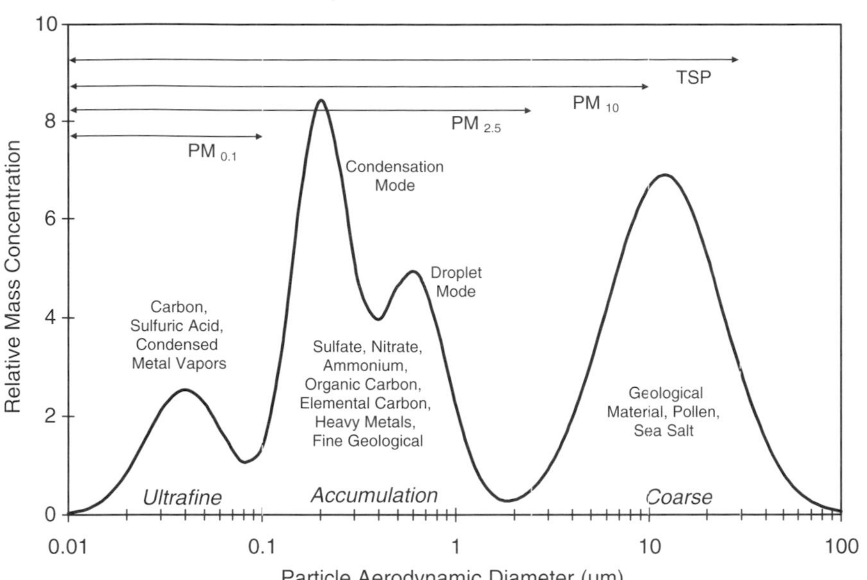

Figure 11.1

Major features of the mass distribution of atmospheric particles by particle size. This idealization was created by weighted sums of four log-normal distributions.

Chow, 1994), and baseline concentrations and compositions are measured within and around US National Parks (Eldred *et al.*, 1997) to support US regional haze regulations (USEPA, 1999). The $PM_{0.1}$ integral, and the large number of ultrafine particles that it contains, may be another indicator of adverse health effects (Oberdörster *et al.*, 1995), as might specific chemical components of the mass (Mauderly *et al.*, 1998). The total suspended particulate (TSP) integral once constituted the US particulate NAAQS (USEPA, 1971) and is still monitored in many parts of the world today.

Ultrafine ($PM_{0.1}$) particles are abundant in fresh combustion emissions, but they rapidly coagulate with each other and with larger particles shortly after emission (Preining, 1998). Buzorius *et al.* (1999) measured 10 000 to 50 000 particles/cm^3 near a heavily traveled street, while levels at a more pristine site were typically 2000 to 3000 particles/cm^3. $PM_{0.1}$ mass is $<10\%$ of $PM_{2.5}$ or PM_{10} mass, but the number of particles in this size range constitutes $>80\%$ of all particles typically found in an urban area (Keywood *et al.*, 1999).

Coarse particles, often reported as TSP minus $PM_{2.5}$ or PM_{10} minus $PM_{2.5}$ mass, result from the dis-aggregation of material, mostly fugitive dust. Concentrations are bounded at larger particle sizes by gravitational settling. This settling results in dustfall nuisances near coarse particle emitters such as heavily-traveled unpaved roads, construction sites, and facilities that handle large amounts of minerals. As a result of settling, the peak of the coarse distribution shifts to smaller particles with transport distance from the point of emissions. Coarse particles can travel over long distances when they are injected high into the atmosphere by severe wind storms (Perry *et al.*, 1997).

The accumulation mode contains most of the $PM_{2.5}$ mass and most of its chemical variability. These particles can be directly emitted 'primary particles' as well as 'secondary particles' that form in the atmosphere from gaseous emissions of sulfur dioxide, oxides of nitrogen, ammonia, and some volatile organic gases. Owing to the similarity between accumulation mode particle diameters and the wavelengths of visible light (0.3 to 0.7 μm), these particles cause most of the haze in urban areas and national parks. The two submodes in Figure 11.1 indicate secondary particle formation processes (John *et al.*, 1990). The condensation mode consists of smaller particles that form from condensation and coagulation, while the larger droplet mode forms from gases adsorbed by and reacted in cloud and fog droplets. When the water in the droplet evaporates, a larger amount of sulfate or nitrate remains than is in the condensation mode. The mode occupied by sulfate in the size spectrum allows inferences about the history of the gases that formed it.

Particle size measurements are made with size selective inlets (Watson and Chow, 2001a) that pass material to a collection filter to be weighed or chemically analyzed. Cascade impactors (Marple *et al.*, 1993; Hering, 1995) use several of these inlets in series to obtain a full size spectrum on substrates that can be analyzed chemically. Continuous detectors quantify the number of particles in discrete ranges for the ultrafine, accumulation, and coarse modes by a combination of their diffusion characteristics, electrical mobility, and optical properties (Cheng, 1993; Rader and O'Hern, 1993).

11.3.2 CHEMICAL COMPOSITION

Chemical composition of particles provides the most information on their origins. The most abundant and commonly measured components of $PM_{2.5}$ or PM_{10} mass are organic carbon, elemental carbon, geological material, sulfate, nitrate, and ammonium (Chow and Watson, 1998). Sodium chloride and other soluble salts are often found near the ocean, open playas, and after road de-icing. Analysis for these basic components indicates the extent to which an excessive mass concentration results from fugitive dust, secondary aerosols, or carbon-generating combustion activities. More specific chemical characterization can then be focused on the components with the largest mass fractions.

For quantitative source apportionment, source profiles must contain chemical abundances (mass fractions) that differ among source types, that do not change appreciably during transport between source and receptor (or that allow changes to be simulated by measurement or modeling), and that are reasonably constant among different emitters and source operating conditions. Minor chemical components, constituting less than 1% of particle mass, are needed for quantitative apportionment as these are more numerous and more likely to occur with patterns that allow differentiation among source contributions.

Figure 11.2 compares profiles from several common sources. These are a few of the hundreds of profiles that have been compiled in various libraries (e.g., Watson, 1979; Sheffield and Gordon, 1986; Cooper *et al.*, 1988; Olmez *et al.*, 1988; Ahuja *et al.*, 1989; Chow and Watson, 1989, 1994; Core, 1989; Houck *et al.*, 1989, 1990; Shareef *et al.*, 1989; Watson and Chow, 1992, 2001b; Watson *et al.*, 2001). Aluminum, silicon, potassium, calcium, and iron are most abundant in geological material, although the abundances vary by type and use of soil. Figure 11.2a is from a paved road, and lead deposited from vehicle exhaust and exhaust systems from prior use of leaded gasoline is evident at low levels. Organic carbon from deposited exhaust, oil drippings, asphalt, and ground up plant detritus is more abundant in this sample than in pristine desert soils. Several of the trace metals such as copper and zinc probably originate from brake and clutch linings as well as metals used in vehicle construction. Water-soluble potassium in dust profiles is about one-tenth of the total potassium abundance. Figure 11.2b from a coal-fired power station contains many of the same elements found in fugitive dust, but in different proportions. Selenium, strontium, and lead are enriched over the dust profile, and organic carbon is depleted. The ratio of sulfur dioxide gas to $PM_{2.5}$ emissions in this sample is large, and this would be a good indicator of fresh emissions from this source type. This ratio would change with time, however, as sulfur deposits more rapidly than $PM_{2.5}$ and augments $PM_{2.5}$ mass as it transforms into secondary sulfate within a day after emission.

The gasoline vehicle exhaust profile in Figure 11.2c shows organic and elemental carbon as its major components, with organic carbon two to three times elemental carbon. A wide variety of different elements is found in these emissions, but they are highly variable. This is consistent with many of these metals originating from exhaust system deterioration and motor oil that will change substantially from vehicle to vehicle. The carbon monoxide to $PM_{2.5}$ ratio is highest for gasoline vehicle exhaust, but it is also highly variable. Figure 11.2d shows that woodburning emissions are also dominated by carbon, but the organic carbon abundance is usually much higher than the elemental carbon fraction. Most of the potassium is water-soluble, in contrast to the insoluble potassium in fugitive dust. Other soluble salts containing sulfate, nitrate, ammonium, and chloride are evident. Trace metals from dust deposited on the burned wood and from stove or fireplace deterioration are often evident in these emissions.

As noted in the discussion of enrichment factors above, trace elements have been used in the past as markers for different source contributions. However, having identified and quantified residual oil combustion, leaded gasoline combustion, and metal and refining processes as contributors to excessive particle

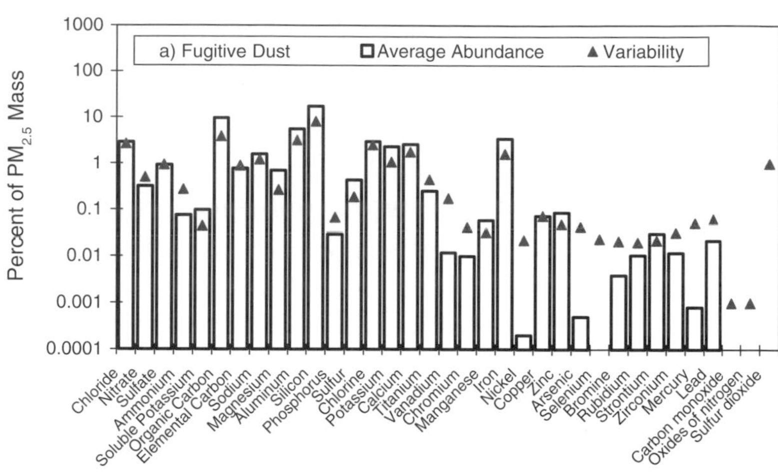

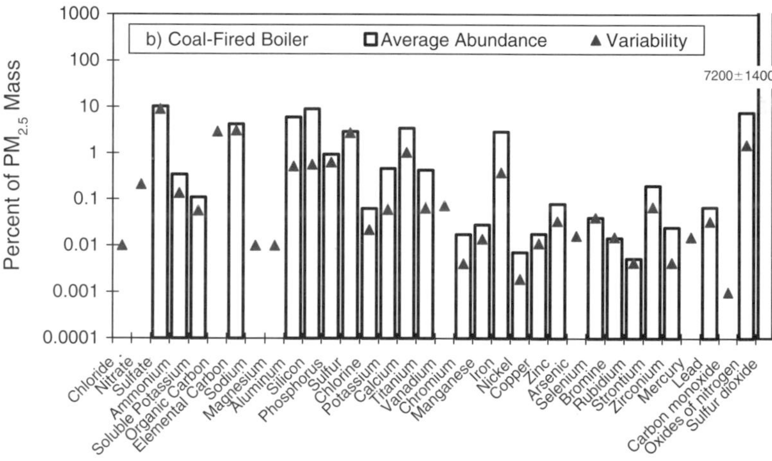

Figure 11.2

Examples of $PM_{2.5}$ source profiles determined from carbon, ion, and elemental analyses of samples from representative sources for (a) fugitive dust, (b) a coal fired power station, (c) hot stabilized gasoline vehicle exhaust, and (d) hardwood burning from typical of Denver, CO during winter, 1997 (Zielinska et al., *1998). The height of the bar represents the average abundance of each species as determined by the average from several source tests and the triangle represents the standard deviation of the average derived from the same tests. Ratios of carbon monoxide, oxides of nitrogen, and sulfur dioxide to $PM_{2.5}$ mass emissions can also be included in source profiles.*

mass and toxic element concentrations, these trace elements have been removed from many fuels and processes. While good for the environment, these modifications have degraded the utility of elemental abundances to distinguish among source contributions in receptor models. Additional chemical and physical properties beyond the commonly measured elements, ions, and carbon need to be examined to determine the ones that are useful in practical source apportionment applications.

Optical microscopy magnifies particles between ~2 and 30 μm so that their shapes, sizes, mineralogy, and optical properties can be related to similar particles from representative sources (McCrone and Delly, 1973; Draftz, 1982). Electron microscopy can be applied to smaller particles (Casuccio *et al.*, 1983, 1989). As shown in Figure 11.3, both shape and elemental content are determined by this

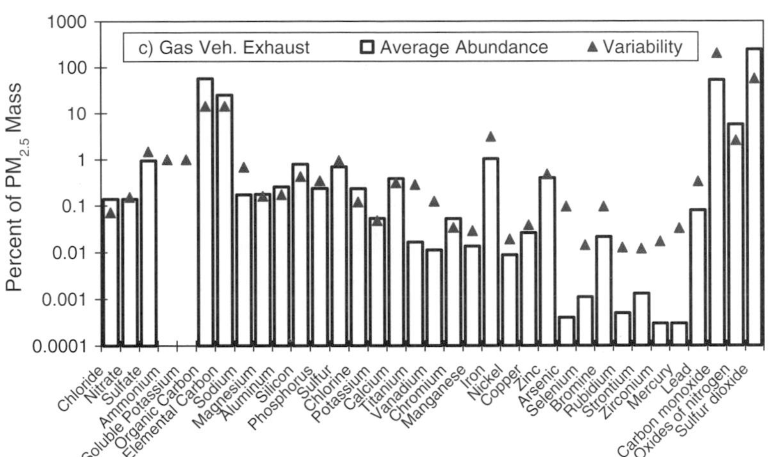

Figure 11.2
Continued

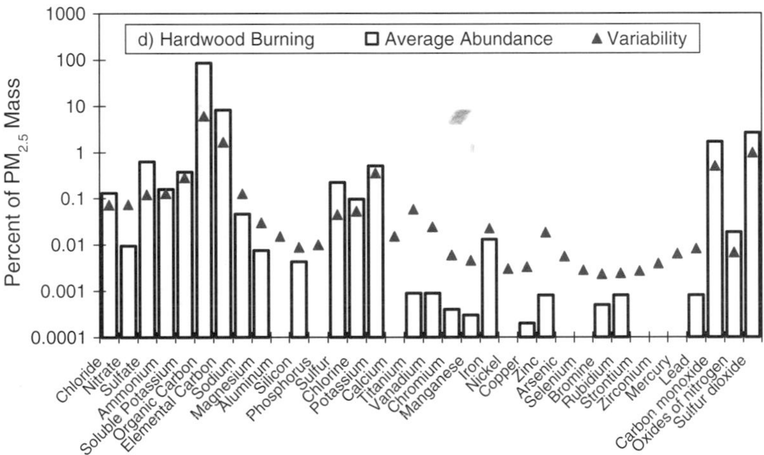

method. In this case, the elemental spectra from a clinker and a cement particle are very similar, but the more polished surface of the cement clearly distinguishes it from the clinker that precedes it in the cement production process (Greer *et al.*, 2000). Results from microscopic examination are semi-quantitative, as thousands of particles must be examined to obtain an adequate statistical representation of the millions deposited on a typical air sampling filter. Computer-automated scanning systems with pattern recognition methods are being developed to perform these analyses automatically (Hopke and Casuccio, 1991).

11.3.2.1 Specific Organic Compounds

The organic carbon components shown in Figure 11.2 are composed of thousands of individual compounds, but the detected compounds amount to only

a)

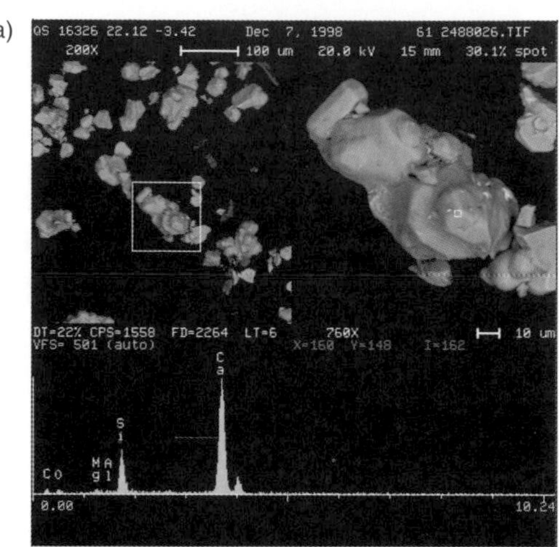

b)

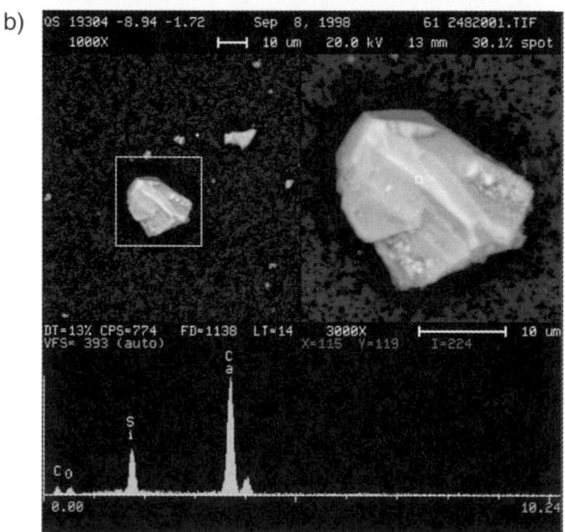

Figure 11.3

Backscattered electron images and elemental spectra illustrating (a) clinker particles and (b) finished cement particles (Watson et al., *1999). The elemental composition of the particles shows them both to be rich in calcium and silicon, but their appearances are quite different.*

10–20% of the organic compound mass. Specific organic compounds may be analyzed by solvent extraction followed by gas or liquid chromatography with different detectors (e.g., Mazurek *et al.*, 1987). Figure 11.4 compares a variety of different organic compound abundances in hardwood burning with those in meat cooking. Meat cooking profiles for elements, ions, and carbon are similar to those for wood burning (e.g., Figure 11.2d). The additional organic compounds shown in Figures 11.4a and 11.4b are quite different, however. Hardwood burning is rich in guaiacols and syringols, but low in sterols such as steroid-m and cholesterol. Just the opposite is true for the meat cooking, where cholesterol is among the most abundant species. Syringols are more abundant in hardwoods, such as oak or walnut, and they are depleted in softwoods such as pine, thereby allowing even greater differentiation to be achieved in source apportionment. Simoneit (1999) cites several examples of odd and even

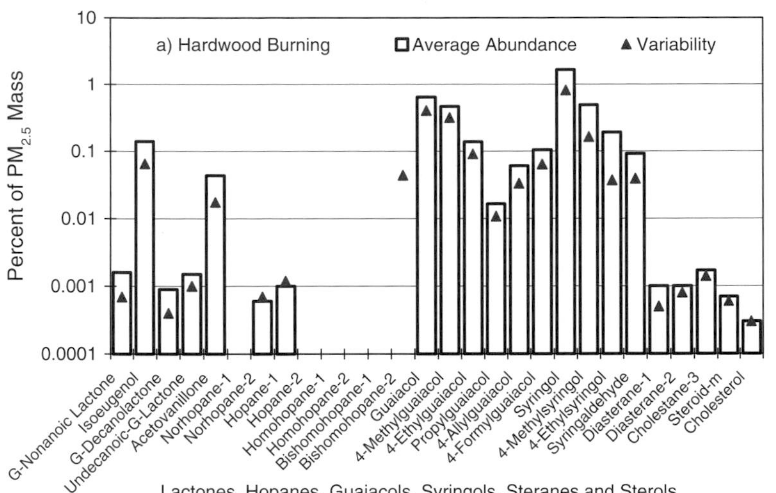

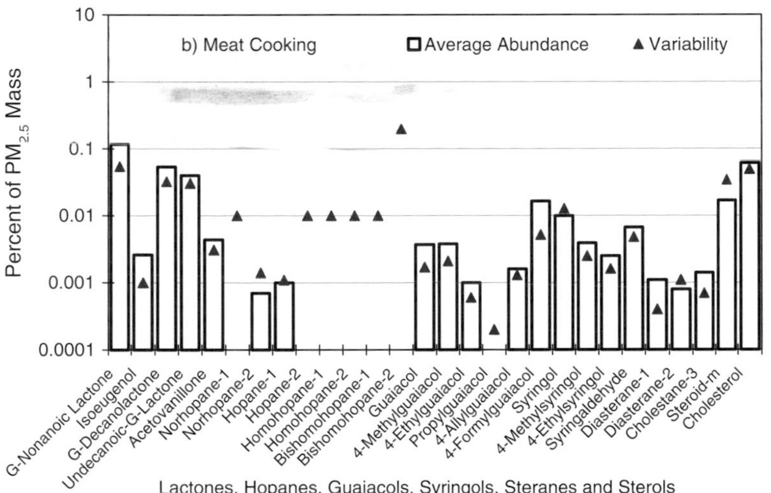

Figure 11.4

Examples of $PM_{2.5}$ source profiles for lactones, hopanes, guaiacols, syringols, steranes, and sterols for (a) hardwood combustion and (b) meat cooking (Zielinska et al., *1998). These are among many different categories of molecular organic markers measured by filter extraction and gas chromatography with mass spectrometric detection.*

numbered carbon molecules in the *n*-alkane series as indicating the presence or absence of organic carbon from manmade sources in the presence of ubiquitous contributions from natural sources. Biogenic material, such as plant waxes or secondary organic aerosols from vegetative hydrocarbon emissions, tends to have more molecules with odd numbers of carbon atoms, whereas carbon from combustion processes has nearly equal quantities of even and odd carbon-numbered molecules. A carbon preference index (CPI), the ratio of odd to even *n*-alkane masses, is used to estimate the relative abundances of odd and even compounds to separate natural from manmade contributions. Some organic compound source profiles have been measured for gasoline and diesel vehicle exhaust (Rogge *et al.*, 1993a; Fraser *et al.*, 1998, 1999; Schauer *et al.*,

1999a; Kleeman *et al.*, 2000), meat cooking (Kleeman *et al.*, 1999; Nolte *et al.*, 1999; Schauer *et al.*, 1999b), natural gas combustion (Rogge *et al.*, 1993b), coal combustion (Oros and Simoneit, 2000), oil combustion (Rogge *et al.*, 1997), wood burning and forest fires (Standley and Simoneit, 1994; Rogge *et al.*, 1998; Schauer *et al.*, 1998; Elias *et al.*, 1999; Simoneit *et al.*, 1999; McDonald *et al.*, 2000), and fugitive dust (Rogge *et al.*, 1993c). Most of these results are from emitters in southern California. Different investigators measure and use different markers for these sources in receptor models. The same measurements, which are often very costly, are also needed at the receptors. Nevertheless, this technology is rapidly evolving and may soon be in widespread use.

The organic and elemental carbon groupings in Figure 11.2 are operationally defined by a thermal evolution method that also monitors the changing reflectance of the sample (Chow *et al.*, 1993, 2001). This is only one of many ways in which carbon can be separated into functional and repeatable categories for which the exact composition is not precisely known. Watson *et al.* (1994) found differences in the quantities of carbon that evolved at different temperatures, with diesel exhaust yielding more carbon at higher temperatures than gasoline vehicle exhaust. Higashi and Flocchini (2000) and Meuzelaar (1992) have experimented with gas chromatographic and mass spectrometric detectors on different air quality samples that yield very complex profiles. These have not yet been quantitatively related to source emissions, and they need to be related to specific organic compounds, but they hold the potential to be much less costly and more convenient to implement than detailed organic compound analysis. Isotopic abundances differ among source material depending on its formation process or geologic origin. Radioactive carbon-14 (^{14}C) is formed by cosmic rays in the atmosphere and is incorporated into living things through respiration and ingestion. Fossil fuels have been depleted in carbon-14 owing to its ~5000 year half-life. ^{14}C abundances allow the separation of biogenic from fossil fuel combustion particles (Currie *et al.*, 1999). Estimates of fuel age must be made, as wood that grew before outdoor nuclear testing in the 1950s has much lower ^{14}C abundances than plant life that grew during and after that period. Other receptor model applications using isotopic abundances include ^{04}He as a marker of continental dust (Patterson *et al.*, 1999), ^{210}Pb to determine sources of lead in house dust (Adgate *et al.*, 1998), and ^{34}S to determine sources of acid deposition (Turk *et al.*, 1993) and to distinguish secondary sulfate contributions from residual oil combustion (Newmann *et al.*, 1975). Isotopes such as non-radioactive ^{34}S or radioactive ^{35}S follow the transformations of sulfur dioxide to sulfate from a source type and offer a way to directly apportion the secondary aerosol in cases where there are large differences in profiles among contributing sources.

11.3.3 TEMPORAL AND SPATIAL VARIABILITY

Sharp spikes of 1 to 60 minute duration indicate contributions from nearby sources. Diurnal patterns show variations over durations of several hours. A morning hump in PM_{10} or $PM_{2.5}$ concentrations is often observed on weekdays in urban areas; disappears on weekends when traffic volumes are more evenly distributed throughout the day (Chow *et al.*, 1999). High PM_{10} levels that correspond to high wind speeds are often indicators of windblown dust, while short-term concentration spikes that correspond to a specific wind direction indicate the direction of the source with respect to the receptor.

Figure 11.5 compares light scattering measurements from nearby locations taken with an inexpensive 'integrating nephelometer.' This nephelometer was equipped with a 'smart heater' that raises the temperature of the airstream when relative humidities (RHs) rise above 70% to maintain it at less than 70% RH. This heating minimizes interferences from liquid water absorbed by soluble particles while retaining volatile particles such as ammonium nitrate that evaporate at temperatures exceeding ~15°C. This system can be battery- or solar-powered for mounting on power poles with minimal expense. It is a cost-effective way to resolve environmental justice concerns in which some communities believe that their exposure to particles is not well represented by a compliance monitor in another neighborhood.

Even though the monitors in Figure 11.5 are separated by more than 1.5 km, the general diurnal pattern appears as if they were right next to each other. Short-duration spikes show the effects of local emissions that affect one monitor but not the others. At the roadway site these represent contributions from nearby road dust or vehicle exhaust. At the residential site during spring, night-time spikes from home heating are evident.

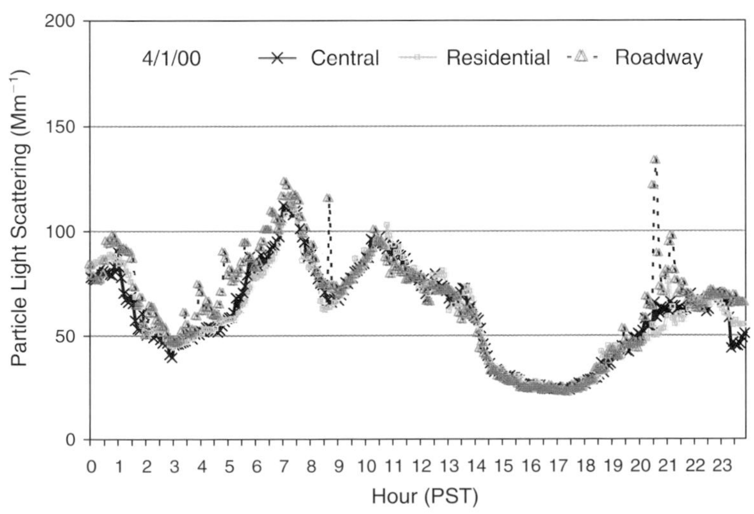

Figure 11.5

Particle light scattering measurements at a central Fresno, California, compliance monitoring sites compared to satellite sites in a residential neighborhood ~0.5 km east and near a heavily traveled road ~1 km west (Watson et al., *2000a). Dividing light scattering by 2 to 5g/m³ provides an estimate of $PM_{2.5}$ mass, a conversion that can be made more precise by comparison with periodic collocated filter sampling with battery-powered monitors. PST, Pacific Standard Time.*

11.4 SOURCE APPORTIONMENT EXAMPLES

The following examples are from actual problems that the authors were commissioned to solve using the methods described here. Each summary provides a statement of the problem, the technical approach, study results, and the decisions that were taken.

11.4.1 SOURCES OF WINTERTIME $PM_{2.5}$ IN DENVER, COLORADO

The Denver Brown Cloud has been the subject of five major studies, and numerous smaller ones, since 1973. The 1987–1988 study (Watson *et al.*, 1988) identified secondary ammonium nitrate and carbon as the largest chemical components, with wood smoke and vehicle exhaust being the largest contributors to the carbon as determined by the CMB model. With the available measurements, diesel contributions could not be distinguished from gasoline vehicle exhaust and wood burning could not be distinguished from cooking contributions. The 1996–1997 study (Watson *et al.*, 1998d) measured organic markers that were sufficient to separate these source contributions, as well as gasoline-exhaust contributions from the cold-starts and poorly maintained vehicles. Relative contributions from the emissions inventory and CMB source apportionment are compared in Figure 11.6.

The 1996–1997 study showed that measures to reduce wood burning emissions implemented after 1988 had been effective. Approximately half of the $PM_{2.5}$ that would have been attributed to wood smoke without the organic markers was found to derive from meat cooking, as evidenced by the ambient sterol levels. Fugitive dust emissions were overestimated by the inventory with

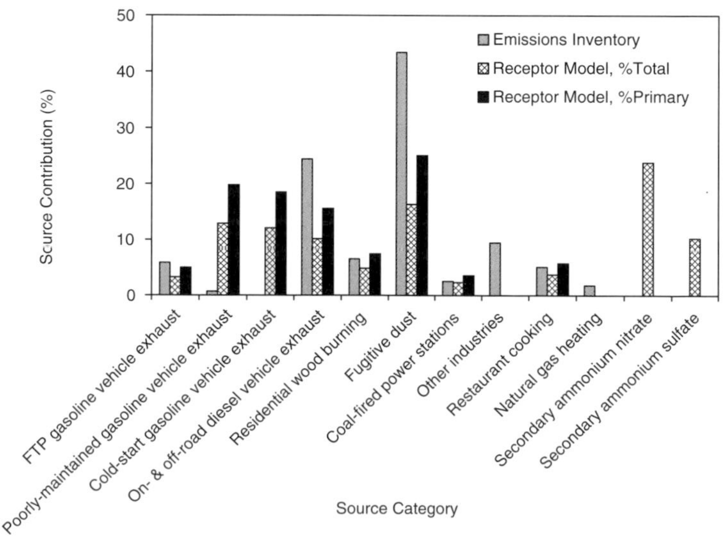

Figure 11.6

Comparison of fractional source contributions to $PM_{2.5}$ during 1996 in Denver, Colorado, (Watson et al., *1998d) from an emissions inventory, receptor model source contributions as fraction of total $PM_{2.5}$ and receptor model source contributions as fraction of primary $PM_{2.5}$ ($PM_{2.5}$ minus secondary ammonium nitrate and ammonium sulfate contributions). FTP, Federal Test Procedure.*

respect to their proportions in ambient air, a feature typical of most inventories that estimate fugitive dust (Watson *et al.*, 2000b).

The most significant result in Figure 11.6 is the large contribution from gasoline vehicle exhaust, and that most of this is from off-cycle operations and a relatively low number of very high emitters. The inventory grossly underestimates these contributions and they were not considered in control strategy development; most of the emphasis was placed on diesel exhaust, which the inventory showed to contribute ~80% of vehicular emissions.

As a result of this study, inventory improvements are intended to better quantify gasoline vehicle exhaust, cooking, and fugitive dust emissions. Although diesel emission reductions are still being considered, additional control measures are being investigated for the other sources.

11.4.2 CAUSES OF HAZE IN THE MOUNT ZIRKEL WILDERNESS AREA

The Mount Zirkel Wilderness Area in northwestern Colorado is a Class I area, as are most National Parks and Wilderness Areas in the US, for which visibility is a protected resource. USEPA regulations require remediation measures when significant visibility impairment in these areas has been certified and is 'reasonably attributable' to an existing industrial source or a group of sources. The Yampa Valley, to the west of the Wilderness, contains two coal-fired power stations with five generating units that were identified as potential contributors to the haze in the Wilderness. Watson *et al.* (1996) conducted a study to determine the frequency and intensity of hazes at the Wilderness caused by contributions from different sources. Many of the approaches described above were used together, including source dispersion modeling and CMB modeling with simulated aging for source profiles.

High time-resolution measurements were also taken for sulfur dioxide, elemental carbon, and light scattering, as shown in Figure 11.7. The 9/17/95 through 9/21/95 episode shows a correspondence between increases in sulfur dioxide concentrations, indicative of local power station emissions, and light scattering. Black carbon, typical of contributions from non-power station combustion, also co-varies with scattering. The early afternoon increases corresponded to the lifting and transport of emissions that had accumulated in the Yampa Valley during overnight fogs that accelerated the transformation of gaseous sulfur dioxide to sulfate. This was confirmed by chemical measurements within the valley as well as measurement of upper air winds and modeling. The phenomenon could also be seen on time-lapsed video.

The 10/13/95 through 10/16/95 episode shows a clear presence of power station emissions at the Wilderness site in the high sulfur dioxide pulses, but it shows no indication of corresponding increases in light scattering. Visibility was

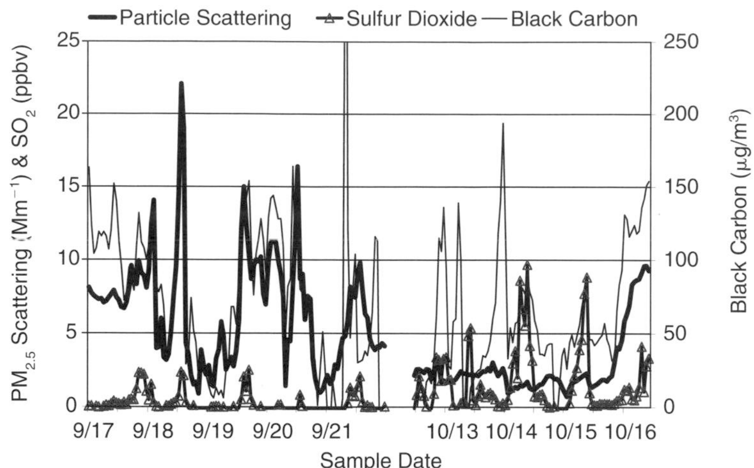

Figure 11.7

Time series of 1-hour average particle light scattering, black carbon, and sulfur dioxide at the Buffalo Pass site near the Mount Zirkel Wilderness area for episodes of 9/17/95 through 9/21/95 and 10/13/95 through 10/16/95 (Watson et al., *1996). mm^{-1}, inverse megameter; ppbv, Parts per billion by volume.*

good for this case and there was insufficient time for gas-to-particle conversion in the absence of aging in clouds and fogs. As a result of this study and other considerations a settlement was negotiated to add state-of-the art sulfur dioxide scrubbers to uncontrolled Yampa Valley power stations.

11.4.3 PARTICLE DEPOSITION NEAR A CEMENT PRODUCTION FACILITY

Residents of a community near a cement production facility along the California coast experienced episodes of deposited dust that adhered to their houses, vehicles, and windows, making it difficult to remove. Several sources of fugitive dust existed in the area, including agricultural tilling and harvesting, traffic suspension from paved and unpaved roads, construction, cement production, and wind erosion from open ground. Previous studies of PM_{10} in the area (Watson, 1998) had found no concentrations exceeding health standards with minor contributions from the cement plant. Deposition plates and scanning electron microscopy were used for long-term monitoring and during events to determine the sources of larger particles that created deposition.

The study showed that the deposited dust consisted of a mixture of substances, including suspended soil, sea salt, limestone, clinker, and cement. Calcium-rich crystalline particles were present in those samples that citizens found to adhere to surfaces. The deposited particles were not similar in appearance or chemical composition to the primary emissions from any of the potential sources, including the clinker and finished cement shown in Figures 11.3a and 11.3b. Samples of the different source materials were deliberately placed on deposition plates and left in a moist outside environment for several weeks. The sea salt dissolved, and most of the mineral materials retained their original shapes and compositions. The cement and clinker samples, however,

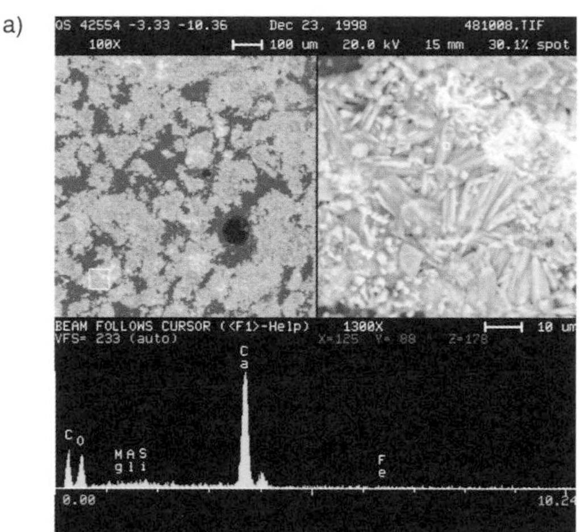

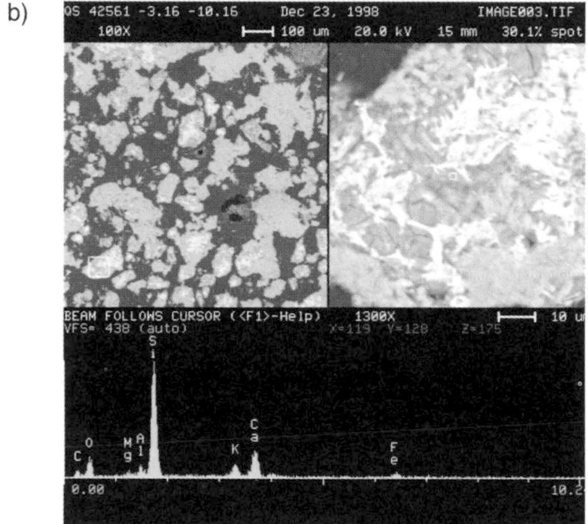

Figure 11.8

Backscattered electron image and elemental spectrum of (a) a polished cross-section of cement particles after exposure to a moisture-enriched environment and (b) the same sample after several microns were removed from the surface (Watson et al.*, 1999). Note that the elemental spectrum of 11.8b is similar to the spectra for clinker and cement particles in Figure 11.3 with a peak distinguished silicon peak.*

reacted with other deposited material to form a coating similar to that of the deposition samples. The silicate portion was missing from the electron-excited X-ray spectrum, as shown in Figure 11.8a. When the surface layer was removed from the test sample, however, it was found that the crystalline calcium surrounded a more complex core of silicon-rich particles that were typical of the clinker and cement, as seen in Figure 11.8b. Although the deposited particles did not look like the primary particles, the changes and interactions with weather and other particles could be simulated to determine the origin.

A thorough survey was conducted of the clinker and cement operations that resulted in: (1) paving and sweeping the road to the clinker storage shed; (2) installing a retaining wall around the base of the storage shed to reduce damage to the shed walls by heavy equipment; (3) creating a curtained airlock with a dust collector at the shed exit; (4) overhauling dust-collectors on the cement storage silos; and (5) installing automatic detection and shutoff systems to prevent overfilling of silos. The deposition surveillance program is being continued by community members to determine improvement trends and to identify additional episodes that can be analyzed to further reduce emissions.

11.5 SUMMARY AND CONCLUSIONS

Particle size, chemical, temporal, and spatial patterns have been used to solve several air quality problems related to public health, visibility, and deposition nuisances. These have employed a variety of measurement technologies and mathematical receptor models that interpret the measurements in terms of source contributions and/or their locations. The advent of *in situ* continuous monitoring for different chemical components as a function of particle size offers new opportunities to combine these different properties into a unified model. Portable and inexpensive monitors, especially for light scattering, offer opportunities to develop spatial and temporal patterns close to sources and in different neighborhoods that might be affected. Challenges remain to quantify most of the molecular compounds that constitute organic carbon, one of the largest particle components; only 10–20% of the organic carbon mass has been identified to date. Although many source types have been characterized, the measurements reported by different investigators often differ from each other and from measurements available at receptors. Particle transformation between source and receptor must be accounted for by mathematical or empirical simulation. These limitations argue for using particle pattern recognition methods in conjunction with other methods, such as source dispersion modeling, that can provide independent source attribution. Reconciling the results of these independent approaches then allows uncertainties to be estimated and for experiments designed to reduce those uncertainties.

REFERENCES

Adgate, J.L., Willis, R.D., Buckley, T.J., Chow, J.C., Watson, J.G., Rhoads, G.G., and Lioy, P.J. (1998) Chemical mass balance source apportionment of lead in house dust. *Environmental Science and Technology* 32, 108–114.

Ahuja, M.S., Paskind, J.J., Houck, J.E., and Chow, J.C. (1989) Design of a study for the chemical and size characterization of particulate matter emissions from selected sources in California. In *Transactions, Receptor Models in Air Resources Management* (Watson, J.G., ed.), pp. 145–158. Air and Waste Management Association, Pittsburgh, PA.

Buzorius, G., Hämeri, K., Pekkanen, J., and Kulmala, M. (1999) Spatial variation of aerosol number concentration in Helsinki City. *Atmospheric Environment* 33, 553–565.

Casuccio, G.S., Janocko, P.B., Lee, R.J., Kelly, J.F., Dattner, S.L., and Mgebroff, J.S. (1983) The use of computer controlled scanning electron microscopy in environmental studies. *Journal of the Air Pollution Control Association* 33, 937–943.

Casuccio, G.S., Schwoeble, A.J., Henderson, B.C., Lee, R.J., Hopke, P.K., and Sverdrup, G.M. (1989) The use of CCSEM and microimaging to study source/receptor relationships. In *Transactions, Receptor Models in Air Resources Management* (Watson, J.G., ed.), pp. 39–58. Air and Waste Management Association, Pittsburgh, PA.

Cheng, Y.S. (1993) Condensation detection and diffusion size separation techniques. In *Aerosol Measurement: Principles, Techniques and Applications* (Willeke, K. and Baron, P.A., eds), pp. 427–448. Van Nostrand Reinhold, New York.

Chow, J.C. (1995) Critical review: Measurement methods to determine compliance with ambient air quality standards for suspended particles. *Journal of the Air and Waste Management Association* 45, 320–382.

Chow, J.C. and Watson, J.G. (1989) Summary of particulate data bases for receptor modeling in the United States. In *Transactions, Receptor Models in Air Resources Management* (Watson, J.G., ed.), pp. 108–133. Air and Waste Management Association, Pittsburgh, PA.

Chow, J.C. and Watson, J.G. (1994) Contemporary source profiles for geological material and motor vehicle emissions. Report No. DRI 2625.2F, Prepared for USEPA, Office of Air Quality Planning and Standards, Research Triangle Park, NC, Desert Research Institute, Reno, NV.

Chow, J.C. and Watson, J.G. (1998) Guideline on speciated particulate monitoring. Prepared for USEPA, Research Triangle Park, NC, Desert Research Institute, Reno, NV.

Chow, J.C., Watson, J.G., Egami, R.T., Frazier, C.A., Lu, Z., Goodrich, A., and Bird, A. (1990) Evaluation of regenerative-air vacuum street sweeping on geological contributions to PM_{10}. *Journal of the Air and Waste Management Association* 40, 1134–1142.

Chow, J.C., Watson, J.G., Pritchett, L.C., Pierson, W.R., Frazier, C.A., and Purcell, R.G. (1993) The DRI Thermal/Optical Reflectance carbon analysis system: Description, evaluation and applications in US air quality studies. Atmospheric Environment 27A, 1185–1201.

Chow, J.C., Watson, J.G., Green, M.C., Lowenthal, D.H., DuBois, D.W., Kohl, S.D., Egami, R.T., Gillies, J.A., Rogers, C.F., Frazier, C.A., and Cates, W. (1999) Middle- and neighborhood-scale variations of PM_{10} source contributions in Las Vegas, Nevada. *Journal of Air and Waste Management Association* 49, 641–654.

Chow, J.C., Watson, J.G., Crow, D., Lowenthal, D.H., and Merrifield, T. (2001) Comparison of IMPROVE and NIOSH carbon measurements. *Aerosol Science and Technology* 34, 23–34.

Cohen, B. and Hering, S.V. (1995) *Air Sampling Instruments for Evaluation of Atmospheric Contaminants*. American Conference of Governmental Industrial Hygienists, Cincinnati, OH.

Cooper, J.A. and Watson, J.G. (1980) Receptor oriented methods of air particulate source apportionment. *Journal of the Air Pollution Control Association* 30, 1116–1125.

Cooper, J.A., Sherman, J.R., Miller, E., Redline, D., Valdonovinos, L., and Pollard, W.L. (1988). CMB source apportionment of PM_{10} downwind of an oil-fired power plant in Chula Vista, California. In *Transactions, PM_{10}: Implementation of Standards* (Mathai, C.V. and Stonefield, D.H., eds), pp. 495–507. Air and Waste Management Association, Pittsburgh, PA.

Core, J.E. (1989) Source profile development for PM_{10} receptor modeling. In *Transactions: Receptor Models in Air Resources Management* (Watson, J.G., ed.), pp. 134–144. Air and Waste Management Association, Pittsburgh, PA.

Currie, L.A., Klouda, G.A., Benner, B.A., Jr., Garrity, K., and Eglinton, T.I. (1999) Isotopic and molecular fractionation in combustion: Three routes to molecular marker validation, including direct molecular 'dating' (GC/AMS). *Atmospheric Environment* 33, 2789–2806.

Draftz, R.G. (1982) Distinguishing carbon aerosols by microscopy. In *Particulate Carbon: Atmospheric Life Cycle* (Wolff, G.T. and Klimisch, R.L., eds), pp. 261–271. Plenum Press, New York.

Eldred, R.A., Cahill, T.A., and Flocchini, R.G. (1997) Composition of $PM_{2.5}$ and PM_{10} aerosols in the IMPROVE Network. *Journal of the Air and Waste Management Association* 47, 194–203.

Elias, V.O., Simoneit, B.R.T., Pereira, A.S., Cabral, J.A., and Cardoso, J.N. (1999) Detection of high molecular weight organic tracers in vegetation smoke samples by high-temperature gas chromatography-mass spectrometry. *Environmental Science and Technology* 33, 2369–2376.

Engelbrecht, J.P., Swanepoel, L., Zunckel, M., Chow, J.C., Watson, J.G., and Egami, R.T. (2000) Modelling PM_{10} aerosol data from the Qalabotjha low-smoke fuels macro-scale experiment in South Africa. *Ecological Modelling* 127, 235–244.

Feeley, J.A. and Liljestrand, H.M. (1983) Source contributions to acid precipitation in Texas. *Atmospheric Environment* 17, 807–814.

Fraser, M.P., Cass, G.R., and Simoneit, B.R.T. (1998) Gas-phase and particle-phase organic compounds emitted from motor vehicle traffic in a Los Angeles roadway tunnel. *Environmental Science and Technology* 32, 2051–2060.

Fraser, M.P., Cass, G.R., and Simoneit, B.R.T. (1999) Particulate organic compounds emitted from motor vehicle exhaust and in the urban atmosphere. *Atmospheric Environment* 33, 2715–2724.

Greer, W.L., Dougherty, A., and Sweeney, D.M. (2000) Portland cement. In *Air Pollution Engineering Manual* 2nd edn (Davis, W.T., ed.), pp. 664–681. John Wiley & Sons, New York.

Henry, R.C. (1986) Fundamental limitations of receptor models using factor analysis. In *Transactions, Receptor Methods for Source Apportionment: Real World Issues and Applications* (Pace, T.G., ed.), pp. 68–77. Air Pollution Control Association, Pittsburgh, PA.

Henry, R.C. (1987) Current factor analysis receptor models are ill-posed. *Atmospheric Environment* 21, 1815–1820.

Henry, R.C. (1991) Multivariate receptor models. In *Receptor Modeling for Air Quality Management* (Hopke, P.K., ed.), pp. 117–147. Elsevier, Amsterdam.

Henry, R.C. (1997a) Receptor model applied to patterns in space (RMAPS) Part I – Model description. *Journal of the Air and Waste Management Association* 47, 216–219.

Henry, R.C. (1997b) Receptor model applied to patterns in space (RMAPS) Part II – Apportionment of airborne particulate sulfur from Project MOHAVE. *Journal of the Air and Waste Management Association* 47, 220–225.

Henry, R.F., Rao, S.T., Zurbenko, I.G., and Porter, P.S. (2000) Effects of changes in data reporting practices on trend assessments. *Atmospheric Environment* 34, 2659–2662.

Hering, S.V. (1995) Impactors, cyclones, and other inertial and gravitational collectors. In *Air Sampling Instruments – For Evaluation of Atmospheric Contaminants* (Cohen, B. and Hering, S.V., eds), pp. 279–321. American Conference of Governmental Industrial Hygienists, Cincinnati, OH.

Hies, T., Treffeisen, R., Sebald, L., and Reimer, E. (2000) Spectral analysis of air pollutants Part 1. Elemental carbon time series. *Atmospheric Environment* 34, 3495–3502.

Higashi, R.M. and Flocchini, R.G. (2000) Fingerprinting sources of dust through analysis of complex organic matter. Report No. ARB 97-309, Prepared for California Air Resources Board, Sacramento, California, Davis, CA, University of California.

Hopke, P.K. (1981) The application of factor analysis to urban aerosol source resolution. In *Atmospheric Aerosol: Source/Air Quality Relationships* (Macias, E.S. and Hopke, P.K., eds), pp. 21–49. American Chemical Society, Washington, DC.

Hopke, P.K. (1985) *Receptor Modeling in Environmental Chemistry.* John Wiley & Sons, New York.

Hopke, P.K. (1988) Target transformation factor analysis as an aerosol mass apportionment method: A review and sensitivity study. *Atmospheric Environment* 22, 1777–1792.

Hopke, P.K. (1991) *Receptor Modeling for Air Quality Management.* Elsevier Press, Amsterdam.

Hopke, P.K. and Casuccio, G.S. (1991) Scanning electron microscopy. In *Receptor Modeling for Air Quality Management* (Hopke, P.K., ed.), pp. 149–212. Elsevier, Amsterdam.

Hopke, P.K., Alpert, D.J., and Roscoe, B.A. (1983) FANTASIA – A program for target transformation factor analysis to apportion sources in environmental samples. *Computers & Chemistry* 7, 149–155.

Houck, J.E., Chow, J.C., and Ahuja, M.S. (1989) The chemical and size characterization of particulate material originating from geological sources in California. In *Transactions, Receptor Models in Air Resources Management* (Watson, J.G., ed.), pp. 322–333. Air and Waste Management Association, Pittsburgh, PA.

Houck, J.E., Goulet, J.M., Chow, J.C., Watson, J.G., and Pritchett, L.C. (1990) Chemical characterization of emission sources contributing to light extinction. In *Transactions, Visibility and Fine Particles* (Mathai, C.V., ed.), pp. 437–446. Air and Waste Management Association, Pittsburgh, PA.

Hsu, K.J. (1997) Application of vector autoregressive time series analysis to aerosol studies. *Tellus* 49B, 327–342.

John, W., Wall, S.M., Ondo, J.L., and Winklmayr, W. (1990) Modes in the size distributions of atmospheric inorganic aerosol. *Atmospheric Environment* 24A, 2349–2359.

Jorquera, H., Palma, W., and Tapia, J. (2000) An intervention analysis of air quality data at Santiago, Chile. *Atmospheric Environment* 34, 4073–4084.

Kelley, D.W. and Nater, E.A. (2000) Source apportionment of lake bed sediments to watersheds in an upper Mississippi basin using a chemical mass balance method. *Catena* 41, 277–292.

Keywood, M.D., Ayers, G.P., Gras, J.L., Gillett, R.W., and Cohen, D.D. (1999) Relationships between size segregated mass concentration data and ultrafine particle number concentrations in urban areas. *Atmospheric Environment* 33, 2907–2913.

Kleeman, M.J., Schauer, J.J., and Cass, G.R. (1999) Size and composition distribution of fine particulate matter emitted from wood burning, meat charbroiling, and cigarettes. *Environmental Science and Technology* 33, 3516–3523.

Kleeman, M.J., Schauer, J.J., and Cass, G.R. (2000) Size and composition distribution of fine particulate matter emitted from motor vehicles. *Environmental Science and Technology* 34, 1132–1142.

Kleinman, M.T., Pasternack, B.S., Eisenbud, M., and Kneip, T.J. (1980) Identifying and estimating the relative importance of sources of airborne particulates. *Environmental Science and Technology* 14, 62–65.

Lawson, D.R. and Winchester, J.W. (1979) A standard crustal aerosol as a reference for elemental enrichment factors. *Atmospheric Environment* 13, 925–930.

Lipfert, F.W. and Wyzga, R.E. (1997) Air pollution and mortality: The implications of uncertainties in regression modeling and exposure measurement. *Journal of the Air and Waste Management Association* 47, 517–523.

Lowenthal, D.H. and Rahn, K.A. (1989) The relationship between secondary sulfate and primary regional signatures in northeastern aerosol and precipitation. *Atmospheric Environment* 23, 1511–1515.

Malm, W.C., Iyer, H.K., and Gebhart, K.A. (1990) Application of tracer mass balance regression to WHITEX data. In *Transactions, Visibility and Fine Particles* (Mathai, C.V., ed.), pp. 806–818. Air and Waste Management Association, Pittsburgh, PA.

Marple, V.A., Rubow, K.L., and Olson, B.A. (1993) Inertial, gravitational, centrifugal and thermal collection techniques. In *Aerosol Measurement: Principles, Techniques and Applications* (Willeke, K. and Baron, P.A., eds), pp. 206–232. Van Nostrand Reinhold, New York.

Mason, B. (1966) *Principles of Geochemistry*, 3rd edn. John Wiley & Sons, New York.

Mauderly, J.L., Neas, L.M., and Schlesinger, R.B. (1998) PM monitoring needs related to health effects. In *Atmospheric Observations: Helping Build the Scientific Basis for Decisions related to Airborne Particulate Matter* (Albritton, D.L. and Greenbaum, D.S., eds), pp. 9–14. Health Effects Research Institute, Cambridge, MA.

Mazurek, M.A., Simoneit, B.R.T., Cass, G.R., and Gray, H.A. (1987) Quantitative high-resolution gas chromatography and high-resolution gas chromatography\mass spectrometry analyses of carbonaceous fine aerosol particles. *International Journal of Environmental and Analytical Chemistry* 29, 119–139.

McCrone, W.C. and Delly, J.G. (1973) *The Particle Atlas,* vol. I: *Principles and Techniques*. Ann Arbor Science Publishers, Ann Arbor, MI.

McDonald, J.D., Zielinska, B., Fujita, E.M., Sagebiel, J.C., Chow, J.C., and Watson, J.G. (2000) Fine particle and gaseous emission rates from residential wood combustion. *Environmental Science and Technology* 34, 2080–2091.

McMurry, P.H. (2000) A review of atmospheric aerosol measurements. *Atmospheric Environment* 34, 1959–1999.

Meuzelaar, H.L.C. (1992) Field evaluation and monitoring of air pollutant levels. Report No. Project No. A2 Thrust Area: Air Quality. Prepared for South West Center for Environmental Research & Policy, University of Utah, Center for Micro Analysis & Reaction Chemistry, Salt Lake City, UT.

Morandi, M.T., Lioy, P.J., and Daisey, J.M. (1991) Comparison of two multivariate modeling approaches for the source apportionment of inhalable particulate matter in Newark, NJ. *Atmospheric Environment* 25A, 927–937.

Newmann, L., Forrest, J., and Manowitz, B. (1975) The application of an isotopic ratio technique to a study of the atmospheric oxidation of sulphur dioxide in the plume from a oil-fired power plant. *Atmospheric Environment*, 9, 969–977.

Nolte, C.G., Schauer, J.J., Cass, G.R., and Simoneit, B.R.T. (1999) Highly polar organic compounds present in meat smoke. *Environmental Science and Technology* 33, 3313.

Oberdörster, G., Gelein, R., Ferin, J., and Weiss, B. (1995) Association of particulate air pollution and acute mortality: Involvement of ultrafine particles? *Inhalation Toxicology* 7, 111–124.

Olmez, I. (1989) Trace element signatures in groundwater pollution. In *Transactions, Receptor Models in Air Resources Management* (Watson, J.G., ed.), pp. 3–11. Air and Waste Management Association, Pittsburgh, PA.

Olmez, I., Sheffield, A.E., Gordon, G.E., Houck, J.E., Pritchett, L.C., Cooper, J.A., Dzubay, T.G., and Bennett, R.L. (1988) Compositions of particles from selected sources in Philadelphia for receptor model applications. *Journal of the Air Pollution Control Association* 38, 1392–1402.

Oros, D.R. and Simoneit, B.R.T. (2000) Identification and emission rates of molecular tracers in coal smoke particulate matter. *Fuel* 79, 515–536.

Paatero, P. (1997) Least squares formulation of robust, non-negative factor analysis. *Chemometrics and Intelligent Laboratory Systems* 37, 23–35.

Patterson, D.B., Farley, K.A., and Norman, M.D. (1999) ^{04}He as a tracer of continental dust: A 1.9 million year record of aeolian flux to the west equatorial Pacific Ocean – The geologic history of wind. *Geochimica et Cosmochimica Acta* 63, 615.

Pena-Mendez, E.M., Astorga-Espana, M.S., and Garca-Montelongo, F.J. (2001) Chemical fingerprinting applied to the evaluation of marine oil pollution in the coasts of Canary Islands (Spain). *Environmental Pollution* 111, 177–187.

Perez, P., Trier, A., and Reyes, J. (2000) Prediction of $PM_{2.5}$ concentrations several hours in advance using neural networks in Santiago, Chile. *Atmospheric Environment* 34, 1189–1196.

Perrier, V., Philipovitch, T., and Basdevant, C. (1995) Wavelet spectra compared to Fourier spectra. *Journal of Mathematical Physics* 36, 1506–1519.

Perry, K.D., Cahill, T.A., Eldred, R.A., and Dutcher, D.D. (1997) Long-range transport of North African dust to the eastern United States. *Journal of Geophysical Research* 102, 11225–11238.

Preining, O. (1998) The physical nature of very, very small particles and its impact on their behaviour. *Journal of Aerosol Science* 29, 481–496.

Pun, B.K. and Seigneur, C. (1999) Understanding particulate matter formation in the California San Joaquin Valley: Conceptual model and data needs. *Atmospheric Environment* 33, 4865–4875.

Rader, D.J. and O'Hern, T.J. (1993) Optical direct-reading techniques: *In situ* sensing. In *Aerosol Measurement: Principles, Techniques and Applications* (Willeke, K. and Baron, P.A., eds), pp. 345–380. Van Nostrand Reinhold, New York.

Rahn, K.A. (1976) Silicon and aluminum in atmospheric aerosols: Crust-air fractionation? *Atmospheric Environment* 10, 597–601.

Rashid, M. and Griffiths, R.F. (1993) Ambient K, S, and Si in fine and coarse aerosols of Kuala Lumpur, Malaysia. *Journal of Aerosol Science* 24, S5–S6.

Reich, S.L., Gomez, D.R., and Dawidowski, L.E. (1999) Artificial neural network for the identification of unknown air pollution sources. *Atmospheric Environment* 33, 3045–3052.

Rogge, W.F., Hildemann, L.M., Mazurek, M.A., Cass, G.R., and Simoneit, B.R.T. (1993a) Sources of fine organic aerosol. 2. Noncatalyst and catalyst-equipped automobiles and heavy-duty diesel trucks. *Environmental Science and Technology* 27, 636–651.

Rogge, W.F., Hildemann, L.M., Mazurek, M.A., Cass, G.R., and Simoneit, B.R.T. (1993b) Sources of fine organic aerosol. 5. Natural gas home appliances. *Environmental Science and Technology* 27, 2736–2744.

Rogge, W.F., Hildemann, L.M., Mazurek, M.A., Cass, G.R., and Simoneit, B.R.T. (1993c) Sources of fine organic aerosol. 3. Road dust, tire debris, and organometallic brake lining dust: Roads as sources and sinks. *Environmental Science and Technology* 27, 1892–1904.

Rogge, W.F., Hildemann, L.M., Mazurek, M.A., Cass, G.R., and Simoneit, B.R.T. (1997) Sources of fine organic aerosol. 8. Boilers burning No. 2 distillate fuel oil. *Environmental Science and Technology* 31, 2731–2737.

Rogge, W.F., Hildemann, L.M., Mazurek, M.A., Cass, G.R., and Simoneit, B.R.T. (1998) Sources of fine organic aerosol. 9. Pine, oak, and synthetic log combustion in residential fireplaces. *Environmental Science and Technology* 32, 13–22.

Schauer, J.J. and Cass, G.R. (2000) Source apportionment of wintertime gas-phase and particle-phase air pollutants using organic compounds as tracers. *Environmental Science and Technology* 34, 1821–1832.

Schauer, J.J., Cass, G.R., and Simoneit, B.R.T. (1998) Characterization of the emissions of individual organic compounds present in biomass aerosol. *Journal of Aerosol Science* 29, S223.

Schauer, J.J., Kleeman, M.J., Cass, G.R., and Simoneit, B.R.T. (1999a) Measurement of emissions from air pollution sources 2. C_1 through C_{30} organic compounds from medium duty diesel trucks. *Environmental Science and Technology* 33, 1578–1587.

Schauer, J.J., Kleeman, M.J., Cass, G.R., and Simoneit, B.R.T. (1999b) Measurement of emissions from air pollution sources. C_1 through C_{29} organic compounds from meat charbroiling. *Environmental Science and Technology* 33, 1566–1577.

Seigneur, C. (1998) $PM_{2.5}$ modeling: Current status and research needs. In *Proceedings, $PM_{2.5}$: A Fine Particle Standard* (Chow, J.C. and Koutrakis, P., eds), pp. 713–724. Air and Waste Management Association, Pittsburgh, PA.

Shareef, G.S., Bravo, L.A., Stelling, J.H.E., Kuykendal, W.B., and Mobley, J.D. (1989) Air emissions species data base. In *Transactions: Receptor Models in Air Resources Management* (Watson, J.G., ed), pp. 73–83. Air and Waste Management Association, Pittsburgh, PA.

Sheffield, A.E. and Gordon, G.E. (1986) Variability of particle composition from ubiquitous sources: Results from a new source-composition library. In *Transactions, Receptor Methods for Source Apportionment: Real World Issues and Applications* (Pace, T.G., ed.), pp. 9–22. Air Pollution Control Association, Pittsburgh, PA.

Shively, T.S. and Sager, T.W. (1999) Semiparametric regression approach to adjusting for meteorological variables in air pollution trends. *Environmental Science and Technology* 33, 3873–3880.

Simoneit, B.R.T. (1999) A review of biomarker compounds as source indicators and tracers for air pollution. *Environmental Science and Pollution Research* 6, 159–169.

Simoneit, B.R.T., Schauer, J.J., Nolte, C.G., Oros, D.R., Elias, V.O., Fraser, M.P., Rogge, W.F., and Cass, G.R. (1999) Levoglucosan, a tracer for cellulose in biomass burning and atmospheric particles. *Atmospheric Environment* 33, 173–182.

Somerville, M.C. and Evans, E.G. (1995) Effect of sampling frequency on trend detection for atmospheric fine mass. *Atmospheric Environment* 29, 2429–2438.

Song, X.H. and Hopke, P.K. (1996) Solving the chemical mass balance problem using an artificial neural network. *Environmental Science and Technology* 30, 531–535.

Spix, C., Heinrich, J., Dockery, D.W., Schwartz, J., Völksch, G., Schwinkowski, K., Cöllen, C., and Wichmann, H.E. (1999) Air pollution and daily mortality in Erfurt, East Germany, from 1980–1989. *Environmental Health Perspectives* 101, 518–526.

Standley, L.J. and Simoneit, B.R.T. (1994) Resin diterpenoids as tracers for biomass combustion aerosols. *Atmospheric Environment* 28, 1–16.

Sturges, W.T. (1989) Identification of pollution sources of anomalously enriched elements. *Atmospheric Environment* 23, 2067–2072.

Thurston, G.D. and Spengler, J.D. (1985) A multivariate assessment of meteorological influences on inhalable particle source impacts. *Journal of Climatology and Applied Meterology* 24, 1245–1256.

Turk, J.T., Campbell, D.H., and Spahr, N.E. (1993) Use of chemistry and stable sulfur isotopes to determine sources of trends in sulfate of Colorado lakes. *Water, Air and Soil Pollution* 67, 415–431.

USEPA (1971) National primary and secondary ambient air quality standards. *Federal Register* 36, 8186.

USEPA (1997) National ambient air quality standards for particulate matter: Final rule. *Federal Register* 62, 38651–38701.

USEPA (1999) Regional haze regulations. *Federal Register* 64, 35714–35774.

U.S. Supreme Court (2001) Whitman v American Trucking Associations Inc., Case No. 99-1257, 531 U.S. http://supremecourtus.gov/opinions/sliplists/s531pt2.html.

Vedal, S. (1997) Critical review – ambient particles and health: Lines that divide. *Journal of the Air and Waste Management Association* 47, 551–581.

Vega, E., García, I., Apam, D., Ruíz, M.E., and Barbiaux, M. (1997) Application of a chemical mass balance receptor model to respirable particulate matter in Mexico City. *Journal of the Air and Waste Management Association* 47, 524–529.

Watson, J.G. (1979) Chemical element balance receptor model methodology for assessing the sources of fine and total particulate matter in Portland, Oregon. PhD Dissertation, Oregon Graduate Center, Beaverton, OR.

Watson, J.G. (1984) Overview of receptor model principles. *Journal of the Air Pollution Control Association* 34, 619–623.

Watson, J.G. (1998) Implications of new suspended particle standards for the cement industry. In *Conference Record, 40th Cement Industry Technical Conference* (Jasberg, M.W., ed.), pp. 331–342. Portland Cement Association, Skokie, IL.

Watson, J.G. and Chow, J.C. (1992) Data bases for PM_{10} and $PM_{2.5}$ chemical compositions and source profiles. In *PM_{10} Standards and Nontraditional Particulate Source Controls* (Chow, J.C. and Ono, D.M., eds), pp. 61–91. Air and Waste Management Association, Pittsburgh, PA.

Watson, J.G. and Chow, J.C. (1994) Clear sky visibility as a challenge for society. *Annual Review of Energy and the Environment* 19, 241–266.

Watson, J.G. and Chow, J.C. (2001a) Ambient air sampling. In *Aerosol Measurement: Principles, Techniques and Applications*, 2nd edn. (Baron, P.A. and Willeke, K., eds), pp. 821–844, Van Nostrand Reinhold, New York.

Watson, J.G. and Chow, J.C. (2001b) Source characterization of major emission sources in the Imperial and Mexicali valleys along the US/Mexico border. *Science of the Total Environment* 277, in press.

Watson, J.G., Cooper, J.A., and Huntzicker, J.J. (1984) The effective variance weighting for least squares calculations applied to the mass balance receptor model. *Atmospheric Environment* 18, 1347–1355.

Watson, J.G., Chow, J.C., Richards, L.W., Andersen, S.R., Houck, J.E., and Dietrich, D. (1988) *The 1987–88 Metro Denver Brown Cloud Air Pollution Study*, Volume III: Data interpretation. Report No. DRI 8810.1, Prepared for 1987–88 Metro Denver Brown Cloud Study, Inc., Greater Denver Chamber of Commerce, Denver, CO, Desert Research Institute, Reno, NV.

Watson, J.G., Robinson, N.F., Chow, J.C., Henry, R.C., Kim, B.M., Pace, T.G., Meyer, E.L., and Nguyen, Q. (1990) The USEPA/DRI chemical mass balance receptor model, CMB 7.0. Environment Software 5, 38–49.

Watson, J.G., Chow, J.C., and Pace, T.G. (1991) Chemical mass balance. In *Receptor Modeling for Air Quality Management* (Hopke, P.K., ed.), pp. 83–116. Elsevier Press, New York.

Watson, J.G., Chow, J.C., Lowenthal, D.H., Pritchett, L.C., Frazier, C.A., Neuroth, G.R., and Robbins, R. (1994) Differences in the carbon composition of source profiles for diesel- and gasoline-powered vehicles. *Atmospheric Environment* 28, 2493–2505.

Watson, J.G., Blumenthal, D.L., Chow, J.C., Cahill, C.F., Richards, L.W., Dietrich, D., Morris, R., Houck, J.E., Dickson, R.J., and Andersen, S.R. (1996) *Mt. Zirkel Wilderness Area Reasonable Attribution Study of Visibility Impairment*, Volume II: *Results of Data Analysis* and Modeling. Prepared for Colorado Department of Public Health and Environment, Denver, CO, Desert Research Institute, Reno, NV.

Watson, J.G., Robinson, N.F., Lewis, C.W., Coulter, C.T., Chow, J.C., Fujita, E.M., Lowenthal, D.H., Conner, T.L., Henry, R.C., and Willis, R.D. (1997) *Chemical Mass Balance Receptor Model Version 8 (CMB) User's Manual*. Prepared for US Environmental Protection Agency, Research Triangle Park, NC, Desert Research Institute, Reno, NV.

Watson, J.G., DuBois, D.W., DeMandel, R., Kaduwela, A.P., Magliano, K.L., McDade, C., Mueller, P.K., Ranzieri, A.J., Roth, P.M., and Tanrikulu, S. (1998a) Field program plan for the California Regional $PM_{2.5}/PM_{10}$ Air Quality Study (CRPAQS).

Prepared for California Air Resources Board, Sacramento, CA, Desert Research Institute, Reno, NV.

Watson, J.G., Robinson, N.F., Lewis, C.W., Coulter, C.T., Chow, J.C., Fujita, E.M., Conner, T.L., and Pace, T.G. (1998b) CMB8 applications and validation protocol for $PM_{2.5}$ and VOCs. Report No. 1808.2D1. Prepared for US Environmental Protection Agency, Research Triangle Park, NC, Desert Research Institute, Reno, NV.

Watson, J.G., Chow, J.C., Moosmüller, H., Green, M.C., Frank, N.H., and Pitchford, M.L. (1998c) Guidance for using continuous monitors in $PM_{2.5}$ monitoring networks. Report No. EPA-454/R-98-012, US Environmental Protection Agency, Research Triangle Park, NC.

Watson, J.G., Fujita, E.M., Chow, J.C., Zielinska, B., Richards, L.W., Neff, W.D., and Dietrich, D. (1998d) Northern Front Range Air Quality Study. Final report. Prepared for Colorado State University, Fort Collins, CO, Desert Research Institute, Reno, NV.

Watson, J.G., Coulombe, W.G., Egami, R.T., Casuccio, G.S., Hanneman, H.G., and Badger, S. (1999) Causes of particle deposition in Davenport, California. Report No. 99-0954.1F2, Prepared for RMC Lonestar Cement, Davenport, CA, Desert Research Institute, Reno, NV.

Watson, J.G., Chow, J.C., Bowen, J.L., Lowenthal, D.H., Hering, S.V., Ouchida, P., and Oslund, W. (2000a) Air quality measurements from the Fresno supersite. *Journal of the Air and Waste Management Association* 50, 1321–1334.

Watson, J.G., Chow, J.C., and Pace, T.G. (2000b) Fugitive dust emissions. In *Air Pollution Engineering Manual* (Davis, W.T., ed.), pp. 117–134. Van Nostrand Reinhold, New York.

Watson, J.G., Chow, J.C., and Houck, J.E. (2001) $PM_{2.5}$ chemical source profiles for vehicle exhaust, vegetative burning, geological material and coal burning in northwestern Colorado during 1996. *Chemosphere*, 43, 1141–1151.

Weatherhead, E.C., Reinsel, G.C., Tiao, G.C., Meng, X.L., Choi, D., Cheang, W.K., Keller, T., DeLuisi, J., Wuebbles, D.J., Kerr, J.B., Miller, A.J., Oltmans, S.J., and Frederick, J.E. (1998) Factors affecting the detection of trends: Statistical considerations and applications to environmental data. *Journal of Geophysical Research* 103, 17149–17162.

Weiss, D., Shotyk, W., Appleby, P.G., Kramers, J.D., and Cheburkin, A.K. (1999) Atmospheric Pb deposition since the Industrial Revolution recorded by five Swiss peat profiles: Enrichment factors, fluxes, isotopic composition, and sources. *Environmental Science and Technology* 33, 1340–1352.

Whitby, K.T., Charlson, R.J., Wilson, W.E., and Stevens, R.K. (1974) The size of suspended particle matter in air. *Science* 183, 1098–1099.

White, W.H. (1986) On the theoretical and empirical basis for apportioning extinction by aerosols: A critical review. *Atmospheric Environment* 20, 1659–1672.

White, W.H. (1999) Phantom spatial factors: An example. *Journal of the Air and Waste Management Association* 49, 345–349.

Wienke, D., Gao, N., and Hopke, P.K. (1994) Multiple site receptor modeling with a minimal spanning tree combined with a neural network. *Environmental Science and Technology* 28, 1023–1030.

Willeke, K. and Baron, P.A. (1993) *Aerosol Measurement: Principles, Techniques and Applications*. Van Nostrand Reinhold, New York.

Zielinska, B., McDonald, J.D., Hayes, T., Chow, J.C., Fujita, E.M., and Watson, J.G. (1998) *Northern Front Range Air Quality Study. Volume B: Source Measurements*. Prepared for Colorado State University, Fort Collins, CO, Desert Research Institute, Reno, NV.

CHAPTER 12

PRINCIPAL COMPONENTS ANALYSIS AND RECEPTOR MODELS IN ENVIRONMENTAL FORENSICS

Glenn W. Johnson, Robert Ehrlich, and William Full

12.1 INTRODUCTION

The identification of chemical contaminant sources is a common problem in environmental forensic investigations. Successful inference of sources depends on sampling plan design, sample collection procedures, chemical analysis methods, and knowledge of historical industrial processes in the study area. However, in complex situations where multiple sources contribute similar types of contaminants, even careful project planning and design may not be enough. If sources cannot be linked to a unique chemical species (i.e., a tracer chemical), then mapping the distributions of individual contaminant concentrations is insufficient to infer source. If, however, a source exhibits a characteristic 'chemical fingerprint' defined by diagnostic proportions of a large number of analytes, source inference may be accomplished through analysis of multiple variables; that is, through use of multivariate statistical methods. The objective of a multivariate approach to chemical fingerprinting is to determine (1) the number of fingerprints present in the system, (2) the multivariate chemical composition of each fingerprint, and (3) the relative contribution of each fingerprint in each collected sample.

Development of numerical methods to determine these parameters has been a major goal in environmental chemometrics and receptor modeling for more than 20 years. The result has been development of a series of procedures designed to accomplish this. As we shall see, procedures developed early in this history are most useful in solving relatively simple problems. Later procedures are designed to solve more general problems, which take into account complications such as bad data, commingled plumes (i.e., mixing of source fingerprints), and the presence of sources not assumed or anticipated at the start of an investigation. The objective of this chapter is to discuss this family of procedures in terms of their strengths and limitations, and in order to guide the working environmental scientist in the use of appropriate procedures.

12.1.1 *PHILOSOPHY AND APPROACH: A CASE FOR EXPLORATORY DATA ANALYSIS*

In terms of experimental design, the source apportionment problem in environmental forensic investigations falls between two extremes. At one extreme, all potential sources are known in terms of their chemical composition, location, history, and duration of activity. At the other extreme, none of these are known with any certainty. Chemicals at the receptor (e.g., estuary sediments, groundwater at a supply well) may be the result of activities long absent from the vicinity of the site.

In the first case (*a priori* knowledge of all sources) the problem is a relatively simple one. Appropriate sampling locations can be determined using a conventional experimental design, which is part of conventional experimental statistics. Determination of contribution of each source can be extracted using a variety of linear methods, such as chemical mass balance receptor models (see Chapter 11 of this volume). However, even when the contributing sources are known, environmental forensic investigations often prove to be more complex than initially anticipated. Chemicals in the environment may not retain their original composition. That is, chemical compositions may change over time by processes such as biodegradation and volatilization. This is true even for relatively recalcitrant contaminants such as polychlorinated biphenyls and dioxins (Bedard and Quensen, 1995; Chiarenzelli *et al.*, 1997). The result of degradation will be resolution of one or more fingerprints, not originally anticipated.

At the other extreme, where nothing is known with certainty, potential sources may be suspected, but samples of the sources (i.e., fingerprint reference standards) may not have been collected, and may not exist in the literature. The industrial history of a region may be imperfectly known. Often, a small, low profile operation may be a major but completely overlooked source of contamination. For cases towards this end of the spectrum, we must take leave of the elegance of conventional experimental statistics, and move into the realm of exploratory data analysis (EDA). The fundamental difference between these two approaches (experimental statistics and EDA) is that former is associated with creation of explicit hypotheses, and evaluation of data in terms of well-defined tests and strong probabilistic arguments. In contrast, the objective of EDA is to find patterns, correlations and relationships in the data itself, with few assumptions or hypotheses (Tukey, 1977). If the fruits of an EDA result in a map where the concentrations of a multivariate fingerprint increase monotonically towards an effluent pipe, and the fingerprint composition is consistent with the process associated with that source, the obvious inference is that the potential source is the actual source. We recognize that we are not working in the realm of classical statistics or formal hypothesis testing, and that EDA is based on less rigorous probabilistic statements. However, such an approach should not be construed as 'second best.' In environmental forensics, an EDA approach may be the only valid option. The cost of highly rigorous probabilistic methods is often a set of narrow assumptions regarding the structure of the data. Such methods cannot be supported if those critical assumptions cannot be safely assumed. Moreover, in environmental forensic investigations, we often lack sufficient information to know what hypotheses to test. Thus, at least at the beginning of a project, an EDA approach is usually most prudent.

Regardless of the data analysis strategy chosen, another important consideration is the presence of bad or questionable data. Common problems with environmental chemical data include the following: (1) chemical analyses performed by different laboratories or by different methods, which may introduce a systematic bias; (2) the presence of data at concentrations at or below method detection limits; (3) the presence of coelution (non-target analytes that elute at the same time as a target analyte); and (4) the ever-present problem of error in data entry, data transcription, or peak integration.

Unfortunately such errors rarely manifest themselves as random noise. More often, they contribute strong systematic variability. If unrecognized, the result may be derivation of 'fingerprints' which have little to do with the true sources. Therefore, a necessary adjunct to any data analysis in environmental forensics is identification of outliers. As we proceed with a discussion of fingerprinting methods, a major consideration in that regard must be inclusion of vigilant outlier identification and data cleaning procedures. If such an effort results in deletion or modification of data, the data must be clearly identified, and justification for the action must be provided in the narrative that accompanies the analyses.

12.1.2 FORMAL DESCRIPTION OF THE RECEPTOR MODELING PROBLEM

Receptor modeling in environmental forensics involves the inference of sources and their contributions through analysis of chemical data from the ambient environment (Gordon, 1988; Hopke, 1991). The objectives are to determine (1) the number of chemical fingerprints in the system; (2) the chemical composition of each fingerprint; and (3) the contribution of each fingerprint in each sample. The starting point is a data-table of chemical measurements in samples collected from the receptor (e.g., estuarine sediments, ambient air in a residential area). These data are usually provided in spreadsheet form where rows represent samples and columns represent chemical analytes. To the multivariate data analyst this table is a matrix. We will refer to the original data table as the m row by n column matrix $\mathbf{X}$, where m is the number of samples and n is the number of analytes. We wish to know the number of fingerprints present (k) and chemical composition of each (objectives 1 and 2 above). This can be expressed as a matrix $\mathbf{F}$, which has k rows and n columns. We also wish to know a third matrix ($\mathbf{A}$), which has m rows and k columns, and represents the contribution of each fingerprint in each sample (objective 3 above). Thus the following linear algebraic equation formally expresses the receptor modeling problem.

$$\underset{(m\times n)}{\mathbf{X}} = \underset{(m\times k)}{\mathbf{A}} \; \underset{(k\times n)}{\mathbf{F}} \tag{12.1}$$

Matrix dimensions

Given our data table we have three knowns (**X**, *m* and *n*), and three unknowns: (1) the number of fingerprints (*k*); (2) the fingerprint compositions (**F**); and (3) the contributions of each fingerprint in each sample (**A**). In the rare case where fingerprints are known *a priori*, both *k* and **F** are known, and only one unknown (**A**) remains. If this is the case, source compositions are contained within a 'training data set' consisting of the compositions of all potential sources, and solving Equation 12.1 for **A** is straightforward. The problem can be solved by regression techniques, which are the basis of chemical mass balance (CMB) approaches (see Chapter 11). Unfortunately, in environmental forensics investigations, we seldom have the luxury of *a priori* knowledge of contributing sources. The use of a training data set constitutes, if not a formal hypothesis test, at least an implicit hypothesis. If at all possible, we would like to derive source patterns directly from analysis of ambient data. That is, we want to employ chemometric methods that are 'self-training.'

While the number of chemometric methods available to us is large, the nature of environmental forensics investigations dictates the use of methods with very special features: they must allow for the conceptual model of mixtures of multiple sources; and they should allow resolution of contributing chemical fingerprints without *a priori* assumption of the number, chemical composition or geographic/temporal distribution. Finally, the results must be interpretable in a scientific context. Several commonly used chemometric methods, which satisfy the above considerations will be presented below.

12.1.3 DEMONSTRATION DATA SETS

The methods demonstrated in this chapter are illustrated using polychlorinated biphenyls (PCB) data. PCBs are a group of chlorinated organic compounds which are commonly the focus of environmental forensic investigations. PCBs were widely used in commercial and industrial settings for much of the twentieth century. Commercial PCB products were marketed under the trade-name Aroclor by Monsanto (the former US manufacturer). Commercial applications of PCBs included their use in fluorescent light ballasts, carbonless copy paper, and as dielectric fluids in electrical transformers and capacitors. PCBs are of concern in environmental studies because they are persistent, toxic and tend to bioaccumulate in tissues of higher predators (Tanabe *et al.*, 1987). PCBs are used for demonstrations in this chapter because they are a group of contaminants that typically require a multivariate approach to data analysis. Commercial PCB formulations (e.g., Aroclors) have unique chemical compositions composed of multiple PCB congeners, but no single congener is a diagnostic tracer for a specific Aroclor.

Two synthetic data sets will be used for demonstration purposes. Each of these represents a three-source system. The congener patterns for the three-source fingerprints were taken from Aroclor standard compositions reported by Frame *et al.* (1996). The source compositions are Aroclor 1248 (Frame sample G3.5) and two variants of Aroclor 1254 (Frame samples A4 and G4). Frame *et al.* (1996) first reported markedly different congener patterns for different lots of Aroclor 1254. Subsequent investigations by Frame indicated that the atypical Aroclor 1254 was the result of a late production change in the Aroclor 1254 manufacturing process that occurred in the early 1970s (Frame, 1999). The congener compositions of these three-source fingerprints are shown as bar-graphs in Figure 12.1.

The two variants of Aroclor 1254 were used for this demonstration because it provides a typical example of the surprises often encountered in environmental forensic investigations. For example, given a situation where (1) the production history of an area was well established; (2) it was known with great certainty that only two PCB formulations were ever used at a site: Aroclor 1248 and Aroclor 1254; and (3) the data analyst was not aware of the two Aroclor 1254 variants; he/she might reasonably make an *a priori* assumption of a two-source system. Clearly, in this case, that assumption would be incorrect. In environmental forensics investigations, even the simplest, well-understood systems can yield surprises.

These three Aroclor compositions were used to create two synthetic data sets. The first of the two (Data Set 1) is quite simple, almost to the point of being unrealistic as an analogue to environmental forensics investigations. However, it is instructive in that it provides a good intuitive understanding of

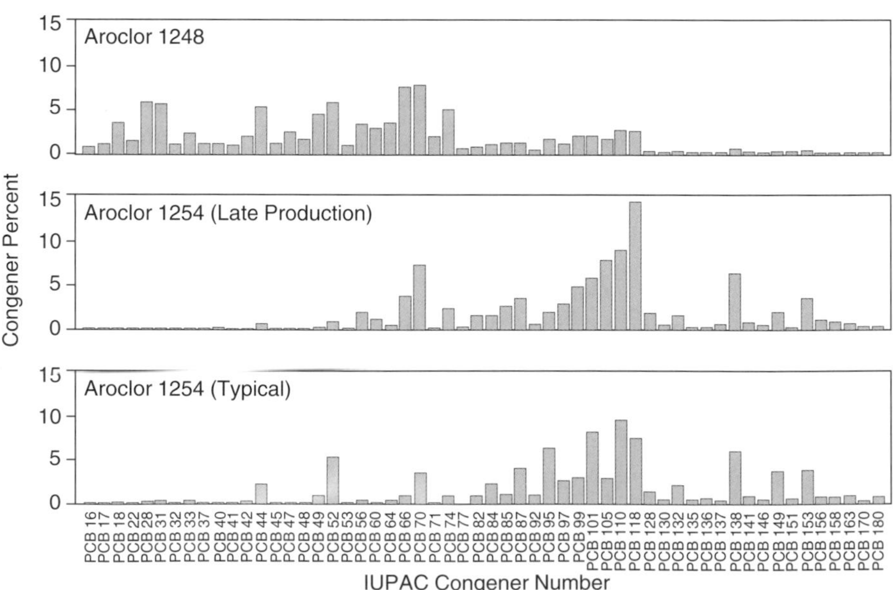

Figure 12.1

Congener compositions of three PCB product formulations (Aroclors) used to create the artificial data sets used for demonstration in this chapter (data from Frame et al., *1996).*

principal components analysis (PCA), which is the mathematical basis of all the methods presented here. Data Set 1 is a 24-sample, 56-congener matrix. The data are simple in that each sample represents a contribution from one and only one source. That is, it is a strongly clustered data set, with no samples representing mixtures of two or of three Aroclors. To create some inter-sample variability within each of the three source categories, random Gaussian noise was added to each sample in the data set. Data Set 1 is shown in Table 12.1.

The second data set (Data Set 2) is considerably more complex, and is more representative of data encountered in environmental forensics investigations. Data Set 2 is shown in Table 12.3, and was created such that:

1 All samples are *mixtures* of the three Aroclor compositions. That is, no sample in the data set represents a 100% contribution from a single source. Varying contributions of multiple sources impacts every sample. The two matrices **A** and **F** (Equation 12.1) used to calculate the original noise free matrix are shown in Table 12.2. For ease of presentation, the transposed 56 × 3 matrix $\mathbf{F}^t$ (rather than the original matrix **F**) is shown in Table 12.2.
2 Ten percent Gaussian noise was added to simulate random error.
3 The data are represented in units of concentration (ng/g), and a method detection limit was established for each sample. As such, many low concentration matrix elements are 'censored' as a function of the detection limits. Non-detects are indicated in Table 12.3 with a 'U' qualifier. For subsequent numerical analyses, we adopt the common practice of replacing non-detect values with half the detection limit.
4 Data transcription errors were added to one sample (Sample 22). The error simulated was a data translation mistake. The reported concentrations of some congeners were offset by one, thus assigning incorrect concentrations to 14 of the 56 spreadsheet columns.
5 For the congener PCB 141, a coelution problem was introduced in 35 of the 50 samples. Coelution of non-PCB peaks with PCB congeners during gas chromatographic analysis is a common problem in PCB chemistry. The coelution simulated here represents one such example. The pesticide *p,p'*-DDT elutes very close to PCB 141 on a Chrompack CP-SIL5-C18 GC column (R. Wagner, personal communication). Therefore, if *p,p'*-DDT is present in a sample undergoing congener specific PCB analysis, the *p,p'*-DDT could erroneously be reported as PCB congener IUPAC 141.

12.2 PRINCIPAL COMPONENTS ANALYSIS

12.2.1 PCA OVERVIEW

Principal components analysis (PCA) is a widely used method in environmental chemometrics, as it is in many scientific disciplines. PCA is used on its own, and as an intermediate step in receptor modeling methods. Before presenting

Table 12.1

Data Set 1 (24 samples, 56 congeners) created by addition of Gaussian noise to three Aroclor compositions reported by Frame et al. *(1996). Units in percent.*

Sample	**PCB 16**	**PCB 17**	**PCB 18**	**PCB 22**	**PCB 28**	**PCB 31**	**PCB 32**	**PCB 33**	**PCB 37**	**PCB 40**	**PCB 41**	**PCB 42**	**PCB 44**
Aroclor 1248	0.76	0.75	2.70	1.16	6.79	4.98	1.05	1.50	1.04	0.74	0.64	1.90	6.34
Aroclor 1248	0.87	1.06	2.61	1.46	5.76	6.11	0.61	2.65	0.89	1.08	0.66	1.80	5.82
Aroclor 1248	0.66	0.99	3.72	1.65	6.46	6.68	0.97	2.64	1.22	0.85	0.63	1.95	5.89
Aroclor 1248	0.66	0.96	3.67	1.72	5.39	5.76	0.93	2.08	0.89	1.02	0.82	1.78	4.77
Aroclor 1248	0.65	1.01	3.95	2.03	6.17	3.78	1.14	2.46	1.05	0.92	0.87	1.85	6.89
Aroclor 1248	0.58	1.18	3.84	1.56	5.05	4.63	0.98	2.26	0.78	1.07	0.88	2.16	5.13
Aroclor 1248	0.70	0.96	3.61	1.30	3.80	5.65	1.07	2.96	1.02	1.07	0.89	1.50	5.93
Aroclor 1248	0.84	1.03	2.72	1.42	4.89	6.62	0.95	2.19	1.20	1.06	0.82	1.99	5.22
Aroclor 1254 (late production)	0.02	0.02	0.09	0.02	0.06	0.11	0.01	0.06	0.01	0.10	0.02	0.08	0.75
Aroclor 1254 (late production)	0.02	0.03	0.09	0.02	0.07	0.14	0.01	0.07	0.01	0.17	0.02	0.06	0.99
Aroclor 1254 (late production)	0.02	0.02	0.08	0.01	0.07	0.11	0.01	0.05	0.01	0.16	0.02	0.10	0.75
Aroclor 1254 (late production)	0.02	0.02	0.09	0.03	0.07	0.16	0.01	0.05	0.01	0.17	0.03	0.12	0.86
Aroclor 1254 (late production)	0.02	0.02	0.07	0.02	0.08	0.10	0.01	0.04	0.01	0.17	0.02	0.08	0.82
Aroclor 1254 (late production)	0.02	0.03	0.08	0.02	0.10	0.14	0.01	0.06	0.01	0.19	0.02	0.10	0.72
Aroclor 1254 (late production)	0.02	0.02	0.04	0.02	0.07	0.14	0.01	0.04	0.01	0.19	0.03	0.14	0.83
Aroclor 1254 (late production)	0.02	0.02	0.09	0.03	0.06	0.15	0.01	0.05	0.01	0.16	0.02	0.06	0.59
Aroclor 1254 (typical)	0.07	0.10	0.24	0.04	0.20	0.32	0.07	0.20	0.07	0.14	0.01	0.14	2.64
Aroclor 1254 (typical)	0.11	0.09	0.30	0.04	0.21	0.30	0.05	0.18	0.07	0.14	0.01	0.14	1.94
Aroclor 1254 (typical)	0.11	0.08	0.21	0.03	0.14	0.22	0.05	0.16	0.08	0.10	0.01	0.16	2.35
Aroclor 1254 (typical)	0.08	0.09	0.27	0.05	0.22	0.31	0.03	0.11	0.08	0.13	0.01	0.24	2.29
Aroclor 1254 (typical)	0.10	0.09	0.35	0.05	0.16	0.33	0.04	0.15	0.08	0.12	0.01	0.12	2.82
Aroclor 1254 (typical)	0.08	0.07	0.23	0.04	0.17	0.30	0.04	0.20	0.08	0.09	0.01	0.13	3.02
Aroclor 1254 (typical)	0.11	0.08	0.33	0.04	0.13	0.24	0.05	0.17	0.07	0.11	0.01	0.14	2.72
Aroclor 1254 (typical)	0.08	0.08	0.31	0.04	0.19	0.27	0.05	0.21	0.08	0.15	0.01	0.18	3.00

PCB 45	PCB 47	PCB 48	PCB 49	PCB 52	PCB 53	PCB 56	PCB 60	PCB 64	PCB 66	PCB 70	PCB 71	PCB 74	PCB 77	PCB 82
0.83	2.83	1.70	4.28	5.45	0.73	2.27	3.76	3.52	6.81	9.70	2.31	5.66	0.70	0.55
1.08	2.70	1.23	4.30	6.28	0.77	2.88	2.49	3.28	9.21	9.01	2.04	4.60	0.49	0.70
1.19	1.87	1.87	4.35	5.01	0.88	3.50	2.41	3.50	7.20	7.40	2.15	4.43	0.57	0.70
1.09	3.14	1.93	4.50	5.08	1.16	3.73	2.71	3.24	6.84	7.48	2.76	5.41	0.66	0.55
1.21	2.80	1.50	4.37	6.43	0.79	4.03	3.16	3.18	7.54	5.85	2.02	4.68	0.67	0.55
0.97	2.80	1.44	4.52	5.77	1.02	2.84	3.33	3.41	8.81	7.60	2.06	4.84	0.61	0.77
0.82	2.43	1.37	4.57	5.54	1.14	3.81	2.72	4.08	7.71	7.67	2.18	5.17	0.46	0.77
1.10	1.77	1.57	4.25	5.71	0.69	4.27	2.80	3.46	8.06	8.19	1.98	4.29	0.51	0.77
0.02	0.06	0.05	0.26	0.69	0.04	1.62	0.95	0.36	3.58	7.92	0.12	2.08	0.23	1.52
0.03	0.10	0.06	0.29	0.83	0.04	1.37	1.06	0.23	4.36	5.63	0.12	2.11	0.23	1.92
0.02	0.06	0.05	0.27	0.78	0.04	1.42	0.62	0.39	3.98	7.45	0.10	2.68	0.23	1.74
0.02	0.09	0.06	0.29	0.93	0.04	2.13	0.96	0.40	3.83	5.65	0.08	2.08	0.24	1.45
0.02	0.07	0.06	0.28	0.96	0.05	1.81	1.18	0.37	4.71	8.06	0.11	2.54	0.21	1.53
0.02	0.07	0.06	0.29	0.85	0.04	2.42	0.96	0.38	3.50	8.58	0.09	2.16	0.26	1.59
0.02	0.08	0.04	0.29	0.83	0.06	1.59	1.12	0.33	3.90	7.38	0.13	1.97	0.23	1.98
0.02	0.08	0.06	0.29	0.74	0.03	1.77	1.06	0.45	3.70	9.32	0.12	2.47	0.16	1.90
0.06	0.16	0.10	1.18	4.12	0.15	0.57	0.19	0.49	1.16	3.69	0.16	0.97	0.04	1.22
0.05	0.17	0.18	1.27	5.64	0.15	0.58	0.17	0.88	1.25	4.16	0.13	0.66	0.04	1.28
0.04	0.15	0.13	1.11	7.11	0.11	0.54	0.18	0.67	0.78	3.69	0.17	0.79	0.02	1.33
0.07	0.15	0.16	1.18	5.13	0.15	0.57	0.20	0.65	0.91	3.60	0.19	0.88	0.03	1.48
0.05	0.12	0.15	1.16	5.71	0.10	0.69	0.23	0.54	1.06	4.37	0.17	0.86	0.03	1.05
0.06	0.12	0.08	1.16	5.70	0.11	0.62	0.24	0.53	1.30	3.15	0.18	1.17	0.03	0.84
0.06	0.15	0.15	1.23	6.77	0.14	0.51	0.20	0.63	0.89	2.96	0.18	0.83	0.03	1.54
0.05	0.16	0.11	1.21	4.39	0.10	0.62	0.16	0.74	1.20	3.38	0.15	0.99	0.03	1.31

Table 12.1
Continued

Sample	PCB 84	PCB 85	PCB 87	PCB 92	PCB 95	PCB 97	PCB 99	PCB 101	PCB 105	PCB 110	PCB 118	PCB 128	PCB 130
Aroclor 1248	0.97	1.05	0.67	0.28	1.55	1.01	1.71	2.07	1.59	3.00	2.41	0.09	0.01
Aroclor 1248	0.98	1.17	0.95	0.25	1.36	1.12	2.00	2.26	1.44	1.87	1.94	0.08	0.01
Aroclor 1248	0.89	0.89	1.14	0.27	1.72	0.98	1.67	2.34	1.48	2.60	2.50	0.08	0.01
Aroclor 1248	1.00	1.15	1.13	0.21	1.57	1.20	1.96	2.43	1.57	2.65	2.00	0.06	0.01
Aroclor 1248	1.04	1.25	1.34	0.28	1.56	1.00	1.79	1.87	1.35	2.82	2.02	0.09	0.01
Aroclor 1248	1.05	1.10	1.43	0.23	1.38	1.10	1.93	1.44	1.77	2.51	3.01	0.09	0.01
Aroclor 1248	0.97	1.29	0.75	0.29	1.78	1.23	2.28	1.73	1.07	2.32	2.98	0.08	0.01
Aroclor 1248	0.80	1.27	1.48	0.23	1.75	0.94	1.88	1.75	1.36	3.14	2.68	0.08	0.01
Aroclor 1254 (late production)	1.83	2.27	3.60	0.63	1.99	2.78	4.81	6.55	8.02	11.26	14.48	1.39	0.50
Aroclor 1254 (late production)	2.02	2.50	3.03	0.62	2.02	2.39	5.32	6.78	9.52	8.60	13.85	2.09	0.64
Aroclor 1254 (late production)	1.67	2.66	3.82	0.62	1.59	2.89	4.53	6.59	7.18	9.10	17.61	1.94	0.34
Aroclor 1254 (late production)	1.72	2.69	3.92	0.64	2.18	2.78	4.30	7.52	5.89	7.90	14.42	2.56	0.65
Aroclor 1254 (late production)	1.47	2.22	2.85	0.64	1.92	3.80	3.68	4.33	7.12	10.01	14.88	2.14	0.50
Aroclor 1254 (late production)	1.98	2.74	3.90	0.45	1.80	2.38	5.50	6.91	6.24	9.93	10.95	2.09	0.52
Aroclor 1254 (late production)	1.95	2.31	4.00	0.48	2.02	2.44	5.43	4.85	7.26	7.78	16.30	1.99	0.55
Aroclor 1254 (late production)	1.39	3.04	3.03	0.57	1.55	3.43	5.26	5.67	7.24	10.00	11.55	2.09	0.60
Aroclor 1254 (typical)	1.98	1.71	4.30	1.27	7.23	2.34	3.34	4.98	3.28	10.74	9.63	1.70	0.66
Aroclor 1254 (typical)	2.15	1.33	4.78	1.75	5.74	3.00	4.40	7.80	3.21	9.91	8.14	1.59	0.51
Aroclor 1254 (typical)	2.74	1.36	4.17	0.88	7.27	3.16	3.51	7.94	2.59	10.14	10.10	1.22	0.48
Aroclor 1254 (typical)	2.99	1.23	4.74	1.19	6.53	3.10	3.42	7.67	2.70	10.28	7.99	1.67	0.69
Aroclor 1254 (typical)	2.22	1.49	3.93	1.17	6.10	2.45	2.64	9.72	3.65	9.61	10.14	1.46	0.75
Aroclor 1254 (typical)	1.82	1.89	3.96	1.41	7.39	2.75	3.81	7.08	3.54	12.26	7.92	1.48	0.62
Aroclor 1254 (typical)	2.91	1.77	4.63	1.02	6.99	2.03	3.25	8.63	2.65	9.86	8.89	1.98	0.68
Aroclor 1254 (typical)	2.02	1.66	4.23	1.02	6.43	2.48	3.54	7.81	3.62	12.39	6.52	1.95	0.58

PCB 132	PCB 135	PCB 136	PCB 137	PCB 138	PCB 141	PCB 146	PCB 149	PCB 151	PCB 153	PCB 156	PCB 158	PCB 163	PCB 170	PCB 180
0.13	0.05	0.06	0.02	0.42	0.09	0.05	0.33	0.08	0.43	0.05	0.04	0.09	0.06	0.24
0.16	0.04	0.05	0.01	0.46	0.09	0.05	0.39	0.08	0.39	0.04	0.04	0.11	0.07	0.17
0.13	0.05	0.06	0.02	0.34	0.10	0.05	0.40	0.08	0.36	0.05	0.04	0.10	0.10	0.21
0.13	0.05	0.05	0.02	0.55	0.10	0.05	0.36	0.10	0.44	0.05	0.04	0.08	0.09	0.25
0.14	0.04	0.07	0.02	0.44	0.10	0.05	0.27	0.07	0.36	0.04	0.04	0.11	0.10	0.19
0.17	0.04	0.05	0.03	0.38	0.08	0.04	0.26	0.06	0.45	0.02	0.04	0.07	0.06	0.29
0.16	0.04	0.06	0.02	0.40	0.11	0.06	0.34	0.08	0.49	0.04	0.05	0.08	0.10	0.29
0.13	0.05	0.06	0.02	0.46	0.11	0.05	0.30	0.10	0.49	0.03	0.04	0.09	0.09	0.23
1.10	0.31	0.19	0.60	6.32	0.59	0.52	1.93	0.26	3.24	1.30	1.25	0.81	0.33	0.30
1.72	0.35	0.28	0.53	7.08	0.78	0.49	2.31	0.29	2.65	0.93	1.36	0.78	0.46	0.53
1.36	0.28	0.29	0.61	5.92	0.64	0.46	1.90	0.31	3.11	0.84	0.91	0.84	0.31	0.43
2.09	0.39	0.32	0.63	7.98	0.74	0.43	2.40	0.26	4.00	1.20	0.93	0.70	0.53	0.38
1.57	0.20	0.26	0.51	6.74	0.74	0.56	2.04	0.22	3.95	1.04	1.35	0.95	0.38	0.48
1.72	0.25	0.23	0.76	6.08	0.58	0.46	2.21	0.26	4.99	1.35	1.31	0.80	0.39	0.47
2.02	0.34	0.30	0.68	6.48	0.63	0.52	1.83	0.21	4.25	1.10	0.94	0.88	0.37	0.55
2.07	0.24	0.30	0.61	6.35	0.63	0.64	2.07	0.28	4.02	1.19	0.94	0.90	0.46	0.42
2.87	0.59	0.69	0.34	6.85	1.22	0.56	4.79	0.81	5.44	1.00	0.87	1.11	0.55	0.67
3.12	0.64	0.96	0.47	5.83	1.35	0.69	3.55	0.87	3.37	0.90	0.88	1.38	0.65	0.84
1.75	0.66	0.79	0.37	6.29	1.16	0.77	3.46	0.76	3.98	1.06	0.91	0.86	0.42	0.65
2.35	0.69	0.74	0.46	7.57	1.30	0.67	3.65	0.82	4.08	0.90	0.64	1.27	0.63	0.77
2.20	0.63	0.78	0.44	6.10	0.92	0.47	3.24	0.75	4.42	0.62	0.96	1.21	0.40	0.79
2.84	0.65	0.85	0.40	4.56	1.17	0.81	3.11	0.64	4.91	1.08	0.96	1.20	0.45	0.68
2.39	0.61	0.61	0.50	5.50	0.80	0.68	3.26	0.62	4.82	0.82	0.76	1.36	0.62	0.89
2.54	0.61	0.72	0.48	6.33	1.22	0.59	4.43	0.75	4.20	0.91	1.23	1.14	0.66	0.72

Table 12.2

*Input matrices for artificial three-source PCB mixture. Multiplication by Equation 12.1 (**X** = **A*****F**) yields error free matrix **X**.*

Source Contributions Matrix (Mixing Proportions) [A]				End-Member Source Compositions Matrix [F^t] (Shown graphically in Figure 12.1)			
Sample Number	**Source 1 Aroclor 1248 (%)**	**Source 2 Late Production Aroclor 1254 (%)**	**Source 3 Typical Aroclor 1254 (%)**	**IUPAC Congener**	**Source 1 Aroclor 1248**	**Source 2 Late Production Aroclor 1254**	**Source 3 Typical Aroclor 1254**
Sample 1	0.03	0.43	0.54	1 PCB 16	0.75	0.02	0.10
Sample 2	0.22	0.26	0.51	2 PCB 17	0.98	0.02	0.09
Sample 3	0.59	0.13	0.28	3 PCB 18	3.46	0.09	0.27
Sample 4	0.71	0.11	0.18	4 PCB 22	1.45	0.02	0.04
Sample 5	0.03	0.38	0.59	5 PCB 28	5.86	0.06	0.21
Sample 6	0.59	0.24	0.17	6 PCB 31	5.76	0.12	0.30
Sample 7	0.45	0.07	0.48	7 PCB 32	0.98	0.01	0.05
Sample 8	0.41	0.02	0.57	8 PCB 33	2.33	0.05	0.17
Sample 9	0.19	0.37	0.44	9 PCB 37	1.00	0.01	0.08
Sample 10	0.60	0.35	0.05	10 PCB 40	0.97	0.16	0.13
Sample 11	0.20	0.74	0.05	11 PCB 41	0.79	0.02	0.01
Sample 12	0.55	0.12	0.33	12 PCB 42	1.88	0.10	0.16
Sample 13	0.48	0.08	0.44	13 PCB 44	5.36	0.72	2.50
Sample 14	0.17	0.72	0.11	14 PCB 45	0.96	0.02	0.05
Sample 15	0.44	0.13	0.43	15 PCB 47	2.54	0.08	0.15
Sample 16	0.44	0.36	0.20	16 PCB 48	1.62	0.05	0.13
Sample 17	0.48	0.50	0.02	17 PCB 49	4.39	0.28	1.19
Sample 18	0.15	0.57	0.29	18 PCB 52	5.87	0.89	5.81
Sample 19	0.47	0.47	0.07	19 PCB 53	0.93	0.04	0.13
Sample 20	0.14	0.01	0.85	20 PCB 56	3.36	1.83	0.59
Sample 21	0.40	0.08	0.52	21 PCB 60	2.81	1.02	0.19
Sample 22	0.46	0.45	0.10	22 PCB 64	3.50	0.39	0.64
Sample 23	0.50	0.22	0.28	23 PCB 66	7.60	3.84	1.09
Sample 24	0.49	0.45	0.06	24 PCB 70	7.78	7.36	3.77
Sample 25	0.63	0.27	0.10	25 PCB 71	1.96	0.12	0.16
Sample 26	0.03	0.67	0.30	26 PCB 74	4.92	2.36	0.91
Sample 27	0.24	0.26	0.50	27 PCB 77	0.55	0.22	0.03
Sample 28	0.57	0.07	0.36	28 PCB 82	0.65	1.65	1.20
Sample 29	0.24	0.12	0.65	29 PCB 84	0.96	1.70	2.51
Sample 30	0.67	0.06	0.27	30 PCB 85	1.20	2.68	1.38
Sample 31	0.50	0.21	0.29	31 PCB 87	1.17	3.68	4.31
Sample 32	0.07	0.79	0.13	32 PCB 92	0.26	0.61	1.39
Sample 33	0.07	0.46	0.47	33 PCB 95	1.51	1.98	6.75
Sample 34	0.03	0.79	0.18	34 PCB 97	1.02	3.00	2.83
Sample 35	0.18	0.38	0.44	35 PCB 99	1.91	4.88	3.26
Sample 36	0.38	0.28	0.34	36 PCB101	1.99	5.92	8.67
Sample 37	0.34	0.30	0.36	37 PCB105	1.53	7.95	3.23
Sample 38	0.51	0.20	0.29	38 PCB110	2.68	9.08	10.04
Sample 39	0.43	0.52	0.06	39 PCB118	2.47	14.65	7.94
Sample 40	0.38	0.00	0.62	40 PCB128	0.08	1.84	1.53
Sample 41	0.07	0.28	0.65	41 PCB130	0.01	0.54	0.65
Sample 42	0.16	0.04	0.80	42 PCB132	0.15	1.62	2.47
Sample 43	0.29	0.50	0.22	43 PCB135	0.04	0.30	0.66
Sample 44	0.21	0.51	0.28	44 PCB136	0.06	0.26	0.76
Sample 45	0.56	0.43	0.01	45 PCB137	0.02	0.56	0.45
Sample 46	0.24	0.42	0.34	46 PCB138	0.43	6.42	6.27

Table 12.2

Continued

Source Contributions Matrix (Mixing Proportions) [A]				End-Member Source Compositions Matrix [F^t] (Shown graphically in Figure 12.1)			
Sample Number	**Source 1 Aroclor 1248 (%)**	**Source 2 Late Production Aroclor 1254 (%)**	**Source 3 Typical Aroclor 1254 (%)**	**IUPAC Congener**	**Source 1 Aroclor 1248**	**Source 2 Late Production Aroclor 1254**	**Source 3 Typical Aroclor 1254**
Sample 47	0.47	0.21	0.32	47 PCB141	0.09	0.74	1.06
Sample 48	0.71	0.06	0.23	48 PCB146	0.05	0.49	0.72
Sample 49	0.37	0.39	0.24	49 PCB149	0.35	1.96	3.94
Sample 50	0.68	0.23	0.08	50 PCB151	0.08	0.24	0.75
				51 PCB153	0.45	3.55	4.07
				52 PCB156	0.04	1.22	0.89
				53 PCB158	0.04	0.97	0.88
				54 PCB163	0.08	0.75	1.11
				55 PCB170	0.08	0.38	0.56
				56 PCB180	0.22	0.45	0.72

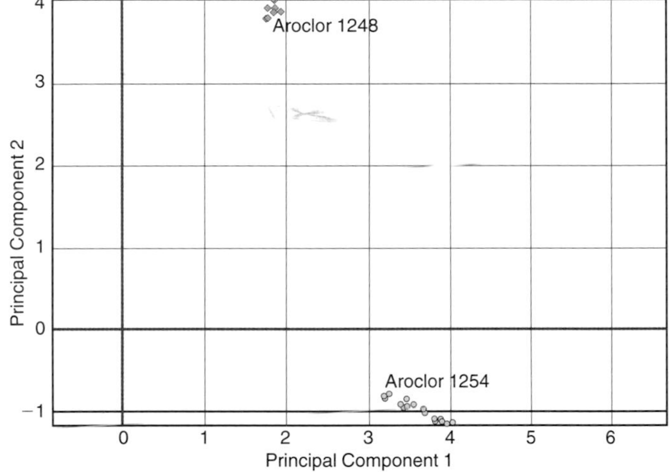

Figure 12.2

Two principal component score plots of Data Set 1. Two PCs account for > 92% of the variance of Data Set 1, but are insufficient to allow distinction of the two Aroclor 1254 variants.

the computational steps in PCA, it is useful to provide a more intuitive discussion, using Data Set 1 as an example. The objective of PCA is to reduce the dimensionality of a data set in which there are a large number of interrelated (i.e., correlated) variables. This reduction in dimension is achieved by transforming the data to a new set of uncorrelated reference variables (principal components or PCs). The PCs are sorted such that each in turn accounts for a progressively smaller percentage of variance within the data set. If nearly all variability between samples can be accounted for by a small number of PCs, then relationships between multivariate samples may be

Table 12.3

Data Set 2 (50 samples by 56 congeners) created by (1) multiplication of matrices **A** *and* **F** *(Table 12.2) as per Equation 12.1; (2) transformation to concentration metric (ng/g); (3) addition of 10% random Gaussian noise; (4) censoring of the data based on a sample specific detection limit (censored data qualified as 'U.' Reported measurement is the detection limit); (5) simulation of data transcription errors in Sample 22; and (6) simulation of DDT coelution with PCB 141 in a subset of samples.*

Sample	PCB 16	Qualifier	PCB 17	Qualifier	PCB 18	Qualifier	PCB 22	Qualifier	PCB 28	Qualifier	PCB 31	Qualifier	PCB 32	Qualifier	PCB 33	Qualifier
Sample 1	0.29		0.30		0.94		0.24	U	0.99		1.2		0.24		0.52	
Sample 2	0.66	U	0.66	U	1.6		0.66	U	2.8		2.8		0.66	U	1.3	
Sample 3	0.59		0.82		2.2		1.16		4.3		3.9		0.79		1.9	
Sample 4	1.5		1.7		6.9		2.70		12.2		9.6		1.6		4.5	
Sample 5	0.64		0.80		2.9		0.64	U	3.1		3.2		0.72		1.7	
Sample 6	1.2		1.7		5.4		2.2		9.2		9.8		1.6		3.2	
Sample 7	3.8		4.3		14.7		6.4		28.0		26.0		4.3		8.8	
Sample 8	2.7		2.6		10.2		3.2		15.4		17.9		2.9		8.2	
Sample 9	1.3	U	1.3	U	1.3	U	1.3	U	1.3	U	1.3	U	1.3	U	1.3	U
Sample 10	1.8		2.0		7.8		3.1		10.9		13.4		2.6		5.6	
Sample 11	1.4		1.3		5.8		2.4		8.6		7.9		1.6		3.4	
Sample 12	1.5		2.0		7.6		3.3		15.6		13.0		2.5		5.0	
Sample 13	2.4		3.2		11.0		4.4		15.7		18.9		3.2		7.5	
Sample 14	1.1		1.3		4.0		1.7		7.1		8.1		1.2		2.9	
Sample 15	1.5		2.2		5.2		2.8		11.3		12.1		2.1		4.9	
Sample 16	1.9		2.9		8.7		4.2		14.8		16.3		2.3		6.7	
Sample 17	1.1	U	1.1	U	1.4		1.1	U	2.3		2.6		1.1	U	1.1	U
Sample 18	0.60		0.76		3.1		1.1		4.2		4.4		0.65		2.02	
Sample 19	1.7		2.4		8.8		3.8		13.9		14.8		2.3		4.9	
Sample 20	3.5		3.6		12.8		3.9		15.9		15.6		2.9		6.9	
Sample 21	1.8		2.1		8.0		2.6		10.8		11.6		2.2		4.7	
Sample 22	3.0		3.7		17.4		5.8		20.3		20.8		4.2		9.0	
Sample 23	1.5		1.8		7.3		3.4		12.2		9.5		2.0		4.8	
Sample 24	2.0		2.3		8.1		3.4		15.2		14.3		2.7		5.0	
Sample 25	4.3		3.7		20.8		9.2		37.2		37.5		5.5		15.9	
Sample 26	0.50	U	0.50	U	1.6		0.50	U	2.0		2.3		0.50	U	1.2	
Sample 27	1.7		2.2		6.8		2.7		12.1		13.3		2.6		5.2	
Sample 28	1.6		2.0		7.2		2.2		11.2		11.0		2.0		4.2	
Sample 29	1.5		1.9		6.6		2.7		9.7		12.1		2.4		4.4	
Sample 30	1.8		2.3		7.5		3.2		14.8		12.7		2.3		5.5	
Sample 31	0.71		0.99		2.14		0.90		4.3		4.6		0.70		2.0	
Sample 32	1.2		1.4		5.2		2.2		8.0		9.3		1.5		3.9	
Sample 33	0.78	U	0.78	U	0.86		0.78	U	1.2		1.2		0.78	U	0.78	U
Sample 34	0.66	U	0.66	U	2.1		0.66	U	2.4		3.0		0.66	U	1.3	
Sample 35	1.8		1.8		8.0		2.8		11.3		10.7		2.1		5.2	
Sample 36	1.3		1.8		6.2		2.4		11.0		7.6		1.7		4.5	
Sample 37	1.0		1.6		4.0		2.4		6.6		8.6		1.5		2.9	
Sample 38	0.78		1.0		2.8		1.3		5.5		6.0		0.76		2.3	
Sample 39	0.72	U	1.1		3.0		1.4		5.5		4.6		0.92		1.8	
Sample 40	1.9		2.5		9.3		3.9		13.1		13.4		1.9		6.7	
Sample 41	0.46		0.44		1.7		0.52		2.2		3.0		0.41		1.1	
Sample 42	0.36	U	0.36	U	0.83		0.36	U	1.2		0.85		0.36	U	0.6	
Sample 43	0.74		0.83		3.3		1.2		6.0		5.1		0.85		2.6	
Sample 44	1.09		1.2		5.3		1.8		6.8		6.3		1.4		2.3	
Sample 45	2.6		2.2		10.3		3.8		20.2		19.2		3.0		8.1	
Sample 46	2.2		2.5		8.4		3.0		12.0		16.5		2.3		4.6	
Sample 47	2.3		2.4		10.0		4.2		13.3		12.5		2.7		6.9	
Sample 48	7.4		9.8		40.3		14.2		60.7		52.6		10.0		23.9	
Sample 49	2.5		3.5		14.1		5.4		21.0		19.7		4.0		8.3	
Sample 50	3.5		4.2		12.9		4.2		17.9		24.0		4.0		9.3	

PCB 37	Qualifier	PCB 40	Qualifier	PCB 41	Qualifier	PCB 42	Qualifier	PCB 44	Qualifier	PCB 45	Qualifier	PCB 47	Qualifier	PCB 48	Qualifier	PCB 49	Qualifier	PCB 52	Qualifier	PCB 53	Qualifier
0.30		0.63		0.24	U	0.66		6.0		0.24	U	0.69		0.50		3.6		12.4		0.41	
0.66	U	0.66	U	0.66	U	0.92		4.0		0.66	U	1.1		0.73		3.0		7.7		0.66	U
0.83		0.74		0.45		1.7		5.2		0.84		2.0		1.3		3.6		6.5		0.82	
1.9		1.8		1.4		3.3		10.8		1.9		4.6		3.8		9.4		14.6		2.0	
0.71		1.5		0.64	U	1.7		16.8		0.64	U	1.9		1.6		9.0		34.6		1.1	
1.6		1.6		1.2		3.1		8.6		1.6		4.6		3.1		7.8		14.2		1.3	
4.4		5.2		3.3		8.7		35.5		4.2		11.7		5.1		23.1		47.7		3.8	
3.0		3.1		2.2		6.1		27.1		2.6		7.5		4.4		14.2		39.2		2.86	
1.3	U	1.3	U	1.3	U	1.3	U	2.2		1.3	U	1.3	U	1.3	U	1.4		3.9		1.3	U
2.2		1.8		1.6		3.8		14.2		1.8		5.9		3.5		9.3		14.7		2.3	
1.4		2.5		1.2		3.2		12.8		1.4		3.9		2.9		8.3		14.9		1.8	
2.4		2.5		1.8		5.2		14.7		2.4		6.0		4.3		10.7		18.2		2.6	
2.9		3.0		2.6		5.0		24.5		3.0		8.1		3.7		16.6		32.2		2.4	
1.4		2.1		1.1		2.8		10.7		1.1		3.4		2.6		6.9		13.7		1.4	
1.8		1.8		1.5		4.5		13.8		1.7		4.9		3.3		10.7		25.3		1.7	
2.5		2.8		1.9		6.0		19.5		2.9		6.4		3.9		13.6		20.5		2.6	
1.1	U	1.1	U	1.1	U	1.1	U	2.4		1.1	U	1.2		1.1	U	1.9		2.4		1.1	U
0.85		1.1		0.61		1.7		8.8		0.92		2.4		1.5		4.4		15.4		0.84	
2.4		2.1		1.9		3.9		15.0		2.3		6.2		3.3		9.3		15.7		2.3	
3.4		3.8		1.9		6.5		45.1		3.3		8.0		5.8		25.5		96.0		3.7	
2.0		2.3		1.4		3.8		15.8		2.1		4.7		3.5		10.7		28.9		2.3	
3.9		4.4		2.8		7.7		21.6		3.2		10.4		5.3		19.8		30.9		3.2	
1.8		2.4		1.6		4.1		14.9		1.8		5.3		3.2		7.6		16.4		1.9	
2.3		2.4		2.0		4.0		14.0		2.3		5.5		4.2		10.7		18.0		2.2	
6.5		6.8		4.8		12.3		38.2		6.5		13.5		10.3		25.6		34.3		6.1	
0.50	U	1.5		0.50	U	0.98		9.0		0.50	U	1.0		0.92		4.8		16.0		0.65	
2.2		2.7		1.7		4.9		19.8		2.2		6.5		3.7		11.5		35.9		2.5	
2.0		1.6		1.5		4.1		13.2		1.9		5.6		3.1		7.9		18.9		1.9	
1.9		2.2		1.4		4.1		23.3		2.2		4.9		3.6		14.2		36.1		2.1	
2.2		2.6		1.8		5.1		11.4		2.6		4.9		3.7		10.5		18.8		2.4	
0.77		0.86		0.58	U	1.6		5.3		0.7		2.3		1.4		4.5		7.3		0.62	
1.4		4.0		1.3		3.8		19.9		1.2		3.6		2.7		11.1		29.6		1.8	
0.78	U	0.78	U	0.78	U	0.78	U	4.2		0.78	U	0.78	U	0.78	U	2.3		8.9		0.78	U
0.66	U	1.6		0.66	U	1.42		13.0		0.66	U	1.4		1.0		4.7		17.7		0.84	
1.9		2.7		1.6		4.7		20.0		1.7		4.3		3.7		14.4		38.1		2.2	
1.6		1.9		1.4		3.2		14.2		1.9		4.4		3.0		10.7		20.0		1.5	
1.2		1.7		1.0		2.4		14.1		1.3		3.9		3.0		8.4		17.1		1.4	
0.88		0.97		0.65		1.7		5.9		0.95		2.2		1.5		4.6		7.6		0.77	
1.08		1.2		0.72	U	2.0		6.0		0.74		2.4		1.4		3.8		5.6		0.81	
2.20		2.6		1.9		4.8		22.4		2.0		6.3		4.3		14.0		34.8		2.2	
0.46		0.64		0.26	U	1.1		9.3		0.43		1.0		0.79		4.2		17.1		0.61	
0.36	U	0.36	U	0.36	U	0.54		3.1		0.36	U	0.49		0.36	U	1.6		5.8		0.36	U
1.1		1.5		0.73		2.0		6.7		0.99		2.6		1.6		5.1		10.9		0.89	
1.0		1.6		0.87		2.6		11.2		1.1		3.2		2.2		6.1		20.1		1.2	
2.8		3.2		2.59		6.59		20.83		3.47		7.37		4.4		15.7		23.5		3.4	
2.0		2.7		1.77		4.55		20.20		2.04		5.57		4.1		10.9		31.3		2.5	
2.6		3.1		2.01		5.54		20.30		2.20		6.03		4.9		14.8		22.9		2.3	
11.8		11.3		8.45		18.67		65.55		9.52		27.48		17.8		47.1		88.7		10.4	
3.4		4.2		2.92		7.14		26.77		3.10		11.34		6.0		18.5		36.5		4.1	
4.1		4.1		3.25		5.79		25.26		4.25		9.21		6.4		19.2		27.6		3.4	

Table 12.3

Continued

Sample	PCB 56	Qualifier	PCB 60	Qualifier	PCB 64	Qualifier	PCB 66	Qualifier	PCB 70	Qualifier	PCB 71	Qualifier	PCB 74	Qualifier	PCB 77	Qualifier
Sample 1	4.1		1.9		1.9		8.7		20.0		0.66		5.2		0.52	
Sample 2	2.8		2.0		2.0		6.3		10.5		0.82		3.9		0.66	U
Sample 3	3.1		2.5		2.8		6.5		7.7		1.3		4.9		0.46	
Sample 4	6.4		5.9		7.3		17.7		21.6		3.2		8.9		1.1	
Sample 5	11.1		5.5		4.5		16.0		53.0		2.0		13.7		1.1	
Sample 6	6.3		5.1		7.0		13.4		20.0		3.8		9.6		1.2	
Sample 7	15.4		14.8		15.9		39.5		42.5		9.5		22.1		2.7	
Sample 8	11.9		8.3		11.6		21.5		35.6		7.0		15.1		1.7	
Sample 9	1.5		1.3	U	1.3	U	2.8		4.8		1.3	U	1.7		1.3	U
Sample 10	9.7		6.2		8.7		20.4		27.7		4.9		15.0		1.5	
Sample 11	12.5		9.8		7.3		32.3		49.7		2.8		18.7		2.1	
Sample 12	7.1		5.9		7.3		21.2		24.5		5.1		12.9		1.5	
Sample 13	13.1		10.3		10.0		25.7		36.7		6.1		17.5		1.6	
Sample 14	13.4		7.8		6.4		27.4		46.8		2.6		18.6		1.2	
Sample 15	8.7		7.6		8.7		16.8		25.1		3.9		11.1		1.2	
Sample 16	12.9		9.9		10.6		27.6		42.8		5.1		15.5		1.7	
Sample 17	2.2		1.9		1.6		4.6		5.4		1.1	U	2.8		1.1	U
Sample 18	8.6		5.1		4.0		14.3		35.1		1.5		12.5		0.83	
Sample 19	12.1		9.4		8.8		24.8		36.7		4.9		19.1		1.4	
Sample 20	16.3		8.3		18.5		28.1		66.5		6.9		28.1		1.7	
Sample 21	8.0		6.4		7.8		20.3		24.6		4.5		15.7		1.1	
Sample 22	18.6		13.3		13.3		36.2		60.9		7.4		27.6		1.6	
Sample 23	9.2		6.3		7.2		17.1		25.1		3.9		11.4		1.1	
Sample 24	10.1		7.3		10.5		28.0		38.0		4.7		18.2		1.8	
Sample 25	25.1		20.6		23.2		49.2		67.8		13.6		33.4		4.2	
Sample 26	10.2		5.9		3.6		22.0		49.7		1.2		15.3		1.1	
Sample 27	13.7		8.0		8.6		31.8		48.8		5.0		20.0		2.0	
Sample 28	7.0		5.7		7.5		18.4		19.7		4.6		11.4		1.1	
Sample 29	9.2		5.9		8.8		22.1		37.8		3.7		16.0		1.1	
Sample 30	9.8		7.2		10.0		17.2		22.0		4.3		13.4		1.5	
Sample 31	4.3		2.8		3.1		8.4		11.0		1.8		5.1		0.58	U
Sample 32	26.1		20.0		10.3		62.3		98.3		3.8		33.6		3.4	
Sample 33	3.2		1.9		1.9		6.0		13.5		0.78	U	3.6		0.78	U
Sample 34	13.5		9.0		5.5		35.2		70.0		1.8		19.6		1.7	
Sample 35	15.4		9.4		11.0		33.0		54.2		4.1		20.0		1.8	
Sample 36	9.1		6.6		8.1		18.9		22.9		3.2		11.4		1.2	
Sample 37	6.7		5.7		5.1		17.0		22.2		3.0		9.4		1.0	
Sample 38	4.1		2.9		3.5		9.6		11.4		2.1		5.5		0.62	U
Sample 39	5.2		3.2		3.3		9.3		15.2		2.1		5.9		0.75	
Sample 40	8.7		7.1		10.8		19.2		29.2		4.7		13.3		1.2	
Sample 41	4.4		2.2		2.7		9.2		19.7		0.87		6.0		0.55	
Sample 42	1.1		0.64		1.1		1.8		4.6		0.36		1.7		0.36	U
Sample 43	6.2		4.9		4.8		11.5		22.4		2.0		8.5		0.8	
Sample 44	11.0		7.1		5.8		19.9		30.0		2.5		10.6		1.2	
Sample 45	14.8		10.0		13.0		28.1		46.0		6.8		20.5		2.9	
Sample 46	16.3		8.9		10.7		34.0		55.3		5.1		21.2		2.3	
Sample 47	11.6		8.9		10.1		22.5		31.1		5.5		17.9		1.6	
Sample 48	32.1		28.3		33.6		93.1		96.4		22.7		67.4		5.5	
Sample 49	20.5		15.1		14.4		48.2		59.9		8.0		31.6		2.9	
Sample 50	13.4		13.1		16.5		32.4		40.0		8.3		23.7		2.3	

PCB 82	Qualifier	PCB 84	Qualifier	PCB 85	Qualifier	PCB 87	Qualifier	PCB 92	Qualifier	PCB 95	Qualifier	PCB 97	Qualifier	PCB 99	Qualifier	PCB 101	Qualifier	PCB 105	Qualifier	PCB 110	Qualifier
5.0		6.7		6.1		14.0		3.0		12.2		9.9		14.3		27.3		17.0		33.2	
1.7		3.3		3.1		6.0		1.5		8.9		5.6		6.1		13.3		8.5		16.4	
1.2		2.0		1.8		3.0		0.80		3.6		2.7		3.7		5.8		3.6		6.9	
2.4		3.1		3.7		4.7		1.2		5.5		4.2		6.3		9.0		6.7		11.0	
10.8		19.8		15.7		40.2		9.1		47.2		23.7		37.6		65.7		51.0		88.3	
3.4		3.7		4.7		5.4		1.6		7.1		4.6		8.9		12.0		8.5		15.2	
9.0		16.3		12.4		24.4		7.2		38.8		17.1		24.4		45.9		27.5		48.6	
7.7		11.1		9.4		20.9		5.4		27.6		13.3		18.5		39.2		19.6		44.4	
1.3		1.9		1.6		3.0		1.3	U	3.4		2.2		3.4		6.3		4.0		6.7	
3.6		4.6		6.7		8.5		1.7		6.1		5.8		11.7		16.1		15.3		19.2	
11.6		10.9		12.3		20.7		3.7		16.6		14.1		25.8		39.7		42.6		47.6	
3.5		6.5		5.5		9.2		3.1		13.5		7.3		10.4		20.0		12.5		27.8	
5.3		12.0		8.2		17.2		4.3		23.0		10.3		17.3		31.8		16.7		37.5	
9.8		12.5		14.3		21.9		4.3		14.7		16.7		27.6		36.9		39.0		56.4	
4.6		8.3		6.1		11.5		3.0		14.5		9.8		10.3		17.3		13.8		27.1	
6.6		8.7		12.2		16.3		3.3		12.6		12.1		17.8		29.1		23.7		35.4	
1.1	U	1.1	U	1.6		2.3		1.1	U	1.6		1.7		3.0		3.5		3.1		5.2	
6.04		8.22		10.74		14.49		3.84		14.25		11.2		17.9		29.7		30.5		43.3	
6.6		6.9		9.9		11.4		2.4		11.7		10.5		21.2		19.5		22.5		31.7	
19.3		42.8		23.4		65.3		20.0		106.9		45.5		57.5		115.2		51.1		128.4	
5.2		8.1		5.9		13.1		3.9		19.1		10.6		11.7		25.9		15.2		32.5	
5.5		12.1		14.3		15.8		4.3		23.2		17.7		24.7		33.1		41.5		27.9	
3.8		5.9		5.5		10.9		2.5		9.9		6.3		10.7		18.7		10.8		18.3	
6.2		8.0		9.6		11.7		2.1		11.2		10.4		13.8		19.3		25.4		28.3	
8.1		11.5		16.7		22.0		4.5		19.5		19.1		27.3		35.0		35.9		51.7	
11.3		14.3		12.5		23.5		6.1		20.2		22.1		31.7		45.0		40.1		67.4	
10.6		16.3		13.7		24.7		6.3		31.8		19.2		30.8		50.1		36.3		74.2	
3.2		5.2		4.1		9.2		2.4		10.3		5.8		8.5		13.0		8.1		19.8	
9.3		16.2		11.6		25.1		7.8		30.6		19.6		23.6		51.0		30.0		51.3	
2.9		5.0		3.3		6.6		2.2		10.0		5.0		8.8		13.1		8.9		18.8	
1.3		2.8		2.7		4.7		1.1		5.7		3.4		5.4		7.7		6.0		8.8	
21.7		26.1		41.5		62.2		12.1		39.5		38.7		68.2		91.4		112.6		137.7	
3.5		3.9		4.8		8.3		1.8		10.0		5.5		9.8		15.8		11.0		19.4	
13.9		19.3		23.3		37.3		7.3		23.3		28.3		44.3		53.2		68.4		83.2	
13.5		16.1		14.4		34.6		8.8		34.3		20.2		38.8		68.4		42.8		72.8	
4.0		7.9		7.0		13.2		3.6		12.5		8.1		12.7		21.6		16.2		33.2	
4.9		6.1		5.8		13.0		3.1		15.1		8.2		12.7		21.2		14.4		26.7	
1.9		2.8		2.9		4.8		1.0		5.6		3.4		5.1		7.6		6.4		11.8	
2.3		2.5		4.1		5.3		1.0		3.9		4.8		8.0		9.5		8.7		13.0	
6.3		11.9		8.9		16.5		4.8		24.2		12.6		12.9		31.6		14.2		36.4	
4.2		7.3		6.2		14.1		3.8		19.9		11.8		12.0		28.6		16.4		30.4	
1.4		2.3		1.5		3.4		1.2		6.1		2.3		3.7		7.6		3.2		8.5	
3.9		6.2		6.1		9.9		2.0		9.6		8.7		9.8		19.4		12.3		27.0	
6.9		7.4		11.9		15.7		4.5		18.7		10.4		19.9		26.2		30.6		27.2	
6.0		8.8		10.1		12.2		3.0		10.6		11.8		19.3		18.8		20.6		33.3	
11.3		12.9		16.8		26.5		5.9		32.1		20.3		33.8		50.4		40.8		65.7	
5.6		8.9		9.8		15.5		4.2		16.0		10.2		15.2		29.8		16.4		40.1	
12.4		17.1		15.9		27.2		7.2		37.2		26.0		33.4		50.4		30.3		57.3	
9.8		17.0		18.4		28.5		5.7		28.0		18.5		38.5		59.5		48.6		70.7	
5.6		6.2		8.0		10.8		2.4		12.4		10.6		14.8		20.3		19.4		26.3	

Table 12.3

Continued

Sample	PCB 118	Qualifier	PCB 128	Qualifier	PCB 130	Qualifier	PCB 132	Qualifier	PCB 135	Qualifier	PCB 136	Qualifier	PCB 137	Qualifier	PCB 138	Qualifier
Sample 1	34.3		4.9		2.1		6.9		1.6		1.5		1.7		24.9	
Sample 2	16.7		2.2		0.93		3.3		0.72		0.81		0.66		6.9	
Sample 3	6.1		0.87		0.44	U	1.3		0.44	U	0.44	U	0.44	U	3.4	
Sample 4	9.6		1.5		0.92	U	1.6		0.92	U	0.92	U	0.92	U	6.0	
Sample 5	90.2		14.5		5.3		17.6		4.6		4.8		3.7		42.5	
Sample 6	18.8		2.1		0.98	U	2.1		0.98	U	0.98	U	0.98	U	8.2	
Sample 7	47.0		7.1		3.4		11.1		3.0		3.9		2.3		34.4	
Sample 8	37.6		5.5		2.3		8.5		2.7		3.3		1.8		26.7	
Sample 9	8.1		1.3	U	1.3	U	1.6		1.3	U	1.3	U	1.3	U	5.0	
Sample 10	27.4		3.0		0.81		2.7		0.63		0.61		0.87		10.0	
Sample 11	78.1		11.2		3.1		11.0		2.0		1.6		3.0		39.1	
Sample 12	26.4		3.1		1.0		4.7		1.2		1.4		0.88		11.8	
Sample 13	29.3		5.2		2.0		6.2		1.9		2.2		1.6		16.8	
Sample 14	78.1		9.7		3.3		8.7		2.1		1.5		2.7		34.4	
Sample 15	20.9		4.4		1.2		6.4		1.3		1.8		1.2		15.4	
Sample 16	44.9		5.6		2.0		5.2		1.5		1.6		1.6		21.7	
Sample 17	5.6		1.1	U	1.1	U	1.1	U	1.1	U	1.1	U	1.1	U	2.6	
Sample 18	55.3		7.2		2.5		6.5		2.0		1.7		2.2		26.5	
Sample 19	40.6		5.6		1.3		5.5		0.96		1.1		1.4		16.2	
Sample 20	126.1		25.1		8.8		39.0		8.1		7.0		7.1		103.7	
Sample 21	27.5		4.4		1.8		8.1		2.0		1.7		1.5		18.7	
Sample 22	9.4		7.7		25.3		7.4		0.29		7		1		2	
Sample 23	26.6		2.8		1.2		4.4		1.1		1.0		1.0		13.1	
Sample 24	41.3		4.7		1.4		4.5		1.0		0.94		1.5		15.6	
Sample 25	60.0		6.0		2.1		8.6		1.7		1.4		2.1		26.3	
Sample 26	94.0		12.0		3.4		11.9		3.1		2.8		3.7		43.4	
Sample 27	77.8		9.8		4.2		16.2		3.6		3.6		2.7		41.4	
Sample 28	16.3		2.1		0.78		3.3		0.8		1.1		0.8		9.4	
Sample 29	53.7		7.3		3.7		15.0		4.7		3.5		2.4		35.1	
Sample 30	17.8		2.0		0.68		2.8		0.75		0.94		0.47		7.8	
Sample 31	11.3		1.5		0.58	U	1.6		0.58	U	0.58	U	0.58	U	5.2	
Sample 32	213.0		23.8		8.6		26.4		4.9		4.9		6.2		87.6	
Sample 33	19.3		3.3		1.4		3.7		1.1		1.1		0.96		11.7	
Sample 34	110.8		13.2		5.4		14.8		3.5		3.5		4.5		59.1	
Sample 35	109.7		14.5		4.5		15.6		4.6		4.1		3.1		57.1	
Sample 36	30.6		5.5		1.5		6.9		1.3		1.6		1.1		19.3	
Sample 37	33.4		5.0		1.5		5.4		1.2		1.5		1.4		14.9	
Sample 38	11.4		1.5		0.62	U	1.86		0.62	U	0.62	U	0.62	U	5.1	
Sample 39	17.8		2.5		0.72	U	2.34		0.72	U	0.72	U	0.72	U	7.4	
Sample 40	34.6		6.5		2.2		9.0		2.5		2.5		1.9		22.2	
Sample 41	36.9		5.7		2.3		8.5		1.9		2.2		1.8		20.1	
Sample 42	8.2		1.3		0.55		1.9		0.58		0.63		0.39		5.55	
Sample 43	32.8		3.8		1.2		4.6		0.92		1.0		1.3		13.2	
Sample 44	52.7		7.5		2.4		9.0		1.7		2.0		2.5		24.2	
Sample 45	38.2		5.3		1.6		5.4		0.84		0.9		1.5		18.3	
Sample 46	74.9		11.0		3.9		13.5		2.9		2.8		3.6		39.4	
Sample 47	40.2		4.9		1.6		6.1		1.6		1.6		1.3		20.5	
Sample 48	67.6		7.1		2.7		10.4		3.1		3.0		2.1		24.7	
Sample 49	94.5		10.8		3.1		12.4		2.9		3.2		3.3		37.2	
Sample 50	31.4		3.6		0.98		4.4		0.88		1.0		1.0		12.8	

PCB 141	Qualifier	PCB 146	Qualifier	PCB 149	Qualifier	PCB 151	Qualifier	PCB 153	Qualifier	PCB 156	Qualifier	PCB 158	Qualifier	PCB 163	Qualifier	PCB 170	Qualifier	PCB 180	Qualifier
10.5		2.0		10.3		1.4		12.1		3.4		3.2		2.7		1.8		2.1	
3.6		0.9		3.8		0.80		5.2		1.2		1.1		1.5		0.76		1.19	
2.2		0.44	U	1.7		0.44	U	1.8		0.57		0.53		0.59		0.44	U	0.46	
4.2		0.92	U	2.5		0.92	U	3.1		0.92	U	0.92	U	0.92	U	0.92	U	0.92	U
26.5		5.4		30.3		4.2		25.4		9.6		6.9		8.5		4.3		5.9	
6.0		0.98	U	4.0		0.98	U	3.94		1.16		1.18		1.30		0.98	U	1.0	
22.4		3.8		19.0		3.4		19.1		5.0		4.6		4.8		2.8		3.7	
15.7		3.2		14.9		3.0		18.2		3.8		3.8		4.6		2.2		3.5	
5.7		1.3	U	2.3		1.3	U	2.4		1.3	U	1.3	U	1.3	U	1.3	U	1.3	U
5.0		0.97		4.8		0.69		7.4		2.1		1.4		1.6		0.76		1.6	
16.9		2.7		13.8		1.6		20.4		6.3		4.6		5.0		2.3		3.3	
10.5		1.0		6.6		1.3		7.7		1.9		1.7		2.1		1.2		1.4	
10.0		2.5		9.7		2.3		15.2		2.9		2.5		3.4		1.7		2.9	
26.1		2.3		12.5		2.1		20.3		7.2		5.2		3.5		2.2		3.0	
9.2		1.6		7.9		1.6		10.6		1.9		2.0		2.5		1.4		1.9	
13.2		1.9		9.3		1.5		14.1		3.4		2.8		3.6		1.4		2.9	
6.0		1.1	U	1.2		1.1	U	1.6		1.1	U	1.1	U	1.1	U	1.1	U	1.1	U
13.8		2.5		11.1		1.6		15.0		4.7		3.8		3.3		1.7		2.6	
8.6		1.6		5.7		0.9		9.8		3.2		2.4		2.7		1.2		1.5	
57.6		10.0		62.8		9.7		59.6		13.9		10.7		11.6		8.5		11.1	
12.1		1.9		11.5		2.2		11.0		2.4		2.8		3.4		1.8		2.4	
7		17		6		0.97		17.76		4.59		5.22		3.59		2.45		2.87	
9.9		1.3		6.6		1.2		8.2		2.0		2.2		2.0		1.0		1.2	
12.2		1.3		7.4		1.1		9.4		3.5		3.0		2.0		1.2		1.5	
12.2		2.4		9.3		1.9		15.4		3.7		3.5		3.2		1.8		3.1	
31.1		3.5		20.0		2.7		25.0		7.6		6.2		6.5		2.6		4.0	
32.6		3.6		17.5		4.2		25.9		6.6		5.3		6.9		3.5		3.8	
4.4		1.1		5.6		0.98		7.0		1.8		1.3		1.9		0.79		1.3	
17.0		3.6		19.9		4.0		23.3		6.4		5.6		6.4		3.5		3.6	
6.8		0.91		4.5		0.93		4.1		1.1		1.1		1.4		0.86		1.3	
2.5		0.58	U	2.7		0.58	U	3.5		0.85		0.82		0.94		0.58	U	0.60	
58.9		8.6		34.4		5.1		51.4		17.0		13.6		12.3		5.5		7.5	
9.1		1.3		5.5		0.93		8.5		2.3		2.0		1.9		1.0		1.4	
35.4		5.1		22.3		2.3		37.7		11.3		8.9		7.3		3.8		3.7	
43.3		4.7		26.3		4.6		30.5		7.9		6.9		8.0		3.5		4.8	
2.8		1.6		8.7		1.5		9.7		3.0		2.7		2.4		1.6		1.9	
2.3		1.8		8.1		1.5		10.7		2.6		2.5		2.4		1.2		1.8	
0.93		0.62	U	2.7		0.62	U	4.9		0.86		0.66		0.94		0.62	U	0.74	
0.98		0.72	U	3.3		0.72	U	4.6		1.2		1.1		0.91		0.72	U	0.80	
4.3		2.4		16.1		2.5		15.8		3.4		3.0		4.2		2.5		3.2	
3.3		2.2		10.1		2.1		14.0		3.0		3.3		3.1		1.9		2.5	
0.95		0.54		3.0		0.76		3.5		0.75		0.75		1.0		0.40		0.66	
1.7		1.2		6.1		0.99		9.0		2.7		2.1		2.0		1.1		1.4	
3.3		2.2		9.1		2.2		14.5		4.8		3.8		3.4		1.9		2.6	
2.1		1.3		5.7		0.95		9.6		3.4		2.5		2.7		1.4		1.8	
5.0		3.6		17.8		3.1		25.0		7.4		6.0		6.0		3.6		3.5	
3.0		2.5		11.5		1.9		11.1		3.2		2.8		2.8		1.7		2.2	
5.2		3.5		20.0		3.0		22.9		4.5		3.5		5.5		2.9		5.6	
5.8		3.4		21.4		2.6		21.5		6.4		5.0		5.6		2.9		3.8	
1.8		1.3		6.2		1.0		9.8		2.4		2.1		1.7		1.2		1.9	

assessed by simple inspection of a two- or three-dimensional plot: a principal components scores plot. Figure 12.2 shows a two-PC scores plot for Data Set 1. Two principal components account for more than 92% of the variance in Data Set 1, and the scores plot clearly divides the samples into two clusters: Aroclor 1248 and Aroclor 1254.

There is an obvious problem though. While this is indeed a two Aroclor system (as Figure 12.2 clearly suggests) it is not a two-source system. The first two PCs do not differentiate between the two Aroclor 1254 variants. This illustrates a common problem in the application of PCA to environmental chemical data. All too often, investigators will present a two-PC scores plot like Figure 12.2, accompanied by a statement of justification indicating that two principal components account for 92.5% of the variance. Such a statement leaves the tacit implication that the residual 7.5% of the variance is random noise, which in this case is clearly not the case.

A three component scores plot for Data Set 1 is shown as Figure 12.3. Three PCs account for 97.5% of the variance; an incremental increase over the percentage accounted for by two PCs. However, that small percentage of total variance is *not* random. The three-PC scores plot clearly distinguishes three clusters, rather than two, and effectively allows the analyst to infer the presence of three sources.

This example highlights several important precautionary notes.

1 For better or for worse, the use of mathematical techniques such as PCA carries with it the aura of precision and exactitude. In the case of the two-PC scores plot (Figure 12.2) the strong clustering into two Aroclor groups, coupled with the statement that two PCs account for 92.5% of the variance, may be sufficiently intimidating to impress skeptics. Moreover, it may even provide a false sense of security to the naïve analyst.

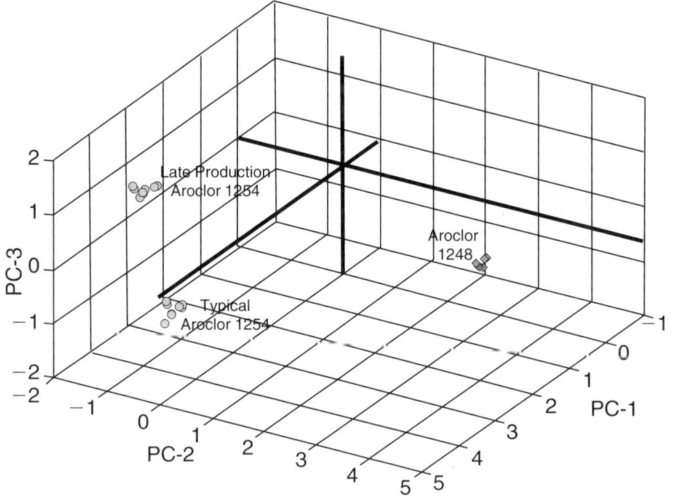

Figure 12.3
Three principal component scores plot of Data Set 1. Three PCs account for 97.5% of the variance, and allow clear distinction of the three PCB sources.

2 In an environmental forensics setting, we seldom have prior knowledge of the true data structure, so we should not be arrogant in our application of rules-of-thumb regarding what percentage of variance should be considered 'significant.'
3 Scientific and legal 'significance' is clearly *not* a function of variance. If a party involved in environmental litigation had used Aroclor 1254, but ceased all operations in the mid-1960s (prior to Monsanto's change in the Aroclor 1254 production process), the small difference in percentage of variance between two and three PCs would have enormous implications.
4 Practitioners of PCA-based methods in environmental forensics must employ more sophisticated goodness-of-fit diagnostics than percentage variance, and they must have a reasonable understanding of how such methods work.

Because PCA is a powerful exploratory data analysis tool on its own, as well as an intermediate step in receptor modeling methods, we will discuss this method in considerable detail below. The key steps in PCA include: (1) data transformations; (2) singular value decomposition (eigenvector decomposition); (3) determination of the number of significant eigenvectors; and (4) visual display of scores and loadings plots. Each of these steps is discussed in turn below.

12.2.2 DATA TRANSFORMATIONS

To the data analyst, the laboratory results received from the chemist are 'raw data.' In environmental chemistry these raw data are usually transmitted in units of concentration. Data Set 2 (Table 12.3: presented in units of ng/g) is a good example. However, seldom is a statistical analysis performed on a matrix in this form. Rather, the matrix is transformed in some manner. A data transformation is the application of some mathematical function to each measurement in a matrix. Taking the square root of every value in a matrix is an example. In analysis of environmental chemical data, data transformations are done either for (1) reasons related to the environmental chemistry of the system; or (2) mathematical reasons, to optimize the analysis.

A common transformation done for chemical or physical reasons is sample normalization. In an environmental system, concentrations can vary widely due to dilution away from a source. For example, in the case of contaminated sediment investigations, concentrations may decrease exponentially away from an effluent pipe. However, if the relative proportions of individual analytes remain relatively constant then we might infer a single source scenario, coupled with dilution away from the source. Inference of source by pattern recognition techniques is concerned more with the relative proportions between analytes than with absolute concentrations. Thus, a transformation is necessary to normalize out concentration/dilution effects. One that is commonly employed is a transformation to a percent metric, where each value is divided by the total concentration of the sample. This percent transformation is also referred to as a 'constant row-sum' transformation, because the sum of analyte concentrations in each sample (i.e., across rows) sums to 100%. Stated mathematically, where **S** is an $m \times m$ diagonal matrix with the m row-sums (total concentrations) of $\mathbf{X}_{(\mathbf{nk})}$ along the diagonal, and zeros on the off-diagonals, the constant row-sum matrix **X** may be calculated as:

$$\mathbf{X} = 100 * \mathbf{S}^{-1} * \mathbf{X}_{(\mathbf{nk})} \tag{12.2}$$

An alternative to the constant row-sum transformation is to normalize the data with respect to a single species or compound (a so-called 'normalization variable'). This transformation involves setting the value of the normalization variable to 1.0, and the values of all other variables to some proportion of 1.0, such that their ratios with respect to the normalization variable remain the same as in the original metric.

The second type of transformation is done more for mathematical/statistical purposes. In any chemical data set, there is usually a strong relationship between the mean value of an analyte and its variance (variance is square of the standard deviation). Therefore, chemicals measured in trace concentrations almost always exhibit smaller variance than those measured at much higher concentrations. Multivariate procedures such as PCA are variance driven, so in the absence of some transformation across variables, the analytes with highest mean and variance usually have the greatest influence on the analysis. Polychlorinated dibenzo-*p*-dioxins provide a particularly instructive example. In most environmental systems, 2,3,7,8-TCDD (dioxin) is typically measured at orders of magnitude lower concentrations than the octa-chlorinated congener OCDD. Thus the mean and variance of 2,3,7,8-TCDD is typically orders of magnitude lower than OCDD. However, this does not imply less precision or accuracy in the chemical measurement of 2,3,7,8-TCDD. Nor does it imply that 2,3,7,8-TCDD is of secondary environmental importance to OCDD. In fact, the opposite is usually the case because 2,3,7,8-TCDD has much higher toxicity than does OCDD. As such, analysis of environmental chemical data almost always requires some sort of 'homogeneity of variance' transformation. A number of transformations may be applied to produce homogeneity of variance. One of the most commonly used transformations in PCA is autoscale transform (also known as the Z transform). Given a matrix **X**, with calculated means ($\bar{x}_j$) and standard deviations (s_j) in each column ($j = 1, 2 \ldots n$). The autoscaled matrix **Z** is calculated:

$$\mathbf{Z}_{ij} = \frac{x_{ij} - \bar{x}_j}{s_j} \tag{12.3}$$

The autoscale transformation guarantees absolutely equal variance in that it sets the mean of each column to 0.0 and the standard deviation to 1.0.

Another common homogeneity of variance transformation is the range-transform (also known as the minimum/maximum transformation). Following the convention of Miesch (1976a), where the original matrix is denoted **X**, the range is denoted as X-prime ($\mathbf{X}' = \{x'_{ij}\}$). The transformation is performed as follows:

$$x'_{ij} = (x_{ij} - x_{\min j})/(x_{\max j} - x_{\min j}) \tag{12.4}$$

This results in a matrix where the minimum value in each column equals 0.0 and the maximum equals 1.0. The range transformation produces variances that are approximately homogeneous (unlike the Z transform, which results in absolute homogeneity of variance). However, the range transform has two advantages: (1) it carries no implication of a standard normal distribution; and (2) because transformed values are within the 0.0 to 1.0 range, all values are non-negative. The latter feature is particularly useful in receptor modeling, where explicit constraints of non-negativity are used to derive source fingerprints. Malinowski (1977, 1991) notes that a disadvantage of the range transform is that it is extremely sensitive to outliers. As we will see, however, in an environmental forensics investigation, detection and evaluation of outliers is a crucial part of the process, so this feature of the range transform is usually desirable.

A third transformation (the equal vector length transform) is often, but not always, employed in multivariate analysis of chemical data. The equal vector length transform is performed to force each of the sample vectors to have equal Euclidean length. If all vectors have equal length, then the differences between samples are a function only of the angles between samples. Thus the similarities and differences between samples can be expressed as a similarity matrix of cosines (Davis, 1986). The cosine between two identical samples is 1.0 – the cosine between two completely dissimilar samples (i.e. vectors at 90°) is 0.0. By Miesch's convention (Miesch, 1976a), the constant-sum input matrix is indicated as **X**, the range transform matrix as $\mathbf{X}'$, and the equal vector length transform as $\mathbf{X}''$. For matrix algebra-based programming languages, a computationally efficient way (Hopke, 1989) is to first define $\mathbf{Y} = \{y_{ij}\}$ (the matrix element of **Y** in row i, column j). **Y** is a $m \times m$ diagonal matrix where each diagonal element equals the inverse of the square root of the sum-of-squares along rows of $\mathbf{X}'$:

$$y_{ij} = \frac{1}{\left(\sum_{i=1}^{m} x'^{2}_{ij}\right)^{1/2}} \tag{12.5}$$

The transformed matrix $\mathbf{X}''$ may then be calculated as follows:

$$\mathbf{X}'' = \mathbf{Y}\mathbf{X}' \qquad (12.6)$$

This transformation has an added advantage for receptor modeling. By definition, if all samples have equal vector lengths, each sample must lie on an $n - 1$ dimensional surface, unit length from the origin. To demonstrate this, a simple three chemical example is shown in Figure 12.4. In this case $n = 3$, because only three variables (chromium, copper and zinc) are present. These data were transformed by Equations 12.5 and 12.6; thus all samples lie on a two-dimensional surface, which is unit length from the origin ($n - 1 = 2$). As we will see in Section 12.3, receptor/mixture modeling involves resolution of a $k - 1$ dimensional geometric figure within k dimensional principal component space. Thus, this transformation has particularly attractive features in that regard.

In summary, normalization across rows (constant row sum or normalization to a marker chemical) is almost always done in environmental chemometric analyses. Some homogeneity of variance transformation (e.g., range transform or autoscale transform) is also typically performed in order to keep high concentration variables from dominating an analysis. The constant vector length transform is sometimes done, usually when (1) there is some advantage to being able to express relationships between samples simply in terms of angles between samples; or (2) when the PCA is an intermediate step in receptor

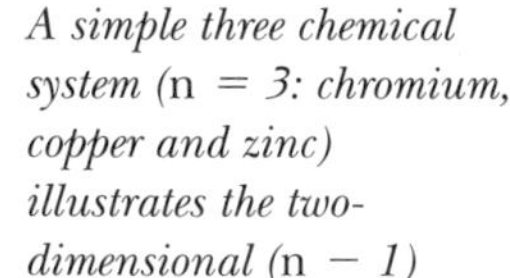

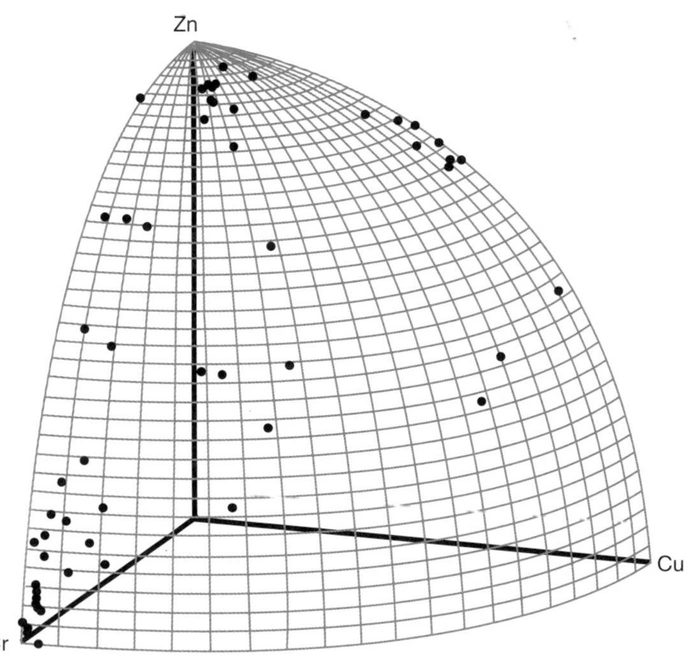

Figure 12.4

A simple three chemical system (n = 3: chromium, copper and zinc) illustrates the two-dimensional (n − 1) geometry of a matrix which has undergone the equal vector length transformation. All sample vectors are unit length from the origin, and lie on the n − 1 dimensional spherical surface.

modeling, which involves resolution of a $k - 1$ dimensional simplex within k dimensional space. The PCA that resulted in the plots shown on Figures 12.2 and 12.3 involved transformation by the constant row sum transformation (Equation 12.2) followed by the range transform (Equation 12.4). The constant vector length transformation was not used.

12.2.3 EIGENVECTOR DECOMPOSITION

Eigenvector decomposition is a simple mathematical procedure that allows a reduction in dimensionality of a data set. This is the core mathematical operation involved in principal components analysis. It is most often accomplished through singular value decomposition (SVD) of the transformed matrix $\mathbf{X}'$ or $\mathbf{X}''$. As shown in Figure 12.4, the transformed $m \times n$ matrix may be thought of as m vectors plotted in n dimensional space. Each variable is represented as one of n orthogonal axes of a cartesian coordinate system. There are, however, an infinite number of sets of n mutually orthogonal basis vectors that may equivalently be used to plot the sample vectors, without loss of information. The set of eigenvectors extracted from a similarity matrix of the original data is one such alternative reference space. The number of eigenvectors (i.e., the number of principal components) will equal m or n, whichever is smaller. However, there are usually correlations between analytes due to common sources. Thus, a relatively small subset of the eigenvectors is typically sufficient to capture the variability in the system, and the interrelationships between samples can be observed without loss of information.

Eigenvector decomposition is a well-established part of the core knowledge of mathematics and is frequently used in the physical and natural sciences. The calculation of eigenvectors and eigenvalues is relatively straightforward, but lengthy and cumbersome. As such, a conceptual discussion of the topic is presented below, and the reader is referred to any number of linear algebra texts for a complete mathematical discussion. Davis (1986) provides a detailed, lucid but less rigorous treatment, using examples from the earth sciences.

Given an error free, noise free matrix of $m > k$ samples and $n > k$ variables, resulting from k sources, only k nonzero eigenvectors and eigenvalues will be extracted. If $k = 3$, the first eigenvector will account for a high percentage of the total variance in the data set. The second eigenvector is constrained in that it must be mutually orthogonal to the first, and accounts for the highest percentage of residual variance (that variance not accounted for by the first eigenvector). The third eigenvector is mutually orthogonal to the first two, and accounts for the remainder of the variance. The data set may equivalently be expressed in this three-dimensional reference space, without loss of information. Given a transformed matrix $\mathbf{X}'$ composed of m samples along the rows, and n variables (chemical analytes) along columns, PCA is accomplished through SVD of $\mathbf{X}'$:

$$\mathbf{X}' = \overbrace{\mathbf{U}}^{\mathbf{F}'_R}\ \overbrace{\mathbf{\Lambda}^{1/2}\ \mathbf{V}^t}^{\mathbf{A}'_R} = \underbrace{\mathbf{U}\ \mathbf{\Lambda}^{1/2}}_{\mathbf{A}'_Q}\ \underbrace{\mathbf{V}^t}_{\mathbf{F}'_Q} \tag{12.7}$$

where $\mathbf{U}$ equals the matrix of principal components scores in R mode (which examines inter-relationships between variables), $\mathbf{V}^t$ equals the matrix of scores in Q mode (which examines inter-relationships between samples), and $\mathbf{\Lambda}$ equals the diagonal matrix of eigenvalues. Principal component scores (**F**) and loadings (**A**) matrices for R and Q modes may result by re-expression of Equation 12.7, as indicated (Zhou *et al.*, 1983). Eigenvectors are abstract in that they usually cannot be interpreted in terms of real world phenomena (although many have tried). The total number of calculated PCs equals m or n, whichever is smaller. A model involving a reduced number of principal components (k) may be represented as follows (Q mode):

$$\text{Matrix dimensions} \qquad \underset{(m\times n)}{\mathbf{X}'} = \underset{(m\times k)}{\mathbf{A}'}\ \underset{(k\times n)}{\mathbf{F}'} + \boldsymbol{\varepsilon} \tag{12.8}$$

where k equals the number of PCs retained for the model, and ε represents error.

12.2.4 DETERMINING THE NUMBER OF SIGNIFICANT PRINCIPAL COMPONENTS

The choice of the number k is equivalent to the decision on the number of 'significant' principal components. Of the many aspects of PCA-based methods, no topic has created more argument or controversy than the criteria used to determine the correct number of eigenvectors (i.e., k, the number of factors, sources, subpopulations or end-members).

Numerous methods have been proposed for determination of k (Cattell, 1966; Exner, 1966; Malinowski, 1977; Miesch, 1976a; Wold, 1978; Ehrlich and Full, 1987; Henry *et al.*, 1999). The spirit and intent of these methods are similar: the estimated data set, as back-calculated from reduced dimensional space (i.e., $\mathbf{X}_{\text{hat}}$ or $\hat{\mathbf{X}}$), should reproduce the measured data ($\mathbf{X}$) with reasonable fidelity.

12.2.4.1 Single Index Methods

Six criteria commonly used in environmental chemometrics were applied to Data Sets 1 and 2, and the results are shown in Tables 12.4 and 12.5, respectively.

The PCA of both data sets involved the constant sum and range transformations. Each of the six indices, and the rationale for their use, is discussed below.

1 *Cumulative percentage variance.* The rationale for this criterion is simple: a reduced dimensional model should account for a large percentage of the variance in the original matrix. However, as discussed in Section 12.2.1, any *a priori* choice regarding what percentage of variance one should consider to be 'significant' is problematic. Workers in multivariate statistics, chemometrics and mathematical geology generally acknowledge that any proposed cutoff criterion is arbitrary (Malinowski, 1991; Deane, 1992; Reyment and Jöreskog, 1993). The lack of a clear criterion makes the cumulative variance method dubious. Nonetheless, in Tables 12.4 and 12.5, we note that the commonly used cutoff of 95% suggests retention of three principal components for both Data Sets 1 and 2.

2 *Scree test.* The scree test of Cattell (1966) is based on the supposition that the residual variance, not accounted for by a k principal component model, should level off at the point where the principal components begin accounting for random error. When residual variance is plotted versus principal component number, the point where the curve begins to level off should show a noticeable inflection point, or 'knee.' The problem with this criterion is that there is often no unambiguous inflection point, and when such is the case, the decision as to the number of significant principal components is arbitrary.

3 *Normalized varimax loadings.* The multivariate statistical algorithms of Klovan and Miesch (1976) included a subroutine that calculates the number of samples with normalized varimax loadings greater than 0.100. Neither Miesch (1976a,b) nor Klovan and Miesch (1976) included explicit discussion of its utility, but Ehrlich and Full (1987) later presented such a discussion. If an eigenvector carries systematic information, then typically, a large number of samples will have high loadings on that varimax factor (loadings in Q-mode terminology). High numbered factors that account for noise and little variance typically have loadings that are small for all samples (i.e., <0.1). Factors with many samples >0.1 indicate principal components that should be retained for a model. This criterion calls for rotation of principal components using the varimax procedure of Kaiser (1958), normalization of the varimax loadings matrix to sum to 1.0 across all sample rows, and tabulation of the number of samples that exceed 0.1 for each factor. The analyst looks for a sharp drop in the index as an indication of the appropriate number of eigenvectors.

4 *Malinowski indicator function.* Malinowski (1977) presented an indicator function, which is calculated as a function of the residual standard deviation (Malinowski, 1977; Hopke, 1989). The function reaches a minimum when the 'correct' number of principal components is retained. The index has worked well with relatively simple data structures but Hopke (1989) reports that it has not proven as successful with complex environmental chemical data.

5 *Cross-validation.* Cross-validation is a commonly used method for determination of number of significant principal components. It involves successive deletion of data points, followed by prediction of the deleted data with increasing numbers of eigenvectors. The PRESS statistic (predicted residual error sum of squares) is then calculated for each number of potential eigenvectors. Many criteria have been proposed based on calculation of some function of PRESS (Wold, 1978; Eastman and Krzanowski, 1982; Deane, 1992; Grung and Kvalheim, 1994). The PRESS value and criterion in Table 12.3 is that presented by Deane (1992).

6 *Signal-to-noise ratio.* This method has been proposed very recently by Henry *et al.* (1999). Henry's NUMFACT criterion involves calculation of a signal-to-noise (S/N) ratio. Henry found that given random data, an S/N ratio as high as 2.0 could be obtained. Based on that, the rule-of-thumb criterion recommended by Henry is that principal components with S/N ratios greater than 2.0 should be retained for a model.

As is evident in Table 12.4, for Data Set 1 (a relatively simple data set with random Gaussian noise) each of these indices provides an accurate estimate of the true number of Aroclor sources. All six indices correctly indicate a three-component system. Table 12.5 reports the values for the same indices, as applied to the more complicated, error-laden Data Set 2. Here, the reproduction indices suggest anywhere between three sources and six sources. Clearly, the complications present in Data Set 2 (which are quite reasonable in terms of common environmental chemistry scenarios) are sufficient to introduce ambiguity between these various indices.

This ambiguity is due in part to the fact that all of the above are 'single-index' methods. Each involves calculation of a single numerical value or statistic, which represents the data set as a whole as a function of the number of principal components retained. The data analyst typically compares the behavior of the index as additional PCs are retained, relative to some rule-of-thumb cutoff criterion. The idea of a rule-of-thumb decision criterion (i.e., a minimum, a change in slope, a threshold) is troublesome in exploratory data analysis, because we have very little information to evaluate the efficacy of these rules. In such situations, we need other tools to gain deeper insight into the chemical system.

12.2.4.2 Variable-by-Variable Goodness of Fit

Miesch (1976) noted and addressed some of these problems. Miesch correctly observed that single index methods (in particular criteria based on percentage of variance) are misleading because they carry the tacit assumption that variability *not* accounted for by a reduced dimensional model is spread evenly across all originally measured variables. Miesch proposed instead, that goodness of fit be evaluated on a variable-by-variable basis. The variance accounted

Table 12.4

Reproduction indices for Data Set 1 (see Table 12.1).

PC No.	Eigenvalue	(1) Cumulative Percentage Variance	(2) Scree Test	(3) Normalized Varimax Loadings	(4) Malinowski Indicator Function	(5) Cross Validation PRESS(i)/ RSS($i-1$)	(6) Signal to Noise Ratio
1	220.430	57.36	42.64	16	0.000674		39.32
2	135.150	92.54	7.46	8	0.000315	0.20	35.04
3	19.212	**97.53**	**2.47**	**16**	**0.000204**	**0.43**	**12.35**
4	1.198	97.85	2.15	1	0.000215	1.26	1.59
5	0.990	98.10	1.90	1	0.000229	1.41	1.35
6	0.861	98.33	1.67	0	0.000246	1.57	1.37
7	0.781	98.53	1.47	1	0.000266	1.76	1.41
8	0.719	98.72	1.28	1	0.000290	1.96	1.29
9	0.634	98.88	1.12	1	0.000318	2.19	1.34
10	0.556	99.03	0.97	0	0.000352	2.50	1.18
11	0.535	99.17	0.83	1	0.000392	2.79	1.27
12	0.475	99.29	0.71	1	0.000442	3.22	1.17
13	0.438	99.41	0.59	1	0.000503	3.67	1.21
14	0.382	99.50	0.50	1	0.000583	4.35	1.03
15	0.371	99.60	0.40	1	0.000681	5.00	1.15
16	0.317	99.68	0.32	1	0.000814	6.01	1.04
17	0.272	99.75	0.25	0	0.001001	7.11	1.09
18	0.213	99.81	0.19	1	0.001296	8.86	0.81
19	0.184	99.86	0.14	0	0.001769	11.31	0.81
20	0.169	99.90	0.10	0	0.002569	14.73	0.83
21	0.143	99.94	0.06	0	0.004157	21.04	0.84
22	0.126	99.97	0.03	1	0.007798	30.60	0.84
23	0.066	99.99	0.01	0	0.027834	308.49	0.64
24	0.043	100.00	0.00	0			0.00

for by *each* of the originally measured variables is evaluated for each potential number of principal components. Given an m sample by n variable data matrix $\mathbf{X}$ of rank m or n (whichever is smaller) the index used by Miesch (1976) was the 'coefficient of determination' (CD) between each variable in the original data matrix ($\mathbf{X}$), and its back-calculated reduced dimensional equivalent ($\hat{\mathbf{X}}$). For each number of potential eigenvectors, $1, 2 \ldots$ *rank*, Miesch calculated an $n \times 1$ vector:

$$r_j^2 \cong \frac{s(x)_j^2 - (d_j)^2}{s(x)_j^2} \tag{12.9}$$

where $s(x)_j^2$ is the variance of values in the jth column of $\mathbf{X}$, and $(d_j)^2$ is the variance of residuals between column j of $\mathbf{X}$ and column j of $\hat{\mathbf{x}}$. Miesch used the '$\cong$' in this equation because he recognized that this was not a conventional r^2 or coefficient of determination ('CD') as defined by least squares linear regression. It is not the variance accounted for by the least

Table 12.5

Reproduction indices for Data Set 2 (see Table 12.3).

PC No.	Eigenvalue	(1) Cumulative Percentage Variance	(2) Scree Test	(3) Normalized Varimax Loadings	(4) Malinowski Indicator Function	(5) Cross Validation PRESS(*i*)/ RSS(*i*−1)	(6) Signal to Noise Ratio
1	585.600	77.92	22.08	50	0.000102	–	73.29
2	115.450	93.28	6.72	47	0.000059	0.33	31.89
3	17.807	**95.65**	4.35	**18**	0.000050	0.73	10.79
4	8.494	96.78	3.22	2	0.000046	**0.97**	6.54
5	4.751	97.41	2.59	1	0.000043	1.21	4.38
6	3.073	97.82	2.18	2	**0.0000421**	1.38	**3.30**
7	1.706	98.05	1.95	2	0.0000422	1.62	1.88
8	1.507	98.25	1.75	1	0.0000424	1.75	1.88
9	1.279	98.42	1.58	1	0.000043	1.88	1.85
10	1.055	98.56	1.44	0	0.000043	2.04	1.49
11	0.973	98.69	1.31	1	0.000044	2.19	1.60
12	0.898	98.81	1.19	1	0.000045	2.33	1.53
13	0.806	98.92	1.08	1	0.000046	2.54	1.51
14	0.782	99.02	0.98	1	0.000047	2.66	1.45
15	0.693	99.11	0.89	1	0.000048	2.81	1.31
16	0.596	99.19	0.81	1	0.000049	3.06	1.25
17	0.570	99.27	0.73	0	0.000050	3.29	1.28
18	0.532	99.34	0.66	1	0.000051	3.57	1.21
19	0.483	99.40	0.60	1	0.000053	3.83	1.16
20	0.434	99.46	0.54	1	0.000055	4.21	1.10
21	0.421	99.52	0.48	1	0.000056	4.52	1.14
22	0.371	99.57	0.43	0	0.000058	5.03	1.10
23	0.364	99.61	0.39	0	0.000060	5.25	1.10
24	0.313	99.66	0.34	0	0.000062	5.72	0.95
25	0.287	99.69	0.31	0	0.000065	6.33	0.94
26	0.274	99.73	0.27	0	0.000067	6.82	1.00
27	0.246	99.76	0.24	0	0.000070	7.74	0.93
28	0.245	99.80	0.20	0	0.000073	8.24	0.94
29	0.215	99.82	0.18	1	0.000076	8.89	0.88
30	0.178	99.85	0.15	0	0.000080	10.12	0.76
31	0.160	99.87	0.13	0	0.000084	11.25	0.72
32	0.142	99.89	0.11	1	0.000089	12.08	0.67
33	0.122	99.90	0.10	1	0.000095	13.78	0.62
34	0.111	99.92	0.08	0	0.000102	16.02	0.61
35	0.105	99.93	0.07	1	0.000109	18.24	0.58
36	0.093	99.95	0.05	0	0.000117	21.61	0.60
37	0.084	99.96	0.04	0	0.000125	22.87	0.54
38	0.060	99.96	0.04	0	0.000138	28.54	0.45
39	0.057	99.97	0.03	0	0.000152	32.49	0.44
40	0.047	99.98	0.02	0	0.000170	39.18	0.41
41	0.040	99.98	0.02	1	0.000191	46.38	0.40
42	0.031	99.99	0.01	0	0.000221	58.76	0.35
43	0.026	99.99	0.01	0	0.000260	73.10	0.32
44	0.020	99.99	0.01	0	0.000317	99.11	0.28
45	0.016	100.00	0.00	0	0.000396	140.67	0.26
46	0.014	100.00	0.00	0	0.000479	172.73	0.26
47	0.007	100.00	0.00	0	0.000697	317.28	0.19
48	0.003	100.00	0.00	0	0.001372	555.71	0.13
49	0.002	100.00	0.00	0	0.005275	–	0.12
50	0.002	100.00	0.00	0	–	–	0.00

squares regression line of $\mathbf{X}_j$ and $\hat{\mathbf{x}}_j$. Rather, the Miesch CD is the r^2 with respect to a line of one-to-one back-calculation between $\mathbf{X}_j$ and $\hat{\mathbf{x}}_j$. For CDs less than 1.0, the analyst must make some decision as to what value may be accepted. That decision is made in context of the analyst's experience, knowledge of measurement error (if available), and scientific context. As an example, if a certain PCB congener is known to be less accurate and precise using a certain gas chromatography (GC) column, the analyst may justifiably accept a lower CD for that congener, than for other congeners.

A graphical extension of Miesch's method has recently been implemented by Johnson, the CD scatter plot (Johnson, 1997; Johnson *et al.*, 2000). The appropriate graphic to illustrate the fit of the Miesch CD is a series of n scatter plots that show the measured value for each variable $\mathbf{X}_j$ plotted against the back-calculated values from the k proposed principal components ($\hat{\mathbf{x}}_j$). A scatter plot series for Data Set 2 is presented as Figure 12.5.

The scatter plot array shows 56 plots (one for each PCB congener) as back-calculated from a three-PC model. The Miesch CD is calculated and reported at the top left corner of each graph. When an insufficient number of principal components are retained there should be evident non-random deviations from the 1:1 fit line. For three principal components a good fit is observed for most congeners, but there is a systematic lack of fit observed for (1) non-detect censored data points (indicated as squares); (2) Sample 22 which had a data transcription error (triangle); and (3) the congener PCB 141, which coelutes with DDT. In particular, note that on many graphs, there are two 'non-detect' samples at high measured concentration. These are Samples 9 and 17 from Table 12.3. Both of these samples had low total PCB concentrations, and yielded non-detects for more than half of the reported PCB congeners.

The important point with regard to determining the number of significant principal components is that these errors, while not related to chemical sources fingerprints, are not random. Rather, they represent systematic signal within the data set, and thus they greatly influence indices shown in Tables 12.4–12.6.

The strength of CD scatter plots is that they allow rapid evaluation of (1) sample-by-sample goodness of fit, (2) variable-by-variable goodness of fit, and (3) outlier detection. Each of these is evaluated simultaneously, as a function of the number of principal components retained. That combination quickly leads the analyst to more insightful and direct questions than are possible based solely on what threshold numerical index one judges acceptable. The questions that we must now ask are:

1 What is the cause of deviation from good fit? Is it random error or systematic variability not accounted for by k PCs?

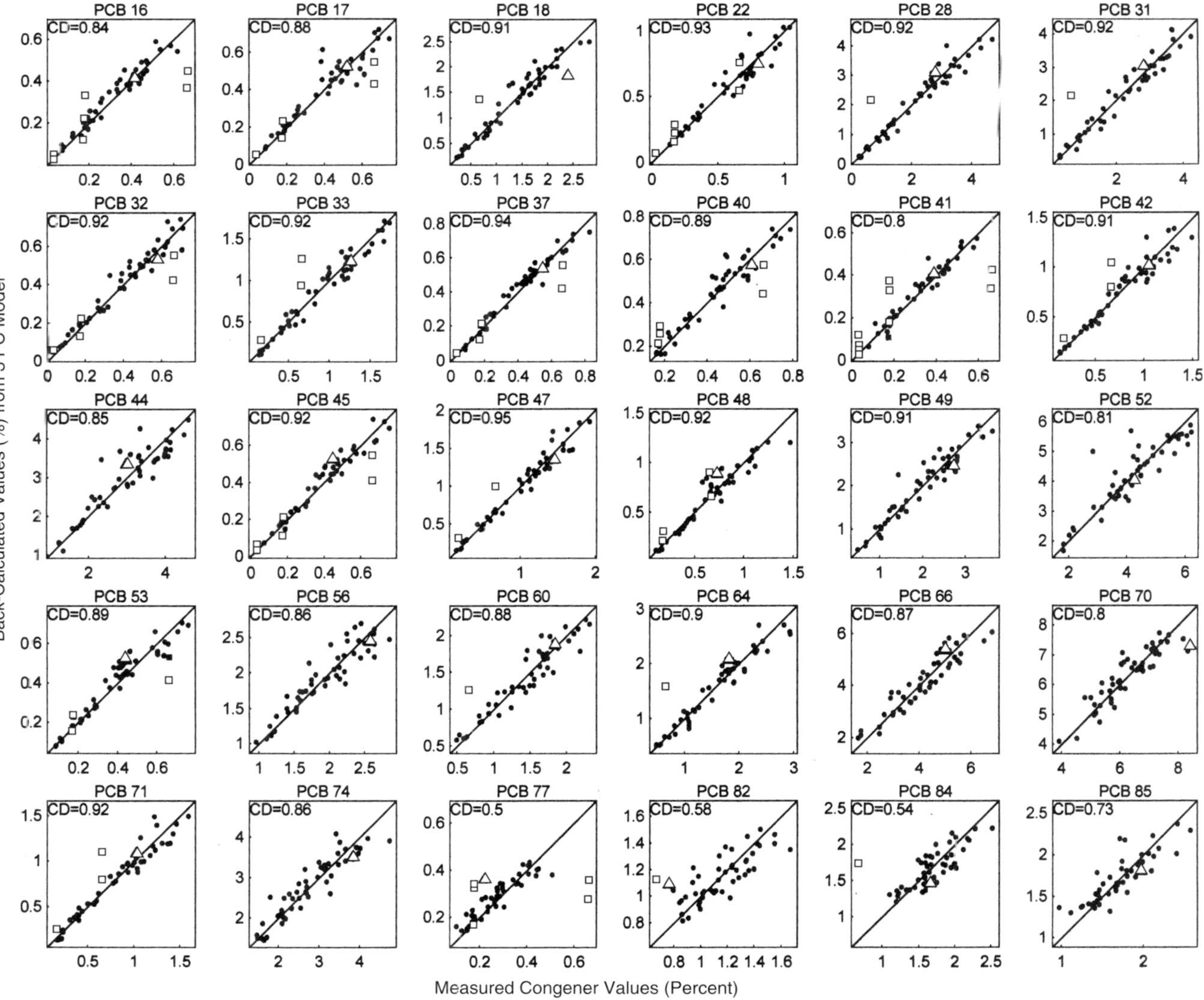

Figure 12.5a

Goodness-of-fit scatter plot array and Miesch coefficients of determination (CDs) for first 30 variables in Data Set 2. The x axis is measured concentration. The y axis is the value back-calculated from a three principal component model. Non-detects (censored data points) are indicated as squares (□). Sample 22, which had a data transcription error, is indicated as a triangle (△).

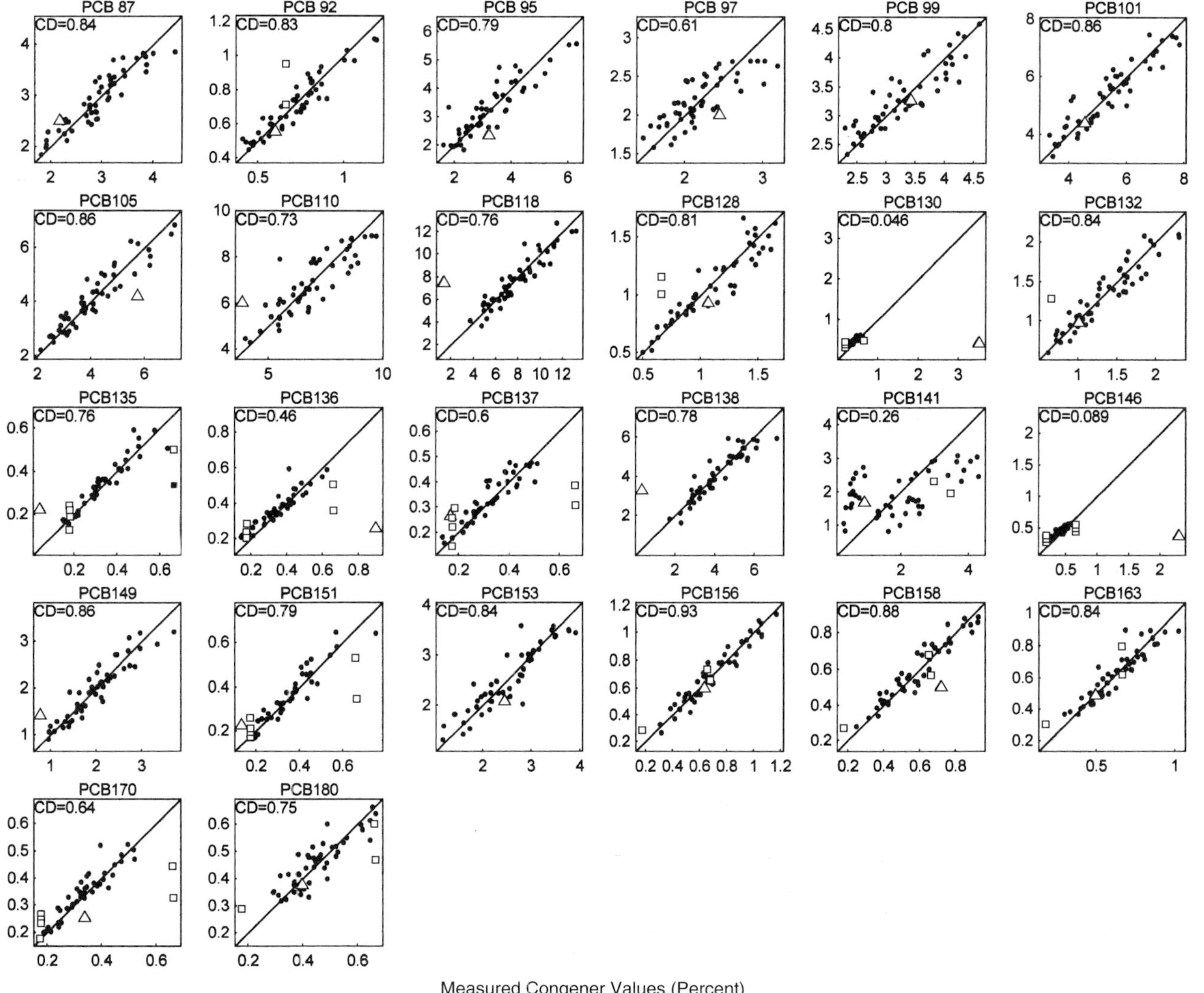

Figure 12.5b

Goodness-of-fit scatter plot array and Miesch coefficients of determination for last 26 variables in Data Set 2. The x *axis is measured concentration. The* y *axis is the value back-calculated from a three principal component model. Non-detects (censored data points) are indicated as squares (□). Sample 22, which had a data transcription error, is indicated as a triangle (△). Note lack of fit for PCB 141 due to DDT coelution.*

2 If the cause of the observed deviation is systematic, is it due to (1) data entry error, (2) analytical error, or (3) presence of an additional source of variability in the field?
3 How does one evaluate the number of significant PCs in the presence of such deviations?

In most cases, these questions cannot be answered in the realm of numerical data analysis. The analyst must now change hats and play the role of forensic scientist. The decision regarding how to deal with outlier samples must be considered in full context of the investigation: geographic/temporal distribution, analytical error, data entry error, method detection limits, as well as the possibility of an additional source, present in only one or a few samples. As discussed earlier, in environmental forensics investigations, decisions of 'significance' are often best made in the scientific context of the investigation, rather than through use of a rule-of-thumb numerical criteria. The use of such diagnostic plots is not new. They are standard in evaluation of linear regression models (Draper and Smith, 1981) but unfortunately are seldom used in evaluation of principal components models.

In the case of Data Set 2, the decision made by the data analyst is different for each type of outlier:

1 In the case of PCB 141 coeluting with DDT, the appropriate decision is to have the chemist go back and reanalyze the chromatograms to ensure that PCB 141 (a shoulder on the DDT peak) is correctly quantified.
2 In the case of the data transcription error, the appropriate decision is to correct the error in the spreadsheet, and rerun the analysis.
3 In the case of the two low concentration samples with multiple censored data points, usually the only realistic solution is to remove those two samples from the data set.
4 In the case of the remaining censored data points, those non-detects are generally at the low end of the measured range, and thus do not adversely affect the accuracy of back-calculation. We usually wish to retain as many samples as possible in the analysis. Therefore, in this case, we would typically leave these remaining non-detect samples in the matrix.

The changes above were made to Data Set 2. The PCA was rerun, and the revised reproduction indices are shown in Table 12.6. As for the much simpler Data Set 1, these indices are now in general agreement with each other, correctly indicating the presence of a three-component system.

12.2.5 PCA OUTPUT

As discussed in Section 12.2.1, the most common way of presenting PCA results is in terms of a PCA scores plot, where the analyst can evaluate relationships

Table 12.6

Reproduction indices for modified Data Set 2. For correctable errors (DDT coelution with PCB 141 and data transcription error in Sample 22) the data were modified accordingly. For uncorrectable problems (low concentration samples with many non-detects in two samples 9 and 17) the samples were removed from the matrix.

PC No.	Eigenvalue	(1) Cumulative Percentage Variance	(2) Scree Test	(3) Normalized Varimax Loadings	(4) Malinowski Indicator Function	(5) Cross Validation PRESS(i)/ RSS($i-1$)	(6) Signal to Noise Ratio
1	561.330	76.26	23.74	46	0.000117	–	38.27
2	135.180	94.63	5.37	44	0.000059	0.25	14.83
3	19.919	**97.34**	**2.66**	**22**	0.000044	**0.57**	**2.55**
4	2.191	97.63	2.37	2	**0.0000434**	1.09	1.87
5	1.642	97.86	2.14	1	0.0000438	1.18	1.90
6	1.416	98.05	1.95	1	0.0000443	1.27	2.14
7	1.321	98.23	1.77	0	0.0000448	1.31	1.98
8	1.147	98.38	1.62	0	0.000046	1.40	1.76
9	1.025	98.52	1.48	1	0.000046	1.50	1.85
10	0.905	98.65	1.35	1	0.000047	1.61	1.59
11	0.839	98.76	1.24	1	0.000048	1.75	1.67
12	0.819	98.87	1.13	0	0.000050	1.77	1.58
13	0.712	98.97	1.03	0	0.000051	1.89	1.54
14	0.656	99.06	0.94	0	0.000052	2.02	1.54
15	0.612	99.14	0.86	1	0.000054	2.16	1.35
16	0.561	99.22	0.78	1	0.000055	2.31	1.37
17	0.511	99.29	0.71	1	0.000057	2.50	1.39
18	0.489	99.35	0.65	1	0.000059	2.74	1.30
19	0.482	99.42	0.58	1	0.000061	2.92	1.30
20	0.439	99.48	0.52	1	0.000063	3.13	1.31
21	0.412	99.53	0.47	0	0.000065	3.26	1.22
22	0.357	99.58	0.42	0	0.000068	3.60	1.16
23	0.339	99.63	0.37	1	0.000071	3.90	1.18
24	0.315	99.67	0.33	0	0.000074	4.15	1.07
25	0.276	99.71	0.29	0	0.000077	4.67	1.10
26	0.269	99.75	0.25	0	0.000081	4.83	0.94
27	0.227	99.78	0.22	0	0.000085	5.47	0.96
28	0.218	99.81	0.19	0	0.000089	6.03	0.89
29	0.199	99.83	0.17	1	0.000094	6.26	0.81
30	0.162	99.86	0.14	0	0.000101	7.05	0.77
31	0.148	99.88	0.12	1	0.000108	7.67	0.71
32	0.129	99.89	0.11	1	0.000116	8.79	0.71
33	0.119	99.91	0.09	1	0.000126	10.20	0.73
34	0.117	99.93	0.07	0	0.000136	11.61	0.68
35	0.106	99.94	0.06	0	0.000147	12.95	0.64
36	0.086	99.95	0.05	0	0.000162	15.57	0.59
37	0.079	99.96	0.04	0	0.000178	15.72	0.45
38	0.053	99.97	0.03	0	0.000204	19.88	0.46
39	0.050	99.98	0.02	0	0.000236	22.65	0.41
40	0.041	99.98	0.02	0	0.000279	28.45	0.40
41	0.037	99.99	0.01	0	0.000335	36.13	0.40
42	0.035	99.99	0.01	1	0.000403	40.03	0.31
43	0.020	99.99	0.01	0	0.000538	59.53	0.30
44	0.019	100.00	0.00	0	0.000740	74.99	0.26
45	0.013	100.00	0.00	0	0.001170	118.92	0.26
46	0.011	100.00	0.00	0	0.002074	175.26	0.19
47	0.006	100.00	0.00	0	0.006138	–	0.13
48	0.002	100.00	0.00	0	–	–	0.00

between samples on a two- or three-dimensional graphic. Scores plots for Data Set 1 were presented as Figures 12.2 and 12.3. For simple, clustered data sets such as this, the scores-plot visualization method is very effective. The problem, however, is that regardless of the complexity of the data or the results of goodness-of-fit diagnostics (Section 12.2.4) graphical limitations dictate a maximum of three principal components. This puts subtle pressure on the analyst to choose, whenever possible, three or fewer components. Therefore, the widespread occurrence of two and three component plots in the literature may be due more to this bias than to the inherent simplicity of environmental chemical systems. Another limitation to PCA score plots is illustrated in Figure 12.6. This is the three-PC scores plot for cleaned-up Data Set 2. The corners of the gray shaded triangle represent the locations of the three Aroclor sources. All Data Set 2 samples plot within a triangle defined by these three vertices. This data cloud geometry is commonly observed when samples are mixtures of multiple sources.

Numerous software packages, including most general-purpose packages, perform PCA and allow the user numerous data transformation and visualization options. A few of these include Statistical Analysis System (SAS: Cary, NC), Number Cruncher Statistical System (NCSS: Kaysville, Utah), Pirouette (Infometrix, Woodinville, WA).

As is evident in Figure 12.6, PCA can be effective in inferring the presence of a two- or three-component mixed system, but by itself PCA is not capable of determining the chemical compositions of the sources, or their relative contributions. In the case of analysis of mixtures, we need another technique in our numerical toolbox. Such tools are discussed in Section 12.3.

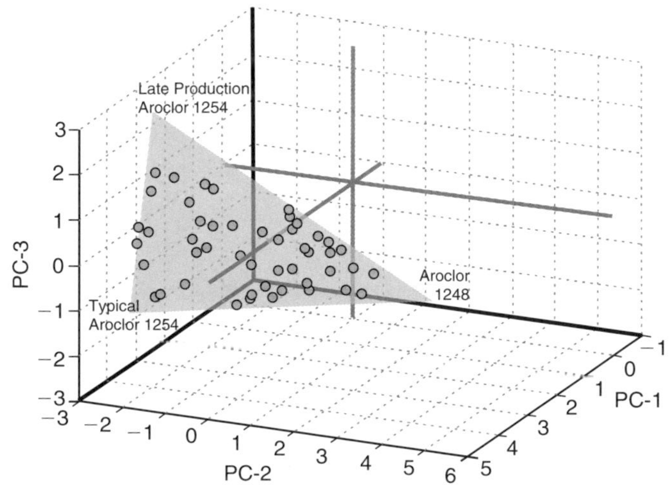

Figure 12.6

Three PC scores plot for Data Set 2, after errors have been corrected. No samples in Data Set 2 (dots) represent pure Aroclor sources. All are mixtures of the three sources. The approximate locations of pure sources are indicated at the vertices (corners) of the gray shaded triangle.

12.3 SELF-TRAINING RECEPTOR MODELING METHODS

An increasingly common approach in environmental forensic investigations involves the use of receptor models. These models are designed to resolve three parameters of concern in a multivariate, mixed chemical system: (1) the number of components in the mixture, (2) the identity (i.e., chemical composition or fingerprints) of each component, and (3) the relative proportions of each component in each sample. These objectives are stated mathematically as the determination of k, **A** and **F** in Equation 12.1. The source apportionment equation is similar to the scores and loadings expression given in Equations 12.7 and 12.8. However, principal component scores and loadings are abstract, orthogonal matrices, and do not represent feasible chemical compositions or source contributions. They are abstract in that when expressed in terms of the original chemical variables, principal component loadings typically contain negative elements.

Receptor modeling methods therefore involve rotation of matrices **A** and **F** to an oblique solution derived within reduced (k-dimensional) principal component space. The rotation is performed per explicit non-negativity constraints imposed on matrices **A** and **F**. For example, a fingerprint cannot have a chemical composition with −10% PCB 138, thus matrix **F** cannot have negative elements. Similarly, a sample cannot have a −35% contribution from a given source, thus matrix **A** cannot have negative elements (Miesch, 1976a; Full *et al.*, 1981, 1982; Gemperline, 1984; Hopke, 1989; Henry and Kim, 1990; Kim and Henry, 1999). Non-negativity constraints are typical requirements for all mixing models. In addition, other explicit restrictions may be imposed, if one has additional knowledge of physical/chemical constraints on the system (Henry and Kim, 1990; Johnson, 1997; Kim and Henry, 1999).

Multivariate receptor modeling methods use PCA as an intermediate step to determine the number of contributing sources, and to provide a reduced dimensional reference space for resolution of the model. In all of the receptor modeling methods discussed here, determining the number of sources (k) essentially reduces to the problem of choosing the number of significant principal components. As such, discussions regarding determination of significant principal components (Section 12.2.4) are equally relevant to receptor modeling.

An assumption of the conceptual mixing model/receptor model is that the system must be over-determined. That is, the data set must contain more variables or samples (whichever number is smaller) than there are sources. If we measure only four chemicals in each sample, and six sources contribute to the contamination, we cannot completely or realistically resolve the model. Another assumption is that of linear mixing. We assume that the relative proportions of variables in each source are fixed, and that source contributions are linearly additive. That is, as we increase the proportion of a source fingerprint in a sample, the variables that are characteristic of that fingerprint will increase proportionally (linearly) in that sample.

After the choice of k (Section 12.2.4), the receptor model then resolves the chemical compositions of sources (**F**) and the contributions of the sources in each of the samples (**A**). Recall, however, that in environmental forensics investigations, we rarely have such *a priori* knowledge of all sources. If possible, we would like to derive source patterns directly from analysis of ambient data. Three such methods have been used in environmental source apportionment investigations: (1) the DENEG algorithm used in polytopic vector analysis (PVA) (Full *et al.*, 1981, 1982); (2) the unique vector iteration method used in target transformation factor analysis (TTFA: Roscoe and Hopke, 1981; Gemperline, 1984; Hopke, 1989; Malinowski, 1991); and (3) source apportionment by factors with explicit restrictions (SAFER) method, used in extended self-modeling curve resolution (ESMCR: Henry and Kim, 1990; Henry *et al.*, 1994; Kim and Henry, 1999).

These three methods are analogous in that (1) they do not require a training data set; (2) they are PCA based methods; (3) they involve solution of quantitative source apportionment equations by development of oblique solutions in PCA space; and (4) each involves the use of non-negative constraints.

The full PVA algorithm has not been set out in any single paper. This chapter provides the opportunity to do so. As such, the mathematics of PVA will be discussed in greater detail than the other two methods. TTFA and ESMCR are presented to provide the reader with an intuitive understanding of how those algorithms operate. The reader is referred to the original papers for specifics of those algorithms. Each of these three methods was applied to Data Set 2 (Table 12.3), modified as described in 12.4.2.2, and yielded source compositions that closely matched the compositions of the Aroclor sources in Table 12.1. While we will focus on these three methods, it should be noted that there are yet other methods with similar objectives, which have been described in the literature (Ozeki *et al.*, 1995; Xie *et al.*, 1998).

12.3.1 POLYTOPIC VECTOR ANALYSIS (PVA)

PVA was developed for analysis of mixtures in the geological sciences, but it has evolved over a period of 40 years, with different aspects of the algorithm presented in a series of publications, by a number of different authors. The roots of PVA are in principal components analysis, pattern recognition, linear algebra, and mathematical geology. Its development in mathematical geology can be traced back to the early 1960s and John Imbrie, a paleontologist

(Imbrie, 1963). Following Imbrie's work, a series of FORTRAN-IV programs were published (Manson and Imbrie, 1964; Klovan, 1968; Klovan and Imbrie, 1971). The resulting software developed by Imbrie (at Brown University) and Ed Klovan (at the University of Calgary) was called CABFAC (*C*algary *a*nd *B*rown *FAC*tor Analysis) and quickly became the most commonly used multivariate analysis algorithm in the geosciences. Subsequent investigators that proved crucial in development of the PVA algorithm included A.T. Miesch and William Full. Miesch, a geochemist with the US Geological survey in the 1970s, was one of the first to take full advantage of Imbrie's oblique vector rotation methods (Miesch 1976a,b). Miesch also developed the variable-by-variable goodness-of-fit criteria (Miesch CDs) discussed in Section 12.2.4.2. William Full, as a PhD candidate at the University of South Carolina in the early 1980s, developed the DENEG algorithm, which allows end-members (sources) to be resolved without *a priori* knowledge of their composition, and without use of a training data set (Full *et al.*, 1981, 1982).

The name 'polytopic vector analysis' follows directly from the jargon of Imbrie (1963) and Full *et al.* (1981, 1982). Imbrie (and many others) referred to eigenvector decomposition models, resolved in terms of orthogonal axes as 'factor analysis.' Solutions resolved in terms of oblique vectors he termed 'vector analysis.' Imbrie's is not the definition of true factor analysis as defined by Harman (1960). Regardless, Imbrie's terminology has held within mathematical geology and a number of other subdisciplines. Because PVA involves resolution of oblique vectors as source compositions, thus the term vector analysis. The term 'polytopic' is due to the fact that PVA involves resolution of a $k-1$ dimensional solid, a 'simplex' or 'polytope,' within k dimensional principal component space. As illustrated in Figure 12.6, the polytope or simplex in three-dimensional space is a two-dimensional triangle. In two space it is a straight line. In four space it is a tetrahedron. Because the objective of PVA is to resolve a $k-1$ dimensional simplex, the constant vector length transform is typically employed because it forces all sample vectors to have unit length, and thus all data are constrained to a $k-1$ dimensional space within k space (see Figure 12.4 – Section 12.2.2).

This section presents the full PVA algorithm running under default conditions; i.e., (1) the EXTENDED CABFAC algorithm (Klovan and Imbrie, 1971) and (2) the iterative oblique vector rotation algorithm originally presented as EXTENDED QMODEL (Full *et al.*, 1981, 1982). While any number of alternative data transformation and calculation options are available and may be implemented, these algorithms represent the core of PVA as it is presently implemented under default options by the commercial version of the SAWVECA (*S*outh *C*arolina *a*nd *W*ichita *Vec*tor *A*nalysis: Residuum Energy, Inc., Dickinson, TX). SAWVECA performs the dimensionality analysis of the

Klovan and Miesch's EXTENDED CABFAC; along with the CD scatter plot goodness-of-fit diagnostics of Johnson (1997; Johnson *et al.*, 2000). SAWVECB is the EXTENDED QMODEL and FUZZY QMODEL algorithms of Full *et al.* (1981, 1982). Readers not interested in the formalism of the PVA algorithm can skip ahead to Section 12.3.1.4.

12.3.1.1 Scaling Functions: Back-Calculation to Original Metric

The transformations presented in Section 2.2.2 serve to optimize the eigenvector decomposition, but interpretation and evaluation of matrices **A**″ and **F**″ in any scientific context is difficult. Calculations are best performed in transform metric, but evaluation/interpretation must be done in measurement metric. Thus the results must be back-calculated. Where double prime (e.g. **X**″) indicates a matrix which has been transformed in turn by the range transform and the constant vector length transform, the mapping functions presented by Miesch (1976a) allow us to translate the equations $\hat{\mathbf{X}}'' = \mathbf{A}''\mathbf{F}''$ back to $\hat{\mathbf{X}} = \mathbf{AF}$ (percent or 'constant row-sum' metric). The mathematics are discussed below. As scaling is called upon in various portions of PVA, the reader is referred back to this section for a description. The first step in back-calculation to measurement metric is the definition of what Miesch termed scale factors: s_k. Given k retained eigenvectors, Miesch defines a $1 \times k$ row vector of scale factors $s = \{s_k\}$ where each element s_k is defined as:

$$s_k = \frac{\mathrm{K} - \sum_{j=1}^{n} x_{\min j}}{\sum_{j=1}^{n} (f''_{kj}(x_{\max j} - x_{\min j}))} \tag{12.10}$$

K is a constant (the sum of each row: usually 100), f''_{kj} is the element of the scores matrix, and $x_{\max j}$ and $x_{\min j}$ are the maximum and minimum values in the jth column of the original data matrix **X**.

The elements of the back-calculated scores matrices **F**′ and **F** are then calculated, in turn, as follows:

$$f'_{kj} = s_k f''_{kj} \tag{12.11}$$

$$f_{kj} = f'_{kj}(x_{\max j} - x_{\min j}) + x_{\min j} \tag{12.12}$$

Similarly, the elements of the back-calculated loadings matrices **A**′ and **A** are calculated as:

$$a'_{ik} = \frac{a''_{ik}}{s_k} \tag{12.13}$$

If $\mathbf{r}$ is then defined as a column vector of the m row-sums of $\mathbf{A}'$, the elements of $\mathbf{A}$ are:

$$a_{ik} = \frac{a'_{ik}}{r_i} \tag{12.14}$$

12.3.1.2 Eigenvector Decomposition and Determining the Number of Sources

A singular value decomposition is carried out on transformed matrix $\mathbf{X}'$ or $\mathbf{X}''$ as presented in Equation 12.7. The results are translated into scores and loading ($\mathbf{A}''_Q$ and $\mathbf{F}''_Q$) again, as per Equation 12.7. The scores and loadings matrices are then translated back to constant sum metric ($\mathbf{A}$ and $\mathbf{F}$) using the scaling functions presented in the preceding section (Equations 12.10 through 12.14). The task then is determination of k, the appropriate number of sources. As discussed in Section 12.2.4, this reduces the problem of choosing the number of significant principal components, and all discussions presented in that section are equally relevant here. The methods used most often in PVA are (1) the normalized loading criteria (Ehrlich and Full, 1987) and (2) the Miesch CDs and scatter plots as described in Section 12.2.4.2. For each number of potential principal components ($k = 1, 2, \ldots n$, if n is smaller; or $k = 1, 2, \ldots m$, if m is smaller), $\hat{\mathbf{X}}$ is calculated as:

Matrix dimensions
$$\underset{(m\times n)}{\hat{\mathbf{X}}} = \underset{(m\times k)}{\mathbf{A}} \; \underset{(k\times n)}{\mathbf{F}} \tag{12.15}$$

The Miesch CDs are calculated and scatter plots constructed using the reduced dimensional scores and loadings, as expressed in percentage metric.

12.3.1.3 Determining End-Member Compositions and Mixing Proportions

Following determination of k (the number of end-members) a set of mathematical procedures are used to resolve the second and third objectives of the receptor modeling problem: (2) determine the composition of the end-members, and (3) determine the relative proportions of each end-member in each sample. This process is termed polytope resolution. The polytope resolution phase is typically conducted within k dimensional varimax space, but can be equivalently performed in unrotated principal component space. The first k columns of $\mathbf{A}''$ are taken, yielding a reduced matrix $\mathbf{A}''$ of size $m \times k$. Because $\hat{\mathbf{X}}'' = \mathbf{A}''\,\mathbf{F}''$, we may determine matrix $\mathbf{F}''$ by the following matrix regression equation:

$$\mathbf{F}'' = (\mathbf{A}''^T\,\mathbf{A}'')^{-1}\,\mathbf{A}''^T\,\mathbf{X}'' \tag{12.16}$$

Using the scaling equations in Section 12.3.1.1, matrices $\mathbf{A}''$ and $\mathbf{F}''$ are then transformed back to the original constant row-sum metric, and the estimate of $\hat{\mathbf{X}}$ is calculated as:

Matrix dimensions
$$\underset{(m\times n)}{\hat{\mathbf{X}}} = \underset{(m\times k)}{\mathbf{A}}\ \underset{(k\times n)}{\mathbf{F}} \tag{12.17}$$

12.3.1.3.1 Selection of Initial Polytope

The first task of the polytope phase is an initial estimate of a polytope. A number of techniques have been proposed. Most commonly employed is the 'extended' method, so named because it was used by Full *et al.* (1981) in the EXTENDED QMODEL Fortran IV algorithm. The formalism of the extended method is described below. Alternative methods, their advantages and disadvantages are subsequently discussed.

The extended method establishes the initial polytope by taking the k most mutually extreme samples in the data set as vertices (the EXRAWC subroutine of Klovan and Miesch, 1976). EXRAWC first picks a good candidate set for these k samples: those with maximum loadings on each of the first k factors of $\mathbf{A}''$. The columns of $\mathbf{A}''$ are scanned, and the maximum absolute value loadings in each column are identified. The rows (samples) of $\mathbf{A}''$ corresponding to the maximum loadings are then put into a new $k \times k$ matrix $\mathbf{O_0}$. This operation is done without duplication (i.e. no two rows of $\mathbf{O_0}$ are the same). Clearly, the samples that make up $\mathbf{O_0}$ are candidates for the k most mutually extreme samples. The PVA algorithm uses these k vectors as oblique reference axes for all samples. The resultant oblique loadings and scores matrices $\mathbf{A}_0''$ and $\mathbf{F}_0''$ are calculated as:

$$\mathbf{F}_0'' = \mathbf{O_0}\,\mathbf{F}'' \tag{12.18}$$

$$\mathbf{A}_0'' = \mathbf{A}''\mathbf{O_0}^{-1} \tag{12.19}$$

Matrices $\mathbf{F}_0''$ and $\mathbf{A}_0''$ are then scaled back to measurement space using the scaling functions described in Section 12.3.1.1, yielding $\mathbf{F_0}$ and $\mathbf{A_0}$.

Matrix $\mathbf{A_0}$ is then inspected to determine if the k samples in $\mathbf{O_0}$ are indeed the k most mutually extreme. Following the method of Imbrie (1963) and Miesch (1976a), if the maximum loadings in matrix $\mathbf{A_0}$ equal 1.0 and correspond to the samples taken for matrix $\mathbf{O_0}$, then the k most mutually extreme samples have been taken and matrix $\mathbf{O_0}$ is then used as initial oblique reference axis for iterative model resolution. If loadings greater than 1.0 are identified in $\mathbf{A_0}$, the sample(s) corresponding to that maximum loading replaces the original sample in $\mathbf{O_0}$, and the process is repeated until a suitable set of k samples is obtained. If the algorithm finds no set of samples with all elements less than 1.0, then the original matrix $\mathbf{O}_0$ is taken as the initial set of oblique vectors. The algorithm for determining extreme samples is termed the EXRAWC procedure.

Outlier samples related to analytical problems or human error should be corrected or omitted from the analysis, as discussed in Section 12.2.4. However, if such errors are not corrected, the extended method can perform poorly. Outliers are extreme samples, and the extended method defines the initial polytope using extreme samples as vertices. In such cases, another method based on the fuzzy *c* means clustering algorithm of Bezdek (Bezdek, 1981; Bezdek *et al.*, 1984) will often produce better results. The mathematics of so-called FUZZY QMODEL are presented by Full *et al.* (1982). Full demonstrated that the *k* fuzzy cluster centers as defined in *k* dimensional eigenspace were always well within the convex hull defined by the sample data cloud, and were minimally affected by the presence of outliers. The vectors that define the *k* fuzzy cluster centers are taken as the row vectors of $\mathbf{O}_0$, and matrices $\mathbf{A}_0$ and $\mathbf{F}_0$ are defined relative to these vectors, as discussed above. The main disadvantage of the fuzzy method is that it is more computationally expensive to run. In the absence of outliers due to error (as we hope would result from diligent outlier identification in the PCA step) no advantage is gained by choosing fuzzy over extended.

A third option is to use the samples with maximum loadings in $\mathbf{A}''$. In instances where the EXRAWC subroutine does not converge, the EXTENDED option will default to the set of samples with maximum loadings.

A final option is external input of end-member compositions. If end-member compositions are known or suspected prior to the analysis, those suspected sample compositions may be plugged into the model as potential end-members. Suppose for instance that Aroclors 1248, and the two 1254 variants *were* suspected as the contributing sources for Data Set 2. If those source compositions were used as external end-members, and the model converged without iteration, the tested end-members would be considered feasible. This method is essentially the same as target testing as described by Hopke (1989) and Malinowski (1991) and is also similar to methods that require use of a training data set (such as chemical mass balance approaches). While this option is included in PVA software, in practice it is seldom used, because it constitutes a tacit hypothesis test, which is contrary to an exploratory data analysis philosophy. Regardless of the method used, the result is ultimately a set of *k* vectors as the rows of matrix $\mathbf{O}_0$. Matrices $\mathbf{A}_0$ and $\mathbf{F}_0$ are then defined relative to these vectors.

12.3.1.3.2 Testing Matrices $\mathbf{A}_0$ and $\mathbf{F}_0$ for Negative Values

Once an initial polytope is defined, the algorithm scans matrices $\mathbf{A}_0$ and $\mathbf{F}_0$ for negative values. The DENEG subroutine described below distinguishes between adjustable negative values and nonadjustable negative values, based on three user-defined numerical criteria. The first, t_1, is the 'mixing proportions cutoff criterion.' The default t_1 is -0.05. In other words, the algorithm will allow up to a -5% mixing proportion in any sample. The purpose of t_1 is to allow for some noise in the model.

If $\mathbf{A_0}$ does contain negative matrix elements less than t_1, the matrix is again scanned using a second criterion, t_2, referred to as the DENEG value. The default DENEG value is -0.25, but may be modified as the user sees fit. The DENEG subroutine recognizes adjustable mixing proportions only if they fall in the range between t_1 and t_2. At first glance, the need for the t_2 is not obvious, but in effect, it serves as our final line of defense against outliers. In early development of the algorithm, Full observed that in the presence of outliers, a model would often converge (i.e., no negatives less than -0.05) with the lone exception of an outlier with an extremely high negative value (less than -25%) in matrix **A**. By using the t_2 criterion, the DENEG subroutine would not iterate in an attempt to fit that single sample into the model.

Finally, the algorithm scans $\mathbf{F_0}$ for negative values using a third criteria, t_3, referred to as the 'end-member composition cutoff criterion.' Again, the default t_3 value is -5%, and serves the purpose of allowing some noise in the model. If there are no adjustable negative elements in matrices $\mathbf{A_0}$ and $\mathbf{F_0}$, the algorithm stops. $\mathbf{A_0}$ is taken as the mixing proportions matrix and $\mathbf{F_0}$ as the end-member compositions matrix.

12.3.1.3.3 The DENEG Algorithm

If $\mathbf{A_0}$ and $\mathbf{F_0}$ contain adjustable negatives, the program starts an iterative process of expansion and rotation of the polytope until two criteria are met: (1) mixing proportions have no adjustable negative values, and (2) end-member compositions have no adjustable negative values.

Geometrically the DENEG procedure is a process of alternate polytope expansion and rotation. The results of DENEG applied to the synthetic three-source PCB data set are shown graphically in Figure 12.7. Recall that data transformations are performed such that each sample vector has unit length (be that in measurement space or reduced dimensional principal components space). Thus, in $K = 3$ dimensional space, all Data Set 2 data points plot on sphere, unit length from the origin (Figure 12.7). DENEG begins by constructing an initial simplex in principal component space (Iteration 0: Figure 12.7). The EXTENDED method (Section 12.3.1.3.1) was used to define the initial polytope (Iteration 0), so the vertices of the Iteration 0 triangle are located at the three most mutually extreme samples in the data set. Had pure end-members been contained within the data set, the non-negativity criteria would have been met at Iteration 0, and the algorithm would have converged without

Frame, G.M., Cochran, J.W., and Bøwadt, S.S. (1996) Complete PCB congener distributions for 17 Aroclor mixtures determined by 3 HRGC systems optimized for comprehensive, quantitative, congener-specific analysis. *Journal of High Resolution Chromatography* 19, 657–668.

Full, W.E., Ehrlich, R., and Klovan, J.E. (1981) Extended Qmodel – objective definition of external end members in the analysis of mixtures. *Journal of Mathematical Geology* 13, 331–344.

Full, W.E., Ehrlich, R., and Bezdek, J.C. (1982) Fuzzy QModel – A new approach for linear unmixing. *Journal of Mathematical Geology* 14, 259–270.

Gemperline, P.J. (1984) A priori estimates of the elution profiles of pure components in overlapped liquid chromatography peaks using target transformation factor analysis. *Journal of Chemical Information and Computer Sciences* 24, 206–212.

Gordon, G.F. (1988) Receptor models. *Environmental Science and Technology* 22, 1132–1142.

Grung, B. and Kvalheim, O.M. (1994) Rank determination of spectroscopic profiles by means of cross validation: the effect of replicate measurements on the effective degrees of freedom. *Chemometrics and Intelligent Laboratory Systems* 22, 115–125.

Harman, H.F. (1960) *Modern Factor Analysis*. University of Chicago Press.

Henry, R.C. and Kim, B.M. (1990) Extension of self-modeling curve resolution to mixtures of more than three components. Part 1: Finding the basic feasible region. *Chemometrics and Intelligent Laboratory Systems* 8, 205–216.

Henry, R.C., Lewis, C.W., and Collins, J.F. (1994) Vehicle related hydrocarbon source compositions from ambient data: the GRACE/SAFER method. *Environmental Science and Technology* 28, 823–832.

Henry, R.C., Spiegelman, C.H., Collins, J.F., and Park, J.F. (1997) Reported emissions of organic gases are not consistent with observations. *Proceedings of the National Academy of Sciences*, USA 94, 6596–6599.

Henry, R.C., Park, E.S., and Spiegelman, C.H. (1999) Comparing a new algorithm with the classic methods for estimating the number of factors. *Chemometrics and Intelligent Laboratory Systems* 48, 91–97.

Hopke, P.K. (1989) Target transformation factor analysis. *Chemometrics and Intelligent Laboratory Systems* 6, 7–19.

Hopke, P.K. (1991) An introduction to receptor modeling. *Chemometrics and Intelligent Laboratory Systems* 10, 21–43.

Imbrie, J. (1963) *Factor and Vector Analysis Programs for Analyzing Geologic Data*. Office of Naval Research. Tech Report No. 6. 83 pp.

Jarman, W.M., Johnson, G.W., Bacon, C.E., Davis, J., Risebrough, R.W., and Ramer, R. (1997) Levels and patterns of polychlorinated biphenyls in water collected from the San Francisco Bay and Esturary, 1993–1995. *Fresnius' Journal of Analytical Chemistry* 359, 254–260.

Johnson, G.W. (1997) Application of Polytopic Vector Analysis to Environmental Geochemistry Problems. PhD Dissertation. University of South Carolina. Columbia, SC.

Johnson, G.W., Jarman, W.J., Bacon, C.E., Davis, J., Ehrlich, R., and Risebrough, R.W. (2000) Resolving polychlorinated biphenyl source fingerprints in suspended particulate matter of San Francisco Bay. *Environmental Science and Technology* 34, 552–559.

Kaiser, H.F. (1958) The varimax criterion for analytic rotation in factor analysis. *Psychometrika* 23, 187–200.

Kim, B.M. and Henry, R.C. (1999) Extension of self-modeling curve resolution to mixtures of more than 3 components Part 2 – finding the complete solution. *Chemometrics and Intelligent Laboratory Systems* 49, 67–77.

Klovan, J.E. (1968) Q-mode factor analysis program in FORTRAN-IV for small computers. Kansas Geological Survey Computer Contribution 20, pp. 39–51.

Klovan, J.E. and Imbrie, J. (1971) An algorithm and fortran-IV program for large-scale Q-mode factor analysis and calculation of factor scores. *Journal of Mathematical Geology* 3, 61–77.

Klovan J.E. and Miesch, A.T. (1976) EXTENDED CABFAC and QMODEL, computer programs for Q-mode factor analysis of compositional data. *Computers and Geosciences* 1, 161–178.

Malinowski, E.R. (1977) Determination of the number of factors and the experimental error in a data matrix. *Analytical Chemistry* 49, 612–617.

Malinowski, E.R. (1991) *Factor Analysis in Chemistry*. John Wiley & Sons, New York.

Manson, V. and Imbrie, J. (1964) FORTRAN program for factor and vector analysis of geological data using an IBM 7090 or 7094 computer system. Kansas Geological Survey Special Distribution Publication 13.

Miesch, A.T. (1976a) Q-mode factor analysis of geochemical and petrologic data matrices with constant row sums. *Geological Survey Prof. Paper*, 574-g, pp. 1–47.

Miesch, A.T. (1976b) Interactive computer programs for petrologic modeling with Extended Q-mode factor analysis. *Computers and Geosciences* 2, 439–492.

Moro, G., Lasagni, M., Rigamonti, N., Cosentino, U., and Pitea, D. (1997) Critical review of the receptor model based on target transformation factor analysis. *Chemosphere* 35, 1847–1865.

Ozeki, T., Koide, K., and Kimoto, T. (1995) Evaluation of sources of acidity in rainwater using a constrained oblique rotational factor analysis. *Environmental Science and Technology* 29, 1638–1645.

Reyment, R.A. and Jöreskog, K.G. (1993) *Applied Factor Analysis in the Natural Sciences*. Cambridge University Press, Cambridge.

Roscoe, B.A. and Hopke, P.K. (1981) Comparison of weighted and unweighted target transformation rotations in factor analysis. *Computers and Chemistry* 5, 1–7.

Tanabe, S., Kannan, N., Subramanian, A., Watanabe, S., and Tatsukawa, R. (1987) Highly toxic coplanar PCBs: Occurrence, source, persistence and toxic implication to wildlife and humans. *Environmental Pollution* 47, 147–163.

Tukey, J.W. (1977) *Exploratory Data Analysis*. Addison-Wesley, Reading, MA.

Wold, S. (1978) Cross-validatory estimation of the number of components in factor and principal components analysis models. *Technometrics* 20, 397–406.

Xie, Y.L., Hopke, P.K., and Paatero, P. (1998) Positive matrix factorization applied to a curve resolution problem. *Journal of Chemometrics* 12, 357–364.

Zhou, D., Chang, T., and Davis, J.C. (1983) Dual extraction of *R*-mode and *Q*-mode factor solutions. *Journal of Mathematical Geology* 15, 581–606.

APPENDIX A

CHEMICAL COMPOSITION OF AUTOMOTIVE GASOLINE AND DIESEL (FUEL #2)

The composition summary for automotive gasoline is excerpted from the California State Water Resources Control Board document titled LUFT Field Manual Revision, Appendix H, dated April 5, 1989. The summary for diesel was obtained from a summary table of laboratory analysis for diesel (fuel #2) compiled by Beth Albertson at Friedman & Bruya in Seattle, Washington for British Petroleum Company in 1996. The intent of including this generalized information regarding the composition of gasoline and diesel is to provide the reader with qualitative information regarding the prominent components of each fuel, and not as a definitive listing of the many components present in any particular brand, grade, or vintage of fuel. As described in Chapter 6, generalizations regarding the composition of a 'typical' automotive gasoline and/or diesel fuel are frequently oversimplified and therefore useful only in a qualitative manner, relative to a specific investigation.

Compound	Number of Carbons	Concentration (Weight %)
Automotive Gasoline		
Straight chain alkanes:		
Propane	3	0.01–0.14
n-Butane	4	3.93–4.70
n-Pentane	5	5.75–10.92
n-Hexane	6	0.24–3.50
n-Heptane	7	0.31–1.96
n-Octane	8	0.36–1.43
n-Nonane	9	0.07–0.83
n-Decane	10	0.04–0.50
n-Undecane	11	0.05–0.22
n-Dodecane	12	0.04–0.09
Branched alkanes:		
Isobutane	4	0.12–0.37
2,2-Dimethylbutane	6	0.17–0.84
2,3-Dimethylbutane	6	0.59–1.55
2,2,3-Trimethylbutane	7	0.01–0.04
Neopentane	5	0.02–0.05

Compound	Number of Carbons	Concentration (Weight %)
Isopentane	5	6.07–10.17
2-Methylpentane	6	2.91–3.85
3-Methylpentane	6	2.4 (volume)
2,4-Dimethylpentane	7	0.23–1.71
2,3-Dimethylpentane	7	0.32–4.17
3,3-Dimethylpentane	7	0.02–0.03
2,2,3-Trimethylpentane	8	0.09–0.23
2,2,4-Trimethylpentane	8	0.32–4.58
2,3,3-Trimethylpentane	8	0.05–2.28
2,3,4-Trimethylpentane	8	0.11–2.80
2,4-Dimethyl-3-ethylpentane	9	0.03–0.07
2-Methylhexane	7	0.36–1.48
3-Methylhexane	7	0.30–1.77
2,4-Dimethylhexane	8	0.34–0.82
2,5-Dimethylhexane	8	0.24–0.52
3,4-Dimethylhexane	8	0.16–0.37
3-Ethylhexane	8	0.01
2-Methyl-3-ehtylhexane	9	0.04–0.13
2,2,4-Trimethylhexane	9	0.11–0.18
2,2,5-Trimethylhexane	9	0.17–5.89
2,3,3-Trimethylhexane	9	0.05–0.12
2,3,5-Trimethylhexane	9	0.05–1.09
2,4,4-Trimethylhexane	9	0.02–0.16
2-Methylheptane	8	0.48–1.05
3-Methylheptane	8	0.63–1.54
4-Methylheptane	8	0.22–0.52
2,2-Dimethylheptane	9	0.01–0.08
2,3-Dimethylheptane	9	0.13–0.51
2,6-Dimethylheptane	9	0.07–0.23
3,3-Dimethylheptane	9	0.01–0.08
3,4-Dimethylheptane	9	0.07–0.33
2,2,4-Trimethylheptane	10	0.12–1.70
3,3,5-Trimethylheptane	10	0.02–0.06
3-Ethylheptane	10	0.02–0.16
2-Methyloctane	9	0.14–0.62
3-Methyloctane	9	0.34–0.85
4-Methyloctane	9	0.11–0.55
2,6-Dimethyloctane	10	0.06–0.12
2-Methylnonane	10	0.06–0.41
3-Methylnonane	10	0.06–0.32
4-Methylnonane	10	0.04–0.26
Cycloalkanes:		
Cyclopentane	5	0.19–0.58
Methylcyclopentane	6	Not qualified
1-Methyl-*cis*-2-ethylcyclopentane	8	0.06–0.11
1-Methyl-*trans*-3-ethylcyclopentane	8	0.06–0.12
1-*cis*-2-Dimethylcyclopentane	7	0.07–0.13
1-*trans*-2-Dimethylcyclopentane	7	0.06–0.20
1,1,2-Trimethylcyclopentane	8	0.06–0.11
1-*trans*-2-*cis*-3-Trimethylcyclopentane	8	0.01–0.25
1-*trans*-2-*cis*-4-Trimethylcyclopentane	8	0.03–0.16

Compound	**Number of Carbons**	**Concentration (Weight %)**
Ethylcyclopentane	7	0.14–0.21
n-Propylcyclopentane	8	0.01–0.06
Isopropylcyclopentane	8	0.01–0.02
1-*trans*-3-Dimethylcyclohexane	8	0.05–0.12
Ethylcyclohexane	8	0.17–0.42
Straight chain alkenes:		
cis-2-Butene	4	0.13–0.17
trans-2-Butene	4	0.16–0.20
Pentene-1	5	0.33–0.45
cis-2-Pentene	5	0.43–0.67
trans-2-Pentene	5	0.52–0.90
cis-2-Hexene	6	0.15–0.24
trans-2-Hexene	6	0.18–0.36
cis-3-Hexene	6	0.11–0.13
trans-3-Hexene	6	0.12–0.15
cis-3-Heptene	7	0.14–0.17
trans-2-Heptene	7	0.06–0.10
Branched alkenes:		
2-Methyl-1-butene	5	0.22–0.66
3-Methyl-1-butene	5	0.08–0.12
2-Methyl-2-butene	5	0.96–1.28
2,3-Dimethyl-1-butene	6	0.08–0.10
2-Methyl-1-pentene	6	0.20–0.22
2,3-Dimethyl-1-pentene	7	0.01–0.02
2,4-Dimethyl-1-pentene	7	0.02–0.03
4,4-Dimethyl-1-pentene	7	0.6 (volume)
2-Methyl-2-pentene	6	0.27–0.32
3-Methyl-*cis*-2-pentene	6	0.35–0.45
3-Methyl-*trans*-2-pentene	6	0.32–0.44
4-Methyl-*cis*-2-pentene	6	0.04–0.05
4-Methyl-*trans*-2-pentene	6	0.08–0.30
4,4-Dimethyl-*cis*-2-pentene	7	0.02
4,4-Dimethyl-*trans*-2-pentene	7	Not qualified
3-Ethyl-2-pentene	7	0.03–0.04
Cycloalkenes:		
Cyclopentene	5	0.12–0.18
3-Methylcyclopentene	6	0.03–0.08
Cyclohexene	6	0.03
Alkyl benzenes:		
Benzene	6	0.12–3.50
Toluene	7	2.73–21.80
o-Xylene	8	0.68–2.86
m-Xylene	8	1.77–3.87
p-Xylene	8	0.77–1.58
1-Methyl-4-ethylbenzene	9	0.18–1.00
1-Methyl-2-ethylbenzene	9	0.19–0.56
1-Methyl-3-ethylbenzene	9	0.31–2.86
1-Methyl-2-*n*-propylbenzene	10	0.01–0.17
1-Methyl-3-*n*-propylbenzene	10	0.08–0.56

Compound	Number of Carbons	Concentration (Weight %)
1-Methyl-3-isopropylbenzene	10	0.01–0.12
1-Methyl-3-*t*-butylbenzene	11	0.03–0.11
1-Methyl-4-*t*-butylbenzene	11	0.04–0.13
1,2-Dimethyl-3-ethylbenzene	10	0.02–0.19
1,2-Dimethyl-4-ethylbenzene	10	0.50–0.73
1,3-Dimethyl-2-ethylbenzene	10	0.21–0.59
1,3-Dimethyl-4-ethylbenzene	10	0.03–0.44
1,3-Dimethyl-5-ethylbenzene	10	0.11–0.42
1,3-Dimethyl-5-*t*-butylbenzene	12	0.02–0.16
1,4-Dimethyl-2-ethylbenzene	10	0.05–0.36
1,2,3-Trimethylbenzene	9	0.21–0.48
1,2,4-Trimethylbenzene	9	0.66–3.30
1,3,5-Trimethylbenzene	9	0.13–1.15
1,2,3,4-Tetramethylbenzene	10	0.02–0.19
1,2,3,5-Tetramethylbenzene	10	0.14–1.06
1,2,4,5-Tetramethylbenzene	10	0.05–0.67
Ethylbenzene	8	0.36–2.86
1,2-Diethylbenzene	10	0.57
1,3-Diethylbenzene	10	0.05–0.38
n-Propylbenzene	9	0.08–0.72
Isopropylbenzene	9	<0.01–0.23
n-Butylbenzene	10	0.04–0.44
Isobutylbenzene	10	0.01–0.08
sec-Butylbenzene	10	0.01–0.13
t-Butylbenzene	10	0.12
n-Pentylbenzene	11	0.01–0.14
Isopentylbenzene	11	0.07–0.17
Indan(e)	9	0.25–0.34
1-Methylindan	10	0.04–0.17
2-Methylindan	10	0.02–0.10
4-Methylindan	10	0.01–0.16
5-Methylindan	10	0.09–0.30
Tetralin	10	0.01–0.14
Polynuclear aromatic hydrocarbons:		
Naphthalene	10	0.09–0.49
Pyrene	16	Not qualified
Benz(a)anthracene	18	Not qualified
Benz(a)pyrene	20	0.19–2.8 mg/kg
Benzo(e)pyrene	20	Not qualified
Benzo(g,h,i)perylene	21	Not qualified
Elements:		
Bromine		80–345 μg/g
Cadmium		0.01–0.07 μg/g
Chlorine		80–300 μg/g
Lead		530–1120 μg/g
Sodium		<0.6–1.4 μg/g
Sulfur		0.10–0.15 μg/g
Vanadium		<0.02–0.001 μg/g
Additives:		
Ethylene dibromide		0.7–177.2 ppm
Ethylene dichloride		150–300 ppm
Tetramethyl lead		
Tetraethyl lead		

Compound	Number of Carbons	Concentration (Weight %)
Diesel (Fuel #2)		
Straight chain alkanes:		
n-Octane	8	0.1
n-Nonane	9	0.19–0.49
n-Decane	10	0.28–1.2
n-Undecane	11	0.57–2.3
n-Dodecane	12	1.0–2.5
n-Tridecane	13	1.5–2.8
n-Tetradecane	14	0.61–2.7
n-Pentadecane	15	1.9–3.1
n-Hexadecane	16	1.5–2.8
n-Heptadecane	17	1.4–2.9
n-Octadecane	18	1.2–2.0
n-Nonadecane	19	0.7–1.5
n-Eicosane	20	0.4–1.0
n-Heneicosane	21	0.26–0.83
n-Docosane	22	0.14–0.44
n-Tetracosane	24	0.35
Branched chain alkanes:		
3-Methylundecane	12	0.09–0.28
2-Methyldodecane	13	0.15–0.52
3-Methyltridecane	14	0.13–0.30
2-Methyltetradecane	15	0.34–0.63
Alkyl benzenes:		
Benzene	6	0.003–0.10
Toluene	7	0.007–0.70
Ethylbenzene	8	0.007–0.20
o-Xylene	8	0.001–0.085
m-Xylene	8	0.018–0.612
p-Xylene	8	0.018–0.512
Styrene	9	<.002
1-Methyl-4-isopropylbenzene	10	0.003–0.026
1,3,5-Trimethylbenzene	9	0.09–0.24
n-Propylbenzene	9	0.03–0.048
Isopropylbenzene	9	<0.01
n-Butylbenzene	10	0.031–0.046
Biphenyl	12	0.01–0.12
Naphtheno-benzenes:		
Huorene	13	0.034–0.15
Fluoranthene	16	0.0000007–0.02
Benz(b)fluoranthene	20	0.0000003–0.000194
Benz(k)fluoranthene	20	0.0000003–0.000195
Indeno (1,2,3cd) pyrene	22	0.000001–0.000097
Alkyl naphthalenes:		
Naphthalene	10	0.01–0.80
1-Methylnaphthalene	11	0.001–0.81
2-Methylnaphthalene	11	0.001–1.49
1,3-Dimethylnaphthalene	12	0.55–1.28
1,4-Dimethylnaphthalene	12	0.110–0.23
1,5-Dimethylnaphthalene	12	0.16–0.36

Compound	Number of Carbons	Concentration (Weight %)
Polynuclear aromatics:		
Anthracene	14	0.000003–0.02
2-Methylanthracene	15	0.000015–0.018
Phenanthrene	14	0.000027–0.30
1-Methylphenanthrene	15	0.000011–0.024
2-Methylphenanthrene	15	0.014–0.18
3-Methylphenanthrene	15	0.000013–0.011
4- and 9-Methylphenanthrene	15	0.00001–0.034
Pyrene	16	0.000018–0.015
1-Methylpyrene	17	0.0000024–0.00137
2-Methylpyrene	17	0.0000037–0.00106
Benz(a)anthracene	18	0.0000021–0.00067
Chrysene	18	0.000045
Triphenylene	18	0.00033
Cyclopenta(cd)pyrene	18	0.000002–0.0000365
1-Methyl-7-isopropylphenanthrene	18	0.0000015–0.00399
3-Methylchrysene	19	<0.001
6-Methylchrysene	19	<0.0005
Benz(a)pyrene	20	0.000005–0.00084
Benz(e)pyrene	20	0.0000054–0.000240
Perylene	20	<0.0001
Benz(g,h,i)perylene	22	0.0000009–0.00004
Picene	22	0.0000004–0.000083

APPENDIX B

APPLICATIONS AND SYNONYMS OF SELECTED CHLORINATED SOLVENTS

Chlorinated Solvent	Applications
Carbon tetrachloride	An azeotropic agent used to dry spark plugs, a delousing agent, an ingredient used in dry/plasma etching, petroleum refining and pharmaceutical manufacturing. An extractant for oils from seeds, flowers, grease from hides and bones and alkaloids from plants (Mellan, 1957). Used as a solvent in the de-inking of paper, in liquid chromatography and in the manufacture of rubber (Environmental Defense Fund, 2000). An ingredient in furniture polishes, floor waxes, paints and varnishes. In the 1950s it was used for Freon production, in fire extinguishers, for metal cleaning and as a seed fumigant (Pankow *et al.,* 1996). Used in the preparation of dichlorodifluoromethane (Montgomery, 1991). *Chemical:* Carbon bisulfide; carbon bisulphide; carbon chloride; carbon disulphide; carbon sulfide; carbon sulphide; dithiocarbonic anhydride; methane tetrachloride; perchloromethane; sulphocarbonic anhydride; tetrachlorocarbon; tetrachloromethane. *Commercial synonyms:* Benzinoform; Carbona (A. Klipstein & Company product used as a dry-cleaning and spot removing agent introduced in about 1898); Carbon chloride; Carbon tet; ENT 4,705; Czterochlorek Wegla (Polish); Fasciolin; Flukoids; Freon 10; Granosan (30% carbon tetrachloride and 70% ethylene dichloride); Halon 104; Necatorina; Necatorine; Refrigerant 10; R10; RCRA waste number U211; Tetrachloorkoolstof (Dutch) Tetrachloormetaan; Tetrachlormethan (German); Tetrachlorure de Carbone (French); Tetracloruro di Carbonio (Italian); Tetrafinol; Tetraform; Tetrasol; UN 1846; Univerm; Vermoestricid. NCI-C04591 UN 1131 Weeviltox.
Chloroform	A solvent used for cleaning electronic circuit boards, a product of chlorination in water treatment, used to prepare fluorocarbon refrigerants and plastics, an ingredient in soil fumigant and used in rubber manufacturing. A solvent for fats, oils, rubber, waxes and resins (Pankow *et al.,* 1996). An ingredient used in toothpaste and liniments. An extractant used in the purification of penicillin and other antibiotics. An intermediate for refrigerants and propellants. A fumigant for soil and prevention of mildew on tobacco seedlings. A general solvent and reaction medium. A solvent for dry cleaning and used to recover fat from waste products. Used in the extraction of essential oils, and of alkaloids from natural substances. In pharmaceutical preparations it is used in cough syrups, expectorants, liniments, sedatives, carminatives, analgesics, anthelmintics and anesthetic preparations. Used in the organic preparation of Freon, dyes and drugs. *Chemical:* Formyl trichloride; methenyl chloride; methyl trichloride; methenyl trichloride; trichloride trichloroform; trichloromethane. *Commercial synonyms:* Chloroforme (French); Choroformio (Italian); methyltrichloride; Methan; Freon 20; R 20; Refrigerant 20; Trichloormethaan (Dutch); Trichlormethan (Czech); Trichlorometano (Italian); UN 1888.
Chloromethane	A natural product in seawater, a coolant and refrigerant, an herbicide and fumigant, used in the manufacture of silicone polymer pharmaceuticals, tetraethyl

Chlorinated Solvent	Applications
	lead, synthetic rubber, methyl cellulose, agricultural chemicals, methylene chloride, carbon tetrachloride, methyl cellulose and chloroform. A fluid used in thermometric and thermostatic equipment. A refrigerant used in household and small commercial refrigerators. A catalyst solvent for low-temperature polymerization. Used in aerosols of insecticides and plant hormones. Used in the extraction of natural food-flavoring materials. An actuating liquid in thermostatic controls. Used in salt baths for annealing metals. Used in the preparation of quaternary ammonium compounds, ethers, esters, hydrocarbons and methyl silicone chlorides. *Chemical:* Methyl chloride; monochloromethane. *Commercial synonyms:* Arctic R40; Freon 40; UN 1063.
Freon 113	Used in fire extinguishers, as a dry cleaning solvent and in the manufacture of chlorotrifluoroethylene. A feedstock in the production of other CFCs and fluoropolymers, a stripper of flux from printed circuit boards (often combined with alcohol), a vapor degreaser and a solvent used in aerosols. Type I CFC-113 is intended for use in the cleaning of space vehicle components, precision assemblies, oxygen systems and electronic equipment and/or precision parts in clean rooms (ASTM, 1995). Type II and IIA designations are typically used in vapor degreasing. *Chemical:* 1,1,2-trifluoro-1,2,2-trichloroethane; trichlorotrifluoroethane; 1,1,2-trichloro-1,2,2-trifluoroethane. *Commercial synonyms:* Arcton 63; Arklone P; Blacotron TF; Daiflon S3; Fluorocarbon 113; F-113; FC-113; Freon 113; Freon TF; Frigen 113a; TR-T; Freon PCA; Genesolv D; Genetron 113; Halocarbon 113; Isceon 113; Isotron 113; Khladeon, Kaiser Chemicals 11; R-113; R113; Refrigerant 113; TTE; 113; Ucon-113; Ucon fluorocarbon (Gorski, 1973).
1,1-Dichloroethene	A hydrolysis product of TCA and daughter product of TCE. Used as a chemical intermediate in vinylidene fluoride synthesis, coating resins, and synthetic fibers and adhesives. Used in the manufacturing of dyes, plastics, perfumes, paints and adhesives. A primary constituent in Saran-type plastics (e.g., copolymerized with vinyl chloride) and as a coating substance. *Chemical:* Dichloroethylene. *Commercial synonyms:* DCE; Sconatex; Chlorure de Vinylidene (French); 1,1 DCE; Chloride II; Vinylidene dichloride; NCI-C54262; Vinylidine chloride; VDC.
Methylene chloride	A secondary blowing agent in the production of low density flexible polyurethane foam used to produce upholstered furniture/bedding and carpet underlay. An extractant for decaffeinated coffee and hops. An ingredient of pill coating in pharmaceuticals (in Western Europe in 1994, accounted for 41% of total usage (European Chlorinated Solvent Association, 1997)). A carrier solvent and reaction medium in the pharmaceutical industry. An inactive ingredient in pesticide formulations, various chemical processing applications, an ingredient in adhesive formulations used to bond contact cements for wood, metal and upholstered furniture. A process solvent for cellulose esters, polycarbonate, triacetate and triacetate ester production. An ingredient in glues for solvent welding of plastic parts. An ingredient in de-waxing solvents, in paint strippers used in the aerospace industry, and as a solvent to clean paint booths, paint lines and spray guns. A vapor degreasing solvent and a vapor pressure depressant aerosol. An ingredient in adhesives for mining applications. A photoresist stripper used in the manufacture of printed circuit boards. A refrigerant used in low pressure ice and air-conditioning machines. A low temperature extractant of essential oils and edible fats. A solvent used to remove oil, wax, paint and as a selective solvent on various cellulose acetates. Methylene chloride is the active ingredient in many

Chlorinated Solvent	Applications
	paint removers, including furniture strippers and home paint removers and is used in aircraft maintenance due to its ability to penetrate, blister, and remove a variety of paint coatings. Methylene chloride was an ingredient in safety solvents used for cold cleaning which included blends of PCE, naphtha and methylene chloride. In the pharmaceutical industry, methylene chloride is used as a reaction and re-crystallization solvent for extraction. In the chemical processing industry, it is used in the production of cellulose triacetate that is used as a base for photographic film. *Chemical:* Methylene bichloride; dichloromethane; methylene chloride; methylene dichloride; methane dichloride; Chlorure de methylene; metylenu chlorek (National Institute of Standards and Technology, 2000). *Commercial synonyms:* Aerothene; Autopride Carburetor Choke and Valve Cleaner (Radiator Specialty Company, 1997a); Autoport Carburetor Choke and Valve Cleaner (Radiator Specialty Company, 1997b); DCM; Freon 30; M-17 solvent; Master Mechanic Carburetor Choke and Valve Cleaner (40–50% by weight) (Radiator Specialty Company, 1996); MM; Narcotil; NCI-C50102; RCRA waste number 84.16; RTECS; GY 4; Turco 5873; #5141 Chlorinated Solvent.
PCE	A dry cleaning fluid of clothing, a metal degreaser, and a solvent for waxes, greases, fats, oil and gums. Used in printing ink manufacturing. An ingredient in paint removers. A feedstock for the production of CFC-113 and of hydrofluorocarbon refrigerant 134a and hydrochlorofluorocarbon 123, 142b and 141b. Small quantities may be present in paper coatings and silicones; brake cleaners (90–100% by weight) and in insulating fluid in some electrical transformers as a substitute for PCBs (Halogenated Solvents Industry Alliance, 1994; Radiator Specialty Company, 1998, 1999a,b). A solvent for rubber, waxes, tar, paraffin and gum. Used in the manufacture of detergents. An extraction solvent for vegetable and mineral oils. Small amounts of PCE are present in printing inks, aerosol specialty products, paper coatings, silicones and maskant formulations used to protect surface from chemical etchants in the aerospace industry (Halogenated Solvents Industry Alliance, 1994). PCE is used as an insulating fluid in some electrical transformers as a substitute for PCBs (Halogenated Solvents Industry Alliance, 1994). Used for sulfur recovery, rubber dissolution, paint removal, printing ink bleeding, soot removal, catalyst regeneration and electroplating pre-cleaning operations (Lowenheim and Moran, 1975). A reforming grade PCE is used to re-activate refinery catalysts where the catalyst is not sensitive to oxygen- and nitrogen-containing compounds. *Chemical:* Acetylene trichloride; 1-chloro-2,2-dichloroethylene; 1,1-dichloro-2-chloroethylene; ethinyl trichloride; ethylene tetrachloride; ethylene trichloride; perchloroethylene; 1,1,2,2-tetrachloroethylene, trichloroethene; trichloroethylene; 1,1,2-trichloroethylene. *Commercial synonyms:* AI3; Algylen; Alk-Tri (Dow Chemical Company); Anameneth; Anamenth; Ankilostin; Antisal 1; Antisol 1; Benzinol; Blacosolv; Blancosolv; Cecolene; Chlorilen; 1-chloro-2,2-dichloroethylene; chlorylea; Chlorylen; Circosolv; Crawhaspol; Czterochloroetylen (Polish); Densinfluat; Didakene Dow-Per; Dow-tri, Dow-TriPhilex; Dukerson; ENT 1860; Ethinyl Tri-plus; Ethinyl trichloride; Ethyl Trichloroethylene; Ex-Tri (Dow Chemical Company); Fedal-un; Fleck-flip; Flock-flip; Fluate; Germalgen; Germalgene; Hi-Tri (Dow Chemical Company); Lanadin; Lethurin; M-17 solvent; Narcogen; Narkosoid; NCI-C04546; Nema; Neu-Tri (Dow Chemical Company); Nialk Trichlor MD (Hooker Chemical Company); Nialk Trichlor MDA (Hooker Chemical Company); Nialk Trichlor X-1 (Hooker Chemical Company); Nialk Trichlor-Extraction (Hooker Chemical Company); Nialk Trichlor-Technical (Hooker Chemical Company); PCE; Per; Perawin; Perc; Perchloorethyleen (Dutch); Perchlor; Perchloraethylen (German); Perclene; Percloroetilene (Italian); Percosolv; Perk; Perklone;

Chlorinated Solvent	Applications
	Perm-a-Chlor (Hooker Chemical Company–Detrex Inc.); Perm-a-Chlor NA (Hooker Chemical Company–Detrex Inc.); Perm-a-Chlor NA-LR (Hooker Chemical Company–Detrex Inc.); Persec; Petzinol; Phillex (industrial grade); Stauffer Trichloroethylene (Stauffer Chemical); Tetlen; Tetracap; Tetrachlooretheen (Dutch); Tetrachloraethen (German); Tetracloroctene (Italian); Tetraguer; Tetraleno; Tetralex; Tetravec; Tetroguer; Tetropil; Threthylene; Trethylene; Tri; Tri-Clene (DuPont de Nemours Company, Diamond Shamrock); Tri-Paint Grade (industrial grade); Triasol; Trichlooretheen (Dutch); Trichlor Type 113, 114, 115 and 112 (industrial grade); Trichloraethen (German); Trichloran; Trichloren; Trichloride Triad E (Hooker-Detrex); Trichloroethilene and Trielina (Italian); Trichloroethylene Dual (industrial grade); Trichloroethylene Extraction Grade (industrial grade); Trichlorretent; Triclene (DuPont, Diamond Shamrock); Triclene D, L, LS, MD, ME, R, Paint Grade and High Alkalinity (DuPont, Diamond Shamrock); Trielene; Triklone (industrial grade); Trilene; Trilene (anesthetic grade); Triline; Triman (anesthetic grade); Trimar; Trisan; Trivec; Tromex; Turco Surjex; tVestrol; UN 1710; V-strol; Vapoclean; Vapoclor; Vestrol; Vitran; Westrosol; Zip Grip Accelerator.
1,1,1-TCA	Used in the production of vinylidene chloride, a primary solvent used for cold cleaning, and used in the photoresist process for developing and stripping electronic circuit boards. An ingredient in septic tank cleaners (*Chemical Week*, 1979), aerosol pesticides, adhesive formulations, coatings for wood furniture, metal substrates, traffic paints for signs and road lines, and an inactive ingredient in pesticide formulations. Used in California after 1988 as a blowing agent in the production of flexible foam used to make upholstered furniture, bedding and carpet underlays. A solvent for fats, resins and waxes. An ingredient in aerosols, textiles, ink, oven cleaners, adhesives and correction fluid formulations (Schober, 1958; Aviado *et al.*, 1976). Used in the manufacture of plastics and metals. Used to clean printing presses, missile components, paint masks, photographic film, plastic molds, motors, generators, appliances, and leather and suede garments. An ingredient in spray and solid pesticides, rodenticides, drain cleaners, carpet glues, and fire ant insecticides (Environmental Defense Fund, 2000). A feedstock for the synthesis of hydrochlorofluorocarbons (United States Environmental Protection Agency, 1994). Paints historically used on high performance aircraft can contain paint-carrier solvents such as *n*-butyl acetate and other acetates, methyl ethyl ketone (MEK), toluene, glycol ethers and 1,1,1-trichloroethane (Evanoff, 1990). *Chemical:* alpha-trichloroethane; alpha-trichloromethane; chloroethene; methyl chloroform; methyltrichloromethane; trichloromethylmethane; trichloro-1,1,1-ethane (French). *Commercial synonyms:* #10 Cleaner; #5141 Chlorinated Solvent; α-T; α-Trichloroethane; Aerothene (1967); Aerothene MM; Aerothene TT; AI3-02061; Algylen; Amsco Solv 5620; Axothene No. 3 (Axton-Cross Company); Baltana; Barcothene Nu; Blaco-thane; Blakeothane; Blakesolve 421; Caswell No. 875; CF2 Film Clean; Chloroethane NU; Chloroethene (1954) (Dow Chemical Company); Chloroethene VG, NU (Dow Chemical Company); Chlorotene; Chloromane; Chlorothane NU; Chlorothane SM; Chlorothene VG; Chlorothene (Inhibited); Chlorothene NU; Chlorten; Chlorthane-NU; Chlorylen; Crack Check Cleaner C-NF; DEV TAP; Devcon; Devon Metal Guard; Dowclene ED (Dow Chemical Company, 1962); Dowclene LS; Dowclene WR (Dow Chemical Company, 1965); Dyno-Sol; ECCO 1550; Ethyl 111 Trichloroethane; FL-20 Flexane primer; Gemalgene; Genklene; ICI-CF 2; Inhibisol; Insolv NU, VG; Kold Phil; Kwik-Solv; Lectrasolv 170; Locquic Primer T; Lube-Lok 4253; M-60; MCF; Methyl Chloroform Tech; Nacon 425; NCI-C04626; NU; One, One, One; PCN UCD 5620; Penolene 643; Perm-Ethane DG (Permathane); Quik Shield; Rapid Tap; RCRA waste number U226; Saf-Sol 20/20; Saf-T-Chlor; SKC-NF/ZC-73; Solvent 111 (Vulcan Chemical Company); Solvent M-50; Solventclean SC-A Aerosol; Sumco 33; TCA; TCEA; Tri; Tri-ethane Type 314, 315,

Chlorinated Solvent	Applications
	324 and 339 (PPG Industries); Trichloran; Trielene; Triple One; Turco Lock; UCD 784; UN 2831; V-301; Vatron 111; VG.
TCE	A metal degreasing solvent and a solvent for waxes, greases, fats, oils and gums. An ingredient in paint removers and septic tanks. A solvent base for metal phosphatizing systems. Used to degrease aluminum and for cleaning sheet and strip steel prior to galvanizing. Used to clean liquid oxygen and hydrogen tanks. An ingredient in grain fumigants (Huff, 1971). Used to degrease bones for making glues. An intermediate compound in the preparation of PCE, polyvinyl chloride, chloroacetic acid, hydrofluorocarbons, and fertilizers (Archer, 1996). A low temperature heat-transfer fluid. A freezing-point depressant in carbon tetrachloride based fire-extinguishing fluids. Used in the manufacture of detergents, dye intermediates, dyestuffs, leather, organic chemicals, paint, perfume, pharmaceuticals, rubber and varnish. A general anesthetic and an analgesic in dental extractions and childbirth short surgical procedures (Doherty, 2000). A dry cleaning spotting agent (Environmental Defense Fund, 2000). *Chemical:* Acetylene trichloride; ethylene trichloride; ethinyl trichloride; trichloroethylene; 1,1,2-trichloroethylene; trichloroethene; 1,1-dichloro-2-chloroethylene; 1-chloro-2,2-dichloroethylene. *Commercial synonyms:* AI3-00052; Algylen; Alk-Tri (Dow Chemical Company); Anameneth; Anamenth; Benzinol; Blacosolv; Blancosolv; Caswell No. 876; Cecolene; Chlorilen; Chlorylea; Chlorylen; Circosolv; Crawhaspol; Densinfluat; Dow-tri, Dow-TriPhilex; Dukeron; Dukerson; DuPont Dry Clean; Dux Water Repellant; EPA Pesticide Chemical Code 081202; Ethinyl Tri-plus; Ethinyl trichloride; Ethyl Trichloroethylene; Ex-Tri (Dow Chemical Company); Fleck-flip; Flock-flip; Fluate; Germalgen; Germalgene; Hi-Tri (Dow Chemical Company-low inhibitor TCE that is ideal for formulations, extractions and catalyst processes that are sensitive to higher inhibitor concentrations); Instant Chimney Sweep Spray (Huff 1971); Lanadin; Lash Bath False Eyelash Cleaner; Lethurin; M-17 solvent; Narcogen; Narkosoid; NCI-C04546; Neu-Tri (Dow Chemical Company – used in formulations, extractions and catalysts); Nialk; Nialk Trichlor MD (Hooker Chemical Company); Nialk Trichlor MDA (Hooker Chemical Company); Nialk Trichlor X-1 (Hooker Chemical Company); Nialk Trichlor-Extraction (Hooker Chemical Company); Nialk Trichlor-Technical (Hooker Chemical Company); NSC 389; Per-A-Clor; Perm-a-Chlor (Hooker Chemical Company – Detrex Inc.); Perm-a-Chlor NA (Hooker Chemical Company – Detrex Inc.); Perm-a-Chlor NA-LR (Hooker Chemical Company – Detrex Inc.); Petzinol; Philex; Phillex (industrial grade); Sears Air Freshener; Sears Odor Neutralizer; Stauffer; Threthylen; Threthylene; Trethylene; Tri; Tri-Clene (DuPont de Nemours Company, Diamond Shamrock); Tri-Paint Grade (industrial grade); Tri-Plus; Tri-Plus M; Triad Metal Cleaner and Polish; Trial; Triasol; Trichlooretheen (Dutch); Trichlor Type 113, 114, 115 and 112 (industrial grade); Trichloran; Trichloraethen (German); Trichloraethylen (German); Trichloren; Trichloretene (Italian); trichloride Triad E (Hooker-Detrex); Trichloroethilene (Italian); Trichloroethylene Dual (industrial grade); Trichloroethylene Extraction Grade (industrial grade); Trichlorretent; Triclene (DuPont, Diamond Shamrock); Triclene D, L, LS, MD, ME, R, Paint Grade and High Alkalinity (DuPont, Diamond Shamrock); Tricloretene (Italian); Tricloroetilene (Italian); Trielene; Trielin; Trielina (Italian); Triklone (industrial grade); Trilene (anesthetic grade); Triline; Triman (anesthetic grade); Trimar; Trisan; Trivec; Tromex; Turco Surjex; tVestrol; UN 1710; V-strol; Vapoclean; Vapoclor; Vestrol; Vitran; Westrosol; Zip Grip Accelerator (Doherty, 2000).
Vinyl chloride	Reduction product of 1,1- and 1,2-DCE. A gas used in the manufacture of polyvinyl chloride (PVC) pipes, wire coatings, automobile upholstery, and copolymers. An ingredient in adhesives for plastics, refrigerants and plastic housewares. Used in the preparation of polyvinyl chloride and other polymers and copolymers such as Flamenol, Koroseal, Vinylite (alone or together with

Chlorinated Solvent	Applications
	vinyl acetate or vinyl acetate and maleic anhydride), Geon (alone or together with vinylidene chloride), Tygon (with vinyl acetate), Velcon (with vinylidene chloride), and Saran (with vinylidene chloride). A refrigerant (American Insurance Association, 1972). *Chemical:* Chloroethylene; chloroethene; ethylene monochloride; monochloroethene; monochloroethylene. *Commercial synonyms:* VC; VCM; 1-Chloroethene; 1-Chloroethylene; Chlorure de Vinyle (French), Chloruro di Vinile (Italian) MVC; Trovidur; UN 1086; Vinile (Cloruro di) (Italian); Vinyl C monomer; Vinylchlorid (German); Vinyle Clorure de (French); Winylu Chlorek (Polish).

REFERENCES

American Insurance Association (AIA) (1972) *Chemical Hazards Bulletin. C-86. Chlorinated Hydrocarbons*. New York City, NY.

American Society for Testing and Materials (ASTM) (1995) Standard specification. Vapor-degreasing grade methylene chloride. ASTM Designation D 4079-95. ASTM, Philadelphia, PA.

Archer, W. (1996) *Industrial Solvents Handbook*. Marcel Dekker, New York.

Aviado, D., Zakhari, S., Simaan, J., and Ulsamer, A. (1976) Review of the literature on trichloroethylene. *Methyl Chloroform and Trichloroethylene in the Environment*, Chapter 7. CRC Press, Cleveland, OH.

Chemical Week (1979) New York seeks to curb solvents in groundwater. 102, 24.

Doherty, R. (2000) A history of the production and use of carbon tetrachloride, tetrachloroethylene, trichloroethylene, and 1,1,1, trichloroethane in the United States. Parts I and II. *Journal of Environmental Forensics* 1(2), 69–93.

Environmental Defense Fund (2000) About the chemicals. Environmental Defense Fund website. www.scorecard.org.

European Chlorinated Solvent Association (ECSA) (1997) Solvents Digest March 1997. Methylene chloride: an update on human and environmental effects. Brussels, Belgium.

Evanoff, S. (1990) Hazardous waste reduction in the aerospace industry. *Chemical Engineering Progress* April, 51–61.

Gorski, R. (1973) Stability of trichlorotrifluoroethane – stainless steel systems. *Cleaning Stainless Steel*. American Society for Testing Materials. STP 538. 65–76.

Halogenated Solvents Industry Alliance, Inc. (1994) *Perchloroethylene*. White Paper, February 1994. Washington, DC.

Halogenated Solvents Industry Alliance, Inc. (1996) *Trichloroethylene*. White Paper, June 1996. Washington, DC.

Huff, J. (1971) New evidence on the old problems of trichloroethylene. *Industrial Medicine* 40(8), 25–33.

Lowenheim, F. and Moran, M. (1975) *Faith, Keyes and Clark's Industrial Chemicals*, 4th edn. John Wiley & Sons, New York.

Mellan, I. (1957) *Source Book of Industrial Solvents*, vol. II: *Halogenated Solvents*. Reinhold Publishing Company, New York.

Montgomery, J. (1991) *Groundwater Chemicals Field Guide*. Lewis Publishers, Chelsea, MI.

National Institute of Standards and Technology (2000) Methylene chloride. webbook.nist.gov/chemistry/.

Pankow, J., Feenstra, S., Cherry, J., and Ryan, C. (1996) Dense chlorinated solvents and other DNAPLs in groundwater: history, behavior, and remediation. In *Dense Chlorinated Solvents and other DNAPLs in Groundwater*. (Pankow, J. and Cherry, J., eds), Chapter 1. Waterloo Press, Portland, OR.

Radiator Specialty Company (1996) Material Safety Data Sheet Mater Mechanic Carburetor Choke and Valve Cleaner CT4814. Radiator Specialty Company, Charlotte, NC.

Radiator Specialty Company (1997a) Material Safety Data Sheet Autopride Carburetor Choke and Valve Cleaner AAF-48-14. Radiator Specialty Company, Charlotte, NC.

Radiator Specialty Company (1997b) Material Safety Data Sheet Brake Cleaner AM7-49C. Radiator Specialty Company, Charlotte, NC.

Radiator Specialty Company (1998) Material Safety Data Sheet Brake Cleaner AM7-49C. Radiator Specialty Company, Charlotte, NC.

Radiator Specialty Company (1999a) Material Safety Data Sheet Brake Cleaner AM7-20 (aerosol). Radiator Specialty Company, Charlotte, NC.

Radiator Specialty Company (1999b) Material Safety Data Sheet Brake Cleaner M7 34, 49. Radiator Specialty Company, Charlotte, NC.

Schober, A. (1958) Chloroethene in hair sprays. *Soap Chemical Specialties* 34, 65.

United States Environmental Protection Agency (1994) Chemical summary for methylchloroform. United States Environmental Protection Agency. Office of Pollution Prevention and Toxics. August 1994. EPA 749-F-014a.

APPENDIX C

CHEMICAL PROPERTIES OF SELECTED CHLORINATED SOLVENTS AT 25°C

Compound	Molecular Weight	Vapor Pressure (p°, torr)	Literature Solubility (mg/L)	Henry's Constant (*H*, atm-m^3/mol)	Relative Vapor Density	Boiling Point (°C)	K_{oc} (mL/g)	Density (g/cm^3)
Dichloromethane	84.9	415	20 000	0.002 12	2.05	41	8.8	1.33
Chloroform	119.4	194	8000	0.003 58	1.80	62	44	1.49
Bromodichloromethane	163.8	64.2	4500	0.002 06	1.39	90	61	1.97
Dibromochloromethane	208.3	17	4000	0.001 15	1.14	119	84	2.38
Bromoform	282.8	6.21	3000	0.000 530	1.07	149	116	2.89
Trichlorofluoromethane	137.4	796	1100	0.0888	4.91	23.8	159	1.49
Carbon tetrachloride	153.8	109	825	0.0298	1.62	76.7	439	1.59
1,1-Dichloroethane	99	221	5100	0.005 43	1.70	57.3	30	1.17
1,2-Dichloroethane	99	82.1	8500	0.0015	1.26	83.5	14	1.25
1,1,1-Trichloroethane	133.4	124.6	1300	0.0167	1.59		152	1.35
1,1,2-Trichloroethane	133.4	24.4	4400	0.001 08	1.12	113.7	56	1.44
1,1,2,2-Tetrachloroethane	167.9	6.36	2900	0.000 459	1.04	146.4	118	1.6
1,1-Dichloroethylene	97	603	3350	0.0255	2.86	31.9	65	1.22
1,2-Dibromoethane (EDB)	187.9	13.8	4200	0.000 680	1.10	131.6	92	2.18
cis-1,2-Dichloroethylene	97	205	3500	0.003 74	1.63	60	86	1.28
trans-1,2-Dichloroethylene	97	315	6300	0.009 16	1.97	48	59	1.26
Trichloroethylene (TCE)	131.5	75	1100	0.009 37	1.35	86.7	126	1.46
Tetrachloroethylene (PCE)	165.8	18.9	200	0.0174	1.12	121.4	364	1.63
1,2-Dichloropropane	113	52.3	2800	0.002 62	1.20	96.8	51	1.16
trans-1,3-Dichloropropylene	110	34	2800	0.0013	1.10	112	48	1.22
bis(chloro)methylether	115	30	22 000	0.000 21	1.09	104	1.2	1.32
bis(2-chloroethyl)ether	143	1.11	10 200	0.000 13	1.004	178	14	1.22
bis(2-chloroisopropyl)ether	171	0.73	1700	0.000 11	1.003	189	61	1.11
2-Chloroethylvinylether	106.6	34.3	15 000	0.000 25	1.010	108	6.6	1.05

Sources:

Montgomery, J. (1991) *Groundwater Chemicals Field Guide*. Lewis Publishers, Chelsea, MI.

Pankow, J. and Cherry, J. (eds) (1996) *Dense Chlorinated Solvents and other DNAPLs in Groundwater*. Waterloo Press, Portland, OR.

Ramamoorthy, S. and Ramamoorthy, S. (1997) *Chlorinated Organic Compounds in the Environment. Regulatory and Monitoring Assessment.* Lewis Publishers, Boca Raton, FL.

AUTHOR INDEX

SUBJECT INDEX

Page numbers in *italics* refer to figures and tables, margin notes by suffix 'n' (e.g. '404n')